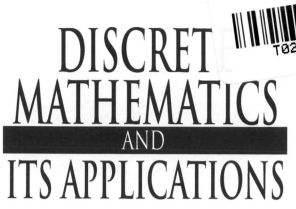

DISCRETE MATHEMATICS AND ITS APPLICATIONS

Series Editor

Kenneth H. Rosen, Ph.D.

Continued Titles

DISCRETE MATHEMATICS AND ITS APPLICATIONS
Series Editor KENNETH H. ROSEN

HANDBOOK OF COMPUTATIONAL GROUP THEORY

DEREK F. HOLT

BETTINA EICK
EAMONN A. O'BRIEN

CRC Press
Taylor & Francis Group
Boca Raton London New York

CRC Press is an imprint of the
Taylor & Francis Group, an **informa** business

A CHAPMAN & HALL BOOK

CRC Press
Taylor & Francis Group
6000 Broken Sound Parkway NW, Suite 300
Boca Raton, FL 33487-2742

First issued in paperback 2020

© 2005 by Taylor & Francis Group, LLC
CRC Press is an imprint of the Taylor & Francis Group, an Informa business

No claim to original U.S. Government works

ISBN-13: 978-1-58488-372-2 (hbk)
ISBN-13: 978-0-367-65944-8 (pbk)

This book contains information obtained from authentic and highly regarded sources. Reasonable efforts have been made to publish reliable data and information, but the author and publisher cannot assume responsibility for the validity of all materials or the consequences of their use. The authors and publishers have attempted to trace the copyright holders of all material reproduced in this publication and apologize to copyright holders if permission to publish in this form has not been obtained. If any copyright material has not been acknowledged please write and let us know so we may rectify in any future reprint.

Except as permitted under U.S. Copyright Law, no part of this book may be reprinted, reproduced, transmitted, or utilized in any form by any electronic, mechanical, or other means, now known or hereafter invented, including photocopying, microfilming, and recording, or in any information storage or retrieval system, without written permission from the publishers.

For permission to photocopy or use material electronically from this work, please access www. copyright. com (http://www.copyright.com/) or contact the Copyright Clearance Center, Inc. (CCC), 222 Rosewood Drive, Danvers, MA 01923, 978-750-8400. CCC is a not-for-profit organization that provides licenses and registration for a variety of users. For organizations that have been granted a photocopy license by the CCC, a separate system of payment has been arranged.

Trademark Notice: Product or corporate names may be trademarks or registered trademarks, and are used only for identification and explanation without intent to infringe.

Library of Congress Cataloging-in-Publication Data

Catalog record is available from the Library of Congress

International Standard Book Number 1-58488-372-3

Visit the Taylor & Francis Web site at
http://www.taylorandfrancis.com

and the CRC Press Web site at
http://www.crcpress.com

Preface

This book is about computational group theory, which we shall frequently abbreviate to CGT. The origins of this lively and active branch of mathematics can be traced back to the late nineteenth and early twentieth centuries, but it has been flourishing particularly during the past 30 to 40 years. The aim of this book is to provide as complete a treatment as possible of all of the fundamental methods and algorithms in CGT, without straying above a level suitable for a beginning postgraduate student.

There are currently three specialized books devoted to specific areas of CGT, namely those of Greg Butler [But91] and Ákos Seress [Ser03] on algorithms for permutation groups, and the book by Charles Sims [Sim94] on computation with finitely presented groups. These texts cover their respective topics in greater detail than we shall do here, although we have relied heavily on some of the pseudocode presented in [Sim94] in our treatment of coset enumeration in Chapter 5,

The most basic algorithms in CGT tend to be representation specific; that is, there are separate methods for groups given as permutation or matrix groups, groups defined by means of polycyclic presentations, and groups that are defined using a general finite presentation. We have devoted separate chapters to algorithms that apply to groups in these different types of representations, but there are other chapters that cover important methods involving more than one type. For example, Chapter 6 is about finding presentations of permutation groups and the connections between coset enumeration and methods for finding the order of a finite permutation group.

We have also included a chapter (Chapter 11) on the increasing number of precomputed stored libraries and databases of groups, character tables, etc. that are now publicly available. They have been playing a major rôle in CGT in recent years, both as an invaluable resource for the general mathematical public, and as components for use in some advanced algorithms in CGT. The library of all finite groups of order up to 2000 (except for order 1024) has proved to be particularly popular with the wider community.

It is inevitable that our choice of topics and treatment of the individual topics will reflect the authors' personal expertise and preferences to some extent. On the positive side, the final two chapters of the book cover applications of string-rewriting techniques to CGT (which is, however, treated in much greater detail in [Sim94]), and the application of finite state automata to the computation of automatic structures of finitely presented groups. On the other hand, there may be some topics for which our treatment is more superficial than it would ideally be.

One such area is the complexity analysis of the algorithms of CGT. During the 1980s and 1990s some, for the most part friendly and respectful, rivalry developed between those whose research in CGT was principally directed towards producing better performance of their code, and those who were more interested in proving theoretical results concerning the complexity of the algorithms. This study of complexity began with the work of Eugene Luks, who established a connection in his 1982 article [Luk82] between permutation group algorithms and the problem of testing two finite graphs for isomorphism. Our emphasis in this book will be more geared towards algorithms that perform well in practice, rather than those with the best theoretical complexity. Fortunately, Seress' book [Ser03] includes a very thorough treatment of complexity issues, and so we can safely refer the interested reader there. In any case, as machines become faster, computer memories larger, and bigger and bigger groups come within the range of practical computation, it is becoming more and more the case that those algorithms with the more favourable complexity will also run faster when implemented.

The important topic of computational group representation theory and computations with group characters is perhaps not treated as thoroughly as it might be in this book. We have covered some of the basic material in Chapter 7, but there is unfortunately no specialized book on this topic. For a brief survey of the area, we can refer the reader to the article by Gerhard Hiss [His03].

One of the most active areas of research in CGT at the present time, both from the viewpoint of complexity and of practical performance, is the development of effective methods for computing with large finite groups of matrices. Much of this material is beyond the scope of this book. It is, in any case, developing and changing too rapidly to make it sensible to attempt to cover it properly here. Some pointers to the literature will of course be provided, mainly in Section 7.8.

Yet another topic that is beyond the scope of this book, but which is of increasing importance in CGT, is computational Lie theory. This includes computations with Coxeter groups, reflection groups, and groups of Lie type and their representations. It also connects with computations in Lie algebras, which is an area of independent importance. The article by Cohen, Murray, and Taylor [CMT04] provides a possible starting point for the interested reader.

The author firmly believes that the correct way to present a mathematical algorithm is by means of pseudocode, since a textual description will generally lack precision, and will usually involve rather vague instructions like "carry on in a similar manner". So we have included pseudocode for all of the most basic algorithms, and it is only for the more advanced procedures that we have occasionally lapsed into sketchy summaries. We are very grateful to Thomas Cormen who has made his LaTeXpackage 'clrscode' for displaying algorithms publicly available. This was used by him and his coauthors in the well-known textbook on algorithms [CLRS02].

Although working through all but the most trivial examples with procedures that are intended to be run on a computer can be very tedious, we have attempted to include illustrative examples for as many algorithms as is practical.

At the end of each chapter, or sometimes section, we have attempted to direct the reader's attention to some applications of the techniques developed in that chapter either to other areas of mathematics or to other sciences. It is generally difficult to do this effectively. Although there are many important and interesting applications of CGT around, the most significant of them will typically use methods of CGT as only one of many components, and so it not possible to do them full justice without venturing a long way outside of the main topic of the book.

We shall assume that the reader is familiar with group theory up to an advanced undergraduate level, and has a basic knowledge of other topics in algebra, such as ring and field theory. Chapter 2 includes a more or less complete survey of the required background material in group theory, but we shall assume that at least most of the topics reviewed will be already familiar to readers. Chapter 7 assumes some basic knowledge of group representation theory, such as the equivalence between matrix representations of a group G over a field K and KG-modules, but it is interesting to note that many of the most fundamental algorithms in the area, such as the 'Meataxe', use only rather basic linear algebra.

A number of people have made helpful and detailed comments on draft versions of the book, and I would particularly like to thank John Cannon, Bettina Eick, Joachim Neubüser, and Colva Roney-Dougal in this regard.

Most of all, I am grateful to my two coauthors, Bettina Eick and Eamonn O'Brien for helping me to write parts of the book. The whole of Chapter 8, on computing in polycyclic groups, was written by Bettina Eick, as were Sections 11.1 and 11.4 on the libraries of primitive groups and of small groups, respectively. Section 7.8 on computing in matrix groups and Section 9.4 on computing p-quotients of finitely presented groups were written by Eamonn O'Brien.

August 2004   Derek Holt

Contents

Notation and displayed procedures

As is usual, $\{a, b, c\}$ will denote the (unordered) set with elements a, b, c. We shall generally use square brackets, as in $[a, b, c, c, a]$, to denote an ordered set or list, which may of course contain repeated elements. There is an unfortunate and occasionally serious conflict here with the notation for commutators of subgroups and elements of a group. If there is a danger of ambiguity in this respect, then we shall write $\mathsf{comm}(g, h)$ rather than $[g, h]$ for the commutator $g^{-1}h^{-1}gh$. Here is some notation used in the book, which might not be completely standard.

$\mathbb{N}$	the set of natural numbers $1, 2, 3, \ldots$ excluding 0
$\mathbb{N}_0$	the set of natural numbers $0, 1, 2, 3, \ldots$ including 0
$\mathbb{F}_q$	the finite field of order q
$[3 \, . \, . \, 8]$	the ordered list $[3, 4, 5, 6, 7, 8]$
$[11 \, . \, . -1 \text{ by } -3]$	the ordered list $[11, 8, 5, 2, -1]$
$\{3 \, . \, . \, 8\}$	the unordered set $\{3, 4, 5, 6, 7, 8\}$
$[1, 3, 2] \mathsf{\ cat\ } [5, 1]$ or $[1, 3, 2] \cup [5, 1]$	
	the concatenation $[1, 3, 2, 5, 1]$ of the two ordered lists

The LaTeX package 'clrscode' has been used for typesetting the displayed procedures in this book. This is the package used in the book [CLRS02], and was downloaded from Thomas Cormen's Web site [Cor].

We shall assume that the meaning of the constructs used in the procedures, like **while, do, if, then, elseif** is sufficiently well-known to require no explanation. Unlike in [CLRS02], however, we use $a := b$ (rather than $a \leftarrow b$) to denote an assignment. By default, a command **break** or **continue** will cause the procedure to break out of or proceed to the next iteration of the current innermost loop. We shall, however, occasionally use "**continue** α" to mean proceed to the next iteration of the loop "**for** $\alpha \in \cdots$".

Procedures may return one or more values. So, if a procedure MULTI-PROC(x) ends with a statement **return** $a, b;$, then we access the two returned values using a statement like $y, z := $ MULTIPROC(x). If an argument of a procedure is preceded by a $\sim$ symbol, as in APPEND$(\sim l, x)$ then, on calling the procedure, the value of the argument may be changed within the procedure.

We shall not display code for a number of basic procedures, of which the meaning is assumed to be obvious. For example, when called with a list l and an element x, the procedure call APPEND$(\sim l, x)$ simply appends x to the end of the list l.

Chapter 1

A Historical Review of Computational Group Theory

We begin the book with a brief review of the history of CGT. Some of the material in this chapter has been taken from Section 2 of the article by J. Neubüser [Neu95].

Group theory has always been a computational subject. For example, there were repeated attempts during the nineteenth century, with varying degrees of accuracy, to compile complete lists of the primitive and transitive permutation groups of low degree. This work began with Ruffini in 1799 and continued until about 1912. It was of course all done painstakingly using hand calculations. Rather than provide citations here, we refer the reader to the extensive bibliography compiled by Short in [Sho92, Appendix A] for details. For whatever reason, very little further work was done on this particular problem until about 1970, at which time computers could be used both to check the results of the older calculations and, over the following years, to extend the lists to groups of much higher degree. We shall be discussing recent progress in this area in Chapter 11 of this book.

The first genuine algorithms proposed for solving problems in group theory predate the days of mechanical computers. As early as 1911, in [Deh11], Dehn formulated the word, conjugacy, and isomorphism problems for finitely presented groups, and he devised an algorithm, the *Dehn algorithm*, for solving the word problem in certain types of small cancellation groups. Chapters 12 and 13 will explore recent computational work on some of these problems. However, problems of this type are usually undecidable in general: it was proved by Novikov [Nov55] and then by Boone in [Boo59] that the word and conjugacy problems are undecidable, and by Adian in [Adi58] and Rabin in [Rab58] that the isomorphism problem is undecidable.

Then, in 1936, Todd and Coxeter described the 'coset enumeration' procedure, which attempts to calculate the index of a finitely generated subgroup of a finitely presented group, on the assumption that this index is finite. While the importance of Dehn's algorithm was more theoretical than practical, the Todd-Coxeter method was intended to be used for real calculations, which indeed it was. Coset enumeration remains one of the most fundamental and important procedures in CGT, and we shall be studying it in detail in Chapter 5.

1

Originally it was used for hand calculations, and there were successful enumerations by hand of the cosets of subgroups of index several hundred. The current most versatile computer implementation, ACE3 [HR99], by George Havas and Colin Ramsay, can routinely handle subgroups of index up to at least 10^8.

Yet another example from the premachine era was an algorithm for the classification of space groups proposed in [Zas48] by Zassenhaus. Some time was to elapse before this could be put to practical use; we shall be discussing this application in Section 11.5 of Chapter 11.

Given that group theory had been recognized as an intrinsically computational branch of mathematics for over a century, it is not surprising that solving problems in group theory was one of the very earliest nonnumerical applications of programmable computers to be proposed. Indeed, some of these early suggestions were surprisingly ambitious, and demonstrated a striking perceptivity for the eventual capabilities of CGT. In 1951, M.H.A. Newman [New51] proposed using computers to investigate groups of order 256 — the complete classification of the 56 092 isomorphism types of groups of this order was eventually completed by O'Brien in 1988 [O'B91]. As another example, there is a quote from A. Turing from 1945, which suggests the idea of writing a computer program to enumerate the groups of order 720. However, the first reported enumeration of the 840 isomorphism types of groups of this order was in 1999, as part of the systematic construction of all groups of order up to 1000 undertaken by Besche and Eick [BE99b].

The first implementation of a group-theoretical algorithm was probably by B. Hazelgrove, who partially implemented Todd-Coxeter coset enumeration on the EDSAC II in Cambridge in 1953. For details of this and other early implementations of coset enumeration, see Leech's article [Lee63]. Possibly the first genuinely useful programs were those of Felsch and Neubüser [Neu60, Neu61] for computing the subgroup lattices of finite permutation groups, and of finite 2-groups given by polycyclic presentations.

A major breakthrough in the art of computing in finite permutation groups came around 1970 with the base and strong generating set methods and the Schreier-Sims algorithm introduced by Charles Sims in [Sim70, Sim71a]. These methods have formed the backbone of this area of CGT ever since, and they now enable detailed structural computations to be carried out routinely in permutation groups of degree up to about 10^7. Indeed, computational group theory made the mathematical headlines for the first time in 1972, when Sims used his methods as the basis for his proof by computer of the existence of the Lyons sporadic simple group [Sim73]. He constructed the group as a permutation group of degree 8 835 156. Later [Sim80], in 1977, he was able to prove the existence of the Baby Monster in a similar fashion; in this case the degree is 13 571 955 000. The groups J_3, He, and ON were also constructed by computer as permutation groups.

Computing in finite permutation groups will be covered in Chapter 4 and also in several later chapters. More recent work in this area has made use of the O'Nan-Scott theorem, which describes the detailed structure of a primitive permutation group, and of the classification of finite simple groups.

Other noteworthy developments in the 1970s include the development of (a number of variations of) a procedure for computing finite quotients of prime power order of finitely presented groups. The resulting algorithm is now known as the *p-quotient* algorithm and will form the main topic of Section 9.4 of Chapter 9. In the mid-1970s, versions were developed and implemented by Macdonald [Mac74], Wamsley [BKW74], and Havas and Newman [New76]. Several new results were discovered using these implementations, including the orders of the restricted Burnside groups $|R(2,5)| = 5^{34}$ and $|R(4,4)| = 2^{422}$ (where $R(d,n)$ denotes the largest finite d-generator group of exponent n).

More recently, in 2002, using a special version of the algorithm that uses techniques involving Lie algebras, O'Brien and Vaughan-Lee [OVL02] proved that $|R(2,7)| = 7^{20416}$. This computation involved about a year of CPU-time!

The p-quotient algorithm outputs finite quotients in the form of a *power-conjugate presentation* or *polycyclic presentation*. Efficient algorithms for computing in finite p-groups, and later in finite solvable and general polycyclic groups, using this type of presentation as the basic data structure for group elements, were first developed around the same period. These methods introduced a number of techniques, such as:

- solving a problem from the top down, by solving it successively in larger and larger quotients G/N of the group G;

- a divide-and conquer approach that uses group homomorphisms to split a problem up into smaller instances of the same problem on the kernel and image of the homomorphism;

which were later generalized to permutation and matrix groups. For example, Neubüser and Felsch [FN79] described and implemented a top-down approach to computing conjugacy classes and centralizers in p-groups, which was used to find a counterexample to an open conjecture, the 'class-breadth' conjecture. Algorithms for computing in polycyclic groups using polycyclic presentations will form the topic of Chapter 8.

Also worthy of mention from this period are:

- Dixon's algorithm for computing character tables of finite groups [Dix67];

- the application of the Zassenhaus algorithm and the subgroup lattice algorithms mentioned above to the determination of the 4-dimensional space groups [BBN+78];

- the implementation by Havas of the Reidemeister-Schreier algorithm for computing and simplifying a presentation of a subgroup of finite index in a finitely presented group;

- a program of Felsch and Neubüser [FN68, FN70] for computing automorphism groups of finite groups.

Computation with characters and representations of finite groups has played a significant rôle in CGT and continues to do so. The Dixon algorithm for computing character tables was based on a method described by Burnside in [Bur11], which was used by John McKay in [McK70] in its original form to compute the character tables of the first two Janko sporadic simple groups J_1 and J_2. It was later refined by Schneider in [Sch90b]. These methods will be described in Section 7.7.

As mentioned already, five of the larger sporadic simple groups were proved to exist by means of computer constructions as permutation groups. Four others, Ru, Th, HN, and J_4 were originally constructed computationally as matrix groups either over an algebraic number field or over a finite field; for example, J_4 was constructed by Norton and others as a subgroup of GL(112, 2) [Nor80]. See Section 2 of [Gor82] or [CCN+85] for details; but be aware that the names used in [Gor82] for several of these groups are different from the ones we are using here, which are as in [CCN+85]. It should be mentioned that several, but not all, of the sporadic simple groups that were originally proved to exist by means of a computer calculation now have alternative existence proofs as subgroups of the largest such group, the Monster, which was constructed by R.L. Griess, without computer assistance, as a group of automorphisms of a certain 196 854-dimensional algebra [Gri82].

The principal computational tool that has been used for working with large groups of matrices over finite fields is a collection of basic linear algebra procedures designed principally by Richard Parker [Par84] which, among other things, can test large dimensional representations of groups over finite fields for irreducibility. This has become known as the 'Meataxe' algorithm on account of the efficiency with which it can chop modules up into their irreducible constituents. These methods will be treated in detail in Chapter 7 of this book. Together with other techniques, they were used extensively in the construction of Brauer trees by Hiss and Lux in [HL89] and of Brauer character tables by Jansen, Lux, Parker, and Wilson in [JLPW95].

These constructions involved an important additional technique, unfortunately beyond the scope of this book, known as *condensation* (see, for example [Ryb90, LMR94, Ryb01]), which can be used to reduce computations in large dimensional modules to equivalent computations over smaller dimensional modules over a different but Morita equivalent algebra.

The very earliest computer programs for working with groups, such as Hazelgrove's 1953 implementation of coset enumeration, were written in machine code. For later programs, Fortran was for many years the most popular choice of language, and more recently 'C' and (less often) 'C++' have been used extensively. There remain a small number of stand-alone programs and packages that are in widespread use, such as ACE3 [HR99] for coset enumeration, which we have mentioned already, and the most recent stand-alone version of the 'Meataxe' programs, written mainly by Michael Ringe [Rin92]. Sims' Rutgers Knuth-Bendix Package (rkbp), and the author's KBMAG package for the Knuth-Bendix algorithm in groups and monoids and computing in automatic groups also spring to mind; the algorithms involved will form the subject of Chapters 12 and 13 of this book.

However, the prevailing philosophy today is that it is nearly always preferable to implement algorithms in CGT as a component of an already existing package or system. The programmer need not then be concerned with much of the underlying functionality required, such as multiplication of permutations, and basic matrix operations, because implementations of these will already be provided by the system. Indeed, the user can make use of or build on top of anything that is already there. So the principal objectives of systems of this kind are:

(i) to provide efficient implementations of a number of basic mathematical procedures and data structures;

(ii) to include the most up-to-date and versatile implementations of the fundamental algorithms of CGT;

(iii) to provide a high-level programming or scripting language to allow the user either to use one of these algorithms to perform a single computation, or to write programs that call one or more of the algorithms, possibly repeatedly.

The programs referred to in (iii) could either be for the purpose of a one-off computation or, at the other extreme, they could be designed to extend the functionality of the system, in which case they might eventually be incorporated into it.

The first such system for CGT, known as 'Group', and written and administered by John Cannon in Sydney and Joachim Neubüser in Aachen, got off the ground in the mid-1970s, and was written in Fortran. By about 1980, this had developed into the Sydney-based CAYLEY system. Around 1993, CAYLEY was superceded by MAGMA, which is written in 'C' (see [CP93] or the Web site and online manual [Mag04]).

A number of more specialized packages, such as CAS [NPP84] for calculating and manipulating characters of finite groups, were developed and maintained in Aachen. They were superceded in 1986 by GAP ('Groups, Algorithms and Programming'); see [GAP04] for detailed information and a list of authors. It has been based in St Andrews, Scotland, since 1997, and is now developed and maintained by an international development team.

GAP and MAGMA are the two systems for CGT that currently enjoy the most widespread usage, although the large computer algebra systems Maple and Mathematica both include some facilities for CGT, including coset enumeration and the Schreier-Sims algorithm for permutation groups. Both GAP and MAGMA aim to cover the whole of CGT and to keep up with all of the latest developments in the area. MAGMA aims for maximal efficiency by implementing many fundamental algorithms at the 'C' level, whereas GAP aims for maximal versatility by having more code written in the GAP language, which allows nonexpert users more flexibility in adapting or extending it. But for both systems, a large proportion of the new code, particularly for algorithms under development, is written in the language of the system.

Until about 1990, CGT was able to exist more or less in isolation from the rest of computer algebra. In more recent years, that situation has been slowly but steadily changing, and many of the newer algorithms in CGT have required facilities from mainstream computer algebra. For example:

- The best performing Smith and Hermite normal form procedures, which will arise in Section 9.2 in connection with computing abelian quotients of finitely presented groups, need the LLL and MLLL (modified) lattice reduction algorithms.

- The flexible version of the Meataxe algorithm described by Holt and Rees in [HR94] uses factorization of polynomials over finite fields.

- The polycyclic quotient algorithm, proposed initially by Baumslag, Cannonito, and Miller [BCM81b, BCM81a], is discussed in Chapters 10 and 11 of [Sim94] for the metabelian case, and has been implemented in the general case by Lo [Lo98]. It uses Gröbner bases over group rings $\mathbb{Z}G$ of polycyclic groups.

- The methods under development by Eick for computing with infinite polycyclic groups, which we shall summarize in Subsection 8.2.2 of this book, require computational facilities from algebraic number theory.

We can expect interactions of this kind to become more and more commonplace in the future.

Both **GAP** and Magma have responded to this situation by expanding the basic facilities that they provide. **GAP** is able to do this by means of its package mechanism, which allows external code written either in **GAP** or in another language, like 'C', to be incorporated into the system by its users.

For example, there is a package of this kind by L. Soicher for computing with combinatorial objects, which includes the 'Nauty' package of Brendan D. McKay for testing finite graphs for isomorphism. A package providing an interface to the KANT system for algebraic number theory is also available.

Since its inception, Magma has been steadily increasing its basic remit. Although it started life as a successor to the CGT system CAYLEY, it is no longer accurate to describe Magma as a CGT system. It now offers a complete and up-to-date range of facilities in computational algebraic number theory, geometry and combinatorics, as well as for computation in groups, rings, modules, etc.; indeed the most exciting developments at the present time are probably in algebraic geometry. The interaction between CGT and other branches of computer algebra has been a two-way affair; for example, some computations in algebraic number theory and algebraic geometry involve computing with cohomology groups.

There are or have been a number of other more specialized systems and packages that deserve a brief mention.

- Derek Holt and Sarah Rees wrote and maintain an interactive graphics package QUOTPIC [HR93], which uses the methods to be described in Chapter 9 together with other basic algorithms from CGT to compute quotients of finitely presented groups.

- The New York Group Theory Cooperative directed by Gilbert Baumslag and others maintain and develop an interactive system MAGNUS (see the Web site [Mag]) for exploring properties of finitely presented groups. It has the interesting feature that several approaches to solving a problem can be tried and run in parallel; for example, if the user asks whether a group is finite, it will simultaneously run one or more programs that attempt to establish finiteness and infiniteness.

- The CHEVIE package of Geck and others for symbolic calculations with generic character tables of groups of Lie type, Coxeter groups, Iwahori-Hecke algebras, and other related structures; see the Web site [Che].

- SYMMETRICA, DISCRETA, and MOLGEN will be mentioned at the end of Section 10.5.

Other recent historical reviews of CGT have been written by Seress [Ser97] and Sims [Sim03]. For a much more detailed account of the early history of CGT and the state of the art around 1970, see the article by Neubüser [Neu70]. Indeed, there are several papers on CGT in the proceedings of the conference on computational problems in abstract algebra, which took place in Oxford in 1967 [Lee70]. The first conference exclusively on CGT was in Durham, England, in 1982 [Atk84]. Since then, there have been four conferences on CGT in Oberwolfach, Germany, but without proceedings. There have also been four organized by DIMACS in the United States, the first three of which have published proceedings [FK93, FK97, KS97]. In addition, a number of special issues of the *Journal of Symbolic Computation* have been devoted to CGT: parts of Volumes 9 (1990), 12 (1991), and 24 (1997).

Chapter 2

Background Material

In this chapter we shall summarize some algebraic background material, generally without proofs or with only outline proofs. We are assuming that the reader is already familiar with the majority of the contents of this chapter, which is included only as a quick reference guide, and to introduce notation. The bulk of this material is from group theory and group representation theory, but we have also included a little field theory, including the basic theory of finite fields, which plays an important rôle in a number of areas of CGT.

2.1 Fundamentals

We start by stating some definitions and results from elementary group theory; see [Rot94], for example, for discussion and proofs. Many readers will want to skip this section entirely, but one of its functions is to introduce notation.

2.1.1 Definitions

DEFINITION 2.1 A group is a set G together with a binary operation $\circ : G \times G \to G$ that satisfies the following properties:

  (i) (*Closure*) For all $g, h \in G$, $g \circ h \in G$;

  (ii) (*Associativity*) For all $g, h, k \in G$, $(g \circ h) \circ k = g \circ (h \circ k)$;

  (iii) There exists a unique element $e \in G$ satsifying:

    (a) (*Identity*) for all $g \in G$, $g \circ e = e \circ g = g$; and

    (b) (*Inverse*) for all $g \in G$ there exists a unique $h \in G$ such that $h \circ g = g \circ h = e$.

This definition assumes more than is necessary; for example, the uniqueness of the identity and inverse elements can be deduced. See Exercise 1 below.

The number of elements in G is called the *order* of G and is denoted by $|G|$. The element $e \in G$ satisfying (iii) of the definition is called the *identity element* of G and, for $g \in G$, the element h that satisfies (iii)(b) of the definition ($h \circ g = e$) is called the *inverse* of g.

DEFINITION 2.2   A group is called *abelian* or *commutative* if it satisfies the additional property: (*Commutativity*) for all $g, h \in G$, $g \circ h = h \circ g$.

We shall now change notation! The groups in this book will either be:

- multiplicative groups, where we omit the $\circ$ sign ($g \circ h$ becomes just gh), we denote the identity element by 1 rather than by e, and we denote the inverse of $g \in G$ by g^{-1}; or

- additive groups, where we replace $\circ$ by $+$, we denote the identity element by 0, and we denote the inverse of g by $-g$.

If there is more than one group around, and we need to distinguish between the identity elements of G and H say, then we will write 1_G and 1_H (or 0_G and 0_H).

Additive groups will always be commutative, but multiplicative groups may or may not be commutative. Multiplicative notation is the default.

Note that in a multiplicative group G, we have $(gh)^{-1} = h^{-1}g^{-1}$ for all $g, h \in G$. The *cancellation laws* $gh = gk \Rightarrow h = k$ and $hg = kg \Rightarrow h = k$ are also easily deduced.

In a multiplicative group G, if $n \in \mathbb{N}$ and $g \in G$, then g^n is defined inductively by $g^1 = g$ and $g^{n+1} = gg^n$ for $n \geq 1$. We also define g^0 to be the identity element 1, and g^{-n} to be the inverse of g^n. Then $g^{x+y} = g^x g^y$ for all $x, y \in \mathbb{Z}$. In an additive group, g^n becomes ng, where $0g = g$ and $(-n)g = -(ng)$.

Let $g \in G$. Then the *order* of g, denoted by $|g|$, is the least $n > 0$ such that $g^n = 1$, if such an n exists. If there is no such n, then g has infinite order, and we write $|g| = \infty$. Note that if g has infinite order, then the elements g^x are distinct for distinct values of x. Similarly, if g has finite order n, then the n elements $g^0 = 1, g^1 = g, \ldots, g^{n-1}$ are all distinct, and for any $x \in \mathbb{Z}$, g^x is equal to exactly one of these n elements. Also $|g| = 1 \Leftrightarrow g = 1$, and if $|g| = n$ then, for $x \in \mathbb{Z}$, $g^x = 1 \Leftrightarrow n|x$.

If G and H are two (multiplicative) groups, then we define the *direct product* $G \times H$ of G and H to be the set $\{(g, h) \mid g \in G,\ h \in H\}$ of ordered pairs of elements from G and H, with the obvious component-wise multiplication of elements $(g_1, h_1)(g_2, h_2) = (g_1 g_2, h_1 h_2)$ for $g_1, g_2 \in G$ and $h_1, h_2 \in H$.

It is straightforward to check that $G \times H$ is a group under this operation. If the groups are additive, then it is usually called the direct sum rather than the direct product, and written $G \oplus H$.

More generally, we can define the direct product G^n of n copies of G for $n \geq 1$, by $G^1 = G$ and $G^n = G^{n-1} \times G$ for $n > 1$. In particular, the direct product of n copies of a cyclic group C_p of order p has order p^n and is called an *elementary abelian p*-group.

In general, if A, B are subsets of a group G, then we define

$$AB := \{\, ab \mid a \in A,\, b \in B \,\} \text{ and } A^{-1} := \{\, a^{-1} \mid a \in A \,\}.$$

2.1.2 Subgroups

DEFINITION 2.3   A subset H of a group G is called a *subgroup* of G if it forms a group under the same operation as that of G.

For any group G, G and $\{1\}$ are subgroups. A subgroup other than G is called a *proper subgroup*, and a subgroup other than $\{1\}$ is called a *nontrivial subgroup*. We often abuse notation and denote the trivial subgroup $\{1\}$ by 1.

We shall use $H \subseteq G$ (respectively, $H \subset G$) to mean that H is a subset (respectively proper subset) of G, and $H \leq G$ (respectively, $H < G$) to mean that H is a subgroup (respectively, proper subgroup) of G.

A subgroup H of G is called *maximal*, if $H < G$ and there are no subgroups K with $H < K < G$. Note that all nontrivial finite groups have maximal subgroups, but an infinite group need not have any.

If $H \leq G$, then it is easy to show that $1_H = 1_G$. The following result gives a commonly used criterion for deciding whether a subset H of G is a subgroup.

PROPOSITION 2.4   *Let H be a nonempty subset of a group G. Then H is a subgroup of G if and only if*

$$h_1, h_2 \in H \Rightarrow h_1 h_2 \in H; \quad \text{and} \quad h \in H \Rightarrow h^{-1} \in H.$$

Let $g \in G$. Then the *right coset* Hg is the subset $\{hg \mid h \in H\}$ of G. Similarly, the *left coset* gH is the subset $\{gh \mid h \in H\}$ of G. (Warning: Of course, by the law of Universal Cussedness, some authors define these the other way round!) For additive groups, we write $H + g$ rather than Hg.

The coset $H1$ is of course H itself. Two cosets Hg, and Hk are equal if and only if $g \in Hk$, which is the case if and only if $gk^{-1} \in H$. Otherwise they are disjoint. So the distinct right cosets partition G. Hence we have what is probably the best-known result in finite group theory:

THEOREM 2.5   (Lagrange's theorem)   *Let G be a finite group and H a subgroup of G. Then the order of H divides the order of G.*

The *index* $|G : H|$ of H in G is defined to be the number of distinct right cosets of H in G, which may be finite or infinite in general, and is equal to $|G|/|H|$ when G is finite. A system of representatives in G of the right cosets of H in G is called a *right transversal* of H in G, and a *left transversal* is defined correspondingly. Note that T is a right transversal if and only if T^{-1} is a left transversal.

An easy consequence of Lagrange's thereom is that, in a finite group G, the order $|g|$ of each $g \in G$ divides $|G|$.

2.1.3   Cyclic and dihedral groups

A group G is called *cyclic* if it consists of the integral powers of a single element. In other words, G is cyclic if there exists an element g in G with the property that, for all $h \in G$, there exists $x \in \mathbb{Z}$ with $g^x = h$. The element g is called a *generator* of G.

The most familiar examples of cyclic groups are additive groups rather than multiplicative. The group $(\mathbb{Z}, +)$ of integers under addition is cyclic, because every integer x is equal to $x1$, and so 1 is a generator. If we fix some integer $n > 0$ and let $\mathbb{Z}_n = \{0, 1, 2, \ldots, n-1\}$ with the operation of addition modulo n, then we get a cyclic group of order n, where 1 is once again a generator, as is any number in $\mathbb{Z}_n$ that is coprime to n. Note that in an infinite cyclic group, any generator has infinite order, and in a finite cyclic group of order n, any generator has order n.

Let $n \in \mathbb{N}$ with $n \geq 2$ and let P be a regular n-sided polygon in the plane. The dihedral group of order $2n$ consists of the isometries of P; that is:

(i) n rotations through the angles $2\pi k/n$ $(0 \leq k < n)$ about the centre of P; and

(ii) n reflections about lines that pass through the centre of P, and either pass through a vertex of P or bisect an edge of P (or both).

Unfortunately, some authors denote this group by D_n and others by D_{2n}, which can be confusing! We shall use D_{2n}.

To study these groups, it is convenient to number the vertices $1, 2, \ldots, n$, and to regard the group elements as permutations of the vertices. Then the n rotations are the powers a^k for $0 \leq k < n$, where $a = (1, 2, 3, \ldots, n)$ is a rotation through the angle $2\pi/n$. Let b be the reflection through the bisector of P that passes through the vertex 1. Then $b = (2, n)(3, n-1)(4, n-2) \cdots$. For example, when $n = 5$, $b = (2, 5)(3, 4)$ and when $n = 6$, $b = (2, 6)(3, 5)$. Notice that there is a difference between the odd and even cases. When n is odd, b fixes no vertex other than 1, but when n is even, b fixes one other vertex, namely $(n+2)/2$. Now we can see, either geometrically or by multiplying permutations, that the n reflections of P are the elements $a^k b$ for $0 \leq k < n$. Thus we have

$$G = \{a^k \mid 0 \leq k < n\} \cup \{a^k b \mid 0 \leq k < n\}.$$

In all cases, we have $ba = a^{n-1}b \ (= a^{-1}b)$.

2.1.4 Generators

Let A be any subset of a group G. The subgroup of G generated by A, which is denoted by $\langle A \rangle$, or by $\langle x_1, x_2, \ldots, x_r \rangle$ if $A = \{x_1, x_2, \ldots, x_r\}$, is defined to be the intersection of all subgroups of G that contain A. In particular, we say that A generates G (or is a generating set for G) if $\langle A \rangle = G$.

An equivalent definition is that $\langle A \rangle$ is equal to the set of all elements of G that can be written as products $x_1 x_2 \cdots x_r$ for some $r \in \mathbb{N}_0$, where each x_i is in A or in A^{-1}. The empty product ($r = 0$) is defined to be 1_G.

In practice, and particularly in computational applications, a generating set $\{x_1, x_2, \ldots, x_r\}$ often has an ordering that is defined implicitly by the subscripts of the x_i. It is also convenient to allow for the possibility that $x_i = x_j$ when $i \neq j$, which can result in conflicts with the standard mathematical convention that sets may not contain repeated elements. We shall therefore frequently refer to generating sequences $[x_1, \ldots, x_k]$, rather than generating sets, where the subgroup generated by the sequence is defined to be the subgroup generated by the set consisting of the elements in the sequence.

2.1.5 Examples — permutation groups and matrix groups

Let Ω be any set, and let $\mathrm{Sym}(\Omega)$ denote the set of permutations of Ω; that is, the bijections from Ω to itself. Then $\mathrm{Sym}(\Omega)$ is a group under composition of maps. It is known as the *symmetric* group on Ω. A subgroup of $\mathrm{Sym}(\Omega)$ is called a *permutation group* on Ω.

For $x \in \Omega$ and $g \in \mathrm{Sym}(\Omega)$, we shall denote the result of applying g to x by x^g rather than by the more usual $g(x)$. The composite gh will mean "first apply g, then apply h", and so $x^{gh} = (x^g)^h$.

Let us recall the cyclic notation for permutations. If $a_1, \ldots, a_r$ are distinct elements of Ω, then the cycle $(a_1, a_2, \ldots, a_r)$ denotes the permutation g of Ω with $a_i^g = a_{i+1}$ for $1 \leq i < r$, $a_r^g = a_1$, and $b^g = b$ for $b \in \Omega \setminus \{a_1, a_2, \ldots, a_r\}$.

When Ω is finite, any permutation of Ω can be written as a product (or composite) of disjoint cycles. Note that a cycle (a_1) of length 1 means that $a_1^g = a_1$, and so this cycle can (and normally is) omitted.

As an example of composition, if $\Omega = \{1, 2, 3\}$, $g_1 = (1, 2, 3)$ and $g_2 = (1, 2)$, then $g_1 g_2 = (2, 3)$, whereas $g_2 g_1 = (1, 3)$. The inverse of a permutation can be calculated by reversing all of the cycles. For example, the inverse of $(1, 5, 3, 6)(2, 8, 7)$ is $(6, 3, 5, 1)(7, 8, 2) = (1, 6, 3, 5)(2, 7, 8)$. (The cyclic representation is not unique: $(a_1, a_2, \ldots, a_r) = (a_2, a_3, \ldots, a_r, a_1)$, etc.)

A cycle of length 2 is called a *transposition*. Since any cycle can be written as a product of transpositions ($(1, 2, 3, \ldots, n) = (1, 2)(1, 3)(1, 4) \cdots (1, n)$, for example), it follows that any permutation on Ω can be written as a product of transpositions. A permutation is called *even* if it is a product of an even number of transpositions, and *odd* if it is a product of an odd number of transpositions. It can be proved that no permutation is both even and odd, so the definition makes sense.

The set H of even permutations forms a subgroup of $\mathrm{Sym}(\Omega)$. This is known as the *alternating group* on Ω, and is denoted by $\mathrm{Alt}(\Omega)$. If (x, y) is some fixed transposition on Ω then, for any $g \in \mathrm{Sym}(\Omega)$, one of the two permutations g and gx is even and the other is odd. Hence $G = H \cup H(x, y)$, so $|G\!:\!H| = 2$.

Let K be a field. Then, for any fixed $d > 0$, the set of $d \times d$ invertible matrices with entries in K forms a group under multiplication. This group is denoted by $\mathrm{GL}(d, K)$ or $\mathrm{GL}_d(K)$.

The $d \times d$ matrices over K with determinant 1 also form a group denoted by $\mathrm{SL}(d, K)$ or $\mathrm{SL}_d(K)$. This is a subgroup of $\mathrm{GL}(d, K)$. When K is a finite field $\mathbb{F}_q$ of order q, we use the alternative notation $\mathrm{GL}(d, q)$, etc. There are a number of important subgroups of $\mathrm{GL}(d, q)$ known as the *classical groups*. These include the symplectic groups $\mathrm{Sp}(d, q)$, the unitary groups $\mathrm{GU}(d, q), \mathrm{SU}(d, q)$, and two types of orthogonal groups $\mathrm{GO}(d, q)$. We refer the reader to [CCN+85] for definitions, essential properties, and further references.

2.1.6 Normal subgroups and quotient groups

DEFINITION 2.6   A subgroup H of a group G is called *normal* in G if the left and right cosets gH and Hg are equal for all $g \in G$.

For example, $\{1\}$ and G are normal subgroups of any group G, and any subgroup of index 2 is normal, but the subgroup $\{1, (1, 2)\}$ of $\mathrm{Sym}(\{1, 2, 3\})$ is not normal.

An equivalent condition, which is often used as the definition, is that H is a normal subgroup of G if and only if $g^{-1}hg \in H$ for all $h \in H$, $g \in G$.

The standard notation for "H is a normal subgroup of G" is $H \vartriangleleft G$ or $H \trianglelefteq G$. We shall use $H \vartriangleleft G$ to mean that H is a proper normal subgroup of G – i.e., $H \neq G$, but this usage is not universal.

A critical property of normal subgroups N of groups G, which is not shared by subgroups in general, is that the product of any element $n_1 g$ of the coset Ng with any element $n_2 h$ of the coset Nh lies in the coset Ngh. Hence $(Ng)(Nh) = Ngh$, and the set of right cosets Ng of N in G forms a group of order $|G\!:\!N|$ under multiplication. This group is denoted by G/N and is called the *quotient group* (or sometimes *factor group*) of G by N.

A group is called *simple* if $|G| > 1$ and the only normal subgroups of G are $\{1\}$ and G. For example, cyclic groups of prime order are simple.

DEFINITION 2.7   Let A be a subset of a group G. The *normal closure* of A in G, which is denoted by $\langle A^G \rangle$, is defined to be the intersection of all normal subgroups of G that contain A.

Since the intersection of any collection of normal subgroups of G is easily shown to be normal in G, $\langle A^G \rangle$ is the smallest normal subgroup of G containing A. An equivalent definition of $\langle A^G \rangle$ is the subgroup of G generated by the set $A^G := \{\, g^{-1}ag \mid g \in G, a \in A \,\}$.

2.1.7   Homomorphisms and the isomorphism theorems

DEFINITION 2.8   Let G and H be groups. A *homomorphism* φ from G to H is a map $\varphi : G \to H$ such that $\varphi(g_1 g_2) = \varphi(g_1)\varphi(g_2)$ for all $g_1, g_2 \in G$.

It is straightforward to prove that any homomorphism $\varphi : G \to H$ maps 1_G to 1_H, and that $\varphi(g^{-1}) = \varphi(g)^{-1}$ for all $g \in G$.

A homomorphism φ is called, respectively, a *monomorphism*, an *epimorphism* or an *isomorphism*, if it is, respectively, an injection, a surjection or a bijection. A homomorphism from G to G is called an *endomorphism* of G, and an isomorphism from G to G is called an *automorphism* of G.

The automorphisms of G form a group, denoted by $\mathrm{Aut}(G)$, under composition. This is of course a subgroup of $\mathrm{Sym}(G)$, and so we follow our convention for permutation groups, and represent the result of applying $\alpha \in \mathrm{Aut}(G)$ to $g \in G$ by g^α rather than by $\alpha(g)$. So, for $\alpha, \beta \in \mathrm{Aut}(G)$, $\alpha\beta$ means first α then β. For any $k \in G$, the map $\alpha_k : G \to G$ defined by $g^{\alpha_k} = k^{-1}gk$ is an automorphism of G. This type of automorphism is known as an *inner automorphism*, and the set of all inner automorphisms forms a normal subgroup $\mathrm{Inn}(G)$ of $\mathrm{Aut}(G)$.

An automorphism that is not inner is called an *outer automorphism* and the outer automorphism group $\mathrm{Out}(G)$ is defined to be the quotient group $\mathrm{Aut}(G)/\mathrm{Inn}(G)$.

Two groups G and H are called *isomorphic*, and we write $G \cong H$, if there is an isomorphism $\varphi : G \to H$. The inverse of an isomorphism is an isomorphism, so $G \cong H$ implies $H \cong G$, and $G \cong H$ and $H \cong K$ imply $G \cong K$.

In many contexts isomorphic groups are regarded as being the same group. It is easy to prove that any two cyclic groups of the same order (finite or infinite) are isomorphic, so we talk about *the* infinite cyclic group, and *the* cyclic group of order n, which we shall denote by C_n in this book. Similarly, if Ω, Δ are sets with $|\Omega| = |\Delta|$, then $\mathrm{Sym}(\Omega) \cong \mathrm{Sym}(\Delta)$ and $\mathrm{Alt}(\Omega) \cong \mathrm{Alt}(\Delta)$, so, for any $n > 0$, we can talk about the symmetric and alternating groups $\mathrm{Sym}(n)$ and $\mathrm{Alt}(n)$ acting on a set of size n. By default, we shall regard these groups as acting on the set $\Omega = \{1 \mathinner{\ldotp\ldotp} n\}$.

Let $\varphi : G \to H$ be any group homomorphism. Then the *kernel* $\ker(\varphi)$ of φ is defined to be the set of elements of G that map onto 1_H; that is,

$$\ker(\varphi) = \{g \mid g \in G,\ \varphi(g) = 1_H\}.$$

It can be shown that $\ker(\varphi)$ is always a normal subgroup of G, whereas the image $\mathrm{im}(\varphi)$ of φ is a subgroup, but not necessarily a normal subgroup, of H. Furthermore, if N is any normal subgroup of G, then there is an epimorphism, known as the *natural epimorphism* $\varphi : G \to G/N$ defined by $\varphi(g) = Ng$. So the set of normal subgroups of G is just the set of kernels of homomorphisms with domain G.

We now state the three isomorphism theorems for groups. The first of these is the most important. For the second, note that if H is any subgroup and K is a normal subgroup of G, then $HK = KH$ is a subgroup of G. It is a normal subgroup if H is.

THEOREM 2.9 (first isomorphism theorem)   *Let $\varphi : G \to H$ be a homomorphism with kernel K. Then $G/K \cong \mathrm{im}(\varphi)$. More precisely, there is an isomorphism $\overline{\varphi} : G/K \to \mathrm{im}(\varphi)$ defined by $\overline{\varphi}(Kg) = \varphi(g)$ for all $g \in G$.*

THEOREM 2.10   (second isomorphism theorem)   *Let H be any subgroup and let K be a normal subgroup of a group G. Then $H \cap K$ is a normal subgroup of H and $H/(H \cap K) \cong HK/K$.*

THEOREM 2.11   (third isomorphism theorem)   *Let $K \subseteq H \subseteq G$, where H and K are both normal subgroups of G. Then $(G/K)/(H/K) \cong G/H$.*

Note that if α_k is the inner automorphism of a group G defined as described above by $k \in G$, then the map $G \to \mathrm{Aut}(G)$ defined by $k \mapsto \alpha_k$ is a homomorphism with image $\mathrm{Inn}(G)$. Its kernel is $\{\, k \in G \mid k^{-1}gk = g \ \forall g \in G \,\}$, which is defined to be the *centre* $\mathbf{Z}(G)$ of G. So, by the first isomorphism theorem, we have $G/\mathbf{Z}(G) \cong \mathrm{Inn}(G)$.

Exercises

1. Let G be a set endowed with an associative binary operation $\circ : G \times G \to G$ that contains an element $e \in G$ such that:

   (a) for all $g \in G$, $e \circ g = g$; and

   (b) for all $g \in G$ there exists $h \in G$ such that $h \circ g = e$.

   Prove that G is a group under the operation $\circ$.

2. Prove that the number $|G{:}H|$ of distinct right cosets of a subgroup H of a group G is equal to the number of distinct left cosets. (This is easy if G is finite, since both numbers are $|G|/|H|$, but less easy when G is infinite.)

3. Let a and b be the generators of the dihedral group D_{2n}, as in Subsection 2.1.3. Let H be a subgroup of D_{2n} that is contained in the cyclic subgroup $\langle a \rangle$. Show that H is normal in D_{2n} and that either $H = \langle a \rangle$, or G/H is isomorphic to a dihedral group.

4. Prove that the two definitions of the subgroup generated by a subset of a group G given in Subsection 2.1.3 are equivalent, and that the two definitions of the normal closure of a subset of a group are equivalent.

5. Prove that a cyclic group C_n of order n has exactly $\Phi(n)$ generators, where $\Phi(n) = |\{ m \in [1..n] \mid \gcd(m, n) = 1 \}|$ is the Euler Φ-function. Deduce that $|\operatorname{Aut}(C_n)| = \Phi(n)$.

6. Prove that the cyclic groups of prime order are the only (finite or infinite) abelian simple groups.

2.2 Group actions

2.2.1 Definition and examples

DEFINITION 2.12 Let G be a group and Ω a set. An *action* of G on Ω is defined to be a homomorphism $\varphi : G \to \operatorname{Sym}(\Omega)$.

So, for each $g \in G$, $\varphi(g)$ is a permutation of Ω. When we are talking about a fixed action φ, we shall usually just write α^g rather than $\alpha^{\varphi(g)}$ for the action of $\varphi(g)$ on $\alpha \in \Omega$. The fact that φ is a homomorphism then translates into the property $\alpha^{gh} = (\alpha^g)^h$ for all $g, h \in G$, $\alpha \in \Omega$.

Notice that it is inherent in our notation α^g that G is acting on Ω from the right. Actions from the left are more common in the literature, largely because this is more consistent with the convention of writing functions on the left of their arguments. However, both GAP and MAGMA use actions from the right, and we shall consistently adopt this convention in this book. Similarly, elements of vector spaces over a field K will be represented by row vectors $v \in K^d$, and a linear map $K^d \to K^d$ will be represented by a $d \times d$ matrix α over K, which acts on the vectors from the right; that is, $v \mapsto v \cdot \alpha$.

The property $\varphi(1_G) = 1_{\operatorname{Sym}(\Omega)}$ translates into $\alpha^1 = \alpha$ for all $\alpha \in \Omega$. In fact, an equivalent way of defining a group action is as a map $\Omega \times G \to \Omega$ with $(\alpha, g) \mapsto \alpha^g$, which satisfies $\alpha^1 = \alpha$ and $\alpha^{gh} = (\alpha^g)^h$ for all $g, h \in G$, $\alpha \in \Omega$. Some authors define a group action like this, and call a homomorphism from G to $\operatorname{Sym}(\Omega)$ a *permutation representation* of G on Ω, but we shall regard the expressions "action of G on Ω" and "permutation representation of G on Ω" as being synonymous. The image $\operatorname{im}(\varphi)$ of the action will be denoted by G^Ω.

Two actions φ_1 and φ_2 of G on sets Ω_1 and Ω_2 are said to be *equivalent*, if there is a bijection $\tau : \Omega_1 \to \Omega_2$ such that $\tau(\alpha^g) = \tau(\alpha)^g$, for all $\alpha \in \Omega_1$ and $g \in G$. This just says that the actions are the same up to a 'relabelling' of the elements of the set on which G is acting.

The kernel $K = \ker(\varphi)$ of an action φ is equal to the normal subgroup

$$\{ g \in G \mid \alpha^g = \alpha \text{ for all } \alpha \in \Omega \}.$$

The action is said to be *faithful* if $K = \{1\}$. In this case, Theorem 2.9 says that $G \cong G/K \cong \mathrm{im}(\varphi)$.

Example 2.1
If G is a subgroup of $\mathrm{Sym}(\Omega)$, then we can define an action of G on Ω simply by putting $\varphi(g) = g$ for $g \in G$. This action is faithful.   □

Example 2.2
Let $G = \{1, a, a^2, a^3, a^4, a^5, b, ab, a^2b, a^3b, a^4b, a^5b\} = D_{12}$ be the group of isometries of a regular hexagon P. In Subsection 2.1.3, we defined a and b to be the permutations $(1,2,3,4,5,6)$ and $(2,6)(3,5)$ of the set $\{1,2,3,4,5,6\}$ of vertices of P, and so we have an action of G on this vertex set.

There are some other related actions, however. We could instead take Ω to be the set $E = \{e_1, e_2, e_3, e_4, e_5, e_6\}$ of edges of P, where e_1 is the edge joining 1 and 2, e_2 joins 2 and 3, etc. The action φ of G on E is then given by $\varphi(a) = (e_1, e_2, e_3, e_4, e_5, e_6)$, $\varphi(b) = (e_1, e_6)(e_2, e_5)(e_3, e_4)$. This action is still faithful, but is not equivalent to the action on vertices.

As a third possibility, let $D = \{d_1, d_2, d_3\}$ be the set of diagonals of P, where d_1 joins vertices 1 and 4, d_2 joins 2 and 5, and d_3 joins 3 and 6. Then the action of G on D is defined by $\varphi(a) = (d_1, d_2, d_3)$, and $\varphi(b) = (d_2, d_3)$. This action is not faithful, and its kernel is the normal subgroup $\{1, a^3\}$ of G. The image is isomorphic to D_6.   □

Example 2.3
There is a faithful action called the *right regular* action, which we can define for any group G. Here we put Ω to be the underlying set of G and simply define α^g to be αg for all $g \in G$, $\alpha \in \Omega$.

More generally, let H be any subgroup of G, and let Ω be the set of right cosets of H in G. Again we can define an action of G on Ω by $\alpha^g := \alpha g$. This is known as a *coset action*. The right regular action corresponds to the case $H = 1$.   □

The right regular action of a group G is closely related to its *Cayley graph* $\Gamma_X(G)$, which is defined with respect to a generating set X of G.

DEFINITION 2.13   If $G = \langle X \rangle$, then the *Cayley graph* $\Gamma_X(G)$ is a directed labelled graph with vertex set $V := \{v_g \mid g \in G\}$ in one-one correspondence with the set of elements of G and, for each $g \in G$ and $x \in X$, an edge labelled x from v_g to v_{gx}. It is customary to identify each such edge with an edge labelled x^{-1} from v_{gx} to v_g.

The edges of the Cayley graph define the action of the generators X in the right regular action of G.

2.2.2   Orbits and stabilizers

DEFINITION 2.14   Let G act on the set Ω. We define a relation $\sim$ on Ω by $\alpha \sim \beta$ if and only if there exists a $g \in G$ with $\beta = \alpha^g$. Then $\sim$ is an equivalence relation. The equivalence classes of $\sim$ are called the *orbits* of G on Ω. In particular, the orbit of a specific element $\alpha \in \Omega$, which is denoted by α^G, is $\{\alpha^g \mid g \in G\}$.

DEFINITION 2.15   Let G act on Ω and let $\alpha \in \Omega$. Then the *stabilizer* of α in G, which is denoted by G_α or by $\mathrm{Stab}_G(\alpha)$, is $\{\, g \in G \mid \alpha^g = \alpha \,\}$.

It is straightforward to check that G_α is a subgroup of G and that the kernel of the action is $\cap_{\alpha \in \Omega}\, G_\alpha$.

Example 2.4
Let G act by right multiplication on the set of right cosets of the subgroup H, as in Example 2.3 above. Since $H^g = Hg$ for any $g \in G$, $H^G = \Omega$ and there is only one orbit. The stabilizer of the coset Hg is the subgroup $g^{-1}Hg$ of G, and so the kernel of the action is equal to $\cap_{g \in G}\, g^{-1}Hg$. This (normal) subgroup of G is called the *core* of H in G. It is denoted by H_G, and is the largest normal subgroup of G that is contained in H.   ☐

THEOREM 2.16   (the orbit-stabilizer theorem)   *Let the finite group G act on Ω, and let $\alpha \in \Omega$. Then $|G| = |\alpha^G|\,|G_\alpha|$.*

PROOF   Let $\beta \in \alpha^G$. Then there exists a $g \in G$ with $\alpha^g = \beta$. If $g' \in G$ with $\alpha^{g'} = \beta$, then $g'g^{-1} \in G_\alpha$, so $g' \in G_\alpha g$, and we see that the elements g' with $\alpha^{g'} = \beta$ are precisely the elements of the coset Hg. But $|Hg| = |H|$, so for each $\beta \in \alpha^G$, there are precisely $|H|$ elements g' of G with $\alpha^{g'} = \beta$, and the total number of such $\beta \in \alpha^G$ must be $|G|/|H|$, which proves the result.
∎

Dually to the stabilizer G_α, for $g \in G$, we define

$$\Omega_g := \{\, \alpha \in \Omega \mid \alpha^g = \alpha \,\}$$

to be the fixed point set of g. The following result, which is often known as Burnside's lemma, but was actually known earlier to Cauchy and Frobenius, says that the number of orbits of an action is equal to the average number of fixed points of the group elements. This result, together with refinements and generalizations, is the basis of countless applications of group theory to the enumeration of combinatorial structures. We shall say some more about this later in the book (Section 10.5).

LEMMA 2.17 (the Cauchy-Frobenius lemma) *Let the finite group G act on Ω. Then the number of orbits of G on Ω is equal to $\frac{1}{|G|} \sum_{g \in G} |\Omega_g|$.*

PROOF

$$\sum_{g \in G} |\Omega_g| = \sum_{g \in G} \sum_{\alpha \in \Omega_g} 1 = \sum_{\alpha \in \Omega} \sum_{g \in G_\alpha} 1 = \sum_{\alpha \in \Omega} |G_\alpha| = |G| \sum_{\alpha \in \Omega} |\alpha^G|^{-1},$$

where the last equality follows from the orbit-stabilizer theorem. But since $\sum_{\alpha \in \Omega} |\alpha^G|^{-1}$ is the total number of orbits of G on Ω, the result follows. ∎

For future use, we introduce some further notation.

DEFINITION 2.18   Let G act on Ω and let $\Delta \subseteq \Omega$. Then

  (i) The *setwise stabilizer* $\{\, g \in G \mid \alpha^g \in \Delta \text{ for all } \alpha \in \Delta \,\}$ of Δ in G is denoted by G_Δ.

  (ii) The *pointwise stabilizer* $\{\, g \in G \mid \alpha^g = \alpha \text{ for all } \alpha \in \Delta \,\}$ of Δ in G is denoted by $G_{(\Delta)}$.

2.2.3   Conjugacy, normalizers, and centralizers

In Example 2.3, a group G was made to act on the set of its own elements by multiplication on the right. There is another important action of G on $\Omega = G$, which is defined by

$$\alpha^g = g^{-1}\alpha g \ \text{ for } \ g, \alpha \in G.$$

This is called *conjugation*. The orbits of the action are called the *conjugacy classes* of G, and elements in the same conjugacy class are said to be *conjugate* in G. So $g, h \in G$ are conjugate if and only if there exists $f \in G$ with $h = f^{-1}gf$. We shall write $\mathrm{Cl}_G(g)$ for the orbit of g; that is, the conjugacy class containing g. Since $f^{-1}gf$ is the image of g under the inner automorphism of G defined by f, conjugate elements must have the same order.

The stabilizer G_g in this action consists of the elements $f \in G$ for which $g^f = g$; that is, $fg = gf$. In other words, it consists of those f that commute with g. It is called the *centralizer* of g in G and is written as $\mathbf{C}_G(g)$. By applying the orbit-stabilizer theorem, we get the well-known formula $|\mathrm{Cl}_G(g)| = |G|/|\mathbf{C}_G(g)|$ for all $g \in G$. The kernel K of the action is the centre $\mathbf{Z}(G)$ of G, which we encountered at the end of Subsection 2.1.7.

Example 2.5

Let $G = \mathrm{Sym}(\Omega)$ and let $f, g \in G$. Write g in cyclic notation, and suppose that one of the cycles of g is $(\alpha_1, \alpha_2, \ldots, \alpha_r)$. Then $\alpha_1^g = \alpha_2$, and so $\alpha_1^{gf} = \alpha_2^f$ and hence $(\alpha_1^f)^{f^{-1}gf} = \alpha_2^f$, and we find that $f^{-1}gf$ has a cycle $(\alpha_1^f, \alpha_2^f, \ldots, \alpha_r^f)$.

In other words, given a permutation g in cyclic notation, we obtain the conjugate $f^{-1}gf$ of g by replacing each element $\alpha \in \Omega$ in the cycles of g by α^f. For example, if $\Omega = \{1, 2, 3, 4, 5, 6, 7\}$, $g = (1, 5)(2, 4, 7, 6)$ and $f = (1, 3, 5, 7, 2, 4, 6)$, then $f^{-1}gf = (3, 7)(4, 6, 2, 1)$.

It follows that two permutations of Ω are conjugate in $\mathrm{Sym}(\Omega)$ if and only if they have the same cycle-type; that is, the same number of cycles of each length. For example, $\mathrm{Sym}(4)$ has five conjugacy classes, with representatives 1, $(1, 2)$, $(1, 2)(3, 4)$, $(1, 2, 3)$, $(1, 2, 3, 4)$. $\quad\square$

We can also take Ω to be the set of subgroups H of G, and use the corresponding action $H^g = g^{-1}Hg$ of G on Ω. Again, subgroups in the same orbit of this action are called *conjugate subgroups*. The stabilizer of a subgroup H is equal to $\{\, g \in G \mid g^{-1}Hg = H \,\}$, which is called the *normalizer* of H in G and denoted by $\mathbf{N}_G(H)$. Then clearly $H \trianglelefteq \mathbf{N}_G(H)$. By the orbit-stabilizer theorem, the number of conjugates of H in G is equal to $|G : \mathbf{N}_G(H)|$.

The *centralizer* $\mathbf{C}_G(H)$ of H in G, on the other hand, is defined to be the subgroup $\{\, g \in G \mid gh = hg$ for all $h \in H \,\}$ of G, and so $\mathbf{C}_G(G) = \mathbf{Z}(G)$.

2.2.4 Sylow's theorems

We state Sylow's theorems as a single theorem. There are many different proofs. See, for example, Chapter 4 of [Rot94] and Section 1.5 of [Cam99] for two different proofs. If G is a finite group, p is a prime, and $|G| = p^m q$ with $p \nmid q$, then a subgroup of G of order p^m is called a *Sylow p-subgroup* of G, and the set of all such subgroups is denoted by $\mathrm{Syl}_p(G)$. In general, a group of order a power of a prime p, including the trivial group of order p^0, is known as a *p-group*.

THEOREM 2.19 (Sylow's theorems.) *Let G be a finite group, let p be a prime number, and let $|G| = p^m q$ where $p \nmid q$. Then:*

(i) *For any l with $0 \le l \le m$, the number of subgroups of G of order p^l is congruent to 1 modulo p. In particular, such subgroups exist for all such l, and $|\mathrm{Syl}_p(G)| \equiv 1 \bmod p$.*

(ii) *Any two subgroups in $\mathrm{Syl}_p(G)$ are conjugate in G, and $|\mathrm{Syl}_p(G)| = |G : \mathbf{N}_G(P)|$ for any $P \in \mathrm{Syl}_p(G)$.*

(iii) *For $0 \le l < m$, any subgroup of G of order p^l is contained in some subgroup of order p^{l+1}. Hence any such subgroup is contained in a Sylow p-subgroup.*

We state without proof a couple of easy consequences of Sylow's theorems.

PROPOSITION 2.20   *Let G be a finite group, p a prime, $P \in \mathrm{Syl}_p(G)$, and H a subgroup of G with $\mathbf{N}_G(P) \le H$. Then $\mathbf{N}_G(H) = H$.*

PROPOSITION 2.21   (the Frattini lemma.)   *Let G be a finite group, p a prime, N a normal subgroup of G and $P \in \mathrm{Syl}_p(N)$. Then $G = N\mathbf{N}_G(P)$.*

2.2.5   Transitivity and primitivity

DEFINITION 2.22   An action of G on a set Ω is called *transitive* if there is a single orbit under the action. That is, for any $\alpha, \beta \in \Omega$, there exists $g \in G$ with $\alpha^g = \beta$. Otherwise, it is called *intransitive*.

Let $n \in \mathbb{N}$ with $n > 0$. The action is *n-fold transitive* (or just *n-transitive* for short) if $|\Omega| \ge n$, and for any two ordered lists $[\alpha_1, \alpha_2, \ldots, \alpha_n]$, $[\beta_1, \beta_2, \ldots, \beta_n]$ of distinct points in Ω, there exists $g \in G$ with $\alpha_i^g = \beta_i$ for $1 \le i \le n$.

Note that 1-transitive is the same as transitive. Also, the condition $|\Omega| \ge n$ for n-fold transitivity ensures that n-transitive actions are $(n-1)$-transitive for $n > 1$. By "G^Ω is transitive" we mean "the action of G on Ω is transitive".

Example 2.6
If $|\Omega| = n$, then $\mathrm{Sym}(\Omega)$ is l-transitive for all l with $1 \le l \le n$, and $\mathrm{Alt}(\Omega)$ is $(l-2)$-transitive for $3 \le l \le n$.   □

The following result says that all transitive actions are equivalent to the action described in Example 2.4 for some subgroup H of G.

PROPOSITION 2.23

(i) *Let H be any subgroup of G, and let φ_H be the action of G by right multiplication on the right cosets of H in G (Example 2.4). Then φ_H is transitive.*

(ii) *If $\varphi : G \to \mathrm{Sym}(\Omega)$ is any transitive action of G and $\alpha \in \Omega$, then φ is equivalent to φ_H, where $H = G_\alpha$.*

(iii) *If $H_1, H_2 \le G$, then φ_{H_1} and φ_{H_2} are equivalent if and only if H_1 and H_2 are conjugate in G.*

PROOF   (i) is clear since, for any cosets Hg_1 and Hg_2 of H, we have $(Hg_1)(g_1^{-1}g_2) = Hg_2$. For (ii), let Σ be the set of right cosets of H in G, with

$H = G_\alpha$. For $g_1, g_2 \in G$, we have $\alpha^{g_1} = \alpha^{g_2}$ if and only if $Hg_1 = Hg_2$, so we can define a bijection $\tau : \Sigma \to \Omega$ by $\tau(Hg) = \alpha^g$, and τ is easily seen to be an equivalence of actions. For any $\beta \in \Omega$, we have $G_\beta = g^{-1}G_\alpha g$, where $\alpha^g = \beta$. It follows from (ii) that φ_{H_1} and φ_{H_2} are equivalent if and only if H_2 is the stabilizer of a point in the action φ_{H_1}, which is the case if and only if H_1 and H_2 are conjugate in G. ∎

DEFINITION 2.24 If G acts on Ω, then G^Ω is *regular* if it is transitive and $G_\alpha = 1$ for some (and hence all) $\alpha \in \Omega$.

The right regular action of G on its own underlying set defined in Example 2.3 is indeed regular. By the orbit-stabilizer theorem, if G is finite and G^Ω is regular, then $|G| = |\Omega|$.

If G acts on Ω, then a nonempty subset Δ of Ω is called a *block* under the action if, for all $g \in G$, we have $\Delta^g = \Delta$ or $\Delta^g \cap \Delta = \phi$. The block is called *nontrivial* if $|\Delta| > 1$ and $\Delta \neq \Omega$. The orbits of any action are clearly blocks, but we normally restrict the consideration of blocks to transitive actions.

DEFINITION 2.25 A transitive action of G on a set Ω is called *primitive* if there are no nontrivial blocks under the action. Otherwise, it is called *imprimitive*.

If Δ is a block under an action, then the distinct translates Δ^g of Δ under G partition Ω. The set of translates is known as a *block system*. So a transitive action is primitive if and only if it does not preserve a nontrivial partition of Ω. Note also that $|\Delta| = |\Delta^g|$, so all blocks in such a partition have the same size. Hence if G^Ω is transitive and $|\Omega|$ is prime, then G^Ω is primitive.

Example 2.7
Let $G = D_{12}$ act on the vertices of a regular hexagon, as in Example 2.2 above. Then G preserves two nontrivial block systems, $\{\{1,3,5\}, \{2,4,6\}\}$ and $\{\{1,4\}, \{2,5\}, \{3,6\}\}$. ⬚

If $G \leq \mathrm{Sym}(\Omega)$ and we say that G is (n-fold) transitive or primitive, then we are referring to the obvious action of G on Ω defined in Example 2.1.

There is an alternative method of describing block systems, which will often be convenient for us to employ.

DEFINITION 2.26 If G acts on Ω, then an equivalence relation $\sim$ on Ω is called a *G-congruence* if $\alpha \sim \beta$ implies $\alpha^g \sim \beta^g$ for all $\alpha, \beta \in \Omega$ and $g \in G$.

Assume that G^Ω is transitive. It is clear that if G^Ω has block system

$\{\Delta^g \mid g \in G\}$, then the equivalence relation on Ω defined by $\alpha \sim \beta$ if and only if α and β lie in the same block is a G-congruence. Conversely, the equivalence classes defined by any G-congruence on Ω form the blocks of a block system. So, for transitive actions, G-congruences and block systems are essentially the same thing.

We remind the reader that, formally speaking, an equivalence relation $\sim$ on a set Ω is a subset $\mathcal{R}$ of $\Omega \times \Omega$, where $\alpha \sim \beta$ means the same as $(\alpha, \beta) \in \mathcal{R}$.

DEFINITION 2.27   If $\mathcal{S} \subseteq \Omega \times \Omega$, then the *$G$-congruence generated by* $\mathcal{S}$ is defined to be the intersection of all G-congruences that contain $\mathcal{S}$.

PROPOSITION 2.28   *Let $\mathcal{S} \subseteq \Omega \times \Omega$, and let $\mathcal{R}$ be the G-congruence generated by $\mathcal{S}$. Call, $\alpha, \beta \in \Omega$ directly equivalent if, for some $g \in G$, we have $\alpha = \gamma^g$ and $\beta = \delta^g$, where either $(\gamma, \delta) \in \mathcal{S}$ or $(\delta, \gamma) \in \mathcal{S}$. Then $(\alpha, \beta) \in \mathcal{R}$ if and only if there exist $\alpha = \alpha_0, \alpha_1, \ldots, \alpha_r = \beta$ in Ω with $r \geq 0$, such that each α_i is directly equivalent to α_{i+1}.*

PROOF   Define the relation $\sim$ on Ω by $\alpha \sim \beta$ if and only if there exists $\alpha = \alpha_0, \alpha_1, \ldots, \alpha_r = \beta$ as in the statement of the proposition. Then clearly $\alpha \sim \beta$ implies that (α, β) is in any G-congruence containing $\mathcal{S}$, and hence $(\alpha, \beta) \in \mathcal{R}$. Conversely, it is straightforward to check that $\sim$ is a G-congruence, and the result follows.   ∎

Let Ω be the disjoint union of sets $\Delta_1, \ldots, \Delta_m$, where each $|\Delta_i| = l$, and let G be the subgroup of $\mathrm{Sym}(\Omega)$ preserving this partition. Then G is the *wreath product* of the permutation groups $\mathrm{Sym}(l)$ and $\mathrm{Sym}(m)$, and we write $G = \mathrm{Sym}(l) \wr \mathrm{Sym}(m)$. It has a normal subgroup $H \cong \mathrm{Sym}(l)^m$ that fixes all of the Δ_i, and $G/H \cong \mathrm{Sym}(m)$. See, for example, Section 5.9 of [Hal59] for details of the theory of wreath products of permutation groups.

We shall now give brief proofs of some standard results pertaining to transitivity and primitivity.

PROPOSITION 2.29

(i) A 2-fold transitive action is primitive.

(ii) Let G act on Ω, and let $N \trianglelefteq G$. Then the orbits of N^Ω are blocks under the action of G.

(iii) If G^Ω is primitive, and N^Ω is nontrivial, then N^Ω is transitive.

PROOF   Let G^Ω be 2-transitive, and let Δ be a block with $|\Delta| > 1$. Let $\alpha, \beta \in \Delta$ and $\alpha \neq \gamma \in \Omega$. Then by 2-transitivity, there exists $g \in G$ with $\alpha^g = \alpha$ and $\beta^g = \gamma$, so $\alpha \in \Delta \cap \Delta^g$, and hence $\Delta = \Delta^g$, and $\gamma \in \Delta$. Thus $\Delta = \Omega$, which proves (i).

Let $\Delta = \alpha^N$ be an orbit of N^Ω, let $g \in G$, and suppose that $\alpha \in \Delta \cap \Delta^g$. So $\alpha = \beta^g$ for some $\beta \in \Delta$. Then for any $\alpha^h \in \Delta$ with $h \in N$, normality of N gives $gh = h'g$ for some $h' \in N$, and $\alpha^h = \beta^{gh} = \beta^{h'g} \in \Delta^g$. So $\Delta = \Delta^g$, proving (ii). (iii) follows easily from (ii).   ∎

PROPOSITION 2.30   *Let G^Ω be a transitive action and let $\alpha \in \Omega$. Then*

(i) *If K is a subgroup with $G_\alpha \leq K \leq G$, then α^K is a block and, conversely, any block containing α is equal to α^K for some K with $G_\alpha \leq K \leq G$.*

(ii) *If $|\Omega| > 1$, then G^Ω is primitive if and only if G_α is a maximal subgroup of G.*

PROOF   Let $G_\alpha \leq K \leq G$ and $\Delta = \alpha^K$. Suppose that $\beta \in \Delta \cap \Delta^g$ for some $g \in G$. So $\beta = \alpha^h = \alpha^{kg}$ for some $h, k \in K$. But then $kgh^{-1} \in G_\alpha \leq K$, so $g \in K$ and $\Delta^g = \Delta$. Conversely, if Δ is a block containing α, then let $K = \{\, g \in G \mid \alpha^g \in \Delta \,\}$. Certainly $G_\alpha \leq K$. For any $k \in K$, we have $\alpha^k \in \Delta \cap \Delta^k$, so $\Delta = \Delta^k$, and hence $K = \{\, g \in G \mid \Delta^g = \Delta \,\}$, which is a subgroup of G. This proves (i), and (ii) follows immediately from (i).   ∎

Exercises

1. Let G act on Ω and let $\alpha \in \Omega$ and $g \in G$. Prove that $G_{\alpha^g} = g^{-1} G_\alpha g$.

2. Find generators of the wreath product $\mathrm{Sym}(l) \wr \mathrm{Sym}(m)$.

3. Let G be a group of order p^n for p prime, and let N be a normal subgroup of G with $N \neq 1$. By adding up the sizes of the conjugacy classes of G that are contained in N, prove that $N \cap \mathbf{Z}(G) \neq 1$. In particular, if $n > 0$, then $\mathbf{Z}(G) \neq 1$.

4. Let H be a subgroup of a group G and let H_G be the core of H in G, as defined in Example 2.4. Prove that $H_G \leq H$ and that, if K is any normal subgroup of G with $K \leq H$, then $K \leq H_G$.

5. Calculate the conjugacy classes of the dihedral group D_{2n}. (The cases when n is odd and even are different.)

6. Prove that a normal subgroup of a group G is a union of conjugacy classes of G.

7. Let $H = \text{Alt}(\Omega)$ and $G = \text{Sym}(\Omega)$, and let $g \in H$. Show that either $\mathbf{C}_H(g) = \mathbf{C}_G(g)$ and $\text{Cl}_G(g)$ is the union of two equally-sized conjugacy classes of H, or $|\mathbf{C}_G(g) : \mathbf{C}_H(g)| = 2$ and $\text{Cl}_G(g) = \text{Cl}_H(g)$. Show that $\text{Alt}(5)$ has five conjugacy classes, with orders 1, 12, 12, 15, 20, and deduce that $\text{Alt}(5)$ is a simple group.

2.3   Series

2.3.1   Simple and characteristically simple groups

Let G be a group and let Ω be a subgroup of $\text{Aut}(G)$. We call G an Ω-*group* for short. A subgroup H of G is called Ω-*invariant* if $g^\alpha \in H$ for all $g \in H, \alpha \in \Omega$. So, if $\Omega = \text{Inn}(G)$, then an Ω-invariant subgroup is the same thing as a normal subgroup. An $\text{Aut}(G)$-invariant subgroup is called a *characteristic* subgroup. For example, the centre $\mathbf{Z}(G)$ of G is characteristic in G. We write $H \operatorname{char} G$ to mean that H is characteristic in G.

If $H \lhd K$ and $K \lhd G$, then it is not necessarily true that $H \lhd K$. However, $H \operatorname{char} K$ and $K \lhd G$ imply $H \lhd G$, and $H \operatorname{char} K$ and $K \operatorname{char} G$ imply $H \operatorname{char} G$. A subgroup H of G is called *subnormal* in G if there exists a finite chain $H = H_0 \lhd H_1 \lhd \cdots \lhd H_r = G$.

An Ω-group G is called Ω-*simple* if $|G| > 1$, and G has no proper nontrivial Ω-invariant subgroups. So a simple group is the same as an $\text{Inn}(G)$-simple group. The celebrated classification of finite simple groups was completed around 1980. The finite nonabelian simple groups comprise a number of infinite families of groups, including $\text{Alt}(n)$ for $n \geq 5$, the classical groups and the exceptional groups of Lie type, together with the 26 sporadic groups. We shall assume that the reader has some basic familiarity with these groups, and refer to [CCN+85] or [Car72] for any detailed information that we may need about them.

An $\text{Aut}(G)$-simple group is called *characteristically simple*. The following result, proved in Theorem 5.20 of [Rot94], describes the finite examples.

THEOREM 2.31   *A finite group is characteristically simple if and only if it is isomorphic to a direct product of one or more isomorphic simple groups.*

The simple groups in this theorem can be either abelian, in which case the group itself is elementary abelian, or nonabelian.

A normal subgroup N of a group G is called a *minimal normal subgroup* of G if $N \neq 1$ and there is no normal subgroup M of G with $1 < M < N$. A minimal normal subgroup must be characteristically simple, and so we have the following corollary.

COROLLARY 2.32   *A minimal normal subgroup N of a finite group G is isomorphic to a direct product of isomorphic simple groups. In particular, if N is abelian, then it is an elementary abelian p-group for some prime p.*

2.3.2   Series

Let G be an Ω-group, and let

$$G = G_0 \triangleright G_1 \triangleright G_2 \triangleright \cdots \triangleright G_r = \{1\}$$

be a finite series of Ω-invariant subgroups in which each group is normal in the preceding group. Such a series is called an (Ω-invariant) *subnormal series* of G. It is called an Ω-*composition series* if each factor G_i/G_{i+1} is Ω-simple. We shall be mainly interested in the cases $\Omega = \{1\}$, which is an ordinary composition series with simple factor groups, and the case $\Omega = \mathrm{Inn}(G)$, where an Ω-composition series is called a *chief series* for G, and the factors G_i/G_{i+1} are characteristically simple groups.

It is clear that finite groups have finite composition series and finite chief series. For an infinite group, such finite series may or may not exist, and there are generalizations which allow infinite series. But for any group G and any $\Omega \leq \mathrm{Aut}(G)$, we have the Jordan-Hölder theorem; see, for example, Section 2.10 of [Sco64].

THEOREM 2.33   *Let G be an Ω-group, and let*

$$G = G_0 \triangleright G_1 \triangleright G_2 \triangleright \cdots \triangleright G_r = \{1\}$$

and

$$H = H_0 \triangleright H_1 \triangleright H_2 \triangleright \cdots \triangleright H_s = \{1\}$$

be two finite Ω-composition series of G. Then $r = s$ and there is a bijection $\varphi : [1..r] \to [1..r]$ such that G_{i-1}/G_i and H_{j-1}/H_j are Ω-isomorphic for $1 \leq i \leq r$, where $j = \varphi(i)$.

We generally refer to the collection of isomorphism types of composition factors or chief factors of a finite group G simply as *the* composition/chief factors of G. This collection is technically a multiset, because it can have repeated elements.

2.3.3   The derived series and solvable groups

For $g, h \in G$, we define the commutator $[g, h] := g^{-1}h^{-1}gh$. If H and K are subgroups of G, then we define $[H, K] := \langle [h, k] \mid h \in H, k \in K \rangle$. If there is a serious danger of confusion with the same notation $[g, h]$ used to mean the ordered list of elements $g, h \in G$, then we shall write $\mathrm{comm}(g, h)$ in place of $[g, h]$. The following lemma lists some elementary properties of commutators.

Recall that, for $x, y \in G$, x^y is defined to be $y^{-1}xy$. We denote the inverse of x^y by x^{-y}.

LEMMA 2.34   *Let x, y, z be elements of a group G and let H, K be subgroups of G. Then:*

(i) *$[xy, z] = [x, z]^y[y, z]$ and $[x, yz] = [x, z][x, y]^z$;*

(ii) *$[H, K]$ is a normal subgroup of $\langle H, K \rangle$;*

(iii) *$[H, K] = [K, H]$;*

(iv) *$H \leq \mathbf{N}_G(K) \Longleftrightarrow [H, K] \leq K$;*

(v) *$[[x, y^{-1}], z]^y[[y, z^{-1}], x]^z[[z, x^{-1}], y]^x = 1$.*

The proof is left as an exercise. For (ii), notice that (i) implies $[x, z]^y = [xy, z][y, z]^{-1}$, and deduce that $H \leq \mathbf{N}_G([H, K])$. Show $K \leq \mathbf{N}_G([H, K])$ similarly. For (iii), use $[x, y]^{-1} = [y, x]$.

The following result will be used for computing commutator subgroups.

PROPOSITION 2.35   *If $H = \langle X \rangle$ and $K = \langle Y \rangle$ are subgroups of a group G, then $[H, K] = \langle Z^{\langle H,K \rangle} \rangle$, where $Z = \{ [x, y] \mid x \in X, y \in Y \}$.*

PROOF   Let $N = \langle Z^{\langle H,K \rangle} \rangle$. By Lemma 2.34 (ii), $[H, K] \trianglelefteq \langle H, K \rangle$, so $N \leq [H, K]$. By Lemma 2.34 (i), we have $[x^{-1}, y] = [x, y]^{-x^{-1}}$ and $[x, y^{-1}] = [x, y]^{-y^{-1}}$ for all $x, y \in G$. So, if $x \in X$, $y \in Y$, then $[x^{-1}, y]$, $[x, y^{-1}]$ and $[x^{-1}, y^{-1}]$ all lie in N. Furthermore, it follows from Lemma 2.34 (i) that $[x, z], [y, z] \in N \Rightarrow [xy, z] \in N$ and $[x, y], [x, z] \in N \Rightarrow [x, yz] \in N$. It then follows by a straightforward induction argument on word length that $[h, k] \in N$ for all $h \in \langle X \rangle = H$ and all $k \in \langle Y \rangle = K$, and so $[H, K] \leq N$, which proves the result.   ∎

In particular, $[G, G]$, the *derived subgroup* or *commutator subgroup* of G, is characteristic in G and is the smallest normal subgroup of G with abelian quotient group.

DEFINITION 2.36   A nontrivial group is called *perfect* if $[G, G] = G$.

So, for example, nonabelian simple groups are perfect. The *derived series* of G is the descending series

$$G = G^{[0]} \trianglerighteq G^{[1]} \trianglerighteq \cdots \trianglerighteq G^{[i-1]} \trianglerighteq G^{[i]} \trianglerighteq \cdots,$$

where $G^{[i]} := [G^{[i-1]}, G^{[i-1]}]$ for $i \geq 1$. (We have used $G^{[i]}$ rather than the more standard $G^{(i)}$ to avoid clashing with our use of $G^{(i)}$ in Chapter 4 in the context of stabilizer-chains of permutation groups.)

Each factor $G^{[i-1]}/G^{[i]}$ in the series is abelian. If there is an r such that $G^{[r]} = \{1\}$ or $G^{[r]}$ is perfect, then the series terminates at $G^{[r]}$ and is finite of length r. If it terminates with $G^{[r]} = \{1\}$, then we say that G is *solvable* (or *soluble*) with derived length r. A group is solvable if and only if it has a normal series with abelian factors. A finite group is solvable if and only if it has a subnormal series with cyclic factors or, equivalently, if all of its composition factors are cyclic of prime order.

If $G^{[i]} = G^{[i+1]}$ for some i, then we define $G^{[\infty]} := G^{[i]}$, in which case $G^{[\infty]}$ is the unique smallest normal subgroup of G with solvable quotient group.

More generally, a group with a finite subnormal series with (possibly infinite) cyclic factors is called *polycyclic*. Equivalently, a polycyclic group is a solvable group in which each factor $G^{[i-1]}/G^{[i]}$ is finitely generated. Polycyclic groups are important in computational group theory because there are efficient algorithms for computing within them. Indeed, algorithms for polycyclic groups is the subject of Chapter 8.

2.3.4 Central series and nilpotent groups

A normal series

$$G = G_0 \trianglerighteq G_1 \trianglerighteq G_2 \trianglerighteq \cdots \trianglerighteq G_r = \{1\}$$

of G is called a *central series* if $G_{i-1}/G_i \leq \mathbf{Z}(G/G_i)$ for $1 \leq i \leq r$. A group is called *nilpotent* if it has such a series, and the *(nilpotency) class* of G is defined to be the length r of the shortest central series for G. Clearly every nilpotent group is also solvable.

The *lower central series* of a group G is the descending series

$$G = \gamma_1(G) \trianglerighteq \gamma_2(G) \trianglerighteq \cdots \trianglerighteq \gamma_i(G) \trianglerighteq \gamma_{i+1}(G) \trianglerighteq \cdots,$$

in which each $\gamma_{i+1}(G) := [G, \gamma_i(G)]$. It is unfortunate that the lower central series begins with $G = \gamma_1(G)$ while the derived series begins with $G = G^{[0]}$ but this is standard notation, so we are stuck with it. The series terminates at $\gamma_r(G)$, and so has length $r - 1$, if $\gamma_r(G) = [G, \gamma_r(G)]$. Since $\gamma_{i-1}(G)/\gamma_i(G) \leq \mathbf{Z}(G/\gamma_i(G))$ for all $i > 1$, it is actually a central series for G if and only if it terminates with $G_r = 1$ for some r. It is not hard to show that this is the case if and only if G is nilpotent of class $r - 1$.

The *upper central series* of a group G is the ascending series

$$\{1\} = \mathbf{Z}_0(G) \trianglelefteq \mathbf{Z}_1(G) = \mathbf{Z}(G) \trianglelefteq \mathbf{Z}_2(G) \trianglelefteq \cdots \trianglelefteq \mathbf{Z}_i(G) \trianglelefteq \mathbf{Z}_{i+1}(G) \trianglelefteq \cdots,$$

where each $\mathbf{Z}_{i+1}(G)$ for $i \geq 1$ is defined by $\mathbf{Z}_{i+1}(G)/\mathbf{Z}_i(G) = \mathbf{Z}(G/\mathbf{Z}_i(G))$. The series terminates at $\mathbf{Z}_r(G)$, and so has length r, if $G/\mathbf{Z}_r(G)$ has trivial centre. It is a central series for G if and only if it terminates with $\mathbf{Z}_r(G) = G$

for some r. It is not hard to show that this is the case if and only if G is nilpotent of class r.

It is straightforward to show by induction that, if

$$G = G_0 \trianglerighteq G_1 \trianglerighteq G_2 \trianglerighteq \cdots \trianglerighteq G_r = \{1\}$$

is any central series of G, then $G_i \geq \gamma_{i+1}(G)$ and $G_i \leq \mathbf{Z}_{r-i}(G)$ for $0 \leq i \leq r$. This accounts for the names lower and upper central series.

There are a number of characterizations of finite nilpotent groups, which we shall list shortly. First we prove:

LEMMA 2.37   *Let H and K be normal subgroups of a group G with $H \cap K = 1$. Then $\langle H, K \rangle \cong H \times K$.*

PROOF   Using Lemma 2.34 (iv), we have $[H, K] \leq H \cap K = 1$, so the map $(h, k) \to hk$ defines an isomorphism from $H \times K$ to $HK = \langle H, K \rangle$.   ∎

It is customary, in the situation described in the lemma, to say that $\langle H, K \rangle$ is equal to (rather than is isomorphic to) the direct product of its subgroups H and K, and we shall follow that custom. Those who like to avoid ambiguities prefer to call $\langle H, K \rangle$ the *internal direct product* of H and K.

THEOREM 2.38   *Let G be a finite group. Then the following are equivalent:*

  (i) G is nilpotent;

  (ii) $|\mathrm{Syl}_p(G)| = 1$ for all primes p dividing $|G|$;

  (iii) G is the direct product of its Sylow p-subgroups;

  (iv) $H < \mathbf{N}_G(H)$ for all proper subgroups $H < G$ of G;

  (v) All maximal subgroups of G are normal in G.

PROOF   By Sylow's theorems, (ii) is equivalent to all Sylow p-subgroups being normal in G and, under this assumption, (iii) can be proved by repeated application of Lemma 2.37. Clearly (iii) implies (ii), so (ii) and (iii) are equivalent. Clearly (iv) implies (v). If $|\mathrm{Syl}_p(G)| > 1$ for some p, and H is a maximal subgroup of G that contains $\mathbf{N}_G(P)$ with $P \in \mathrm{Syl}_p(G)$, then Proposition 2.20 yields $H = \mathbf{N}_G(H)$. Hence (v) implies (ii).

By Exercise 1 of Subsection 2.2.3, a nontrivial finite group P of prime power order has nontrivial centre. It follows easily by induction on $|P|$ that the upper central series of P reaches P, and hence that P is nilpotent. It is straightforward to show that a direct product of nilpotent groups is nilpotent, so (iii) implies (i).

(i) implies (iv) is also proved by induction on $|G|$. For $H < G$, either $\mathbf{Z}(G) \leq H$ and we apply induction to $G/\mathbf{Z}(G)$, or $H < H\mathbf{Z}(G) \leq \mathbf{N}_G(H)$. This completes the proof. ∎

2.3.5   The socle of a finite group

The socle of a finite group plays an important rôle in the more advanced structural computations in finite groups that will be discussed in Chapter 10, and we present its basic properties here. These are well-known, but it was not easy to find a convenient reference containing all of the properties that we require, so we include proofs here.

DEFINITION 2.39   The *socle* $\mathrm{soc}(G)$ of a group G is the subgroup generated by all minimal normal subgroups of G.

PROPOSITION 2.40   *Let G be a finite group. Then $\mathrm{soc}(G)$ is the direct product $M_1 \times \cdots \times M_m$ of some minimal normal subgroups M_i of G, where each M_i is itself a direct product of isomorphic simple groups. Furthermore, the nonabelian M_i are the only nonabelian minimal normal subgroups of G.*

PROOF   Let $N \trianglelefteq G$ be maximal subject to being a direct product of minimal normal subgroups of G. If $N \neq \mathrm{soc}(G)$, then there is a minimal normal subgroup M of G not contained in N. Then minimality of M gives $M \cap N = 1$ and hence, by Lemma 2.37, $\langle M, N \rangle = M \times N$, which contradicts the maximality of N. So $N = \mathrm{soc}(G)$, which proves the first assertion, and the fact that the M_i are direct products of isomorphic simple groups follows from Theorem 2.31.

Let M be any minimal normal nonabelian subgroup of G. If M is not equal to any of the M_i then, by Lemma 2.37 again, $[M, M_i] = 1$ for all i, so $[M, \mathrm{soc}(G)] = 1$. Since M is nonabelian, this contradicts $M \leq \mathrm{soc}(G)$. ∎

COROLLARY 2.41   *If G, M_i are as in the proposition, and M_k is nonabelian for some k, then the simple direct factors of M_k are permuted transitively under the action of G by conjugation on $\mathrm{soc}(G)$.*

PROOF   By the proposition applied to M_k in place of G, the simple direct factors are the only minimal normal subgroups of M_k, so they are permuted under conjugation by G. If S is one of these factors, then $\langle S^G \rangle = M_k$ by minimality of M_k, so the action on the simple factors is transitive. ∎

COROLLARY 2.42   *The simple direct factors of the nonabelian minimal normal subgroups of a finite group G are precisely the nonabelian simple subnormal subgroups of G.*

PROOF   Clearly each simple factor of a minimal normal subgroup is subnormal in G, so suppose that S is a nonabelian simple subnormal subgroup of G. If $S \trianglelefteq G$ then the result is clear. Otherwise G has a proper normal subgroup N containing S and, by induction, we may assume that S is a direct factor of a minimal normal subgroup of N. Now $\mathrm{soc}(N) \operatorname{char} N$ implies that $\mathrm{soc}(N) \trianglelefteq G$, so the conjugates of S under G are all minimal normal subgroups of $\mathrm{soc}(N)$. Hence $\langle S^G \rangle$ is the direct product of these conjugates and is a minimal normal subgroup of G, which proves the result.   ∎

2.3.6   The Frattini subgroup of a group

DEFINITION 2.43   The *Frattini subgroup* $\Phi(G)$ of a group G is the intersection of its maximal subgroups.

Clearly $\Phi(G)$ is a characteristic subgroup of G. Its principal property is that it consists of the so-called *nongenerators* of G. We shall only be interested in this result for finite groups; see, for example, Section 10.4 of [Hal59] for a more general treatment.

PROPOSITION 2.44   *If X is a subset of a finite group G with $G = \langle X, \Phi(G) \rangle$, then $G = \langle X \rangle$.*

PROOF   Otherwise $\langle X \rangle$ is contained in a maximal subgroup M of G, but then $\Phi(G) \leq M$ and so $G = \langle X, \Phi(G) \rangle \leq M$, contradiction.   ∎

We can describe $\Phi(G)$ very precisely in the case when G is a finite p-group. This topic will arise in Section 9.4 in connection with the p-quotient algorithm.

Let G be a finite p-group for some prime p. By Theorem 2.38, all maximal subgroups M of G are normal in G and so they must have index p in G. So, for all $x, y \in G$, we have $x^p, [x,y] \in M$, and hence $x^p, [x,y] \in \Phi(G)$. In other words, $G/\Phi(G)$ is elementary abelian.

On the other hand, if $N \trianglelefteq G$ with G/N elementary abelian, then the intersection of the maximal subgroups of G/N is trivial, so $\Phi(G) \leq N$. Hence we have:

PROPOSITION 2.45   *If G is a finite p-group, then $\Phi(G)$ is equal to the smallest normal subgroup N of G such that G/N is elementary abelian.*

Exercises

1. Prove that subgroups, quotient groups, and direct products of solvable, nilpotent, or polycyclic groups are solvable, nilpotent, or polycyclic, respectively.

2. Prove that if $N \trianglelefteq G$ and N and G/N are both solvable, then G is solvable. Does this remain true with 'nilpotent' in place of 'solvable'?

3. Prove that if $N \leq \mathbf{Z}(G)$ and G/N is nilpotent then G is nilpotent.

4. Use Lemma 2.34 (v) to prove the *three subgroups lemma*: if $H, K, L \leq G$ with $[[H, K], L] = [[K, L], H] = 1$ then $[[L, H], K] = 1$. Deduce that $[\gamma_i(G), \gamma_j(G)] \leq \gamma_{i+j}(G)$ for all $i, j \geq 1$.

5. Show that the final assertion in Proposition 2.40 may be false for abelian direct factors M_i of $\mathrm{soc}(G)$.

6. Use the Frattini lemma (Proposition 2.21) together with Proposition 2.44 to prove that, if G is a finite group and $P \in \mathrm{Syl}_p(\Phi(G))$, then $P \trianglelefteq G$. In particular, the Frattini subgroup of a finite group is nilpotent.

7. Let G be a finite p-group, and suppose that $G/\Phi(G)$ has order p^d. Prove that $|X| = d$ for any irredundant generating set X of G. (*Hint:* Since X generates $G/\Phi(G)$ modulo $\Phi(G)$, we have $|X| \geq d$. If $|X| > d$, then some element $x \in X$ is redundant as a generator of $G/\Phi(G)$, and so we can write $x = wu$, where w is a word over $X \setminus \{x\}$ and $u \in \Phi(G)$. Now use Proposition 2.44 to show that $G = \langle X \setminus \{x\} \rangle$.)

2.4 Presentations of groups

We include here a rapid treatment of the basic theory of free groups and group presentations. The reader could consult [Joh98] for a more leisurely account of the same material.

2.4.1 Free groups

A free group on a set X is roughly speaking the largest possible group that is generated by X. This idea is made precise by the universal property described in the following definition.

DEFINITION 2.46 A group F is *free* on the subset X of F if, for any group G and any map $\theta : X \to G$, there is a unique group homomorphism $\theta' : F \to G$ with $\theta'(x) = \theta(x)$ for all $x \in X$.

It follows directly from the definition that any group G that is generated by a set X is an epimorphic image of, and hence isomorphic to a quotient of, a free group on X. (We shall see shortly that a free group on X exists.)

Uniqueness (up to isomorphism) of a free group on a set X is straightforward to prove, as is the fact that it depends only on $|X|$.

PROPOSITION 2.47
(i) Two free groups on the same set X are isomorphic.
(ii) Free groups on X_1 and X_2 are isomorphic if and only if $|X_1| = |X_2|$.

PROOF   (i) follows from the 'if' part of (ii), so we just need to prove (ii). Suppose that F_1, F_2 are free on X_1, X_2 with $|X_1| = |X_2|$, let $i_i : X_i \to F_i$ be the inclusion maps, for $i = 1, 2$, and let $\tau : X_1 \to X_2$ be a bijection. By the definition applied to the maps $i_2\tau : X_1 \to F_2$ and $i_1\tau^{-1} : X_2 \to F_1$, we find group homomorphisms $\theta_1 : F_1 \to F_2$ and $\theta_2 : F_2 \to F_1$ that restrict to the $i_2\tau$ and $i_1\tau^{-1}$ on X_1 and X_2, respectively. So $\theta_2\theta_1 : F_1 \to F_1$ restricts to the identity on X_1, and by the uniqueness part of the definition, $\theta_2\theta_1$ is the identity on F_1. Similarly $\theta_1\theta_2$ is the identity on F_2, so θ_1, θ_2 are mutually inverse and hence are isomorphisms.

Conversely suppose that F_1, F_2 are free on X_1, X_2, and F_1 and F_2 are isomorphic. Let G be a group of order 2. Then there are $2^{|X_i|}$ distinct maps from X_i to G for $i = 1, 2$, and hence by the definition there are also $2^{|X_i|}$ homomorphisms from F_i to G. So, for finite X_1 and X_2, F_1 isomorphic to F_2 implies that $|X_1| = |X_2|$. The result is also true for infinite generating sets because, for infinite X, the cardinality of a free group on X is equal to that of X, but we shall only be interested in finite X in this book, so we omit the details.   ∎

The cardinality $|X|$ of X is known as the *rank* of a free group F on X.

To prove existence, we need to construct a free group on a set X. For any set X, we define the set X^{-1} to be equal to $\{ (x, -1) \mid x \in X \}$, but we shall write x^{-1} rather than $(x, -1)$. For $y = x^{-1} \in X^{-1}$, we define y^{-1} to be equal to x. We define A_X (or just A when X is understood) to be $X \cup X^{-1}$.

DEFINITION 2.48   For a set A, we define A^* to be the set of all strings $x_1 x_2 \cdots x_r$ with each $x_i \in A$. The number r is the *length* of the string, and for $\sigma \in A^*$, we denote the length of σ by $|\sigma|$. The empty string over A of length 0 is denoted by ε_A, or just by ε.

Elements of A^* are also known as *words* in A^* or words over A. A *subword* or *substring* of $w = x_1 \cdots x_r \in A^*$ is the empty word, or any word of the form $x_i x_{i+1} \cdots x_j$, with $1 \leq i \leq j \leq r$. It is called a *prefix* of w if it is empty or $i = 1$, and it is called a *suffix* of w if it is empty or $j = r$.

Let us fix our set X and set $A = A_X$. Call two words v, w over A *directly equivalent* if one can be obtained from the other by insertion or deletion of a subword xx^{-1} with $x \in A$. For example, if $X = \{x, y\}$, then $xxx^{-1}y$ and xy are directly equivalent, as are xy and $y^{-1}yxy$. Let $\sim$ be the equivalence relation on A^* generated by direct equivalence. That is, for $v, w \in A^*$, $v \sim w$ if and only if there is a sequence $v = v_0, v_1, \ldots, v_r = w$ of elements of A^*, with $r \geq 0$, such that v_i and v_{i+1} are directly equivalent for $0 \leq i < r$.

We denote the equivalence class of $w \in A^*$ under $\sim$ by $[w]$. Let F_X be the set of equivalence classes of $\sim$. It is clear that $u_1 \sim v_1$ and $u_2 \sim v_2$ imply $u_1 u_2 \sim v_1 v_2$ so we can define a well-defined multiplication on F_X by setting $[u_1][u_2] = [u_1 u_2]$ and, since the operation of concatenation of words is clearly associative, this multiplication is associative. Furthermore, $[\varepsilon]$ is an identity element and $[x_1 x_2 \cdots x_r]$ multiplied by $[x_r^{-1} \cdots x_2^{-1} x_1^{-1}]$ is equal to $[\varepsilon]$, so F_X is a group under this multiplication.

THEOREM 2.49 *For any set X, F_X is a free group on the set $[X] := \{[x] \mid x \in X\}$, and the map $x \mapsto [x]$ defines a bijection from X to $[X]$.*

PROOF Let $F = F_X$ and $A = A_X$. Let G be a group and let $\varphi : X \to G$ be a map. We extend φ to a map $\varphi' : A^* \to G$ as follows. Let $w = x_1^{\epsilon_1} x_2^{\epsilon_2} \cdots x_r^{\epsilon_r} \in A^*$, where each $x_i \in X$ and $\epsilon_i = \pm 1$, and let $g_i = \varphi(x_i)$ for $1 \leq i \leq r$. Then we define $\varphi'(w) = g_1^{\epsilon_1} g_2^{\epsilon_2} \cdots g_r^{\epsilon_r}$ and of course $\varphi'(\varepsilon) = 1_G$. Since $v \sim w$ implies $\varphi'(v) = \varphi'(w)$, φ' induces a well-defined map $\theta' : F_X \to G$, which is clearly a homomorphism.

To show that F_X is free on $[X]$, for a given map $\theta : [X] \to G$, we define $\varphi : X \to G$ by setting $\varphi(x) = \theta([x])$, and then the map θ' defined above is a homomorphism extending θ, and is clearly the unique such homomorphism.

To show that the map $x \mapsto [x]$ defines a bijection from X to $[X]$, let G be any group with $|G| \geq |X|$, and let $\varphi : X \to G$ be an injection. Then, since $\theta'([x]) = \varphi(x)$ for $x \in X$, $x \mapsto [x]$ must define a bijection, as claimed. ∎

It is customary to write w rather than $[w]$ for elements of $F := F_X$, and we shall do that from now on. This renders the equality of words ambiguous. To resolve this, we shall always write $w_1 =_F w_2$ to mean that $[w_1] = [w_2]$, whereas $w_1 = w_2$ will mean that w_1 and w_2 are equal as words. The second part of Theorem 2.49 enables us to identify X with $[X]$ in F_X, which we shall do, and we shall also follow the custom of referring to F_X as *the* free group on X. Since F_X is generated by X, it follows from Proposition 2.47 and Theorem 2.49 that any free group on X is generated by X. This can also be proved directly from the definition of a free group; see Exercise 1, below.

A word in A^* is called *reduced*, if it does not contain xx^{-1} as a substring for any $x \in A$. We have not used quite the same definition of free groups as that given in Chapter 1 of [Joh98], for example, which defines the elements of F_X to be reduced words over A rather than equivalence classes of words. We chose our definition because, as we shall see later in Chapter 12, it enables us to regard group presentations as a special case of monoid presentations. The following proposition shows that the two definitions are essentially the same.

PROPOSITION 2.50   *If X is any set, then each equivalence class in F_X contains exactly one reduced word.*

PROOF   Let $\sim_X$ be defined as above. Since $xx^{-1} \sim_X \varepsilon$ for all $x \in A = A_X$, it is clear that each equivalence class contains a reduced word. Suppose that a class contains two distinct reduced words, u and v. Then there is a sequence $u = u_0, u_1, \ldots, u_r = v$ of elements of A^* such that u_i and v_i are directly equivalent for $0 \le i < r$. Choose this sequence to make the sum of the $|u_i|$ as small as possible. It is clear that directly equivalent elements of A^* differ in length by two, and they cannot both be reduced words, so we must have $r > 1$. Choose i such that $|u_i|$ is maximal. Then $0 < i < r$, $|u_{i-1}| = |u_{i+1}| = |u_i| - 2$, and both u_{i-1} and u_{i+1} are derived from u_i by deleting a substring of the form xx^{-1}. If these two substrings of u_i are disjoint, then we can reverse the order of the substitutions and obtain another sequence with $|u_i| = |u_{i-1}| - 2$, contrary to the minimality condition on the sequence. On the other hand, if the two substrings xx^{-1} are not disjoint, then either they are equal, or they are the substrings xx^{-1}, $x^{-1}x$ of a substring $xx^{-1}x$ of u_i. In both of these cases, we have $u_{i-1} = u_{i+1}$, so we can shorten the sequence and again contradict its minimality condition.   ∎

If $w = x_1 x_2 \cdots x_r \in A^*$ then, for any i with $1 \le i \le r$, w is conjugate in F_X to the cyclic permutation $w_i := x_i x_{i+1} \cdots x_r x_1 x_2 \cdots x_{i-1}$ of w. The words w and w_i are said to be *cyclic conjugates* of one another. The word w is said to be *cyclically reduced* if it is reduced and $x_1 \ne x_r$. This is equivalent to all of its cyclic conjugates being reduced words. Clearly every w is conjugate in F to a cyclically reduced word.

2.4.2   Group presentations

DEFINITION 2.51   Let X, $A = A_X$ be defined as above, and let R be a subset of A^*. We define the group presentation $\langle\, X \mid R\,\rangle$ to be equal to the quotient group F/N, where $F = F_X$ is the free group on X, and N is the normal closure $\langle R^F \rangle$ of R in F.

Informally, $\langle\, X \mid R\,\rangle$ is the 'largest' group G that is generated by X in which all of the strings $v \in R$ represent the identity element. This idea is expressed more precisely by the universal property proved next. This property could be used as an alternative definition of a group presentation.

Notice that this definition does not distinguish properly between the presentation itself, which can be thought of as just a set or sequence X together with a set R of words in A^*, and the group G that is defined by the presentation. If we wish to emphasize this distinction, then we shall denote the presentation by $\{\, X \mid R \,\}$, and the group defined by it by $\langle\, X \mid R \,\rangle$.

THEOREM 2.52   *Let X be a set, let $A = A_X$, let R be a subset of A^*, and let G be a group. For any map $\theta : X \to G$, we extend θ to a map $\theta : A \to G$ by putting $\theta(x^{-1}) = \theta(x)^{-1}$ for all $x \in X$. Suppose that $\theta : X \to G$ is a map with the property that, for all $w = x_1 \cdots x_r \in R$, we have $\theta(x_1) \cdots \theta(x_r) = 1_G$. Then there exists a unique group homomorphism $\theta' : \langle\, X \mid R \,\rangle \to G$ for which $\theta'(xN) = \theta(x)$ for all $x \in X$, where $\langle\, X \mid R \,\rangle = F/N$ as in the definition above.*

Conversely, if there is a group homomorphism $\theta' : \langle\, X \mid R \,\rangle \to G$ for which $\theta'(xN) = \theta(x)$ for all $x \in X$, then we must have $\theta(x_1) \cdots \theta(x_r) = 1_G$ for all $w = x_1 \cdots x_r \in R$.

PROOF   Since $\langle\, X \mid R \,\rangle$ is generated by the elements xN for $x \in X$, the homomorphism θ' is certainly unique if it exists. By the definition of the free group $F = F_X$, there is a homomorphism $\theta'' : F \to G$ with $\theta''(x) = \theta(x)$ for all $x \in X$. The hypothesis of the theorem then says that $w \in \ker(\theta'')$ for all $w \in R$, and hence $N = \langle R^F \rangle \leq \ker(\theta'')$, and so θ'' induces a homomorphism $\theta' : \langle\, X \mid R \,\rangle = F/N \to G$ with the required property.

Conversely, if θ' exists as described, then we must have $\theta'(wN) = \theta''(w)$ for all $w \in F$, and so $\theta'(w) = 1_G$ for all $w \in N$. In particular,

$$\theta(x_1) \cdots \theta(x_r) = \theta'(x_1) \cdots \theta'(x_r) = \theta'(x_1 \cdots x_r) = 1_G$$

for all $w = x_1 \cdots x_r \in R$.   ∎

We shall further abuse notation by using v to represent the image vN of v in $\langle\, X \mid R \,\rangle$ for $v \in A^*$. We shall use $u =_G v$ to mean that $u, v \in A^*$ represent the same element of G.

In general, if G is a group generated by a set X, $A = A_X$, and $w \in A^*$ with $w =_G 1_G$, then we call the word w a *relator* of G. The elements of R in a group presentation $G = \langle\, X \mid R \,\rangle$ are clearly relators of G, and these are known as *defining relators*. If $w_1, w_2 \in G$ with $w_1 =_G w_2$, then we call $w_1 =_G w_2$ a *relation* in G. Of course, this is equivalent to $w_1 w_2^{-1}$ being a relator of G. It is often convenient to write down group presentations in the alternative form $G = \langle\, X \mid \mathcal{R} \,\rangle$, where $\mathcal{R}$ is a subset of $A^* \times A^*$, but the element (w_1, w_2) of $\mathcal{R}$ is usually written as $w_1 = w_2$. We formally define this group to be equal to $\langle\, X \mid R \,\rangle$, where the elements of R are obtained from those of $\mathcal{R}$ by replacing $w_1 = w_2$ by $w_1 w_2^{-1}$. The elements of $\mathcal{R}$ are called *defining relations* of G.

For example, the group $\langle\, x, y \mid x^2, y^3, (xy)^2 \,\rangle$, which is dihedral of order 6, could be written as $\langle\, x, y \mid x^2 = 1,\ y^2 = y^{-1},\ xyx = y^{-1} \,\rangle$.

If the hypotheses of Theorem 2.52 are satisfied and the resulting map θ' : $\langle X \mid R \rangle \to G$ is an isomorphism, then we often, in practice, identify the set X with its image in G under θ, and say that $\langle X \mid R \rangle$ is a presentation of G on the generators X of G. For example, if $G = \langle (1,2), (1,2,3) \rangle = \text{Sym}(3)$, then we might say that $\langle x, y \mid x^2 = 1, y^2 = y^{-1}, xyx = y^{-1} \rangle$ is a presentation of G on the generators $x = (1,2)$, $y = (1,2,3)$ of G.

Using that slight abuse of notation, we can formulate the following version of Theorem 2.52, which is computationally useful for testing whether a map φ from a group G to a group H is a homomorphism.

THEOREM 2.53   *Let $\langle X \mid R \rangle$ be a presentation of a group G, and let $\theta : X \to H$ be a map from X to a group H. Extend θ to $\theta : X^{-1} \to H$ by putting $\theta(x^{-1}) = \theta(x)^{-1}$ for all $x \in X$.*

Then θ extends to a homomorphism $\theta' : G \to H$ if and only $\theta(x_1) \cdots \theta(x_r) = 1_H$ for all $w = x_1 \cdots x_r \in R$. This extension is unique if it exists.

PROOF   The uniqueness of the extension follows from the fact that X generates G. The statement that $\langle X \mid R \rangle$ is a presentation of a group G really means that there is an isomorphism $\varphi : \langle X \mid R \rangle \to G$, and that we are identifying $\langle X \mid R \rangle$ with its image under φ. Similarly, the statement $\theta(x_1) \cdots \theta(x_r) = 1_H$ for all $w = x_1 \cdots x_r \in R$ means $\theta(\varphi(x_1)) \cdots \theta(\varphi(x_r)) = 1_H$ for all for all $w = x_1 \cdots x_r \in R$. By Theorem 2.52, this is true if and only if $\theta\varphi$ extends to a homomorphism $\psi : \langle X \mid R \rangle \to H$. Since φ is an isomorphism, the existence of ψ is equivalent to the existence of the required homomorphism $\psi\varphi^{-1} : G \to H$.   ∎

2.4.3   Presentations of group extensions

DEFINITION 2.54   A group G with a normal subgroup N is called a *(group) extension* of N by G/N. It is called a *split extension* if there is a subgroup C of G with $NC = G$ and $N \cap C = 1$. In that case, C is called a *complement* of N in G/N.

Suppose that the group G has a normal subgroup N, and that we have presentations $\langle Y \mid S \rangle$ of N and $\langle \overline{X} \mid \overline{R} \rangle$ of G/N on generating sets Y and $\overline{X}$, respectively. Here we shall describe a general recipe for constructing a presentation of G as an extension of N by G/N.

For each $\overline{x} \in \overline{X}$, choose $x \in G$ with $xN = \overline{x}$, and let

$$X := \{\, x \mid \overline{x} \in \overline{X} \,\}.$$

Then, for any word $\overline{w} \in (\overline{X} \cup \overline{X}^{-1})^*$, we can define $w \in (X \cup X^{-1})^*$ with $wN = \overline{w}$, by substituting x or x^{-1} for each $\overline{x}$ or $\overline{x}^{-1}$ occurring in $\overline{w}$.

In particular, for each $\overline{r} \in \overline{R}$ there is a corresponding word r, and then $\overline{r} = 1_{G/N}$ implies that $r \in N$, so in the group G we have $r =_G w_r$, for some word $w_r \in (Y \cup Y^{-1})^*$. Let R be the set $\{ rw_r^{-1} \mid \overline{r} \in \overline{R} \}$.

For each $y \in Y$ and $x \in X$, we have $x^{-1}yx \in N$, so $x^{-1}yx =_G w_{xy}$ for some word $w_{xy} \in (Y \cup Y^{-1})^*$. Let T be the set $\{ x^{-1}yxw_{xy}^{-1} \mid x \in X, y \in Y \}$.

PROPOSITION 2.55   *With the above notation, $\langle X \cup Y \mid R \cup S \cup T \rangle$ is a presentation of G.*

PROOF   Let F be the group defined by the above presentation. To avoid confusion, let us denote the generators of F mapping onto $x \in X$ or $y \in Y$ by $\hat{x}$ and $\hat{y}$, respectively. We have chosen the sets R, S, T to be words that evaluate to the identity in G, so Theorem 2.52 tells us that the mapping $\hat{x} \to x, \hat{y} \to y$ induces a homomorphism $\theta : F \to G$. In fact θ is an epimorphism, because clearly G is generated by $X \cup Y$.

Let K be the subgroup $\langle \hat{y} \mid y \in Y \rangle$ of F. Then $\theta(K) = N$. Since, by assumption, $\langle Y \mid S \rangle$ is a presentation of N, and the relators of S are also relators of K, Theorem 2.52 implies that the map $y \to \hat{y}$ induces a homomorphism $N \to K$. But this homomorphism is then an inverse to θ_K, so $\theta_K : K \to N$ is an isomorphism.

We now wish to assert that the relators in T imply that $K \trianglelefteq F$. This is not quite as clear as it may at first sight appear! For it to be clear, we would need to know that $\hat{x}\hat{y}\hat{x}^{-1} \in K$ for all $x \in X, y \in Y$, in addition to $\hat{x}^{-1}\hat{y}\hat{x} \in K$. But the fact that $N \trianglelefteq G$ tells us that each $x \in X$ induces an automorphism of N by conjugation, which implies that, for each $x \in X$, we have $N = \langle w_{xy} \mid y \in Y \rangle$. Since N is isomorphic to K via θ_K, the corresponding statement is true in K for each $\hat{x}$. So the fact that $\hat{x}^{-1}$ conjugates each word w_{xy} into K implies the desired property $\hat{x}\hat{y}\hat{x}^{-1} \in K$ for all $x \in X, y \in Y$.

So we do indeed have $K \trianglelefteq F$. Now, by a similar argument to the one above for θ_K, the fact that $\langle \overline{X} \mid \overline{R} \rangle$ is a presentation of G/N implies that the induced homomorphism $\theta_{F/K} : F/K \to G/N$ is an isomorphism. So, if $g \in \ker(\theta)$, then $gK \in \ker(\theta_{F/K})$ implies that $g \in K$, but then $g \in \ker(\theta_K) = 1$, so θ is an isomorphism, which proves the result.   ∎

In the special case of a split extension, we can choose our inverse images x of $\overline{x}$ to lie in a complement of N in G, in which case we get $w_r = 1$ for all $r \in R$. In the even more special case of trivial action of the complement on N, we have a direct product, and we get the following well-known result as a corollary.

COROLLARY 2.56   *Let $\langle X \mid R \rangle$ and $\langle Y \mid S \rangle$ be presentations of groups G and H. Then $\langle X \cup Y \mid R \cup S \cup T \rangle$ is a presentation of $G \times H$, where $T = \{ [x,y] \mid x \in X, y \in Y \}$.*

2.4.4   Tietze transformations

There are a number of simple manipulations of a group presentation that do
not change the group or at least the isomorphism class of the group defined
by the presentation. These are known as *Tietze transformations* of the presentation. There are four of these, which come in mutually-inverse pairs. As
before, we denote $X \cup X^{-1}$ by A.

TT1 Adjoin an extra relator in $\langle X \mid R \rangle$ to R;

TT2 Remove a redundant relator w from R where $\langle R^F \rangle = \langle (R \setminus \{w\})^F \rangle$;

TT3 For any word $w \in A^*$, adjoin an extra element x, not already in X, to
X, and adjoin the element wx^{-1} to R.

TT4 If there is a generator $x \in X$ and a relator $w \in R$ such that there is
exactly one occurrence of x or x^{-1} in w, and neither x nor x^{-1} occurs
in any $v \in R \setminus \{w\}$, then remove x from X and w from R.

It is almost obvious that transformations of type (i) or (ii) do not change
the group defined by the presentation, and it is not hard to prove that (iii) and
(iv) replace this group by an isomorphic group. We leave this as an exercise
for the reader, who could alternatively refer to [Joh98].

There are some combinations of these transformations that are used very
frequently, and we shall mention a few of these here.

We can always replace a relator in a presentation by its inverse or by a
conjugate of itself. This is immediately clear from the definition of a group
presentation, but it could also be regarded as an application of TT1 followed
by TT2. We shall feel free to carry out such replacements without comment. In particular, we often need to replace a relator $x_1 x_2 \cdots x_r$ by a cyclic
conjugate. This facility also enables us to assume that defining relators are
cyclically reduced words.

A slightly less trivial instance is when we have two relators $uv, uw \in R$,
where $u, v, w \in A^*$ and $u \neq \varepsilon$. The common substring u is called a *piece*. If
$G = F/N$ as before, then $uv, uw \in N$ implies $v^{-1}w \in N$, and so we can apply
TT1 to adjoin $v^{-1}w$ to R. But $uv, v^{-1}w \in N$ implies $uw \in N$, so we can apply
TT2 to remove uw from R. The combined effect is to replace uw by $v^{-1}w$.
Let us call this combination a *common substring replacement*. In practical
applications, we might want to do this if $|v^{-1}w| < |uw|$, because this shortens
the presentation. Since we can always replace relators by cyclic conjugates, we
can apply common substring replacement whenever two elements of R have a
common substring u, regardless of whether or not u is a prefix of the relators.

TT4 is more useful when applied in combination with common substring
replacement. Suppose that we have $x \in X$ and $w \in R$ such that there is
exactly one occurrence of x or x^{-1} in w. By replacing r by its inverse and a
cyclic conjugate if necessary, we can assume that $w = xv$, where v does not
contain x or x^{-1}. If there are occurrences of x or x^{-1} in any other relators

in R, then we can apply common substring replacements to replace them by v^{-1} or v, thereby eliminating them. We can then apply TT4 and eliminate x from X and w from R. We shall refer to this combination of transformations as *eliminating a generator*. One practical problem with doing this is that if v is long, then the common substring replacements may render the total relator length of the presentation considerably greater than it was before the elimination.

Exercise

1. Prove from the definition of a free group F on a set X that F is generated by X.

2.5 Presentations of subgroups

Let G be a group generated by a set X, and let $A = X \cup X^{-1}$. Then $G \cong F/N$ for some N, where $F = F_X$ is the free group on X, and we shall assume that $G = F/N$. Let $H = E/N$ be a subgroup of G, and let T be a set of reduced words in A^* that form a right transversal of E in F. We shall always assume that T contains the empty word ε as the representative of the coset E. Hence the images of T in G form a right transversal of H in G. For a word $w \in A^*$, we denote the unique element of $T \cap Ew$ by $\overline{w}$.

2.5.1 Subgroup presentations on Schreier generators

Let $w = x_1 x_2 \cdots x_r \in A^*$ and, for $0 \le i \le r$, let $t_i := \overline{x_1 x_2 \cdots x_i}$, where $t_0 = \varepsilon$. Then we have

$$w =_F (t_0 x_1 t_1^{-1})(t_1 x_2 t_2^{-1}) \cdots (t_{r-1} x_r t_r^{-1}) t_r. \qquad (*)$$

Define the subset $\hat{Y}$ of E by

$$\hat{Y} := \{\, t x \overline{tx}^{-1} \mid t \in T, x \in X, tx \ne_F \overline{tx}^{-1} \,\}.$$

Note that the final condition $tx \ne_F \overline{tx}^{-1}$ simply has the effect of omitting the identity element from $\hat{Y}$. If $x \in X$, $t \in T$, and $u = \overline{tx^{-1}}$, then $t = \overline{ux}$, so $tx^{-1}u^{-1} = (uxt^{-1})^{-1} = (ux\overline{ux}^{-1})^{-1}$, which is an element of $\hat{Y}^{-1}$ (unless it equals 1_E). Since $t_i = \overline{t_{i-1} x_i}$ for $1 \le i \le r$, we see that each of the bracketed terms $t_i x_{i+1} t_{i+1}^{-1}$ in the equation $(*)$ is either equal to 1_E, or is an element of $\hat{Y} \cap \hat{Y}^{-1}$. If w represents an element of H, then $t_r = \overline{w} = \varepsilon$, and so we have:

THEOREM 2.57 (Schreier) *The subgroup E of F is generated by $\hat{Y}$. Similarly the subgroup H of G is generated by the images of $\hat{Y}$ in G.*

The set $\hat{Y}$ is called the set of *Schreier generators* of E with respect to the group F.

We now let Y be a set of new symbols corresponding to $\hat{Y}$, where we denote the element of Y corresponding to $tx\overline{tx}^{-1} \in \hat{Y}$ by y_{tx}. To ease the notation, we will let y_{tx} denote the empty string ε when $tx =_F \overline{tx}$. The map $\varphi : Y \to F$ defined by $\varphi(y_{tx}) := tx\overline{tx}^{-1}$ extends to a homomorphism from the free group F_Y on Y to F, and the theorem above says that $\mathrm{im}(\varphi) = E$. For $t \in T$, $x \in X$, we denote the element of Y^{-1} that φ maps to $tx^{-1}\overline{tx^{-1}}^{-1}$ by $y_{tx^{-1}}$.

We define a map $\rho : A^* \to F_Y$ by

$$\rho(w) := y_{t_0 x_1} y_{t_1 x_2} \cdots y_{t_{r-1} x_r},$$

where $w = x_1 x_2 \cdots x_r$, as in the equation $(*)$ above. Of course, ρ and φ depend on the choice of the transversal T. It is straightforward to check that $\rho(uxx^{-1}v) =_{F_Y} \rho(uv)$ for any $u, v \in A^*$, $x \in A$, and so ρ induces a well-defined map, also denoted by ρ, from F to F_Y. This map is called a *rewriting map*.

More generally, for $t \in T$ and $w = x_1 x_2 \cdots x_r \in A^*$, we define $\rho(t, w)$ in the same way as for $\rho(w)$, but with $t_0 = t$; that is,

$$\rho(t, w) := y_{t_0 x_1} y_{t_1 x_2} \cdots y_{t_{r-1} x_r},$$

where $t_i = \overline{tx_1 x_2 \cdots x_i}$ for $0 \le i \le r$; so $\rho(w) = \rho(\varepsilon, w)$.

LEMMA 2.58　　(i) $\rho(uv) = \rho(u)\rho(\overline{u}, v)$ for all $u, v \in A^*$.
(ii) $\rho(u)^{-1} =_{F_Y} \rho(\overline{u}, u^{-1})$ for all $u \in A^*$.

PROOF　　(i) is clear from writing down the definition of $\rho(uv)$. For (ii), note that $\rho(uu^{-1}) =_{F_Y} 1_{F_Y}$ and use (i). ∎

It follows from (i) of this lemma that, if $u \in E$, then $\overline{u} = \varepsilon$ and $\rho(uv) = \rho(u)\rho(v)$, so ρ restricts to a homomorphism from E to F_Y.

The transversal T is called a *Schreier transversal* if it is prefix-closed; that is, if all prefixes of all elements of T lie in T. To prove that Schreier transversals exist, let $<$ be a well-ordering of A^* with the property that $u < v \Rightarrow uw < vw$ for all $u, v, w \in A^*$, and choose the elements of T to be the least elements of their cosets under this ordering (see the exercises below). Examples of orderings of A^* with this property are shortlex orderings, which are defined below. They are defined with respect to a given well-ordering of A and so, if we assume the axiom of choice, then they exist for all sets A; but in this book, we are only concerned with finite sets A.

DEFINITION 2.59   Let $<$ be a given well-ordering of a set A. Then, the associated *lexicographical* (dictionary) ordering $<_L$ of A^*, is defined by $a_1 a_2 \cdots a_m < b_1 b_2 \cdots b_n$ (with $a_i, b_j \in A$) if, for some $k \geq 0$, $a_i = b_i$ for $1 \leq i \leq k$, and either $k = m < n$ or $a_{k+1} < b_{k+1}$.

The lexicographical ordering satisfies the $u < v \Rightarrow uw < vw$ property, but it is not a well-ordering for $|A| > 1$ because, if $a, b \in A$ with $a > b$, then $a^i b >_L a^{i+1} b$ for all $i \geq 0$. The shortlex ordering, however, which we shall now define, is a well-ordering.

DEFINITION 2.60   Let $<$ be a given well-ordering of a set A. Then, the associated *shortlex* (also called *lenlex*) ordering $<_S$ of A^* is defined by $u <_S v$ if either $|u| < |v|$, or if $|u| = |v|$ and $u <_L v$.

If T is a Schreier transversal and $w = t \in T$, then all of the bracketed terms in the equation $(*)$ are equal to 1_E, and so $\rho(t) = \varepsilon$. If $w = tx$ with $t \in T$, $x \in X$ and $tx \neq_F \overline{tx}$, then all of the bracketed terms are trivial except for the last, and so $\rho(tx) = y_{tx}$ and, using Lemma 2.58, we have

$$\rho(tx\overline{tx}^{-1}) = y_{tx}\rho(\overline{tx}, \overline{tx}^{-1}) =_{F_Y} y_{tx}\rho(\overline{tx})^{-1} = y_{tx}.$$

So the composite $\rho\varphi$ is equal to the identity homomorphism on F_Y. But this implies that φ is a monomorphism, so we have proved the well-known Nielsen-Schreier theorem:

THEOREM 2.61   *If T is Schreier transversal, then the subgroup E of F is a free group on the generating set $\hat{Y}$.*

Now suppose that $G = \langle X \mid R \rangle$ is a presentation of G, and $G = F/N$ with $N = \langle R^F \rangle$. The following theorem will be used later, in Subsection 5.3.1, in an algorithm to compute a presentation of the subgroup H of G.

THEOREM 2.62   *Assume that T is a Schreier transversal of E in F, and define $S := \{ \rho(twt^{-1}) \mid t \in T, w \in R \}$. Then the group $H' := \langle Y \mid S \rangle$ is isomorphic to H, and an isomorphism is induced by the map $\varphi : F_Y \to E$ defined above.*

PROOF   We saw above that φ and $\rho|_E$ are inverse isomorphisms. Since each twt^{-1} with $t \in T$, $w \in R$ lies in $N \leq E$, we have $\varphi\rho(twt^{-1}) =_F twt^{-1} \in N$, and so by Theorem 2.52 φ induces an epimorphism $\overline{\varphi}$ from H' to H.

Now $N = \langle R^F \rangle$ is generated by conjugates uwu^{-1} of elements $w \in R$ by elements $u \in A^*$. For any such u, we have $u =_F vt$, with $v \in E$ and $t \in T$, and then $uwu^{-1} =_F v(twt^{-1})v^{-1}$ and so $\rho(uwu^{-1}) =_{F_Y} \rho(v)\rho(twt^{-1})\rho(v)^{-1}$

is a conjugate of an element of S. It follows that $\rho(z) \in \langle S^{F_Y} \rangle$ for all $z \in N$, and hence $\rho|_E$ induces a homomorphism $\overline{\rho} : H = E/N \to H'$.

We saw above that $\rho\varphi$ is the identity homomorphism on F_Y, and hence $\overline{\rho\varphi}$ is the identity on H'. So $\overline{\varphi}$ is an isomorphism, which proves the result. ∎

As a final observation on the above theorem, note that by Lemma 2.58 we can express the elements $\rho(twt^{-1})$ of S as

$$\rho(twt^{-1}) = \rho(t)\rho(t,w)\rho(\overline{tw}, t^{-1}) = \rho(t)\rho(t,w)\rho(t, t^{-1}) = \rho(t)\rho(t,w)\rho(t)^{-1}.$$

We saw earlier that the fact that T is a Schreier transversal implies that $\rho(t) = \varepsilon$, so in fact we have

$$S = \{ \rho(t, w) \mid t \in T, w \in R \}.$$

This form of S is more convenient for use in the algorithm in Subsection 5.3.1 for computing subgroup presentations.

2.5.2 Subgroup presentations on a general generating set

We shall now describe a generalization of the results in Subsection 2.5.1, which can be used to find a presentation of a subgroup on a given generating set, rather than on the set of Schreier generators. This subsection could be omitted on a first reading.

As before, we assume that $G = \langle X \mid R \rangle$ and $G = F/N$ with $F = F_X$ and $N = \langle R^F \rangle$, and that $H = E/N$ is a subgroup of G. Let T be any right transversal of E in F (not necessarily a Schreier transversal) consisting of reduced words and containing ε and, as before, denote the unique element of $T \cap Ew$ by $\overline{w}$ for $w \in A^*$. Let Y be a set disjoint from X, and suppose that we have an epimorphism $\varphi : F_Y \to E$, and that, for $t \in T$ and $x \in X$, we are given elements $y_{tx} \in F_Y$ satisfying $tx =_G \varphi(y_{tx})\overline{tx}$ for all $t \in T$, $x \in X$. Note that these conditions are satisfied by the φ and y_{tx} defined in Subsection 2.5.1. For $t \in T$, $x \in X$, we use $y_{tx^{-1}}$ to denote the inverse of the element y_{ux} of F_Y that satisfies $ux =_G \varphi(y_{ux})t$. We define the rewriting map $\rho : A^* \to F_Y$ exactly as before: for any $w \in A^*$, we put

$$\rho(w) = y_{t_0 x_1} y_{t_1 x_2} \cdots y_{t_{r-1} x_r}$$

where $w = x_1 x_2 \cdots x_r$ as in the equation $(*)$ in Subsection 2.5.1.

Again the images of equivalent words under ρ are equal, so we can think of ρ as mapping F to F_Y, and Lemma 2.58 is true as before, so again $\rho|_E$ is a homomorphism. In Subsection 2.5.1, we had $\rho\varphi(y_{tx}) = \rho(tx\overline{tx}^{-1}) = y_{tx}$ for all $y_{tx} \in Y$, but this is no longer necessarily the case. So, in Theorem 2.62, the relators in the set S_2 in the theorem below freely reduce to the empty word, and can be omitted.

THEOREM 2.63   *With the notation just defined, let* $S_1 := \{\rho(twt^{-1}) \mid t \in T, w \in R\}$, *and let* $S_2 := \{\rho\varphi(y)y^{-1} \mid y \in Y\}$. *Then the group* $H' := \langle Y \mid S \rangle$ *with* $S = S_1 \cup S_2$ *is isomorphic to* H, *and an isomorphism is induced by the map* $\varphi : F_Y \to E$.

PROOF   Let $w = x_1 x_2 \cdots x_r \in A^*$ and $\rho(w) = y_{t_0 x_1} y_{t_1 x_2} \cdots y_{t_{r-1} x_r}$ with the t_i defined as in the equation $(*)$. Since $t_i = \overline{t_{i-1} x_i}$ for $1 \le i \le r$, our assumption $tx =_G \varphi(y_{tx})\overline{tx}$ tells us that $\varphi(y_{t_{i-1} x_i}) =_G t_{i-1} x_i t_i^{-1}$ for $1 \le i \le r$, and so we have

$$\varphi\rho(w)t_r =_G \varphi(y_{t_0 x_1} y_{t_1 x_2} \cdots y_{t_{r-1} x_r})t_r =_G w.$$

In particular, if $w \in E$, then $\varphi\rho(w) =_G w$. From this equation, it follows that $\varphi(s) \in N$ for all $s \in S_1 \cup S_2$ and so φ induces a epimorphism $\overline{\varphi} : H' \to H$.

It follows exactly as in Theorem 2.62 that $\rho|_E$ induces a homomorphism $\overline{\rho} : H = E/N \to H'$. The inclusion of S_2 in the set of relators of H' ensures that $\rho(\varphi(y)) =_{H'} y$ for all $y \in Y$, and so $\overline{\rho\varphi}$ is the identity on H', and hence $\overline{\varphi}$ is an isomorphism, which proves the result.   ∎

As in the final paragraph of Subsection 2.5.1, the element $\rho(twt^{-1})$ of S_1 is equal to $\rho(t)\rho(t, w)\rho(t)^{-1}$ in F_Y. It is no longer necessarily true that $\rho(t) = \varepsilon$, but since $\rho(t)\rho(t, w)\rho(t)^{-1}$ is a conjugate of $\rho(t, w)$, the theorem still remains true if we replace S_1 by

$$\{\rho(t, w) \mid t \in T, w \in R\}.$$

Exercises

1. Show that the transversal T of E in F defined as the set of least elements in their cosets under a well-ordering of A^* satisfying $u < v \Rightarrow uw < vw$ for all $w \in A^*$ is indeed a Schreier transversal.

2. Show that the shortlex ordering $<_S$ of A^* associated with a given well-ordering $<$ of A is a well-ordering and satisfies $u < v \Rightarrow uw < vw$ and $wu < wv$ for all $w \in A^*$.

3. Let $A = \{x, x^{-1}, y, y^{-1}\}$ and write down the first 20 elements of A^* in the shortlex ordering of A^* defined with respect to the ordering $x < y < y^{-1} < x^{-1}$ of A.

4. Show that the normal closure of the subgroup $\langle y \rangle$ of the free group on $\{x, y\}$ is free on $\{y^{x^i} \mid i \in \mathbb{Z}\}$.

2.6   Abelian group presentations

The theory that we have just developed for group presentations can be recast in the framework of abelian groups rather than general groups, but much of the material becomes significantly easier in the abelian case. In particular, a free abelian group is just a direct sum of copies of an infinite cyclic group. We shall quickly run through this basic theory here. We shall, however, delay our treatment of the structure theorem for finitely generated abelian groups until Section 9.2, because it is more convenient to discuss this at the same time as we describe the related algorithms.

DEFINITION 2.64   An abelian group F^{ab} is *free abelian* on the subset X of F^{ab} if, for any abelian group G and any map $\theta : X \to G$, there is a unique group homomorphism $\theta' : F^{\mathrm{ab}} \to G$ with $\theta'(x) = \theta(x)$ for all $x \in X$. The set X is known as a *free basis* of F^{ab}.

Again it follows directly from the definition that any abelian group generated by a set X is an epimorphic image of, and hence isomorphic to a quotient of, a free abelian group on X.

Exactly as in Proposition 2.47, we can show that free abelian groups on X_1 and X_2 are isomorphic if and only if $|X_1| = |X_2|$, and we call $|X|$ the *rank* of a free abelian group on X.

We generally prefer to use additive rather than multiplicative notation when discussing abelian groups, and we shall do so here.

For an arbitrary set X, we define F_X^{ab} to be the direct sum $\oplus_{x \in X} \mathbb{Z}_x$ of infinite cyclic groups $\mathbb{Z}_x$ generated by x. As usual, we identify the groups $\mathbb{Z}_x$ with the corresponding component subgroups of the direct sum, and so an element $g \in F_X^{\mathrm{ab}}$ has the form $\sum_{x \in X} \alpha_x x$ for $\alpha_x \in \mathbb{Z}$, where all but finitely many of the α_x are zero. (In this book, X will virtually always be finite, but there is no reason not to consider general X at this stage.)

Many books define the free abelian group on X to be the group F_X^{ab}, and we shall now show that this is equivalent to our definition.

PROPOSITION 2.65   *If X is a set, then the group F_X^{ab} is free abelian on X.*

PROOF   For an abelian group G and a map $\theta : X \to G$, we define $\theta' : F_X^{\mathrm{ab}} \to G$ by $\theta'(g) = \sum_{x \in X} \alpha_x \theta(x)$, where $g = \sum_{x \in X} \alpha_x x$ with all but finitely many of the α_x equal to zero. It is easy to see that θ' is a homomorphism extending θ.   ∎

It follows that a finitely generated free abelian group is isomorphic to $\mathbb{Z}^n$ for some $n \in \mathbb{N}_0$, and we shall generally use $\mathbb{Z}^n$ as our prototype free abelian group. Similarly, any finitely generated abelian group with n generators is isomorphic to a quotient group of $\mathbb{Z}^n$.

There is an alternative construction of the free abelian group on X as an ordinary group presentation. For a set X, we define $[X, X]$ to be the set of formal commutators

$$\{\, x^{-1}y^{-1}xy \mid x, y \in X,\ x \neq y \,\}.$$

This is a subset of the set A_X^* of words over A_X, where $A_X := X \cup X^{-1}$.

PROPOSITION 2.66   *If F is a free group on a set X, then $\langle\, X \mid [X, X] \,\rangle$ is a free abelian group on X.*

PROOF   Let $\theta : X \to G$ be a map, with G an abelian group. By the definition of the free group on X, θ extends to a homomorphism $\theta' : F \to X$. Since G is abelian, $[X, X]$ and hence $N := \langle [X, X]^F \rangle$ is contained in the kernel of θ', so θ' induces a homomorphism $\langle\, X \mid [X, X] \,\rangle = F/N \to G$, and the result follows.   ∎

We can now define an abelian group presentation.

DEFINITION 2.67   Let X be a set, let R be a subset of F_X^{ab}, and let N be the subgroup of F_X^{ab} generated by R. Then the abelian group presentation $\mathrm{Ab}\langle\, X \mid R \,\rangle$ is defined to be equal to the quotient group F_X^{ab}/N.

The analogous result to Theorem 2.52 holds for presentations of abelian groups, but it is not used so frequently as Theorem 2.52 itself.

Next we relate abelian group presentations to the abelianizations of ordinary group presentations. For a set of words $R \subseteq A_X^*$, we let R^+ denote the corresponding set of elements of F_X^{ab} obtained by writing each element of R additively. So, for example, if $R = \{\, x^2yx^{-1}y^3x^4,\ y^3x^{-4}xy \,\}$, then $R^+ = \{\, 5x + 4y,\ -3x + 4y \,\}$.

PROPOSITION 2.68   *Let $G = \langle\, X \mid R \,\rangle$ be a group presentation. Then*

(i) $G/[G, G] \cong \langle\, X \mid R \cup [X, X] \,\rangle$ *and*

(ii) $G/[G, G] \cong \mathrm{Ab}\langle\, X \mid R^+ \,\rangle.$

PROOF   Let F be the free group on X. Then $G = F/N$ with $N = \langle R^F \rangle$, and $\langle\, X \mid R \cup [X, X] \,\rangle = F/M$ with $M = \langle (R \cup [X, X])^F \rangle$. Since each pair of generators commutes in F/M, F/M is abelian, and so $(F/N)/(M/N) \cong$ F/M is an abelian quotient of G and hence $[G,G] \le M/N$. But $M = \langle R^F, [X,X]^F \rangle = NL$, where $L = \langle [X,X]^F \rangle$, so M/N is generated by commutators, and hence $[G,G] = M/N$, which proves (i).

By Proposition 2.66, F/L is a free abelian group on the set X, and so $F/M \cong (F/L)/(M/L)$. An element of R^F has the form u^w for $u \in R$, $w \in F$, and $u^w = u[u,w]$ with $[u,w] \in [F,F] \subset L$ (since F/L is abelian), so $M/L = NL/L$ is generated by the images of the elements $u \in R$, which proves (ii).   ∎

Exercise 3 in Section 2.5 shows that subgroups of finitely generated groups are not themselves finitely generated in general. The following result shows that this property does hold for finitely generated abelian groups, from which it follows that all finitely generated abelian groups have finite abelian group presentations.

PROPOSITION 2.69   *Any subgroup H of a finitely generated abelian group G is finitely generated.*

PROOF   Let G be generated by $x_1, \ldots, x_n$. We use induction on n, the result being clear for $n = 0$. Let $K := \langle x_2, \ldots, x_n \rangle$. By induction, $K \cap H$ is finitely generated, whereas $H/(K \cap H) \cong HK/K \le G/K$ is cyclic, and hence finitely generated. The result follows.   ∎

2.7   Representation theory, modules, extensions, derivations, and complements

This section concerns *group representation theory*, together with the basic theory of group extensions of M by G in the case when M is abelian.

Computation with group representations is a significant subtopic within CGT. Some of the methods in this area, particularly those related to group representation theory for its own sake involve some rather advanced theory. Even for computations that are concerned only with the group-theoretical structure of finite groups, some of the more sophisticated algorithms require some familiarity with representation theory.

The basic reason for this is that if a finite group G has normal subgroups $N < M$ for which M/N is an elementary abelian p-group for some prime p, then the conjugation action of G on M gives rise to a representation of G/M over the field of order p, and properties of that representation translate into group-theoretical properties of G. For example, the representation is irreducible if and only if M/N is a chief factor of G.

In this book, mainly in Chapter 7, we shall attempt to cover only those parts of computational representation theory that have applications to the analysis of the structure of finite groups. This does not require that the reader have a very deep knowledge of representation theory, but it does involve certain topics that are not always covered by the most elementary books on the subject, such as representations over finite fields rather than over $\mathbb{C}$, so it is not easy to find references at the appropriate level. Two possibilities are the books by Isaacs [Isa76] and Rotman [Rot02], which we shall use here for references to theoretical results.

2.7.1 The terminology of representation theory

Let us briefly review the basic definitions and results from the representation theory of finite groups. Let K be a commutative ring with 1, and let G be a finite group. The *group ring KG* of G over K is defined to be the ring of formal sums

$$\{ \sum_{g \in G} r_g g \mid r_g \in K \}$$

with the obvious addition and multiplication inherited from that of G. In fact KG is an associative algebra with 1 and thus it is a ring with 1 and a module over K. It is also known as the *group algebra* of G over K.

Let M be a right (unital) KG-module. We shall write $m \cdot x$ ($m \in M$, $x \in KG$) to represent the module product in M, but when $x \in K$ and we are thinking of M primarily as a K-module, then we may write xm rather than $m \cdot x$. Since K is commutative, this does not cause any problems. From the module axioms, and the fact that $(m \cdot g) \cdot g^{-1} = m$ for $m \in M$, $g \in G$, we see that multiplication by a group element $g \in G$ defines an automorphism of M as a K-module. So we have an associated action $\varphi : G \to \mathrm{Aut}_K(M)$, and we shall sometimes use the group action notation m^g as an alternative to $m \cdot g$. Conversely, if M is a K-module, then any action $\varphi : G \to \mathrm{Aut}_K(M)$ can be used to make M into an KG-module.

We shall always assume that M is finitely generated and free as a K-module, and so, after fixing on a free basis of M, we can identify M with K^d for some d. Then, using the same free basis of M, $\mathrm{Aut}_K(M)$ can be identified with the group $\mathrm{GL}(d, K)$ of invertible $d \times d$ matrices over K. So the action homomorphism φ is $\varphi : G \to \mathrm{GL}(d, K)$, which is the standard definition of a representation of G of degree d over K.

According to basic results from representation theory, two KG-modules are isomorphic if and only if the associated representations φ_1, φ_2 are *equivalent*, which means that they have the same degree and there exists $\alpha \in \mathrm{GL}(d, K)$ with $\alpha \cdot \varphi_2(g) = \varphi_1(g) \cdot \alpha$ for all $g \in G$.

When K is a field, a KG-module and its associated representation is called *simple* or *irreducible* if it has no proper nonzero KG-submodules. (A slightly different definition of irreducibility is normally used when $K = \mathbb{Z}$, for example.)

Since we are dealing only with finite-dimensional modules, any such module M has a *composition series*, (that is, an ascending series of submodules with simple factor modules), and the Jordan-Hölder theorem (Thereom 8.18 of [Rot02]) asserts that any two such series have isomorphic factors, counting multiplicity, and so we can refer to the *composition factors* of M.

For any KG-module M, we can define the K-algebra $\mathrm{End}_{KG}(M)$ of endomorphisms of M ($= KG$-homomorphisms from M to M). This is also known as the *centralizing algebra* of M and its associated representation. It contains the scalar automorphisms, which form a subalgebra isomorphic to K. When K is a field and M is a simple KG-module, then Schur's lemma (Theorem 8.52 [Rot02] or Lemma 1.5 of [Isa76]) says that $\mathrm{End}_{KG}(M)$ is division ring. This can be noncommutative in general, but in this book we shall be particularly concerned with the case when K is a finite field, in which case a well-known theorem of Wedderburn (Theorem 8.23 of [Rot02]) tells us that $\mathrm{End}_{KG}(M)$ is a field, and it can be regarded as an extension field of K.

When K is a field, the KG-module M is called *absolutely irreducible* if it is irreducible and remains irreducible when regarded as an LG-module for any extension field L of K. By Theorem 9.2 of [Isa76], M is absolutely irreducible if and only if $\mathrm{End}_{KG}(M)$ consists of scalars only.

So, if K is a finite field and $L = \mathrm{End}_{KG}(M)$, then we can use the action of L on M to make M into an LG-module with $\dim_L(M)|L : K| = \dim_K(M)$, and M is absolutely irreducible as an LG-module. In particular, there is a finite extension L of K for which all irreducible LG-modules are absolutely irreducible, and such an L is called a *splitting field* for G.

2.7.2 Semidirect products, complements, derivations, and first cohomology groups

This and the following subsection contain a very brief description of the first and second cohomology groups of groups acting on modules, insofar as they are relevant to the (computational) study of group extensions. For a more complete treatment, the reader may consult Chapter 10 of [Rot02], particularly Sections 10.2 and 10.3. But, in common with the majority of published material on this topic, the account in [Rot02] is in terms of left modules, whereas ours uses right modules, so there will be some differences, such as in the definitions of cocycles.

We defined the notion of a (split) extension of one group by another in Subsection 2.54. Let G and M be groups, and suppose that we are given a homomorphism $\varphi : G \to \mathrm{Aut}(M)$. We define the *semidirect product* of M by G using φ to be the set $G \times M$ endowed with the multiplication $(g, m)(h, n) = (gh, m^h n)$, for $g, h \in G$, $m, n \in M$, where, as usual, m^h is an abbreviation for $m^{\varphi(h)}$. The standard notation for a semidirect product is $G \ltimes M$ or $G \ltimes_\varphi M$.

The semidirect product is an extension of M by G, using the maps $M \to G \ltimes M$ and $G \ltimes M \to G$ defined by $m \mapsto (1_G, m)$ and $(g, m) \mapsto g$. It is a split extension, with complement $\{ (g, 1_M) \mid g \in G \}$.

Conversely, if the group E has a normal subgroup M with a complement G then any $e \in E$ can be written uniquely as $e = gm$ for $g \in G$, $m \in M$, and $gmhn = ghm^h n$, so we have:

PROPOSITION 2.70 *Any split extension E of M by G is isomorphic to the semidirect product $G \ltimes_\varphi M$, where the action φ of G on M is defined by the conjugation action of a complement of M in E on M.*

In general, different complements could give rise to different actions φ. However, if M is abelian, then the actions coming from different complements are the same. We shall assume for the remainder of this subsection that M is abelian and use additive notation for M.

We shall also assume that M is a K-module for some commutative ring K with 1. This is no loss of generality, because any abelian group can be regarded as a $\mathbb{Z}$-module just by defining $n \cdot m = nm$ for $n \in \mathbb{Z}$, $m \in M$. In the case when M is an elementary abelian p-group, we can take K to be the field $\mathbb{F}_p$.

As we saw in Subsection 2.7.1, an action $\varphi : G \to \mathrm{Aut}_K(M)$ of G on the K-module M corresponds to endowing M with the structure of a KG-module, and so we can talk about the semidirect product $G \ltimes M = G \ltimes_\varphi M$ of the KG-module M with G. The multiplication rule in $G \ltimes M$, using additive notation in M, is $(g, m)(h, n) = (gh, m^h + n)$.

A general left transversal of the subgroup $\hat{M} := \{ (1_G, m) \mid m \in M \}$ isomorphic to M in $G \ltimes M$ has the form $T_\chi = \{ (g, \chi(g)) \mid g \in G \}$, for a map $\chi : G \to M$. Then T_χ is a complement of $\hat{M}$ in $G \ltimes M$ if and only if $(g, \chi(g))(h, \chi(h)) = (gh, \chi(gh))$ for all $g, h \in G$ or, equivalently,

$$\chi(gh) = \chi(g)^h + \chi(h) \quad \forall g, h \in G. \tag{$\dagger$}$$

If M is a KG-module, then a map $\chi : G \to M$ is called a *derivation* or a *crossed homomorphism* or a *1-cocycle* if ($\dagger$) holds. Notice that by putting $h = 1_G$ in ($\dagger$), we see that $\chi(1_G) = 0_M$ for any derivation χ.

We denote the set of such derivations by $Z^1(G, M)$. By using the obvious pointwise addition and scalar multiplication, we can make $Z^1(G, M)$ into a K-module. We have proved:

PROPOSITION 2.71 *If M is a KG-module, then the set T_χ defined above is a complement of $\hat{M}$ in $G \ltimes M$ if and only if $\chi \in Z^1(G, M)$.*

Notice that for a fixed $m \in M$, $\{ (g, 0_M)^{(1, m)} = (g, m - m^g) \mid g \in G \}$ is a complement of $\hat{M}$ in $G \ltimes M$, and so $g \mapsto m - m^g$ is a derivation. Such a map is called a *principal derivation* or *1-coboundary*. The set of all principal derivations is denoted by $B^1(G, M)$ and forms a K-submodule of $Z^1(G, M)$.

DEFINITION 2.72   The *first cohomology group* $H^1(G, M)$ is the quotient K-module $Z^1(G, M)/B^1(G, M)$.

From the discussion above, it follows that $H^1(G, M)$ is in one-one correspondence with the set of conjugacy classes of complements of $\hat{M}$ in $G \ltimes M$.

The following result tells us that derivations are uniquely determined by their action on a generating set of a group.

PROPOSITION 2.73   *Suppose that $G = \langle X \rangle$ and let $\chi : X \to M$ be a map. If χ extends to a derivation $\chi : G \to M$, then:*

(i) $\chi(x^{-1}) = -\chi(x)^{x^{-1}}$ for all $x \in X$.

(ii) Let $g \in G$ with $g = x_{k_1}^{\varepsilon_1} \cdots x_{k_l}^{\varepsilon_l}$, where each $\varepsilon_i = \pm 1$. Then

$$\chi(g) = \sum_{i=1}^{l} \varepsilon_i \chi(x_{k_i})^{g_i}$$

where $g_i = x_{k_{i+1}}^{\varepsilon_{i+1}} \cdots x_{k_l}^{\varepsilon_l}$ or $x_{k_i}^{\varepsilon_i} \cdots x_{k_l}^{\varepsilon_l}$, when $\varepsilon_i = 1$ or -1, respectively.

PROOF   By (†), we have $0_M = \chi(xx^{-1}) = \chi(x)^{x^{-1}} + \chi(x^{-1})$, which proves (i). (ii) is proved by repeated use of (†).   ∎

In general, given any map $\chi : X \to M$, we can use (i) and (ii) of the above proposition to extend χ to a derivation $F_X \to M$, where F_X is the free group on X.

2.7.3   Extensions of modules and the second cohomology group

Let E be any extension of an abelian group M (regarded as subgroup of E) by a group G. So we have an epimorphism $\rho : E \to G$ with kernel M. For $g \in G$, choose $\hat{g} \in E$ with $\rho(\hat{g}) = g$ and, for $m \in M$, define $m^g := m^{\hat{g}}$. Since M is abelian, this definition is independent of the choice of $\hat{g}$, and it defines an action of G on M. In general, this action makes M into a $\mathbb{Z}G$-module, but if M happens to be a module over a commutative ring K with 1, and the conjugation actions of $g \in G$ define K-automorphisms of M, then M becomes a KG-module. In particular, this is true with $K = \mathbb{F}_p$ in the case when M is an elementary abelian p-group.

DEFINITION 2.74   Let G be a group and M a KG-module for some commutative ring K. We define a *KG-module extension of M by G* to be a group extension E of M by G in which the given KG-module M is the same as the KG-module defined as above by conjugation within E.

Given E as above, the elements $\{\hat{g} \mid g \in G\}$ form a transversal of M in G. For $g, h \in G$, we have $\hat{g}\hat{h} = \widehat{gh}\,\tau(g, h)$, for some function $\tau : G \times G \to M$, where the associative law in E implies that, for all $g, h, k \in G$,

$$\tau(g, hk) + \tau(h, k) = \tau(g, h)^k + \tau(gh, k).$$

A function $\tau : G \times G \to M$ satisfying this identity is called a *2-cocycle*, and the additive group of such functions forms a K-module and is denoted by $Z^2(G, M)$.

Conversely, it is straightforward to check that, for any $\tau \in Z^2(G, M)$, the group $E = \{(g, m) \mid g \in G, \, m \in M\}$ with multiplication defined by

$$(g, m)(h, n) = (gh, \tau(g, h) + m^h + n)$$

is a KG-module extension of M by G that defines the 2-cocycle τ when we choose $\hat{g} = (g, 0)$.

A general transversal of M in E has the form $\hat{g} = (g, \chi(g))$ for a function $\chi : G \to M$, and it can be checked that this transversal defines the 2-cocycle $\tau + c_\chi$, where c_χ is defined by $c_\chi(g, h) = \chi(gh) - \chi(g)^h - \chi(h)$. A 2-cocycle of the form c_χ for a function $\chi : G \to M$ is called a *2-coboundary*, and the additive group of such functions is a K-module and is denoted by $B^2(G, M)$.

Two KG-module extensions E_1 and E_2 of a KG-module M by G are said to be *equivalent* if there is an isomorphism from E_1 to E_2 that maps M_1 to M_2 and induces the identity map on both M and on G. From the above discussion, it is not difficult to show that the extensions corresponding to the 2-cocycles τ_1 and τ_2 are equivalent if and only if $\tau_1 - \tau_2 \in B^2(G, M)$ and, in particular, an extension E splits if and only if its corresponding 2-cocycle $\tau \in B^2(G, M)$.

DEFINITION 2.75   The *second cohomology group* $H^2(G, M)$ is the quotient K-module $Z^2(G, M)/B^2(G, M)$.

So $H^2(G, M)$ is in one-one correspondence with the equivalence classes of KG-module extensions of M by G.

Checking directly whether a given 2-cocycle is a 2-coboundary can be difficult, but there is an alternative approach to deciding whether a KG-module extension E splits, which we shall now describe.

Suppose that we have a finite presentation $\langle X \mid R \rangle$ of G, and let us identify G with the group F/N defined by the presentation, where F is the free group on X and $N = \langle R^F \rangle$. For each $x \in X$, choose an element $\hat{x} \in E$ with $\rho(\hat{x}) =_G x$. Then there is a unique homomorphism $\theta : F \to E$ with $\theta(x) = \hat{x}$ for all $x \in X$ and, since $\rho\theta(x) =_G x$, $\rho\theta$ is the natural map from F to

G. Hence $\ker(\rho\theta) = N$. It follows that $\ker(\theta) \le N$ in any case and, by Theorem 2.52, $\ker(\theta) = N$ if and only if θ induces $\hat{\theta} : G \to E$, in which case $\rho\hat{\theta}$ is an isomorphism and $\mathrm{im}(\theta) = \langle\, \hat{x} \mid x \in X \,\rangle$ is a complement of M in E. Furthermore, $\ker(\theta) = N$ if and only if $\theta(w) = 1_E$ for all $w \in R$, so we have proved the following lemma.

LEMMA 2.76   *If $\hat{x} \in E$ are chosen with $\rho(\hat{x}) =_G x$ and $\theta : F \to E$ is defined by $\theta(x) = \hat{x}$, then the elements $\hat{x}$ generate a complement of M in E if and only if $\theta(w) = 1_E$ for all $w \in R$.*

The elements $\theta(w)$ for $w \in R$ always lie in $\ker(\rho) = M$. Let $\{\, \hat{x}\chi(x) \mid x \in X \,\}$ be another choice of the inverse images of x under ρ, where $\chi : X \to M$ is a map, and let $\theta_\chi : F \to E$ be the associated homomorphism with $\theta_\chi(x) = \hat{x}\chi(x)$ for $x \in X$. If we use (i) and (ii) of Proposition 2.73 to extend χ to $\chi : F \to M$, then a simple calculation shows that $\theta_\chi(w) = \theta(w)\chi(w)$ for all $w \in R$. Since the elements in these equations all lie in M, we can switch to additive notation and write them as $\theta_\chi(w) = \theta(w) + \chi(w)$. So we have the following result.

PROPOSITION 2.77   *With the above notation, the elements $\{\, \hat{x}\chi(x) \mid x \in X \,\}$ generate a complement of M in E if and only if $\chi(w) = -\theta(w)$ for all $w \in R$.*

We shall use this result later, in Section 7.6, to help us determine computationally whether an extension splits.

In particular, from the case $E = G \ltimes M$, and $\hat{x} = (x, 0)$ for all $x \in X$, Proposition 2.71 yields:

THEOREM 2.78   *If M is a KG-module with $G = \langle\, X \mid R \,\rangle$, then a map $\chi : X \to M$ extends to a derivation $G \to M$ if and only if $\chi(w) = 0_M$ for all $w \in R$, where $\chi : F \to M$ is defined by (i) and (ii) of Proposition 2.73.*

This last result is the analogue of Theorem 2.52 for derivations. It can of course be proved directly from the definition of derivations, without involving semidirect products. It will be used later in Section 7.6 to help us to compute $Z^1(G, M)$.

2.7.4   The actions of automorphisms on cohomology groups

Two KG-module extensions E_1 and E_2 of M by G can be isomorphic as groups without being equivalent as extensions. This remains true even if we restrict our attention to isomorphisms that map M to M.

Example 2.8

Let $M = C_2 = \langle t \rangle$, $G = C_2 \times C_2 = \langle x, y \rangle$ with G acting trivially on M. It is not difficult to check that there are eight equivalence classes of extensions of M by G (in fact $H^2(G, M) \cong C_2 \times C_2 \times C_2$), in which each of x^2, y^2 and $[x, y]$ can be equal to 1_M or t. However, there are only four isomorphism classes of groups E that arise, namely $C_2 \times C_2 \times C_2$, $C_4 \times C_2$, D_8, and Q_8. ☐

Suppose that $\alpha : E_1 \to E_2$ is a group isomorphism mapping M to M, and inducing $\mu \in \mathrm{Aut}_K(M)$. Then, since $E_1/M \cong E_2/M \cong G$, α also induces $\nu \in \mathrm{Aut}(G)$. We have $m^{g\mu} = m^{\mu g^{\nu}}$ for all $m \in M$, $g \in G$, because both expressions result from applying α to $\hat{g}^{-1} m \hat{g}$, where $\hat{g} \in E_1$ maps onto $g \in G$.

DEFINITION 2.79 Let M be a KG-module. If $\mu \in \mathrm{Aut}_K(M)$ and $\nu \in \mathrm{Aut}(G)$ are isomorphisms satisfying $m^{g\mu} = m^{\mu g^{\nu}}$ for all $m \in M$ and $g \in G$, then (ν, μ) is called a *compatible pair*.

This expression was introduced by Robinson [Rob81]. It was first used in computational group theory by M.J. Smith in Section 4.2 of [Smi94], in connection with the computation of the automorphism groups of solvable groups; we shall return to that theme in Section 8.9.

The set $\mathrm{Comp}(G, M)$ of compatible pairs forms a group under composition, and is a subgroup of $\mathrm{Aut}(G) \times \mathrm{Aut}_K(M)$. If $\tau \in Z^2(G, M)$ and (ν, μ) is a compatible pair, then we can define $\tau^{(\nu, \mu)}$ by the rule

$$\tau^{(\nu, \mu)}(g, h) = \tau(g^{\nu^{-1}}, h^{\nu^{-1}})^{\mu}$$

for all $g, h \in G$. It is straightforward to check that $\tau^{(\nu, \mu)} \in Z^2(G, M)$. Indeed, the mapping $\tau \mapsto \tau^{(\nu, \mu)}$ defines an automorphism of $Z^2(G, M)$ that fixes $B^2(G, M)$ setwise, and so it induces an automorphism of $H^2(G, M)$.

So $\mathrm{Comp}(G, M)$ induces a group of automorphisms of $H^2(G, M)$. From the above discussion, it can be shown that the isomorphism classes of KG-module extensions of M by G (where we are restricting attention to isomorphisms that fix M) correspond to the orbits of $\mathrm{Comp}(G, M)$ on $H^2(G, M)$.

Exercises

1. If $G = C_n$ and $K = \mathbb{F}_q$, then show that the smallest splitting field for G containing L is $\mathbb{F}_{q^r}$ where r is minimal with $q^r \equiv 1 \pmod n$.

2. Let M be an KG-module, where G is finite with $|G| = n$.

   (i) Let $\chi \in Z^1(G, M)$. By considering $\sum_{g \in G} \chi(gh)$ for $h \in G$, show that $n\chi \in B^1(G, M)$.

   (ii) Let $\tau \in Z^2(G, M)$. By considering $\sum_{g \in G} \tau(g, hk)$ for $h, k \in G$, show that $n\tau \in B^2(G, M)$.

(iii) Deduce that if the map $m \mapsto nm$ of M is an automorphism of M, then $H^1(G, M) = H^2(G, M) = 0$. This holds, for example, when M is finite with $|M|$ coprime to n.

3. Calculate $\mathrm{Comp}(G, M)$ and its action on $H^2(G, M)$ in Example 2.8.

4. Show that $\mathrm{Comp}(G, M)$ induces a naturally defined action on $H^1(G, M)$.

5. Let M be a KG-module defined via $\varphi : G \to \mathrm{Aut}_K(M)$.

   (i) If φ is a faithful action, then show that $\mathrm{Comp}(G, M) \cong \mathbf{N}_{\mathrm{Aut}_K(M)}(\varphi(G))$.

   (ii) If φ is trivial, then show that $\mathrm{Comp}(G, M) = \mathrm{Aut}(G) \times \mathrm{Aut}_K(M)$.

6. If M is a KG-module defined via $\varphi : G \to \mathrm{Aut}_K(M)$, and $\nu \in \mathrm{Aut}(G)$, then we can define a KG-module M^ν via the action $g \mapsto \varphi(g^\nu)$.

   (i) Verify that $M \mapsto M^\nu$ defines an action of $\mathrm{Aut}(G)$ on the set of isomorphism classes of KG-modules.

   (ii) Check that $(\nu, \mu) \in \mathrm{Comp}(G, M)$ exactly when μ is a KG-module isomorphism from M to M^ν. Hence, for $\nu \in \mathrm{Aut}(G)$, there exists $(\nu, \mu) \in \mathrm{Comp}(G, M)$ if and only if $M \cong_{KG} M^\nu$.

2.8   Field theory

For a detailed treatment of the material in this section, see any book on abstract algebra.

2.8.1   Field extensions and splitting fields

If F is a subfield of K, then K is said to be a (field) *extension* of F. The *degree* $[K : F]$ of the extension is defined to be the dimension of K as a vector space over F.

If $F \leq K \leq L$, then we have $[L : F] = [L : K][K : F]$. This is proved by showing that for bases $[u_i \mid 1 \leq i \leq [K : F]]$ and $[v_j \mid 1 \leq j \leq [L : K]]$ of K over F and L over K, $[u_i v_j \mid 1 \leq i \leq [K : F], 1 \leq j \leq [L : K]]$ is a basis of L over F.

An element $\alpha \in K$ is said to be *algebraic* over F if $f(\alpha) = 0$ for some polynomial $f \in F[x]$. Otherwise α is *transcendental* over F. If $[K : F]$ is finite, then all elements of K satisfy polynomials of degree at most $[K : F]$ over F, and so are algebraic over F.

In any case, $F(\alpha)$ is defined to be the subfield of K generated by F and α; that is, the intersection of all subfields of K that contain F and α. The elements of $F(\alpha)$ are all quotients $f(\alpha)/g(\alpha)$ with $f, g \in F[x]$, $g \neq 0$.

If α is *algebraic* over F, then the set of all $f \in F[x]$ with $f(\alpha) = 0$ forms an ideal, and hence a principal ideal (p) of $F[x]$, where p can be chosen to be monic. Then p is called the *minimal polynomial* of α over F. Since F has no zero divisors, p must be irreducible over F.

If p is the minimal polynomial over F of $\alpha \in K$, then (p) is the kernel of the ring homomorphism $\tau : F[x] \to K$ in which $\tau(\gamma) = \gamma$ for all $\gamma \in F$ and $\tau(x) = \alpha$. Hence $\mathrm{im}(\tau) \cong F[x]/(p)$. Since p is irreducible, (p) is a maximal ideal of $F[x]$ and so $F[x]/(p)$ is a field and

$$\mathrm{im}(\tau) = F(\alpha) = \{ f(\alpha) \mid f \in F[x], \ \deg(f) < \deg(p) \},$$

with $[F(\alpha):F] = \deg(p)$.

On the other hand, if F is any field and $p \in F[x]$ is an irreducible polynomial, then $K := F[x]/(p)$ is a field, which can be thought of as an extension of F of degree $\deg(p)$ by identifying F with its natural image in K. If we define α to be the image $x + (p)$ of x in K, then $p(\alpha) = 0$. Hence α is a root of p in K, and p factorizes in K as $(x - \alpha)q(x)$ for some $q \in K[x]$.

More generally, if $f \in F[x]$ has an irreducible factor p of degree greater than 1, then p factorizes nontrivially in the extension field $K = F[x]/(p)$. By repeating this construction, we can show by a straightforward induction argument on the largest degree of an irreducible factor of f, that there is an extension field K of F, with $[K:F] \leq n!$, in which f factorizes into $\deg(f)$ linear factors.

If K has this property, and no proper subfield of K containing F has this property, then K is called a *splitting field* of f over F.

It is an important result that if K and K' are two splitting fields of f over F, then there is a field isomorphism from K to K' that fixes every element of F. Here is a very brief outline of the proof. Use induction on the minimum of $[K:F]$ and $[K':F]$, let $\alpha \in K$, $\alpha' \in K'$ be roots of the same irreducible factor p of f with $\deg(p) > 1$, observe that $F(\alpha) \cong F[x]/(p) \cong F(\alpha')$, and then apply the inductive hypothesis to K and K' regarded as splitting fields of f over $F[x]/(p)$.

The *characteristic* $\mathrm{char}(F)$ of a field F is defined to be the smallest integer $n > 0$ such that $n1_F = 0_F$, or zero if there is no such integer $n > 0$. So the familiar fields $\mathbb{Q}$, $\mathbb{R}$ and $\mathbb{C}$ all have characteristic zero. If $\mathrm{char}(F) > 0$ then, since F has no zero divisors, $\mathrm{char}(F)$ must be a prime p.

It is easily shown that a polynomial $f \in F[x]$ has repeated roots (that is, repeated linear factors) if and only if $\gcd(f, f') \neq 1$, where f' is the derivative of f. If $F \leq K$, then $\gcd(f, f') = 1$ in $F[x]$ if and only if $\gcd(f, f') = 1$ in $K[x]$, so we can use this condition in $F[x]$ to check whether f has repeated roots in its splitting field.

If f is irreducible then $\gcd(f, f') = 1$ if and only if $f' \neq 0$, which is certainly the case when $\mathrm{char}(F) = 0$. If $\mathrm{char}(F) = p > 0$, then $f' = 0$ if and only if f is a polynomial in x^p.

2.8.2 Finite fields

The main result on finite fields is that all finite fields have prime power order and that, for each positive prime power $q = p^n$, there is, up to isomorphism, a unique finite field of order q. We shall outline the proof of this fact in this subsection. Although it is only defined up to isomorphism, and can be constructed in different ways, it is customary to regard 'the' finite field of order q as a fixed object, and to denote it by $\mathbb{F}_q$.

Let K be a finite field. Then we must have $\text{char}(K) = p > 0$, and the subset $F := \{n1_K \mid n \in \mathbb{Z}, 0 \le n < p\}$ forms a subfield of K, which is isomorphic to the field $\mathbb{F}_p$ of integers modulo p. In particular, all fields of order p are isomorphic to $\mathbb{F}_p$.

The degree $n := [K : F]$ must be finite, and then $q := |K| = p^n$ is a prime power, so all finite fields have prime power order.

The multiplicative group $K^{\#}$ of $K \setminus \{0_K\}$ has order $q - 1$, and hence $\alpha^{q-1} = 1_K$ for all $\alpha \in K^{\#}$, and $\alpha^q - \alpha = 0_K$ for all $\alpha \in K$. So K contains q distinct roots of the polynomial $x^q - x \in F[x]$. Clearly no proper subfield of K can have this property, so K is a splitting field of $x^q - x$ over F. It now follows from the uniqueness of splitting fields that all fields of order q are isomorphic.

To prove the existence of a field of order q for any prime power $q = p^n$, let K be the splitting field of $x^q - x$ over $F := \mathbb{F}_p$. The derivative of $x^q - x$ is -1, which is nonzero so, as we saw earlier, $x^q - x$ has q distinct roots in K. It is easily checked that, if α, β are roots of $x^q - x$ in K then so are $\alpha \pm \beta$, $\alpha\beta$, and α/β if $\beta \ne 0_K$, so the set of these roots form a subfield of K which, by minimality of the splitting field, must be equal to K itself. So $|K| = q$, as required.

PROPOSITION 2.80   *Any finite subgroup of the multiplicative group of a field K is cyclic. In particular, the multiplicative group of a finite field is cyclic.*

The proof of this depends on the following result from group theory. The *exponent* of a group G is defined to be the least common multiple of the orders of its elements or, equivalently, the least $n > 0$ such that $g^n = 1$ for all $g \in G$.

LEMMA 2.81   *If G is a finite abelian group of exponent n, then G has an element of order n.*

PROOF   Let $n = p_1^{e_1} p_2^{e_2} \cdots p_r^{e_r}$ for distinct primes p_i. Then G must have elements g_i of order $p_i^{e_i}$ for each i. Since G is abelian, $g_1 g_2 \cdots g_r$ has order n, as required.   ∎

Note that the final step in the proof of the lemma is not necessarily true for nonabelian groups G, and indeed the lemma itself is not true in general.

To prove the proposition, let the finite subgroup H in question have order n. If H had exponent $m < n$, then the polynomial $x^m - 1$ would be satisfied by all elements of H and hence have $n > m$ distinct roots, which is impossible. So H has exponent n, and the result follows from the lemma.

An element α of multiplicative order $q-1$ in $\mathbb{F}_q$ is called a *primitive* element of $\mathbb{F}_q$. Clearly $\mathbb{F}_q = F(\alpha)$ with $F = \mathbb{F}_p$, and so the minimal polynomial f of α over F must be of degree n, where $q = p^n$. An irreducible polynomial of degree n over $\mathbb{F}_p$ of which the roots are primitive elements of $\mathbb{F}_q$ is called a *primitive* polynomial.

(This meaning is distinct from and unconnected with the meaning of a primitive polynomial over $\mathbb{Z}$ as one in which the greatest common divisors of the coefficients is 1. This clash of meanings is unfortunate, but since the concept of a greatest common divisors of field elements is trivial, there is probably little danger of confusion.)

It is easily verified that the map $x \to x^p$ defines an automorphism of $\mathbb{F}_q$ of order n (it is called the *Frobenius automorphism*, and generates the automorphism group of $\mathbb{F}_q$, but we shall not need that fact). So, if f is a primitive polynomial of degree n over F with root $\alpha \in \mathbb{F}_q$, then the n elements in the set $\{\alpha^{p^i} \mid 0 \le i < n\}$ are all roots of f in $\mathbb{F}_q$. Hence $\mathbb{F}_q$ is a splitting field of f.

If w is a primitive element of $\mathbb{F}_q$ then, for $0 \le k < q-1$, w^k is primitive if and only if $\gcd(k, q-1) = 1$, and so the total number of primitive elements is $\Phi(q-1)$, where Φ is the Euler Phi-function. Each primitive polynomial f has n roots in $\mathbb{F}_q$ so there are a total of $\Phi(q-1)/n$ primitive polynomials.

2.8.3 Conway polynomials

Although $\mathbb{F}_q$ is unique up to isomorphism, it can, and often does, arise as the splitting field of many different irreducible polynomials of degree n over $\mathbb{F}_p$. For computational purposes, it is useful to agree on a standard primitive polynomial, so that different computer algebra systems can use the same representation of the elements of $\mathbb{F}_q$. Unfortunately, there appears to be no natural mathematical way of choosing such a standard polynomial.

The standard that has been generally agreed upon is known as the *Conway polynomial* for $\mathbb{F}_q$. (This is an unfortunate choice of name, because there is another meaning of Conway polynomial in knot theory!) They were originally introduced by Richard Parker, who also computed many examples. To define them, we first need to define an ordering on the set of all polynomials of degree n over $F = \mathbb{F}_p$, and it is here that an apparently arbitrary choice had to be made.

We order $\mathbb{F}_p$ itself by $0 < 1 < 2 < \cdots < p-1$. Then the polynomial

$$x^n - \alpha_{n-1} x^{n-1} + \alpha_{n-2} x^{n-2} - \cdots + (-1)^n \alpha_0$$

is mapped onto the word $\alpha_{n-1} \alpha_{n-2} \cdots \alpha_1 \alpha_0$, and the resulting words are ordered lexicographically using the above ordering of $\mathbb{F}_p$.

The Conway polynomial for $\mathbb{F}_p$ is defined to be the least primitive polynomial of degree 1 under this ordering. In other words, it is $x - \alpha$, where α is the smallest primitive element in $\mathbb{F}_p$. For nonprime fields, there is an extra condition.

It turns out (see exercises below) that $\mathbb{F}_{p^m}$ is a subfield of $\mathbb{F}_{p^n}$ if and only if m divides n. For compatibility between the Conway polynomial f of the field $\mathbb{F}_{p^n}$ and its subfields, it is required that if α is a root of the Conway polynomial f of $\mathbb{F}_{p^n}$ then, for all proper divisors m of n, α^t with $t := (p^n - 1)/(p^m - 1)$ should be a root of the Conway polynomial of $\mathbb{F}_{p^m}$.

It is a slightly tricky exercise in modular arithmetic to show that a primitive polynomial f exists that satisfies this property, where we may assume by induction that the Conway polynomials are already defined and satisfy the property for all proper subfields of $\mathbb{F}_{p^n}$. A proof can be found in the thesis of W. Nickel [Nic88]. We can then define the Conway polynomial f of $\mathbb{F}_{p^n}$ to be the least polynomial under the ordering defined above that has the property.

For example, for $q = 2, 4, 8, 16, 32, 3, 9, 27, 5, 25$, the Conway polynomials are respectively $x+1$, x^2+x+1, x^3+x+1, x^4+x+1, x^5+x^2+1, $x+1$, x^2+2x+2, x^3+2x+1, $x+3$, x^2+4x+2.

One disadvantage of this definition is that there is nothing much better than brute-force algorithms available for calculating Conway polynomials, which become rapidly impractical as the field grows larger. On the other hand, they only have to be computed once, and some of the calculations can be done in parallel. Some recent computations are described by Heath and Loehr in [HL04]. For example, they have been computed for fields up to order p^{97} for $p \leq 11$, p^{43} for $p \leq 23$, 67^{31}, 127^6. Complete lists are available on [Con].

Exercises

1. Find an example of a nonabelian group having no element of order equal to its exponent.

2. If q and r are prime powers, show that $\mathbb{F}_r$ occurs as a subfield of $\mathbb{F}_q$ if and only if q is a power of r, and in that case there is a unique subfield of $\mathbb{F}_q$ isomorphic to $\mathbb{F}_r$.

3. Prove that Conway polynomials can be defined for all finite fields $\mathbb{F}_q$. You may want to use the fact that $\gcd(p^m - 1, p^n - 1) = p^{\gcd(m,n)} - 1$ for all positive integers p, m, n.

Chapter 3

Representing Groups on a Computer

We are now ready to embark upon our study of computational group theory! Many of the methods in CGT depend on whether the group is represented as a group of permutations, a group of matrices, or by means of a presentation using generators and relators. The most fundamental algorithms in CGT generally apply just to one specific representation type: for example, coset enumeration can be applied only to a group defined by means of a finite presentation. Most of the later chapters in the book will be concerned with algorithms for one particular representation type of groups.

In the current chapter, however, we shall start with some general considerations about the different ways of representing groups on computers, and about the properties of a group and its elements that we might hope to be able to compute in various situations. We shall then go on to introduce a few basic algorithms and techniques, such as choosing random elements of groups, which can be studied independently of the representation type.

3.1 Representing groups on computers

3.1.1 The fundamental representation types

There are three methods commonly used to represent groups on a computer, namely, as groups of permutations of a finite set, groups of matrices over a ring, and as groups defined by a finite presentation. Most of the algorithms discussed in this book will concern groups defined using one of these three methods. Chapter 4 and various sections of later chapters will be devoted to computing in permutation groups, Section 7.8 will deal specifically with matrix groups, and Chapters 5, 9, 12, and 13 will be concerned with groups given by a finite presentation. In addition, Chapter 8 will be concerned with groups defined by a particular type of finite presentation, namely a polycyclic presentation.

We generally prefer our permutation groups of degree n to act on the set $\{1 .. n\}$; that is, we prefer to compute with subgroups of $\text{Sym}(n)$. In cases when the action is given or arises naturally on some different set, we would normally choose to rename the points as $\{1 .. n\}$ before attempting any further computations. In the case of matrix groups over a ring, we shall only be concerned with the case where we can carry out the basic operations of addition and multiplication exactly within the ring. So the ring might be a finite field, the integers or the rational numbers, or perhaps an algebraic number field, but not the real or complex numbers.

We may also occasionally want to consider other groups arising from these basic types, such as quotient groups of permutation or matrix groups, which cannot always be easily represented in the same way themselves. Examples are given by P.M. Neumann in [Neu87] of permutation groups of order 8^n and degree $4n$, for $n \geq 1$, which have quotients for which the smallest degree of a faithful permutation representation is 2^{2n+1}. See the exercise below.

Virtually all of the techniques that have been built up for computing within groups take as a starting point a finite generating set for the group or groups concerned, and we shall assume in this book that all of our groups are defined by means of a finite generating set (or sequence). So, for example, if we were trying to develop an algorithm to find a Sylow p-subgroup of a finite group, then our aim would be to output generators for that subgroup.

We want to be able to compute in very large finite groups, and so we try very hard to avoid algorithms that would require computing a list of all elements of the group, or even looping over all such elements. This is not always possible. As we saw in Example 2.3, any finite group whatsoever has a faithful permutation representation acting by right multiplication on its own elements, which we called the *(right) regular* action. Occasionally, we may need to resort to using this representation, which is tantamount to listing all of the group elements, although the group elements would be represented by natural numbers in this situation. For example, if we need a faithful permutation representation of a proper quotient of an existing permutation or matrix group, then the only default choice is the regular representation.

3.1.2   Computational situations

As a rough guideline, there are three basic situations that occur in practice, and which affect our ability to extract information about a group G computationally. Of course, variations and combinations of them are also possible, but they serve to give an approximate idea of the possible states of affairs.

1. *Situation A.* We can represent every element of G on the computer, and we have methods to compute representations of inverses and products of group elements. But we may not be able to decide whether two representations of group elements define the same element.

   Example: a group G defined by a finite presentation $\langle\, X \mid R \,\rangle$.

2. *Situation B.* As in Situation A and, additionally, we can decide whether two representations of group elements define the same element.

   Examples: a permutation group G on a finite set defined by an arbitrary generating set; a matrix group G over a finite field or an algebraic number field defined by an arbitrary generating set.

3. *Situation C.* As in Situation B and, additionally, we have a special generating set $g_1, \ldots, g_r$ of G and an algorithm that can compute for every element $g \in G$ a unique word w_g with $g = w_g(g_1, \ldots, g_r)$.

   Example: a permutation group G with a so-called 'strong generating set'. This 'strong generating set' is used as the special generating set for G. For its definition and the corresponding algorithm to compute unique words w_g, we refer to Section 4.4.

   In general, the algorithm for computing w_g is known as a *rewriting algorithm*.

If a certain group is given to the computer, then our first and principal aim before doing anything else with the group is usually to move into the best possible computational situation for this group. The methods to move into a better situation depend (again) on the representation of the given group. For example:

1. A group G is defined by a finite presentation $\langle X \mid R \rangle$ and, additionally, it is known that G is finite. Then for most computational purposes we would first try to determine a faithful permutation representation for G and thereby move from Situation A to B. The Todd-Coxeter coset enumeration algorithm to be discussed in detail in Chapter 5 is most commonly used for this purpose.

2. A group G is given by a generating set of permutations. Then for most computational purposes we would first compute a strong generating set for G and thereby move from Situation B to C. Algorithms for achieving this will be described in Section 4.4 and in Chapter 6.

Situation C is our preferred situation for computational purposes. It has various computational advantages. For example, the words w_g used in Situation C play the role of a normal form for group elements. They allow us to read off the order of the group G (which may be finite or infinite) as the number of possible normal forms. Also, in Situation C we can store and compare elements using their corresponding normal form words, which is often a computational advantage. Often it enables us also to test a given element of some superstructure (such as $\mathrm{Sym}(\Omega)$ for permutation groups, or the general linear group for matrix groups) for membership in G.

In some situations, including the base and strong generating set framework for permutation and matrix groups, and the polycyclic series for polycyclic groups to be described in Chapter 8, this normal form is derived from a chain of subgroups

$$G = G(1) \supset G(2) \supset \cdots \supset G(k+1) = 1$$

of G, and the normal form itself is $g = g_k \cdots g_2 g_1$, where g_i lies in a fixed transversal U_i of $G(i+1)$ in $G(i)$.

3.1.3 Straight-line programs

The most convenient generating set for the normal form words w_g that we discussed above in connection with Situation C is often not the same as the initial generating set on which the group is defined. For example, for a permutation group, a strong generating set used for w_g is generally a superset of the initial generators. However, in many situations we need to be able to relate the generators used for the w_g to the initial generators.

For that purpose, we shall introduce a data structure known as a a *straight-line program* (SLP). It also provides a space-efficient method for storing elements in groups such as permutation or matrix groups. The terminology was introduced in computer science as a certain type of program. Here we shall use it in the following form.

We define an *SLP group* S of rank k to be a free group of rank k on generators $\hat{x}_1, \ldots, \hat{x}_k$. We shall call the elements of S *SLPs* or *SLP elements*. From a purely mathematical viewpoint, there is no difference between an SLP group and any other free group. The difference lies in the way that the SLP elements are stored. They are not, in general, stored as words in the original $\hat{x}_i$, but as words in SLPs that have been previously defined. So, for example, if $k = 2$, then

$$w_1 := \hat{x}_1, \ w_2 := \hat{x}_2, \ w_3 := (w_1 w_2)^2, \ w_4 := w_2^2 w_3^{-1} w_2$$

defines a sequence of SLP elements w_1, w_2, w_3, w_4. To avoid the words growing too long, they are not evaluated as words in the original generators of S, but stored exactly as they are defined. So, in particular, a product of two existing SLP elements is stored as that product, and the product is not expanded.

An *evaluation* φ of an SLP group S in a group G is an assignment of the generators $\hat{x}_i$ of S to elements x_i of G. By definition of a free group, this defines a unique homomorphism $\varphi : S \to G$ in which each $\hat{x}_i$ is mapped to x_i. The storage method renders the evaluation of $\varphi(w)$ for an SLP element $w \in S$ very easy, and this is the motivation behind these ideas. Since w is stored as a (usually short) word in SLP elements w_i that were defined earlier, we first calculate the evaluation $\varphi(w_i) \in G$ of the w_i used to define w, and then we can calculate $\varphi(w)$ itself.

So, in the example above, if we take $G = \mathrm{Alt}(6)$ and $x_1 = (1,2,3)$, $x_2 = (2,3,4,5,6)$, then w_1, w_2, w_3 and w_4 evaluate in G respectively to x_1, x_2, $x_3 := (2,5)(4,6)$, and $x_4 := (4,5,6)$.

Now it turns out to be much easier to write an arbitrary element of Alt(6) as a word in x_1, x_2, x_3, x_4 than in the original generators x_1, x_2. (The reason for this will become clear in Section 4.4, when we define bases and strong generating sets. In fact x_1, x_2, x_3, x_4 is a strong generating set with respect to the base $[1, 3, 2]$.) For example, $(1,5)(2,6) = x_4 x_3 x_2^{-1} x_1 x_3$. We could now, if we wanted to, use the definitions of x_3 and x_4 coming from the corresponding SLP elements to evaluate this element as a word in the original generators. In this example, $x_3 = (x_1 x_2)^2$ and

$$x_4 = x_2^2 x_3^{-1} x_2 = x_2^2 (x_1 x_2)^{-2} x_2 = x_2 x_1^{-1} x_2^{-1} x_1^{-1} x_2,$$

so we have

$$(1,5)(2,6) = x_2 x_1^{-1} x_2^{-1} x_1^{-1} x_2 x_1 x_2 x_1^2 (x_1 x_2)^2.$$

Incidentally, this is not the shortest word possible for $(1,5)(2,6)$, which is $x_2 x_1 x_2 x_1^{-1} x_2^{-2}$, but in large groups there is no satisfactory algorithm for finding shortest words for elements in the original generators. We shall discuss this problem later, in Subsection 4.8.3.

In general, we proceed by defining an SLP group on generators $\hat{x}_i$ corresponding to our initial generators x_i $(1 \le i \le k)$ of G, and the corresponding evaluation in which $\hat{x}_i$ maps to x_i. We then define SLP elements $w_1, \ldots, w_r$, usually with $w_i = \hat{x}_i$ for $1 \le i \le k$, together with their evaluations $x_1, \ldots, x_r$ in G, where these elements are chosen in such a way that arbitrary elements $g \in G$ can be written easily as reasonably short words in the extended generating set $\{x_1, \ldots, x_r\}$. In the case of permutation and matrix groups, we extend the initial generating set to a strong generating set (see Section 4.4), which enjoys this particular property. The definitions within the SLP could, in principal, be used for rewriting group elements as words in the original generators, but we would avoid doing that if possible. We might choose to store the elements and their inverses from this extended generating set explicitly as permutations or matrices, but that is not essential, because they can always be evaluated when required.

As we shall explain shortly, straight-line programs are of vital importance in computations involving group homomorphisms.

3.1.4 Black-box groups

We shall not be devoting too much attention to black-box groups in this book, but the topic is important enough to warrant a mention, if only a brief one.

The term 'black-box group' can be confusing, because it does not describe a particular type of group; it refers rather to a method of representing group elements within a computer, together with some assumptions on the availability of algorithms for composing, inverting, and comparing group elements.

It was introduced by Babai and Szemerédi in [BS84]. The definition is as follows. A finite alphabet A and an integer $N > 0$ are given. Group elements are represented by strings in A^* of length at most N. If we have strings representing $g, h \in G$, then we can compute in constant time (using an 'oracle') strings that represent g^{-1} and gh, and we can decide whether $g =_G h$.

Note that we cannot in general decide whether a given string of length at most N in A^* represents a group element. (Deciding that is similar to the membership testing problem.)

In fact a black-box group is really just a description of Situation B for a group G, but with a significant additional restriction: the stipulation that strings have length at most N means that a black-box group is necessarily finite, with an upper bound of $\sum_{i=0}^{N} |A|^i$ on the group order. Finite permutation groups and matrix groups over finite fields are important examples of black-box groups.

An algorithm for black-box groups lends itself readily to complexity analysis in terms of both time and space, where the time complexity will involve the number of calls to the oracle.

There is a significant research project in CGT to develop a full collection of algorithms for finite groups given as black-box groups. The advantage is that they are independent of the particular type of representation of the group. The disadvantage is that they cannot make any use of this representation. For example, this automatically rules out many of the most important algorithms for permutation groups, which typically rely heavily on the orbit structure of the group and its subgroups.

There is one other important attraction of black-box group algorithms, which is that if K is a normal subgroup of G, and we have an algorithm for testing strings for membership in K, then this algorithm can be used as the oracle for testing whether an element of G/K is the identity, and then G/K becomes a black-box group, where the representatives of $g \in G$ become representatives of $gK \in G/K$.

Unfortunately, the black-box setting is too restrictive for most purposes. One major problem is that there is no efficient black-box algorithm for determining the order of an element g; the only way is to test $g^k = 1$, for $k = 1, 2, \ldots$, which is likely to be too slow if $|g|$ is large. Since there are efficient ways of finding the order of an element in a permutation group or in a matrix group (a method for the latter is described by Celler and Leedham-Green in [CLG97]), a common recourse is to assume that the group is equipped with a further oracle for determining element orders. But then the problem of passing to quotient groups reasserts itself!

Perhaps the most important achievement in this area to date is the development of methods for the constructive recognition of the finite nonabelian simple groups given as black-box groups. We shall return briefly to this topic later in the book, in Section 10.3.

Exercises

1. Let G be the direct product of n copies of the dihedral group of order 8. Show that G has a faithful permutation representation of degree $4n$. Let z_i be the central element of order 2 in the i-th copy of D_8, and let K be the subgroup of G generated by $\{z_1 z_i^{-1} \mid 2 \le i \le n\}$. So K has order 2^{n-1}, and G/K is a central product of n copies of D_8. Show that the smallest degree of a faithful permutation representation of G/K is 2^{n+1} (see [Neu87].)

   (*Hint*: Let the i-th copy of D_8 be generated by x_i, y_i. Then the image in G/K of the subgroup generated by the x_i is abelian of order 2^n and is disjoint from $Z(G/K)$. Any larger subgroup must include two noncommuting elements and hence must contain $Z(G/K)$.)

2. Let $G = \langle x \mid x^n \rangle$ be cyclic of order n for some large positive integer n. Devise a method of representing elements of G efficiently using an SLP group on the single generator $\hat{x}$.

3.2 The use of random methods in CGT

Many algorithms in CGT, and indeed in computer algebra in general, depend critically on making some random choices, such as choosing random elements of groups. In the first subsection we describe the two principal types of randomized algorithms that are used in CGT, and in the second we present one specific method of choosing random elements from a finite group.

3.2.1 Randomized algorithms

A *deterministic algorithm* is one that depends on its input data alone. So, if it is run repeatedly with the same input data, it will always proceed in the identical fashion. A complexity analysis of such an algorithm will provide an accurate and consistent estimate of its time and space requirements.

A *randomized algorithm* is one that makes use of a random number generator during its execution, typically in order to choose random elements of some group G. Such an algorithm does not depend on its input data alone, and its performance may vary, sometimes dramatically, from one run to another with the same input.

It may still be possible to carry out complexity analyses of such algorithms, but the results will typically merely estimate the average running time and space requirements of the algorithm. An analysis of this kind will need to assume that the random number generator being used is working properly, and is capable of choosing genuinely random integers within a given range $[a \mathinner{.\,.} b]$. In practice, of course, there is no such thing as a perfect random number generator, but this is not the place to explore this interesting question, and we refer the interested reader to the lengthy treatment by Knuth in Chapter 3 of [Knu69].

There are two distinct types of randomized algorithm that arise in CGT; they have become generally known within the trade as *Monte Carlo* and *Las Vegas* algorithms.

A *Monte Carlo algorithm* is one which may sometimes output a wrong answer! It is, however, a requirement that one of the input parameters of such an algorithm should be a real number ϵ with $0 < \epsilon < 1$, and that the probability of the answer being wrong should then be less than ϵ for any values of the remaining input data. The performance will depend on ϵ, with smaller values of ϵ resulting in longer running times.

Although the first reaction of many traditional mathematicians (who may in any case be generally suspicious of the use of computers in mathematics) is to recoil in horror at the idea of a process that has an intrinsic possibility of generating a false result, these algorithms do have their uses. The best-known example is primality testing of integers, for which there are Monte Carlo methods that run significantly faster that the best-known deterministic methods, and which are adequate, and essential in practice, for their applications to cryptography.

There are other mathematicians, more in the habit of relying on the computer to help with their work, who go to the opposite extreme, and argue that an answer delivered with an error estimate of 10^{-20}, for example, should simply be regarded as proven correct, because even traditional mathematical proofs written out in full detail might struggle to acquire a comparable degree of reliability. Indeed, with any computer calculation, there are small possibilities of error, such as hardware or compiler error, which are completely outside the control of the typical mathematical user.

For a mathematical treatment of algorithms, such as the one we are attempting in this book, it seems desirable, however, to make suitable simplifying assumptions, such as the absence of hardware or compiler errors, and the ability to generate genuinely random numbers, and then to treat the output of a Monte Carlo algorithm exactly for what it is; namely, an answer with a small, known probability of being wrong.

In most, if not all, of the important Monte Carlo algorithms in CGT, certain answers are in fact guaranteed correct, whereas other answers come with the possibility of error. For example, in Section 4.2 we shall describe a Monte Carlo algorithm for testing whether an input permutation group on Ω is $\mathrm{Alt}(\Omega)$ or $\mathrm{Sym}(\Omega)$. The answer 'yes' is guaranteed correct, whereas the answer 'no' has a small chance of being wrong. In practice, such algorithms are often used as filters; we want to check quickly whether the group is very large ($\mathrm{Alt}(\Omega)$ or $\mathrm{Sym}(\Omega)$) and, if not, then we plan to analyze it further and perhaps to calculate its precise order.

A *Las Vegas algorithm* is one that never delivers an incorrect answer, but has a probability of at most ϵ of not returning an answer at all. It is again required that ϵ should be an input parameter.

These are much less controversial than Monte Carlo algorithms, because in practice they can always be made to return a correct answer just by running them repeatedly with the same ϵ, and implementations will often do this automatically. The only problem is that, if one is unlucky, then one may have an unexpectedly long wait for an answer. The majority of randomized algorithms that we shall encounter in this book are of this type.

3.2.2 Finding random elements of groups

Again we shall assume that we can choose random integers within a given range. This clearly enables us to choose random elements of any indexed list l, and we shall assume the availability of a function RANDOM(l) that does this for us.

Many algorithms in CGT depend critically on the ability to generate quickly uniformly distributed random elements of a group G. There are some cases where this is relatively easy; for example if $G = \text{Sym}(n)$ (see exercise below) and, more generally, we shall see in Chapter 4 that it is easy for finite permutation groups G for which a base and strong generating set are known. However, in some situations we need to be able to generate a sequence of random elements from a group G for which we are in Situation A or B, as defined in Subsection 3.1.2 above. For finitely presented groups (Situation A), all that we can sensibly do is to find a random word for which the length lies within a given range.

Unfortunately, even for black-box groups, there is no known method of generating random elements that is completely satisfactory. A method proposed by Babai in [Bab91] does generate elements that are guaranteed to be genuinely uniformly distributed, but it is far too slow to be useful in practice. Here, we shall describe the *product replacement algorithm*, proposed in [CLGM$^+$95], which generates a sequence of elements that is not guaranteed to be uniformly distributed, but which is very fast, and appears to behave satisfactorily in those algorithms in which it has been used to date. (However, there are some serious reservations. For example, the behaviour can be unsatisfactory when the group is a direct product of a large number of copies of the same finite simple group. See the article of I. Pak [Pak01] for a theoretical investigation.) We refer the reader to Section 2.2 of [Ser03] for a detailed discussion and theoretical analysis of both Babai's method and the (original) Product Replacement Algorithm.

We maintain a list $[x_1, x_2, \ldots, x_r]$ of elements of G, which generate G. A suitable value of r has been found in practice to be about 10, but if the given initial generating sequence X of G has more than this number of elements, then we must take $r = |X|$. It is argued in [Pak01] that a larger value of r would be necessary for guaranteed good theoretical behaviour. If $|X| < r$, then we initialize the list to contain the generators in X repeated as many times as required. In the original version, we repeatedly choose random distinct integers $s, t \in [1 \mathbin{..} r]$, and replace x_s by $x_s x_t^{\pm 1}$ or $x_t^{\pm 1} x_s$. Notice that the list still generates G after making this change. After doing this a certain number of times to get us away from the initial generating set (50 times has been used as a default in implementations), we start returning the new values of x_s as our random elements.

It is proved in [CLGM+95] that the elements returned will eventually be uniformly distributed random elements of G, but there is no reasonable estimate known on how long we will have to wait for this, and it is likely that the *ad hoc* default waiting time of 50 changes is not enough. Another possible problem with the algorithm is that the group elements g_1, g_2 returned on successive calls of the algorithm will never be uniformly distributed in $G \times G$.

The version that we shall present here is a modification due to Leedham-Green, which involves the use of an accumulator x_0 that is always used as the return value. (As a cricket enthusiast, Leedham-Green likes to take $r = 11$ and to call x_0 the 'twelfth man', but we prefer to avoid the gender bias inherent in that terminology!) There is heuristic evidence that this converges more quickly than the original version.

To use the method, we must first call the following initialization function on our generating set X. This sets up the list $\mathcal{X} = [x_0, x_1, \ldots, x_r]$, performs the product replacement process PrRandom n times as initialization, and returns $\mathcal{X}$. In some applications, we shall need to know exactly how the random elements returned were derived from the original generators, and so we shall also maintain SLP elements w_i, which express each element x_i as a word in the original generators $x_1, \ldots, x_k$ of G. As usual, we denote the generators of the SLP group that evaluate to the x_i by $\hat{x}_i$.

PrInitialize(X, r, n)

> **Input**: $X = [x_1, \ldots, x_k]$ generating a black-box group, $r, n \in \mathbb{N}$
> **Output**: List $\mathcal{X}$ and associated SLP elements $\mathcal{W}$
> 1   Let $\hat{x}_i$ $(1 \le i \le k)$ be generators of an SLP group;
> 2   **for** $i \in [1 \mathbin{..} k]$ **do** $w_i := \hat{x}_i$;
> 3   **for** $i \in [k{+}1 \mathbin{..} r]$ **do** $x_i := x_{i-k}$; $w_i := w_{i-k}$;
> 4 $x_0 := 1_G$; $w_0 := \varepsilon$;
> 5 $\mathcal{X} := [x_0, x_1, \ldots, x_r]$; $\mathcal{W} := [w_0, w_1, \ldots, w_r]$;
> 6   **for** $i \in [1 \mathbin{..} n]$ **do** PrRandom$(\sim\!\mathcal{X}, \sim\!\mathcal{W})$;
> 7   **return** $\mathcal{X}, \mathcal{W}$;

We remind the reader that we precede a variable by a $\sim$ symbol in a function or procedure call when the value of that variable may be be altered by the function or procedure.

The function PrRandom, which outputs random group elements, should be called for the first time using the lists $\mathcal{X}, \mathcal{W}$ returned by PrInitialize. Subsequently, it should be called using the altered values of $\mathcal{X}, \mathcal{W}$ resulting from the previous call of PrRandom.

The reader should be aware that we are omitting the details of the implementation of SLP groups in the code for PRRANDOM; each SLP element w_i is being defined as a word in existing SLP elements, and so all of these existing elements would need to be stored in order to allow later evaluation of the w_i. In situations later in the book in which we do not need the SLP-elements, we shall call PRRANDOM with only one input parameter, $\mathcal{X}$.

PRRANDOM($\sim \mathcal{X}, \sim \mathcal{W}$)

Input: List $\mathcal{X} = [x_0, x_1, \ldots, x_r]$, $\mathcal{W} = [w_0, w_1, \ldots, w_r]$
Output: An element of G and corresponding SLP element

```
1   s := RANDOM([1 .. r]);
2   t := RANDOM([1 .. r] \ [s]);
3   x := RANDOM([1 .. 2]);
4   e := RANDOM({−1, 1});
5   if x = 1
6       then x_s := x_s x_t^e;   x_0 := x_0 x_s;
7               w_s := w_s w_t^e;   w_0 := w_0 w_s;
8       else  x_s := x_t^e x_s;   x_0 := x_s x_0;
9               w_s := w_t^e w_s;   w_0 := w_s w_0;
10  return x_0, w_0;
```

The list $[1 .. r] \setminus [s]$ in this function can be easily constructed by removing k from $[1 .. r]$ and, if $s \neq r$, replacing s by r.

Exercises

1. Find a method for constructing uniformly distributed random elements of $\text{Sym}(n)$ in time $O(n)$, on the assumption that random elements from a list of length up to n can be found in constant time. (This assumption is false for large n, but provided that n is small enough for the list to be represented on a computer in RAM, then it is a reasonable practical assumption to make.) It is clear how to start: choose the image 1^g under our random permutation g to be a random element of $\{1 .. n\}$. But now we have to choose 2^g from $\{1 .. n\} \setminus \{1^f\}$, so how can we manage this by doing only a small constant amount of reorganization of our data?

2. If you have access to **GAP** or **MAGMA**, then implement the procedures PRINITIALIZE and PRRANDOM (but leave out the calculations with SLP elements if you prefer). Try out your functions on the groups $\text{GL}(d, q)$ for various parameter settings, and various d and q. How many iterations of PRRANDOM in PRINITIALIZE are needed to ensure that the elements returned by PRRANDOM appear to be random? How dependent is this number on q?

3.3   Some structural calculations

In this section, we present examples of some structural computations that can be carried out in any class of finitely generated groups for which we can find random group elements, and in which we can test membership of group elements in finitely generated subgroups. This is typically the case when we are in Situation C (see Subsection 3.1.2).

3.3.1   Powers and orders of elements

Suppose that we are in Situation B for a group G, and that we need to calculate g^n for some $g \in G$ and $n \in \mathbb{N}_0$. The obvious thing to do is just to multiply g by itself $n - 1$ times. This is fine for small n, but for larger n, there is a much quicker way, which has complexity $O(\log(n))$ rather than $O(n)$. We calculate g^n as a product of elements g^{2^i}, where each individual g^{2^i} can be calculated by repeated squaring of g. This is carried out in the following function.

Of course, if we knew in advance that we were going to need to calculate g^n for the same g and for many different values of n, then we could save time by storing the elements g^{2^i} rather than recalculating them each time.

$\text{POWER}(g, n)$

    Input: $g \in G$, $n \in \mathbb{N}_0$
    Output: g^n
1   $x := 1_G$;
2   **if** $n \bmod 2 = 1$ **then** $x := xg$;
3   **while** $n > 1$
4   **do** $g := g^2$;
5           $n := n$ div 2;
6   **if** $n \bmod 2 = 1$ **then** $x := xg$;
7   **return** x;

A well-designed computer algebra system would use this method by default whenever the user types g^n, at least for large values of n.

As we mentioned in Subsection 3.1.4, finding the order of an element in a black-box group is not so easy in general and, without further information, all that we can do is to test whether $g^n = 1_G$ for each power g^n in turn. In some situations, however, such as for finite field elements, or matrices over finite fields, it is possible to find an n such that the order $|g|$ of g divides n. Provided that we can factorize n (which, unfortunately, may not be completely straightforward when n is very large), the following function will find the order of g in time at most $O(\log(n)^3)$.

ORDERBOUNDED(g, n)

    Input: $g \in G$, $n \in \mathbb{N}_0$ with $|g|$ dividing n
    Output: $|g|$
1   **if** $n = 1$ **then return** 1;
2   **for** $p \in$ PRIMEDIVISORS(n)
3   **do if** POWER($g, n/p$) $= 1_G$
4   **then return** ORDERBOUNDED($g, n/p$);
5   **return** n;

3.3.2   Normal closure

Let $H = \langle Y \rangle$ be a subgroup of a group $G = \langle X \rangle$ with X and Y finite, and suppose that we want to find the normal closure $N := \langle H^G \rangle$ of H in G. The obvious way to proceed is to start with $N = H$, test each y^x with $y \in Y$, $x \in A := X \cup X^{-1}$ for membership in N, and adjoin it to Y as an extra generator if the test fails.

Of course, in general, N might not be finitely generated, in which case this process would not terminate, but usually we will want to do this when at least $|G:H|$ is finite, in which case termination is guaranteed, and the above procedure provides us with a deterministic algorithm.

In fact, when $|G:H|$ is large and finite, it is usually faster to adjoin a number of random conjugates of H by G to the generating set of N, and then to test whether N is the full normal closure. The procedure NORMALCLOSURE does this. The random conjugates are added in batches of size n. In Section 5.1.4 of [Ser03], Seress suggests the value $n = 10$ for permutation groups.

3.3.3   The commutator subgroup, derived series, and lower central series

Let $H = \langle X \rangle$ and $K = \langle Y \rangle$ be subgroups of a group G, where X and Y are finite. By Proposition 2.35, $[H, K]$ is the normal closure in $\langle H, K \rangle$ of the set $C := \{\, [h, k] \mid h \in X, k \in Y \,\}$. This can be computed by the normal closure algorithm described in Subsection 3.3.2 provided that the appropriate assumptions on membership testing in subgroups are satisfied.

However, if the generating sets X and Y are large, then it may be expensive to construct the set C, which has size $|X||Y|$, explicitly. In that case, we might prefer to construct our random elements of the group H in NORMALCLOSURE as random words in a smaller set of commutators $[h, k]$, where h and k are themselves random elements in H and K. If we do this, then we must keep changing the commutators $[h, k]$ used, as well as the words in these commutators.

These techniques immediately enable us to calculate the derived series and the lower central series of a group.

NORMALCLOSURE(X, Y, n)

    Input: Generating sequences X, Y of groups G, H, $n \in \mathbb{N}$
    Output: Generators of H^G
1 $Z := Y$; $C := \texttt{false}$;
2 $\mathcal{X} := $ PRINITIALIZE$(X, 10, 20)$;
3 **while not** C
4   **do** $\mathcal{Z} := $ PRINITIALIZE$(Z, 10, 10)$;
          $(*$ Add some new random conjugates to Z $*)$
5   **for** $i \in [1 .. n]$
6   **do** $g := $ PRRANDOM$(\sim \mathcal{X})$;
7                  $h := $ PRRANDOM$(\sim \mathcal{Z})$;
8   **if** $h^g \notin \langle Z \rangle$ **then** APPEND$(\sim Z, h^g)$;
          $(*$ Test whether $\langle Z \rangle = H^G$ $*)$
9          $C := \texttt{true}$;
10   **for** $g \in X$, $h \in Z$
11   **do if** $h^g \notin \langle Z \rangle$
12   **then** $C := \texttt{false}$;
13   **break**;
14 **return** Z;

3.4   Computing with homomorphisms

The ability to compute effectively with group homomorphisms is one of the most important aspects of CGT. In particular, it allows us to move from one representation of a group to another, possibly more convenient, representation. For example, if a group is given initially as a matrix group or finitely presented group, then we might prefer to perform our computations in an isomorphic permutation group.

    We shall consider some particular types of homomorphisms, such as mapping to the induced action of a permutation group on a nontrivial block system in later chapters, but there are some general ideas that are independent of the particular representation of the group, which we can consider at this point. The reader should bear in mind that, in certain specific situations, there may happen to be more efficient solutions to these problems than the general ones described here.

3.4.1   Defining and verifying group homomorphisms

There are two methods that are commonly used for defining group homomorphisms $\varphi : G \to H$. The first is by using a general rule for calculating $\varphi(g)$ for $g \in G$. The induced action of a permutation group on a block system is an example of this. The second method is to specify the images $\varphi(x) \in H$ for

x in a finite generating set X of G. This will of course uniquely determine $\varphi : G \to H$ if such a φ exists, but unless F happens to be free on X, an arbitrary assignment like this will not usually define a homomorphism at all.

In a particular situation, we might know already that our φ does define a homomorphism, but in general we need a method for testing whether this is the case. For example, if we are trying to compute the automorphism group $\mathrm{Aut}(G)$ of a finite group G, then we will certainly need to be able to carry out such a test.

If we know a presentation $\langle X \mid R \rangle$ of G on X, then Theorem 2.53 provides us with a simple test for this. We assume that φ does extend, and then evaluate its image in H on each of the group relators in R. If these all evaluate to 1_H then φ extends, and otherwise it does not.

In some situations, such as when G is a large permutation group, it is not easy to find a group presentation on the original generators $X = [x_1, \ldots, x_k]$, but one can be found efficiently on an extended sequence $X' = [x_1, \ldots, x_r]$, where the new generators are defined from the existing ones via an SLP, as described in Subsection 3.1.3 above. In that case, if we are given $\varphi(x_i)$ for $1 \leq i \leq k$, then we can simply use the expressions for each x_j with $j > k$ as SLP elements in the earlier x_i, in turn, to compute $\varphi(x_j)$. We can then carry out our test for φ being a homomorphism using the presentation on X'.

3.4.2 Desirable facilities

Ideally, for a group homomorphism $\varphi : G \to H$, we would like to be able to perform the following operations.

(i) Compute the image $\varphi(g)$ of any $g \in G$.

(ii) For $h \in H$, test whether $h \in \mathrm{im}(\varphi)$ and, if so, find a $g \in G$ with $\varphi(g) = h$.

(iii) Find $\varphi(K)$ for a subgroup K of G.

(iv) Find $\varphi^{-1}(K)$ for a subgroup K of H.

(v) Find $\ker(\varphi)$.

In some of these cases, it may be unclear what we mean by "Find". It might be relatively easy, for example, to find generators of $\varphi(K)$ in (iii), which would place us in Situation B for $\varphi(K)$, but we would prefer to be in Situation C, providing us with a rewriting algorithm and membership testing for $\varphi(K)$.

What is immediately clear is that, if we can solve the rewriting problem in G, possibly by using extra generators and a SLP, then we can solve (i), and we can at least find generators of $\varphi(K)$ for finitely generated subgroups K of G. Similarly, if we can solve the rewriting problem in $\mathrm{im}(\varphi)$, then we can solve (ii), provided that we know inverse images for our chosen generating set of $\mathrm{im}(\varphi)$. To solve (iv), we need to be able to solve (ii) and (v), because, if

$K \leq H$ is generated by X, then $\varphi^{-1}(K)$ is generated by $\ker(\varphi)$ and any set of inverse images in G of the elements of X.

Let us assume now that we can solve the rewriting, order, and membership problems in finitely generated subgroups of both G and H. So we can solve (i) and (ii) efficiently. The procedure IMAGEKERNEL displayed below is a very useful general method of finding $\ker(\varphi)$ and $\mathrm{im}(\varphi)$ which, taken together with the remarks in the preceding paragraph, allows us to solve all of (i) to (v).

Although we have not indicated it in the code, we also need to remember the $g \in G$ for which $h = \varphi(g)$ is appended to Z, in order to be able to solve the inverse image problem (ii).

IMAGEKERNEL(φ, X)

    Input: $G = \langle X \rangle$, group homomorphism $\varphi : G \to H$
    Output: Generators of $\ker(\varphi)$, $\mathrm{im}(\varphi)$
    ($*$ Initialize generating sequences Y and Z of kernel and image $*$)
1   $Y := []$; $\ Z := []$;
2   $\mathcal{X} := $ PRINITIALIZE$(X, 10, 20)$;
3   **while** $|\langle Y \rangle|\,|\langle Z \rangle| < |G|$
4   **do** $g := $ PRRANDOM$(\sim \mathcal{X})$;
5         $h := \varphi(g)$;
6   **if** $h \notin \langle Z \rangle$
7   **then** APPEND$(\sim Z, h)$;
8   **else** $k := \varphi^{-1}(h)$; ($* \ gk^{-1} \in \ker(\varphi) \ *$)
9   **if** $gk^{-1} \notin \langle Y \rangle$
10   **then** APPEND$(\sim Y, gk^{-1})$;

It is clear why this method works. If g is a random element in G, then $\varphi(g)$ is a random element of $\mathrm{im}(\varphi)$. We also need gk^{-1} to be a random element of $\ker(\varphi)$. Usually, an algorithm for solving (ii) will always produce the same inverse image of a given element of $\mathrm{im}(\varphi)$, so k depends only on h, and then gk^{-1} will indeed be random in $\ker(\varphi)$. This would also be the case if the solution of (ii) resulted in a random inverse image.

Exercise

1. Show that ORDERBOUNDED has complexity $O(\log(n)^3)$.

Chapter 4

Computation in Finite
Permutation Groups

This lengthy chapter is devoted to computations within finite permutation groups. With a few exceptions, which include orbit computations, finding systems of imprimitivity, and testing whether a group is alternating or symmetric, all of the algorithms involved rely on first computing a base and strong generating set (BSGS) for the group G. This topic will be treated in detail in Section 4.4.

As we shall see in Section 4.6, there is an important subdivision of algorithms for finite permutation groups into those that run in polynomial-time, and those that do not; typically, the latter involve backtrack searches through the elements of the group. The polynomial-time algorithms include transitivity and primitivity testing, finding a BSGS, finding normal closures of subgroups, computing commutator subgroups, and finding Sylow subgroups, although the last of these has not yet been fully implemented as a polynomial-time algorithm. Polynomial-time algorithms are not currently known for computing intersections of subgroups, centralizers of elements, and normalizers of subgroups, for example, and it is conjectured that none such exist.

The reader's attention is directed to two existing books on computation in finite permutation groups. The first, by Butler [But91], provides an elementary and leisurely introduction to the topic. The second and more recent book, by Seress [Ser03], is more advanced, particularly in its treatment of complexity questions, and we shall frequently refer the reader to this book for details of more advanced topics.

4.1 The calculation of orbits and stabilizers

In this section, we shall suppose that we are given a finite group G defined by an explicit finite generating sequence $X = [x_1, \ldots, x_r]$, and that G acts on a finite set Ω. We shall assume that, for each $\alpha \in \Omega$ and $x \in X$, the image α^x under the action is stored or can be computed, and that we can decide whether two given elements of Ω are equal.

The most common situation that we shall encounter is where $\Omega = \{1 \ldots n\}$ for some $n \in \mathbb{N}$, and the actions of the generators on Ω are stored as simple arrays of length n. But there are other possibilities. For example, Ω could be the set of all subgroups of G where the action is by conjugation, in which case testing for equality of elements of Ω is a nontrivial process.

We first write down a straightforward function for computing the orbit α^G of a point $\alpha \in \Omega$ under G.

ORBIT(α, X)

    Input: $\alpha \in \Omega$, $X = [x_1, \ldots, x_r]$, $x_i \in \mathrm{Sym}(\Omega)$ with $\langle X \rangle = G$
    Output: α^G

1   $\Delta := [\alpha]$;
2   **for** $\beta \in \Delta$, $x \in X$ **do if** $\beta^x \notin \Delta$
3   **then** APPEND$(\sim \Delta, \beta^x)$;
4   **return** Δ;

To show that the above procedure performs correctly, first observe that all elements appended to Δ are images under an element of X of an element already in Δ, and so by induction we always have $\Delta \subseteq \alpha^G$. Conversely, if $\beta = \alpha^g$ for some $g \in G$ then, since X generates G, we have $g = x_{i_1} x_{i_2} \cdots x_{i_t}$ with each $x_{i_j} \in X \cup X^{-1}$. But since G is finite, the x_i all have finite order, so we can assume that each $x_{i_j} \in X$. We can now show by induction on t that $\beta \in \Delta$ at the end of the procedure. This is true if $t = 0$, since then $\beta = \alpha$, and Δ is initialized to $[\alpha]$. Otherwise, by inductive hypothesis, we have $\gamma := \alpha^{g'} \in \Delta$, where $g' := x_{i_1} x_{i_2} \cdots x_{i_{t-1}}$, and so $\beta = \gamma^{x_{i_t}}$ will be appended to Δ during the loop in the procedure.

On the assumption that images α^x can be looked up or computed in constant time, and that membership of elements of Ω in Δ can be tested in constant time, it is not hard to see that the complexity of ORBIT is $O(|\Delta|r)$. From the simple code above, it is not clear that the membership testing can be achieved in constant time. In the standard situation, where $\Omega = \{1 \ldots n\}$, this can be done by maintaining the current characteristic function of Δ as a subset of Ω. This is best accomplished by means of a *Schreier vector*, which we shall discuss shortly.

In practice, we shall often require not only the orbit itself, but, for each $\beta \in \alpha^G$, an element $u_\beta \in G$ with $\alpha^{u_\beta} = \beta$. By the Orbit-Stabilizer theorem, the set $\{u_\beta \mid \beta \in \alpha^G\}$ is a right transversal of the stabilizer G_α of α in G. By Theorem 2.57, we have

$$G_\alpha = \langle \{u_\beta x u_{\beta^x}^{-1} \mid \beta \in \alpha^G, x \in X\} \rangle.$$

We shall call the elements of this generating set the *Schreier generators* of G_α. The function ORBITSTABILIZER returns a list Δ of pairs (β, u_β) for $\beta \in \alpha^G$, and also a generating sequence Y of G_α.

ORBITSTABILIZER(α, X)

> **Input**: $\alpha \in \Omega$, $X = [x_1, \ldots, x_r]$, $x_i \in \text{Sym}(\Omega)$ with $\langle X \rangle = G$
> **Output**: Δ, Y, as described above

1    $\Delta := [(\alpha, 1_G)]$;
2    $Y := []$;
3   **for** $(\beta, u_\beta) \in \Delta$, $x \in X$ **do if** $\beta^x \notin \Delta$
4   **then** APPEND$(\sim\Delta, (\beta^x, u_\beta x))$;
5   **else** APPEND$(\sim Y, u_\beta x(u_{\beta^x})^{-1})$;
6   **return** Δ, Y;

Observe that, if an orbit member is initially introduced into Δ as β^x, then $u_{\beta^x} = u_\beta x$, and so the associated Schreier generator $u_\beta x u_{\beta^x}^{-1}$ is necessarily trivial, and does not need to be appended to Y. In this case, we shall say that the Schreier generator is trivial by definition, and write $u_\beta x \equiv u_{\beta^x}$. (It is of course possible to have $u_\beta x = u_{\beta^x}$ but not $u_\beta x \equiv u_{\beta^x}$!)

4.1.1   Schreier vectors

We now turn to an important alternative approach to handling the transversal elements u_β when $\Omega = \{1..n\}$ for some $n \in \mathbb{N}$. In many situations, we would like to avoid storing the u_β explicitly. We want to be able to handle permutations acting on as many as 10^6 or even 10^7 points. For a transitive group action, we would have $\Delta = \Omega$, and so it would not be practical to store $|\Delta|$ permutations of this degree. Even for groups of smaller degree, where storage is not critical, we may wish to avoid the $O(n^2 r)$ time complexity that results from computing all of the u_β.

DEFINITION 4.1   A *Schreier vector* for $\alpha \in \Omega$ is a list v of length n with the following properties.

(i) $v[\alpha] = -1$.

(ii) For $\gamma \in \alpha^G \setminus \{\alpha\}$, $v[\gamma] \in \{1..r\}$; more precisely, $v[\gamma] = i$, where γ is appended to Δ as β^{x_i} in the function ORBIT.

(iii) $v[\beta] = 0$ for $\beta \notin \alpha^G$.

So, in particular, v can be used as a characteristic function of the orbit. The function ORBITSV displayed below is an amended version of ORBIT, which computes v. (It could of course be easily modified to compute generators of G_α as in ORBITSTABILIZER.)

ORBITSV(α, X)

    Input: $\alpha \in [1 .. n]$, $X = [x_1, \ldots, x_r]$, $x_i \in \text{Sym}(n)$, $\langle X \rangle = G$
    Output: α^G and Schreier vector v for α
1   **for** $i \in [1 .. n]$ **do** $v[i] := 0$;
2   $\Delta := [\alpha]$; $v[\alpha] := -1$;
3   **for** $\beta \in \Delta$, $i \in [1 .. r]$ **do if** $\beta^{x_i} \notin \Delta$
4   **then** APPEND($\sim \Delta, \beta^{x_i}$); $v[\beta^{x_i}] := i$;
5   **return** Δ, v;

Example 4.1

Let $n = 7$ and $X = [x_1, x_2]$ with $x_1 = (1, 3, 7)(2, 5)$ and $x_2 = (3, 4, 6, 7)$, and let us apply ORBITSV with $\alpha = 1$. Initially $\Delta = [1]$ and $v = [-1, 0, 0, 0, 0, 0, 0]$. Starting the main loop with $\beta = 1$, we have $1^{x_1} = 3$, so we append 3 to Δ and set $v[3] = 1$. Since $1^{x_2} = 1 \in \Delta$, we proceed to $\beta = 3$. Then $3^{x_1} = 7 \notin \Delta$, so we append 7 to Δ and put $v[7] = 1$. Also $3^{x_2} = 4 \notin \Delta$, so we append 4 to Δ and put $v[4] = 2$. Moving on to $\beta = 7$, we have $7^{x_1} = 1$ and $7^{x_2} = 3$ are already in Δ, so we go on to $\beta = 4$. Then $4^{x_1} = 4 \in \Delta$ but $4^{x_2} = 6 \notin \Delta$, so we append 6 to Δ and put $v[4] = 2$. Both 6^{x_1} and 6^{x_2} are in Δ, so the process completes and returns $\Delta = [1, 3, 7, 4, 6]$ and $v = [-1, 0, 1, 2, 0, 2, 1]$.   □

ORBITSV has complexity $O(nr)$. The function U-BETA can then be used to compute a particular u_β if required. It returns false if β is not in α^G. We leave it to the reader to prove that it performs correctly.

U-BETA(β, v, X)

    Input: Schreier vector v for some $\alpha \in \Omega$ and $\beta \in \Omega$
    Output: u_β with $\alpha^{u_\beta} = \beta$ or false if $\beta \notin \alpha^G$
1   **if** $v[\beta] = 0$ **then return** false;
2   $u := 1_G$; $k := v[\beta]$;
3   **while** $k \neq -1$ **do** $u := x_k u$; $\beta := \beta^{x_k^{-1}}$; $k := v[\beta]$;
4   **return** u;

If we call U-BETA with $\beta = 6$ in Example 4.1, then we find $v[6] = 2$, so u is set equal to x_2 and β replaced by $6^{x_2^{-1}} = 4$. Now $v[4] = 2$, so u becomes x_2^2 and β becomes $4^{x_2^{-1}} = 3$. Now $v[3] = 1$, so u becomes $x_1 x_2^2$ and β becomes $3^{x_1^{-1}} = 1$. We now have $v[1] = -1$, so the process halts and returns $u = (1, 6, 3, 4, 7)(2, 5)$, which maps 1 to 6 as required.

    Given an orbit α^G, a method of computing the associated transversal elements u_β, and also a method for computing random elements of G, we can easily generate random elements of G_α. We simply choose g random in G, find $\beta := \alpha^g$, and then $g u_\beta^{-1}$ is random in G_α. The following function implements this process using a Schreier vector for the orbit, and the function

RANDOMSTAB$(\alpha, v, X, \sim \mathcal{X})$

    Input: $G = \langle X \rangle \leq \text{Sym}(n)$, Schreier vector v for orbit α^G
    Output: Random element of G_α
1   $g := \text{PRRANDOM}(\sim \mathcal{X})$;
2   $h := \text{U-BETA}(\alpha^g, v, X)$;
3 **return** gh^{-1};

PRRANDOM from Section 3.2.2. We assume that the random element list $\mathcal{X}$ has already been initialized by a call of PRINITIALIZE.

4.2   Testing for Alt(Ω) and Sym(Ω)

In this section, we shall describe a fast and easy Monte Carlo algorithm for testing whether a transitive permutation group on a set Ω is equal to $\text{Alt}(\Omega)$ or $\text{Sym}(\Omega)$ when $|\Omega| \geq 8$. (For smaller values of $|\Omega|$, there is no need for this, because the order of the group can be computed easily by the Schreier-Sims algorithm to be described in Subsection 4.4.2.) A positive answer is guaranteed to be correct, whereas a negative answer has a small probability of being wrong. Note that in the event of a positive answer, it easy to distinguish between $\text{Alt}(\Omega)$ and $\text{Sym}(\Omega)$; we simply check whether all of the group generators are even permutations.

When attempting to analyze an unknown permutation group G on Ω, a sensible approach is first to test G for transitivity using the methods of the preceding section and then, if G^Ω is indeed transitive, to check whether it is one of the large groups $\text{Alt}(\Omega)$ and $\text{Sym}(\Omega)$. If, as a result of carrying out this test, we become virtually certain that it is not one of these two groups, then we can proceed with further analysis, as described in the following sections. If the result of the Alt/Sym test had actually been wrong, then we would expect to discover this error later, although we might have wasted considerable computational effort before this discovery!

The method is based on the following result.

THEOREM 4.2   *Let $G \leq \text{Sym}(\Omega)$ be transitive on Ω with $|\Omega| = n$, and let p be a prime number with $n/2 < p < n-2$. If G contains an element with a cycle of length p, then $G = \text{Alt}(\Omega)$ or $\text{Sym}(\Omega)$.*

We shall not prove this theorem here. It follows from an old result of Jordan (see [Jor73] or Theorem 13.9 of [Wie64]), which says that a primitive group that contains an element consisting of a single p-cycle for a prime p with $p < n-2$ is alternating or symmetric. The extra condition $n/2 < p$ in our theorem ensures primitivity (exercise below).

The algorithm itself, which we shall call IsAltSym, is so simple that we do not need to display its code. It takes two input parameters: G, and our required upper bound ϵ on the error probability. We use PrRandom to choose some number $N(n, \epsilon)$ random elements of G. If one of these elements contains a p-cycle for a prime p with $n/2 < p < n-2$, then we stop immediately and return the answer true. If not, then we return false. Notice that this will only work for $|\Omega| = n \geq 8$, since there are no primes with $n/2 < p < n-2$ when $n < 8$.

By Theorem 4.2, we know that the answer true is correct. It remains for us to calculate the number $N(n, \epsilon)$ that will ensure that an answer false has a probability of less than ϵ of being wrong. More precisely, we want the probability of the answer false being returned when the input group is $\mathrm{Alt}(\Omega)$ or $\mathrm{Sym}(\Omega)$ to be less than ϵ.

For this, we need to estimate the proportion of elements of $\mathrm{Alt}(\Omega)$ and $\mathrm{Sym}(\Omega)$ that contain a p-cycle for a prime p with $n/2 < p < n-2$. It is proved in Lemma 10.2.3 of [Ser03] that this proportion is at least

$$\sum_{\substack{p \text{ prime,} \\ n/2 < p < n-2}} \frac{1}{p}$$

which, by standard results in number theory, is approximately $\log(2)/\log(n)$. Choose $c(n)$ such that this proportion is at least $c(n)\log(2)/\log(n)$ and let $d(n) := c(n)\log(2)/\log(n)$. (For example, we could choose $c(n) = 0.34$ for $8 \leq n \leq 16$ and $c(n) = 0.57$ for $n > 16$.)

Now, if $G = \mathrm{Alt}(\Omega)$ or $\mathrm{Sym}(\Omega)$, then the probability of choosing k random elements of G and not encountering an element containing a p-cycle is at most $(1 - d(n))^k < e^{-d(n)k}$. Therefore the choice of $N(n, \epsilon) := -\log(\epsilon)/d(n)$ random elements of G will satisfy the required condition on the error probability.

4.3 Finding block systems

4.3.1 Introduction

Let $G = \langle x_1, \ldots, x_r \rangle$ be a subgroup of $\mathrm{Sym}(n)$. The algorithms described in Section 4.1 enable us to decide in time $O(nr)$ whether G is transitive. In this section, we assume that G is transitive. It is now natural to ask whether G is primitive (see Subsection 2.2.5), and we address this problem in this section. Deciding primitivity and finding block systems for G is one of the few computational problems with permutation groups that can be solved efficiently without knowing a base and strong generating set for G.

The basic algorithm to be described finds the smallest block of G that contains $k > 1$ given points $\alpha_1, \ldots, \alpha_k \in \Omega = \{1 \ldots n\}$, together with the other blocks in the associated block system. This is equivalent to finding the G-congruence generated by $\{ (\alpha_1, \alpha_i) \mid 2 \leq i \leq k \}$ (see Definition 2.27). This algorithm is originally due to Atkinson [Atk75]. It runs in time that is just a tiny bit slower than $O(kn)$, and for all practical purposes it can be regarded as being $O(kn)$. It relies on one of the standard algorithms from computer science, the UNION-FIND algorithm; see, for example, Sections 4.6 and 4.7 of [AHU74]. We shall encounter what is essentially the same algorithm again in Section 5.1 in connection with coincidence processing in coset enumeration.

If, as is usually the case, we want to find all minimal block systems preserved by G (and hence test G for primitivity), it appears at first sight that we need to run this test $n - 1$ times, on the pairs of points $\{1, \alpha\}$ for $2 \leq \alpha \leq n$, thereby rendering the complexity of primitivity testing $O(kn^2)$.

If α and β lie in the same orbit of the point stabilizer $H := G_1$ on $\{2 \ldots n\}$, then there exists $h \in H$ mapping $\{1, \alpha\}$ to $\{1, \beta\}$, and so a block for G contains $\{1, \alpha\}$ if and only if it contains $\{1, \beta\}$. We can therefore reduce the number of tests by testing only the pairs $\{1, \alpha\}$ for orbit representatives α of H. If we already know a base and strong generating set for G then, assuming that the first base point is 1, we will already know generators for H, and so we can easily calculate the orbits of H. Even if we do not know H already, then we can quickly generate a subgroup of H using repeated calls of RANDOMSTAB.

But primitivity testing by this method still has quadratic complexity, and this will be noticeable when n is large and H is small. A nearly-linear-time algorithm is described by Beals in [Bea93a]. An alternative version of complexity $O(n \log^3 |G| + nk \log |G|)$ due to Schönert and Seress is described in [SS94] and in [Ser03]; this may be faster for small base groups of large degree. Here, we shall content ourselves with describing the Atkinson method, which does seem to perform satisfactorily in practice for degrees in the millions. It has the advantage of relative simplicity.

4.3.2 The Atkinson algorithm

The displayed function MINIMALBLOCK manipulates an equivalence relation $\sim$ on $\{1 \ldots n\}$. We start by putting the k given points in the same equivalence class, and all other points in separate classes. During the function some of these classes are merged, and the final returned value of $\sim$ is the G-congruence generated by $\{ (\alpha_1, \alpha_i) \mid 2 \leq i \leq k \}$. At any stage in the algorithm, each equivalence class will have a stored representative.

We shall first present and prove the correctness of the algorithm in this basic form, and then discuss how the equivalence class merging process can be implemented efficiently. Representatives of classes that are merged with other classes and then cease to be representatives are queued in a list q of length l, and this queue is initialized to contain $\{ \alpha_i \mid 2 \leq i \leq k \}$. The main 'While' loop of the procedure processes the points in the queue in turn. The equivalence class of $\alpha \in [1\mathinner{.\,.}n]$ under $\sim$ is denoted by $\mathrm{CLASS}(\alpha)$, and the representative of the class containing α by $\mathrm{REP}(\alpha)$. As usual, we denote the Given generating set of G by X.

$\mathrm{MINIMALBLOCK}(G, \{\alpha_1, \ldots, \alpha_k\})$

  Input: $G = \langle X \rangle \leq \mathrm{Sym}(n)$, G transitive, $\alpha_1, \ldots, \alpha_k \in [1\mathinner{.\,.}n]$
  Output: G-congruence $\sim$ generated by $\{(\alpha_1, \alpha_i) \mid 2 \leq i \leq k\}$

1 Initialize $\sim$ with classes $\{\alpha_i \mid 1 \leq i \leq k\}$ with representative α_1,
    and $\{\{\gamma\} \mid \gamma \neq \alpha_i \ (1 \leq i \leq k)\}$.
2 **for** $i \in [1\mathinner{.\,.}k{-}1]$ **do** $q[i] := \alpha_{i+1}$;
3 $l := k{-}1$; $i := 1$;
4 **while** $i \leq l$
5   **do** $\gamma := q[i]$; $i := i+1$;
6   **for** $x \in X$
7   **do** $\delta := \mathrm{REP}(\gamma)$; $\kappa := \mathrm{REP}(\gamma^x)$; $\lambda := \mathrm{REP}(\delta^x)$;
8   **if** $\kappa \neq \lambda$
9   **then** Merge classes $\mathrm{CLASS}(\kappa)$ and $\mathrm{CLASS}(\lambda)$, and
10   make κ representative of combined class.
11        $l := l+1$; $q[l] := \mathrm{REP}(\lambda)$;
12 **return** $\sim$;

Example 4.2

Let $n = 6$ and $G = \langle x_1, x_2 \rangle$ with $x_1 = (1,2,3,4,5,6)$, $x_2 = (2,6)(3,5)$ (Example 2.2), and let us run through the algorithm with $\alpha_1 = 1$, $\alpha_2 = 3$. We start by making 1 the representative of the class $\{1,3\}$ and putting 3 on the queue. Going into the main loop with $\gamma = 3$, with $x = x_1$ we get $\delta = 1$, $\kappa = 4$, $\lambda = 2$, so we amalgamate classes $\{2\}$ and $\{4\}$, make 4 the representative, and add 2 to the queue. With $x = x_2$, we get $\delta = 1$, $\kappa = 5$, $\lambda = 1$, so we amalgamate $\{1,3\}$ and $\{5\}$, make 5 the representative, and add 1 to the queue. Moving onto the next element, $\gamma = 2$, in the queue, with $x = x_1$ we get $\delta = 4$, $\kappa = \lambda = 5$, but with $x = x_2$ we get $\delta = 4$, $\kappa = 6$, $\lambda = 4$, so we amalgamate $\{2,4\}$ and $\{6\}$, make 6 the representative, and add 4 to the queue. Then, $\gamma = 1$, $x = x_1$ gives $\delta = 5$, $\kappa = \lambda = 6$; $\gamma = 1$, $x = x_2$ gives $\delta = 5$, $\kappa = \lambda = 5$; $\gamma = 4$, $x = x_1$ gives $\delta = 6$, $\kappa = \lambda = 5$; and $\gamma = 4$, $x = x_2$ gives $\delta = 6$, $\kappa = \lambda = 6$. So the equivalence classes of the relation returned are $\{1,3,5\}$, $\{2,4,6\}$, and these are the blocks of the corresponding block system preserved by G. $\square$

Let us now prove that the algorithm does what is claimed.

THEOREM 4.3   *The equivalence relation $\mathcal{R}$ returned by $\mathrm{MINIMALBLOCK}$ is the G-congruence generated by $\{(\alpha_1, \alpha_i) \mid 2 \leq i \leq k\}$.*

PROOF   We will denote the final returned value of the equivalence relation $\sim$ by $\sim_F$, and the G-congruence generated by $\{\,(\alpha_1, \alpha_i) \mid 2 \le i \le k\,\}$ by $\equiv$. We have to show that $\sim_F$ and $\equiv$ are equal. We first show that, at any point in the algorithm, we have $\mu \sim \nu$ implies $\mu \equiv \nu$, for any $\mu, \nu \in \Omega$. This is certainly true after the initialization stage, when $\{\,\alpha_i \mid 1 \le i \le k\,\}$ is the only nontrivial $\sim$-class. Whenever two classes are merged during the course of the algorithm, they contain the images γ^x and δ^x under the group element x of two points γ and δ for which $\gamma \sim \delta$ before the merging. We may therefore assume inductively that $\gamma \equiv \delta$. Since $\equiv$ is a G-congruence, we have $\gamma^x \equiv \delta^x$, and so the property "$\mu \sim \nu$ implies $\mu \equiv \nu$" remains true after the merging. Hence it is true at all times, so $\sim_F \; \subseteq \; \equiv$.

Conversely, to show that $\equiv \; \subseteq \; \sim_F$, it is sufficient to show that $\sim_F$ is a G-congruence, because it certainly contains $\{\,(\alpha_1, \alpha_i) \mid 2 \le i \le k\,\}$. So we must prove that $\mu \sim_F \nu$ implies $\mu^g \sim_F \nu^g$ for any $g \in G$. Clearly it suffices to prove this for the generators $g = x$ of G. (Since G is finite, we need not worry about inverses of generators.) Notice that whenever a point λ is added to the queue q, we have $\text{REP}(\lambda) = \lambda$ before adding λ to the queue q, and $\text{REP}(\lambda) \ne \lambda$ immediately afterwards. It is possible that $\text{REP}(\lambda)$ will be redefined again later in the procedure, but it will always be put equal to an existing class representative, so it will never equal λ again. Hence, at all times, λ is in the queue if and only if $\text{REP}(\lambda) \ne \lambda$. (This argument shows that each element is added to the queue at most once, and so the procedure must terminate!)

Let l_F be the length of the queue at the end of the algorithm. For $\lambda \in [1..n]$, define $w(\lambda) = k$ if $q[k] = \lambda$, and $w(\lambda) = l_F + 1$ if λ is not in the queue at the end of the algorithm. We use induction on $z := 2l_F + 2 - w(\mu) - w(\nu)$. If $z = 0$, then neither μ nor ν is in the queue so, by the remark above, $\text{REP}(\mu) = \mu$ and $\text{REP}(\nu) = \nu$. But then $\mu \sim_F \nu$ implies $\mu = \nu$ and there is nothing to prove. If $z > 0$, then at least one of μ, ν, is in the queue, say $\mu = q[k]$. Then, when we reach $i = k$ in the main loop of the algorithm and consider the action of generator x, we have $\gamma = \mu$ and $\delta = \text{REP}(\gamma) \ne \gamma$. At this point, δ is a $\sim$-class representative, so is not in the queue. Hence we must have $w(\delta) > w(\gamma)$. The classes of γ^x and δ^x are now amalgamated, so $\gamma^x \sim_F \delta^x$. Since $w(\delta) > w(\gamma)$, we have $\delta^x \sim_F \nu^x$ by inductive hypothesis, and the result follows.   ∎

4.3.3   Implementation of the class merging process

We now move on to the question of how we implement the $\sim$–equivalence class merging process. The algorithm for this process (UNION-FIND), which we shall now describe, is analyzed in detail in Sections 3.6 and 3.7 of [AHU74] and it follows from that analysis that MINIMALBLOCK will run in time that is microscopically longer than $O(kn)$.

The information on the classes under $\sim$ is stored in an array p, defined on $[1..n]$, which will always satisfy the properties:

$$p[\alpha] \sim \alpha, \quad \text{and} \quad p[\alpha] = \alpha \iff \alpha = \text{REP}(\alpha),$$

REP$(\kappa, \sim p)$

> **Input:** $\kappa \in [1 \mathbin{.\,.} n]$, array p
> **Output:** Representative of class containing κ

1    $\lambda := \kappa;\ \ \rho := p[\lambda];$
2   **while** $\rho \neq \lambda$ **do** $\lambda := \rho;\ \ \rho := p[\lambda];$
     $(*$ Now perform path compression $*)$
3    $\mu := \kappa;\ \ \rho := p[\mu];$
4   **while** $\rho \neq \lambda$ **do** $p[\mu] := \lambda;\ \ \mu := \rho;\ \ \rho := p[\mu];$
5   **return** $\lambda;$

MERGE$(\kappa, \lambda, \sim c, \sim p, \sim q, \sim l)$

> **Input:** $\kappa, \lambda \in \Omega,\ c,\ p,\ q,\ l$

1    $\varphi := \mathrm{REP}(\kappa, \sim p);\ \ \psi := \mathrm{REP}(\lambda, \sim p);$
2   **if** $\varphi \neq \psi$
3   **then if** $c[\varphi] \geq c[\psi]$
4   **then** $\mu := \varphi;\ \ \nu := \psi;$
5   **else** $\mu := \psi;\ \ \nu := \varphi;$
6        $p[\nu] := \mu;\ \ c[\mu] := c[\mu] + c[\nu];$
7        $l := l+1;\ \ q[l] := \nu;$

MINIMALBLOCK$(G, \{\alpha_1, \ldots, \alpha_k\})$

> **Input:** $G = \langle X \rangle \leq \mathrm{Sym}(n)$, G transitive, $\alpha_1, \ldots, \alpha_k \in [1 \mathbin{.\,.} n]$
> **Output:** G-congruence $\sim$ generated by $\{\, (\alpha_1, \alpha_i) \mid 2 \leq i \leq k \,\}$

1   **for** $i \in [1 \mathbin{.\,.} n]$ **do** $p[i] := i;\ \ c[i] := 1;$
2   **for** $i \in [1 \mathbin{.\,.} k-1]$ **do** $p[\alpha_{i+1}] := \alpha_1;$
3   **for** $i \in [1 \mathbin{.\,.} k-1]$ **do** $q[i] := \alpha_{i+1};$
4    $c[\alpha_1] := k;\ \ i := 1;\ \ l := k-1;$
5   **while** $i \leq l$
6   **do** $\gamma := q[i];\ \ i := i+1;$
7   **for** $x \in X$
8   **do** $\delta := \mathrm{REP}(\gamma, \sim p);\ \ \mathrm{MERGE}(\gamma^x, \delta^x, \sim c, \sim p, \sim q, \sim l);$
9   **for** $i \in [1 \mathbin{.\,.} n]$ **do** $\mathrm{REP}(i, \sim p);$
10   **return** $p;$

for all $\alpha \in [1 \mathbin{.\,.} n]$. So, for $\alpha \in [1 \mathbin{.\,.} n]$, we can find REP$(\alpha)$ by continually replacing α by $p[\alpha]$ until it becomes constant. It increases the efficiency of later class representative calculations if, whenever we find REP$(\alpha) = \beta$ in this way, we set $p[\gamma]$ equal to β for all γ in the p-chain from α to β. This process is known as *path compression* and is done in the function REP below, which takes the array p as a (mutable) second argument.

Merging of the two classes with representatives κ and λ can then be accomplished simply by putting $p[\kappa] := \lambda$. It turns out to be more efficient to choose the new representative from the larger of the two classes being merged.

To do this, we maintain another array c (cardinality) such that, for class representatives α, $c[\alpha]$ is equal to the size of that class. The procedure MERGE performs the merging operation, updates the array c, and adds the deleted class representative to the queue q. The revised MINIMALBLOCK function uses REP and MERGE.

The purpose of line 9 in MINIMALBLOCK is to force path compression and thereby ensure that the array p returned satisfies $p[\alpha] = \text{REP}(\alpha)$ for all α.

There is a further simple improvement that we could easily build in. Since we are assuming that G is transitive, we know that the any block system preserved by G has blocks of size dividing n. Let q be the smallest prime dividing n. Then if ever a class grew larger than n/q, we could immediately abort the process and make all elements of $[1 \mathbin{..} n]$ equivalent.

Exercises

1. Prove Theorem 4.2 assuming the theorem of Jordan.

2. Work through MINIMALBLOCK with G as in Example 2.2, first with input points $\{1, 4\}$ and then with input points $\{1, 2\}$.

3. Why is the assignment $\delta := \text{REP}(\gamma)$; in line 7 of the first version of MINIMALBLOCK (and in line 8 of the second version) inside rather than outside of the loop over the generators in X? Would the algorithm still work if it were moved outside?

4.4 Bases and strong generating sets

4.4.1 Definitions

In 1970 [Sim70], Sims introduced a chain of subgroups in a finite permutation group, which plays a very similar role to that played by the chain of subspaces defining the echelon form representation of a subspace of a vector space.

We shall now introduce some notation that we shall retain throughout this section, and indeed throughout the remainder of the book. We assume that G is a finite permutation group acting on $\{1 \mathbin{..} n\}$ and generated by a finite sequence S of elements of $\text{Sym}(n)$. Let $B = [\beta_1, \ldots, \beta_k]$ be a sequence of distinct elements of Ω. Define $G^{(i)} := G_{\beta_1, \ldots, \beta_{i-1}}$ for $1 \leq i \leq k+1$ (so $G^{(1)} = G$).

Now, for $1 \leq i \leq k+1$, let $S^{(i)} := S \cap G^{(i)}$, $H^{(i)} := \langle S^{(i)} \rangle$, and $\Delta^{(i)} := \beta_i^{H^{(i)}}$. We let $U^{(i)}$ be a right transversal of $H_{\beta_i}^{(i)}$ in $H^{(i)}$; so, by the Orbit-Stabilizer theorem, we have $U^{(i)} = \{u_\beta^{(i)} \mid \beta \in \Delta^{(i)}\}$, where $u_\beta^{(i)}$ maps β_i to β.

Note that, for given B and S, it is straightforward to compute the subsets $S^{(i)}$, and then the sets $\Delta^{(i)}$ and $U^{(i)}$ can be computed using the orbit algorithms described in Section 4.1. The transversals $U^{(i)}$ can either be computed explicitly and stored, or their corresponding Schreier vectors can be computed. If they are stored explicitly, then we still need to store a characteristic function of $\Delta^{(i)}$ in $\{1 .. n\}$, because we need to be able to test points for membership in $\Delta^{(i)}$. For convenience, we shall use Δ^* to denote the list of orbits $\Delta^{(i)}$, together with the chosen data structures for computing the $u_\beta^{(i)}$ and testing points for membership in $\Delta^{(i)}$.

The sequence B is said to be a *base* for G if the only element in G that fixes each of $\beta_1, \ldots, \beta_k$ is the identity. That is, if $G^{(k+1)} = 1$, and hence

$$1 = G^{(k+1)} \leq G^{(k)} \leq \cdots \leq G^{(2)} \leq G^{(1)} = G.$$

The sequence S is said to be a *strong generating set* for G relative to the base B if it includes generators for each stabilizer $G^{(i)}$ in the chain above; that is, for $i = 1, 2, \ldots, k+1$, $G^{(i)} = \langle S^{(i)} \rangle = H^{(i)}$. Note that this is true by definition for $i = 1$.

(The expression 'strong generating set' is now firmly entrenched in the CGT literature but, for reasons outlined in Subsection 2.1.4, we prefer to regard S as an ordered set or sequence.)

For brevity, we shall say that (B, S) is a *BSGS* for G, if B is a base and S is a strong generating set relative to B. In that case, $G^{(i)}$ is called the i-th *basic stabilizer*, and $\Delta^{(i)} = \beta_i^{G^{(i)}}$ the i-th *basic orbit*.

If (B, S) is a BSGS, then the sequence of transversals $U^{(1)}, \cdots, U^{(k)}$ provides us with a convenient normal form for the elements of G, since every element $g \in G$ has a unique representation $g = u_k u_{k-1} \cdots u_1$ with $u_i \in U^{(i)}$.

Note that the order of the group can be read off from the transversals, since

$$|G| = |U^{(k)}| \cdot |U^{(k-1)}| \cdots |U^{(1)}|.$$

If $B = [\beta_1, \ldots, \beta_k]$ is a base, and $g \in G$, then $B^g := [\beta_1^g, \ldots, \beta_k^g]$ is called the *base image* of g (relative to B).

LEMMA 4.4   *The base image of g relative to a base B of G uniquely determines the element $g \in G$.*

PROOF   If $B^g = B^h$ for $g, h \in G$, then $B^{gh^{-1}} = B$, and so $gh^{-1} = 1$ by the definition of a base.   ∎

One of the advantages of representing elements by means of base images arises from the fact that, for many interesting groups, the size of a base B may be rather small compared to the degree of the group. Clearly a permutation group G of degree n with a base of length k satisfies $|G| \leq n^k$. It turns out that the groups in certain important families have permutation representations with a base of length polylogarithmic in their degree.

Such families are called families of *short base groups*. They have associated constants c and d such that $\log |G| \leq d \log^c |\Omega|$ for all groups G in the family. The nonalternating nonabelian finite simple groups, and the primitive permutation groups that do not have an alternating group as a composition factor are examples of families of small base groups.

The worst case occurs for the symmetric group S_n: it is clear that a base must contain $n - 1$ points (and that any $n - 1$ will do); choosing $B := [n, n - 1, \ldots, 2]$, one finds that $G^{(i)} \cong S_{n-i+1}$, the full symmetric group acting on the first $n - i + 1$ points (which form the i-th basic orbit).

The usefulness of the concept of a BSGS is supported by the following (empirical) observations.

Universality: For the great majority of fundamental permutation group algorithms, a BSGS appears to be the most appropriate way to represent the group.

Inheritability: Almost all algorithms that are used to construct subgroups and homomorphic images of a permutation group G have the property that the subgroup or homomorphic image inherits a BSGS from that of G.

Availability: Effective algorithms can be designed for constructing BSGS for groups G.

Before turning to the central problem of how we construct a BSGS of a finite permutation group, we shall assume for the moment that these are given, and explain how we use them to test membership in our group G of an arbitrary permutation in $\text{Sym}(\Omega)$. This is accomplished by the function STRIP.

STRIP(g, B, S, Δ^*)

    Input: $g \in \text{Sym}(\Omega)$, B, S, Δ^* as described above
    Output: $h \in \text{Sym}(\Omega)$ and $i \in [1 .. k{+}1]$
1   $h := g$;
2   **for** $i \in [1 .. k]$
3   **do** ($*$ h fixes base points $\beta_1, \ldots, \beta_{i-1}$ $*$)
4          $\beta := \beta_i^h$;
5   **if** $\beta \notin \Delta^{(i)}$ **then return** h, i;
6   Let $u_i \in U^{(i)}$ with $\beta_i^{u_i} = \beta$;
7          $h := hu_i^{-1}$;
8 **return** $h, k{+}1$;

If the element h returned is the identity, then $g = u_k u_{k-1} \cdots u_1$, and so $g \in G$. Conversely, if $g \in G$ and (B, S) is a BSGS for G, then g has a unique expression as $g = u_k u_{k-1} \cdots u_1$ with $u_i \in U^{(i)}$, and the function will find these u_i and return $h = 1_G$. Hence, if (B, S) is a BSGS for G, then $g \in G$ if and only if $h = 1_G$. If an integer $i \leq k$ is returned, then the membership test $\beta \in \Delta^{(i)}$ failed and the function aborted. The output in that situation will be used later. The function could also be made to return the elements u_i if the decomposition of g was required. As we shall see later, STRIP can be usefully applied even when (B, S) is not known to be a BSGS for G.

4.4.2 The Schreier-Sims algorithm

We now present a basic version of the Schreier-Sims algorithm for determining a base and strong generating set of a permutation group. It is based on the following simple result.

LEMMA 4.5 *Let B, S, $S^{(i)}$ and $H^{(i)}$ be as defined above. Then (B, S) is a BSGS for G if and only $H^{(k+1)} = 1$ and $H^{(i)}_{\beta_i} = H^{(i+1)}$ for $1 \leq i \leq k$.*

PROOF By definition, (B, S) is a BSGS if and only if $G^{(k+1)} = 1$ and $H^{(i)} = G^{(i)}$ for $1 \leq i \leq k$. Since $H^{(1)} = G^{(1)}$ and $G^{(i)}_{\beta_i} = G^{(i+1)}$ for $1 \leq i \leq k$, the condition in the lemma is certainly necessary for (B, S) to be a BSGS. Conversely, if the condition holds, then we see by induction on i that $H^{(i)} = G^{(i)}$ for $1 \leq i \leq k + 1$, and then the assumption $H^{(k+1)} = 1$ implies that $G^{(k+1)} = 1$. ∎

The Schreier-Sims algorithm displayed in SCHREIERSIMS takes an initial sequence B (which may be empty) and generating sequence S for G as input, and extends B and S to a BSGS for G. It starts by extending B if necessary to ensure that no element of S fixes each point in B, and then sets up the basic orbits and associated transversals. At this stage we certainly have $H^{(k+1)} = 1$, and the algorithm proceeds by testing the conditions $H^{(i)}_{\beta_i} = H^{(i+1)}$ for $i = k, k-1, \ldots, 1$. This is done in the main loop of the algorithm, starting at line 6. Since $H^{(i+1)} \leq H^{(i)}_{\beta_i}$ by definition, we only need to test that $H^{(i)}_{\beta_i} \leq H^{(i+1)}$, which we do by testing that generators of $H^{(i)}_{\beta_i}$ lie in $H^{(i+1)}$.

These generators of $H^{(i)}_{\beta_i}$ are found using Schreier's theorem (Theorem 2.57), as described in the algorithm ORBITSTABILIZER in Section 4.1. In an implementation, we should be careful to consider only those Schreier generators that are not trivial by definition, as defined in Section 4.1. It is convenient to do the tests in the order $i = k, k-1, \ldots, 1$ because, by doing that, we know that, when we are testing a particular i, the required condition is satisfied for larger i, and so we can use the algorithm STRIP to test the generators of $H^{(i)}_{\beta_i}$ for membership in $H^{(i+1)}$.

If this membership test fails for some Schreier generator g, then we set the Boolean variable y to `false`, and adjoin a new generator h to $S^{(i+1)}$, thereby replacing $H^{(i+1)}$ by a larger subgroup of $G^{(i+1)}$. (The new generator h appended is not necessarily g itself, but is rather the element output by STRIP applied to g.) If h fixes all existing base points, then we adjoin a new

SCHREIERSIMS($\sim B, \sim S$)

    Input: B and S as defined above.

1   **for** $x \in S$ **do if** $x \in G_{\beta_1, \ldots, \beta_k}$

2   **then** Let $\gamma \in \Omega$ with $\gamma^x \neq \gamma$; APPEND($\sim B, \gamma$); $k := k+1$;

3   **for** $i \in [1 \ldots k]$

4   **do** $S^{(i)} := S \cap G_{\beta_1, \ldots, \beta_{i-1}}$; $H^{(i)} := \langle S^{(i)} \rangle$; $\Delta^{(i)} := \beta_i^{H^{(i)}}$;
    ($*$ Now $H^{(k+1)} = 1$ $*$)

5    $i := k$;

6   **while** $i \geq 1$

7   **do** ($*$ Test condition $H_{\beta_i}^{(i)} = H^{(i+1)}$ $*$)

8   **for** $\beta \in \Delta^{(i)}$

9   **do** Find $u_\beta \in H^{(i)}$ with $\beta_i^{u_\beta} = \beta$;

10   **for** $x \in S^{(i)}$ **do if** $u_\beta x \not\equiv u_{\beta^x}$

11   **then** ($*$ Test if Schreier gen. $u_\beta x u_{\beta^x}^{-1} \in H^{(i+1)}$ $*$)

12                   $y := \texttt{true}$;

13                   $h, j := $ STRIP($u_\beta x u_{\beta^x}^{-1}, B, S, \Delta^*$);

14   **if** $j \leq k$

15   **then** ($*$ New strong gen. h at level j $*$)

16                         $y := \texttt{false}$;

17   **elseif** $h \neq 1$

18   **then** ($*$ h fixes all base points $*$)

19                         $y := \texttt{false}$;

20   Let $\gamma \in \Omega$ with $\gamma^h \neq \gamma$;

21   APPEND($\sim B, \gamma$);

22                         $k := k+1$; $S^{(k)} := []$;

23   **if** $y = \texttt{false}$

24   **then for** $l \in [i+1 \ldots j]$

25   **do** APPEND($\sim S^{(l)}, h$);

26                             $H^{(l)} := \langle S^{(l)} \rangle$;

27                             $\Delta^{(l)} := \beta_l^{H^{(l)}}$;

28                     $i := j$;

29   **continue** i;
                   ($*$ $u_\beta x u_{\beta^x}^{-1} \in H^{(i+1)}$ verified $*$)
              ($*$ Condition $H_{\beta_i}^{(i)} = H^{(i+1)}$ is verified $*$)

30           $i := i - 1$;

31    $S := \cup_{i=1}^{k} S^{(i)}$;

base point moved by h to B, thereby maintaining the condition $H^{(k+1)} = 1$ of the lemma.

For a particular i in SCHREIERSIMS, each new generator that we append to $S^{(i)}$ strictly increases the group $H^{(i)}$, so we can only append generators finitely many times, and the procedure must terminate. When it does terminate, all of the conditions of the lemma have been verified, and so (B, S) is a BSGS for G.

Notice that we do not maintain the condition $S^{(i)} = S \cap G^{(i)}$ throughout the procedure. It is considerably more efficient if we adjoin new generators found for a particular i only to $S^{(l)}$ for $l > i$, because they would be redundant as generators of $H^{(l)}$ for $l \leq i$, and would significantly increase the time taken by the loop over the elements of $S^{(l)}$ for $l \leq i$.

Example 4.3

Let $n = 5$ and $S = [a, b]$, with $a = (1, 2, 4, 3)$, $b = (1, 2, 5, 4)$. The default is to start with B empty, and then the procedure will begin by adjoining 1 to B, so we have $B = [1]$, $k = 1$, and no element of S fixes B pointwise. Using ORBIT, we have $S^{(1)} = S$, and calculate $\Delta^{(1)} = [1, 2, 4, 5, 3]$, with

$$U^{(1)} = [1, a, a^2 = (1, 4)(2, 3), ab = (1, 5, 4, 3, 2), a^3 = (1, 3, 4, 2)].$$

We now enter the main loop of the algorithm at line 6 with $i = 1$. Looping through $\Delta^{(1)} = [1, 2, 4, 5, 3]$, we start with $\beta = 1$, $u_\beta = 1_G$. Then $\beta^a = 2$ and $u_\beta a \equiv u_{\beta^a}$, but $\beta^b = 2$ and the associated Schreier generator is $u_\beta b u_{\beta^a}^{-1} = ba^{-1} = (2, 5)(3, 4)$. Since $k = 1$, STRIP applied to $(2, 5)(3, 4)$ will return $(2, 5)(3, 4)$ and 2, so we put $c := (2, 5)(3, 4)$, append the point 2, moved by c, to B, increment k to 2, and set $S^{(2)}$ to $[c]$. We also calculate $\Delta^{(2)} = [2, 5]$, with $U^{(2)} = [1, c]$. (Note that we do not adjoin c to $S^{(1)}$, because $H^{(1)}$ is unchanged, and c would be redundant as a generator of $H^{(1)}$.)

We now jump back to the beginning of the main loop at line 6 with $i = 2$, and find that the two associated Schreier generators cc^{-1} and c^2 are both trivial (the first by definition), so we return to line 6 with $i = 1$. The Schreier generator $u_\beta b u_{\beta^a}^{-1}$ that caused the problem before now strips to the identity, so we move on to $\beta = 2$, $u_\beta = a$. Here both Schreier generators are trivial by definition, so we can proceed to $\beta = 4$, $u_\beta = a^2$. We have $u_\beta a \equiv u_{\beta^a}$, but $\beta^b = 1$, so the associated Schreier generator is $u_\beta b = a^2 b = (2, 3, 5, 4)$. Since $3 = 2^{a^2 b}$ is not in $\Delta^{(2)}$, STRIP applied to $a^2 b$ returns $a^2 b$ and 2. So we put $d := a^2 b$ and append d to $S^{(2)}$. We now have $S^{(2)} = [c, d]$ and we recalculate $\Delta^{(2)} = [2, 5, 3, 4]$ with $U^{(2)} = [1, c, d, cd]$.

Again we jump back to line 6 with $i = 2$. The Schreier generators not trivial by definition are c^2 (for $\beta = 5$), $dc(cd)^{-1}$ and $d^2 c^{-1}$ (for $\beta = 3$), and $cdcd^{-1}$ and cd^2 (for $\beta = 4$), and these are all trivial, so we go back to line 6 with $i = 1$ for the third time. On the previous pass, we had got as far as $\beta = 4$, and the Schreier generator $a^2 b$ now strips successfully to $a^2 b d^{-1} = 1$. Moving on to $\beta = 5$, $u_\beta = ab$ we find that the associated Schreier generators are $aba(ab)^{-1} = (2, 3, 5, 4)$, which strips to $aba(ab)^{-1} d^{-1} = 1$ and $ab^2 a^{-2} = (2, 3, 5, 4)$, which again strips to 1. Finally, with $\beta = 4$, $u_\beta = a^3$, the Schreier generators are $a^4 = 1$, and $a^3 b a^{-3} = (2, 4, 5, 3)$, which strips to $a^3 b a^{-3} (cd)^{-1} = 1$. So the algorithm terminates, and we have shown that G is a group of order $|\Delta^{(1)}||\Delta^{(2)}| = 20$. $\quad\square$

The observant reader will have noticed an inefficiency in SCHREIERSIMS. When we return to line 6 with a value of i that has occurred previously then, if $S^{(i)}$ has not changed since the previous occurrence, we do not need to repeat the checks for those Schreier generators that have already been tested. Indeed, even a Schreier generator that failed the test previously will now pass it because of the new generator that has been appended. It is routine to modify the code to remember where we have got to on any previous occurrences of each i. Indeed, in the version given in Algorithm 3, page 136 of [But91], this repetition is avoided even when $S^{(i)}$ has changed, because the existing orbit and transversal are extended rather than recalculated.

4.4.3 Complexity and implementation issues

We shall refer the reader to other sources for detailed complexity analyzes of the Schreier-Sims algorithm. The first major implementation decision is whether to calculate and store the transversals $U^{(i)}$ explicitly, or whether to use Schreier vectors. The former method generally results in lower complexity and faster running times, but the space requirement is much greater. If we wish to work with groups of degree more than about 1000, then storing the $U^{(i)}$ becomes impractical.

The running time of a version of Sims' algorithm that stores the transversals $U^{(i)}$ was shown by Furst, Hopcroft Luks in [FHL80] to be $O(n^6 + kn^2)$. Using the so-called *labelled branching* data structure for the $U^{(i)}$, Jerrum [Jer82] devised a variant with running time $O(n^5 + kn^2)$. Knuth [Knu81] describes another version with running time $O(n^5 + kn^2)$. These results are all presented in Chapter 14 of [But91]. Chapter 4 of [Ser03] contains detailed analyzes of several variations, including those that use Schreier vectors.

We shall now discuss some features of the algorithm for small base groups. This is an important case, because many computations with permutation groups in practice take place in groups with large degree but with a small base, typically with size less than 10, say. In this situation, we certainly use Schreier vectors for calculating transversal elements and, what is more, we must avoid multiplying complete permutations together so far as possible.

To help us avoid multiplying permutations, we introduce a data structure, which we shall call a *permutation word*. Theoretically, this is just a list $[g_1, g_2, \ldots, g_r]$ of permutations, which represents the product $g_1 g_2 \cdots g_r$. The elements of the list are, however, stored as pointers or index numbers of permutations, so that lists can be manipulated without copying complete permutations. We adopt the policy of storing inverses of all stored permutations, including the permutations in the initial generating sequence S of the group, and any new generators appended during the Schreier-Sims algorithm. Then a permutation word can be easily inverted by reversing the list and replacing each entry by its inverse, and two permutation words can be multiplied simply by concatenating them. The image of a single point in Ω under a permutation word can be computed in time proportional to the length r of the word.

Many computations with small base groups, including most of the steps in the Schreier-Sims algorithm, can be carried out by looking at images of base points under permutation words. In particular, the algorithm U-BETA can be easily adapted to output a permutation word with entries in S rather than a permutation. In Example 4.1, the modified version of U-BETA applied to $\beta = 6$ would return $[x_1, x_2, x_2]$ as a word representing the permutation $(1, 6, 3, 4, 7)(2, 5)$, which maps 1 to 6. Then STRIP can be modified so that it takes a permutation word as input and returns a possibly modified permutation word as output. The Schreier generators in SCHREIERSIMS can be computed as permutation words. If the integer returned by STRIP for a Schreier generator is less than $k + 1$, then we need to append the group element output to S, and in that case we do need to multiply out the word and store it and its inverse as complete permutations.

The above discussion also shows that each new strong generator that is appended to S during SCHREIERSIMS arises as a word in the existing strong generators. It is therefore possible to associate with each strong generator an SLP element that expresses it as a word in the earlier generators. An implementation of the algorithm should have the option of returning these SLP elements because, as we shall see in Section 4.5 below, they may be needed for calculations involving group homomorphisms. Note that the modified version of STRIP that we have just described can be used to express an arbitrary permutation in g as a word in the strong generators and their inverses; this is also important for homomorphism calculations.

Unfortunately, we also need to multiply out the word when STRIP returns $k+1$, because we have to test whether the word represents the identity. Surprisingly, it is this innocent-looking test at line 17 of SCHREIERSIMS that takes up nearly all of the time when running it on a small base group. One consequence is that the complexity of the algorithm is worse than $O(n^2)$ even if the base size is regarded as a constant.

For this reason, it is worth noting that if, for some reason, we already know that B is base for G, and STRIP returns h and $k+1$, then h fixes all base points, so we must have $h = 1$, and we do not need to multiply out the permutation word for h to check this. A 'known-base' version of SCHREIERSIMS differs only from the standard version in that the test for $h = 1$ is omitted, but it runs faster by a factor of about n/k in practice. One common situation where this occurs is when G is a subgroup of some larger group $K \leq \mathrm{Sym}(\Omega)$, and we already know a small base B for K. Then B is also a base of G. This observation enables many computations with subgroups of a given group to run very fast.

In particular, an SGS for the join $\langle H, K \rangle$ of two subgroups $H, K \leq G$ for which a base of G is known is easily computable. We just apply the 'known-base' version of SCHREIERSIMS to the union of the generating sets of H and K. Unfortunately, the intersection $H \cap K$ is much harder to compute in general. We shall discuss that in Subsection 4.6.6 below.

4.4.4 Modifying the strong generating set — shallow Schreier trees

(This subsection may be omitted on a first reading.) We are now going to discuss two apparently contradictory ideas. First we shall consider the problem of removing redundant strong generators, and then we shall discuss situations under which it might be a good idea to adjoin some new redundant ones! In fact it is perfectly sensible to carry out the first of these processes followed by the second. Redundant strong generators tend to be removed from $S^{(i)}$ for low values of i, whereas the new ones that we adjoin are more often for high values of i. We assume throughout the subsection that a BSGS (B, S) has already been computed for G.

What often happens in SCHREIERSIMS is that new generators appended to $S^{(i)}$ during the algorithm render some generators of $S^{(j)}$ for $j < i$ redundant as generators of $S^{(j)}$. If we wish to save some space by removing these, then we can proceed as in the following outline algorithm:

REMOVEGENS$(B, \sim S, \sim \Delta^*)$

  Input: BSGS data structure for permutation group
1 **for** $i \in [k \mathinner{.\,.} 1 \text{ by } -1]$, $g \in S^{(i)} \setminus S^{(i+1)}$
2   **do if** $\beta_i^{\langle S^{(i)} \setminus \{g\} \rangle} = \Delta^{(i)}$ **then** remove g from S;

It is probably advisable in practice to modify the above procedure so that it does not eliminate the original generators of G. One reason for this is that images of the original generators are often used to define homomorphism $\varphi : G \to H$. If the original generating set is inordinately large (which can sometimes happen for groups computed as Sylow subgroup of larger groups, for example), then it is preferable to reduce it, redefine G, and recalculate the strong generating set using a known-base version of SCHREIERSIMS.

Now we turn to the situation where we might adjoin new strong generators. There is a problem that occasionally arises with Schreier vector calculations that can adversely affect running times. Typically, for long orbits Δ, the length of the permutation word returned by ORBITSV (that is, the length of the coset transversal element returned as a word in S) is of order $\log(|\Delta|)$, but it can occasionally be as long as Δ itself. This occurs, for example, when the group is a large cyclic group and $|S| = 1$, a situation that can arise during the Schreier-Sims algorithm: whenever we adjoin a new base point, we initially have a single generator that fixes the other base points but moves the new one, and so we are potentially in that bad situation for the orbit at the bottom level. This can result in an extra factor $O(n)$ rather than $O(\log(n))$ in the complexity of the complete algorithm, and so it is important in practice to find a way of avoiding it.

The following function seeks to solve this problem by adjoining extra random chosen generators of the input group G. This approach was first described and analyzed by Cooperman, Finkelstein, and Sarawagi in [CFS90]. We have modified it by keeping the original generators X of G, which would help to reduce storage requirements in an implementation. It requires the size m of the orbit to be calculated to be known in advance, but in an implementation we would only wish to use this method for large orbits with small initial generating sets, and so we would want to precompute m by using the basic ORBIT algorithm in any case.

SHALLOWORBITSV$(\alpha, \sim X, m)$

      Input: $\alpha \in [1 \mathinner{\ldotp\ldotp} n]$, $X = [x_1, \ldots, x_r] \subseteq \mathrm{Sym}(n)$ with $\langle X \rangle = G$, $m = |\alpha^G|$
      Output: α^G and Schreier vector v for α
      ($*$ The generating sequence X of G may be enlarged $*$)

1   For $i \in [1 \mathinner{\ldotp\ldotp} n]$ **do** $v[i] := 0$;
2   $v[\alpha] := -1$;
3   $\Delta := [\alpha]$;
4   **for** $x_i \in X$ **do if** $\alpha^{x_i} \notin \Delta$
5   **then** APPEND$(\sim \Delta, \alpha^{x_i})$; $v[\alpha^{x_i}] := i$;
6   $\mathcal{X}, \mathcal{W} := \mathrm{PRINITIALIZE}(X, 10, 20)$;
7   **while** $|\Delta| \neq m$
8   **do repeat** $g := \mathrm{PRRANDOM}(\sim \mathcal{X}, \sim \mathcal{W})$
9   **until** $|\Delta \cup \Delta^g|/|\Delta|$ is large;
10     $r := r+1$; $x_r := g$; $\Delta' := \Delta$;
11   **for** $x_i \in X$, $\beta \in \Delta$ **do if** $\beta^{x_i} \notin \Delta'$
12   **then** APPEND$(\sim \Delta', \beta^{x_i})$; $v[\beta^{x_i}] := i$;
13     $\Delta := \Delta'$;
14 **return** Δ, v;

Of course, we have not yet specified what we mean by the condition "$|\Delta \cup \Delta^g|/|\Delta|$ is large" in this function. The idea is that we only want to adjoin the new random element as a new generator if it is effective in increasing the current orbit size. We could use the following specific criterion from Lemma 4.4.5 of [Ser03]. Let $p = 0.46$. Then 'large' means:

$$|\Delta \cup \Delta^g|/|\Delta| \geq 2 - \frac{|\Delta|}{(1-p)m} \text{ if } |\Delta| \leq m/2;$$

$$|\Delta \cup \Delta^g|/|\Delta| \geq (m - |\Delta|)\frac{(m - |\Delta|)}{(1-p)m} \text{ if } |\Delta| \geq m/2.$$

It is proved in [Ser03] that, with this choice, the probability of a random element $g \in G$ satisfying this condition is at least p, and so we can expect to find suitable elements g quickly. Furthermore, we will have $|X| \leq 2\log_2(m) + 4 + r$ after a call of SHALLOWORBITSV, where r is the initial value of $|X|$, and the length of the words output by calls of U-BETA using the Schreier vector output by SHALLOWORBITSV will also be at most $2\log_2(m) + 4 + r$.

If we were using this function on the basic orbits $\Delta^{(i)}$ of a group G, and appending new strong generators to $S^{(i)}$, then we would also need to store any new generators as SLP elements in the existing generators. Since the function PRRANDOM returns the required SLP element together with each random element g that it returns, this presents no difficulty, and we omit the technical details.

A different method for calculating shallow Schreier trees, which was originally proposed in [BCFS91], is discussed in Section 4.4 of [Ser03].

4.4.5 The random Schreier-Sims method

Let (B, S) be as before, and suppose that we are given a random element $g \in G$ using a uniform distribution on G. If (B, S) is *not* a BSGS for G, then it is not hard to prove that STRIP(g, B, S, Δ^*) returns a nonidentity element h with probability at least $1/2$. We can then use the output of STRIP to extend (B, S), as we did in the Schreier-Sims algorithm.

Furthermore, if (B, S) is not a BSGS and we apply STRIP to w elements of G, then the probability that the identity is returned each time is at most 2^{-w}. In other words, if STRIP returns the identity on each of, say, 10 random elements of G, then we can be highly confident that (B, S) *is* a BSGS for G. This observation is the basis of the following algorithm, which runs much more quickly than SCHREIERSIMS and, with high confidence level, yields a BSGS for G.

It is generally known as the 'Random Schreier-Sims algorithm', which is really a misnomer, because it uses random elements rather than Schreier generators! The name dates from the first implementation, which was described by Leon in [Leo80b].

Although we have omitted the details in the displayed procedure RANDOMSCHREIER, the SLP elements returned as a second value by PRRANDOM should be stored for the new strong generators appended to S.

One common situation in which a variant of RANDOMSCHREIER is most useful is when we know the order of G in advance. In that case, we do not use c and w to determine when we halt; rather, we can stop as soon as the product of the orders of the basic orbits is equal to $|G|$. This occurs, for example, if we start with known generators of a standard group, such as $\mathrm{PGL}(n, q)$, or if G was constructed as a wreath product of two smaller groups.

In general, however, although we can be highly confident that (B, S) is a BSGS after a call of RANDOMSCHREIER, the consequences of an error tend to be dire, so we prefer to verify the result. (Our confidence level may also be decreased by any doubts that we may have about the product replacement algorithm for producing random elements.) The verification can be done by calling SCHREIERSIMS on the (B, S) produced by RANDOMSCHREIER: a run of RANDOMSCHREIER followed by a verification by SCHREIERSIMS is usually significantly faster than calling SCHREIERSIMS on the original generating set S with B empty. But, in an implementation, we would arrange for RANDOMSCHREIER to return the sets $S^{(i)}$ individually, and pass these on to SCHREIERSIMS, in order to avoid using redundant generators of $H^{(i)}$.

RANDOMSCHREIER($\sim B, \sim S, w$)

    Input: B and S as in SCHREIERSIMS, $w \in \mathbb{N}$.

1   $\mathcal{X}, \mathcal{W} := \text{PRINITIALIZE}(S, 10, 20)$;
2   **for** $x \in S$ **do if** $x \in G_{\beta_1, \dots, \beta_k}$
3   **then** Let $\gamma \in \Omega$ with $\gamma^x \neq \gamma$; APPEND($\sim B, \gamma$); $k := k+1$;
4   **for** $i \in [1 .. k]$
5   **do** $S^{(i)} := S \cap G_{\beta_1, \dots, \beta_{i-1}}$; $H^{(i)} := \langle S^{(i)} \rangle$; $\Delta^{(i)} := \beta_i^{H^{(i)}}$;
     ($*$ Now $H^{(k+1)} = 1$ $*$)
6   $c := 0$;
     ($*$ c counts the random elements sifted without change to (B, S) $*$)
7   **while** $c < w$
8   **do** $g := \text{PRRANDOM}(\sim \mathcal{X}, \sim \mathcal{W})$;
9         $h, j := \text{STRIP}(g, B, S, \Delta^*)$;
10      $y := \texttt{true}$;
11   **if** $j \leq k$
12   **then** ($*$ New strong generator h at level j $*$)
13             $y := \texttt{false}$;
14   **elseif** $h \neq 1$
15   **then** ($*$ New strong generator h fixes all base points $*$)
16             $y := \texttt{false}$;
17   Let $\gamma \in \Omega$ with $\gamma^h \neq \gamma$; APPEND($\sim B, \gamma$);
18             $k := k+1$; $S^{(k)} := []$;
19   **if** $y = \texttt{false}$
20   **then for** $l \in [2 .. j]$
21   **do** APPEND($\sim S^{(l)}, h$);
22             $H^{(l)} := \langle S^{(l)} \rangle$; $\Delta^{(l)} := \beta_l^{H^{(l)}}$;
23             $c := 0$;
24   **else** $c := c + 1$;
25   $S := \cup_{i=1}^{k} S^{(i)}$;

For small base groups of large degree, however, there are much quicker ways of verifying the correctness of (B, S), which we shall discuss in Chapter 6.

4.4.6 The solvable BSGS algorithm

We shall now describe an alternative method of computing a base and strong generating set in a solvable permutation group G; this method was described by Sims in [Sim90a]. See also Section 7.1 of [Ser03] for a complexity analysis; it turns out that, if we use shallow Schreier trees as described in Subsection 4.4.4, then the algorithm is nearly-linear-time. Sims observed that his implementation ran significantly faster than the standard Schreier-Sims. It can safely be used on groups that are not known for certain to be solvable, because it quickly halts and reports failure when the input group is not solvable.

A further advantage of the algorithm is that the strong generating sequence for G that it computes can be reordered to form a *polycyclic sequence* for G; that is, a generating sequence $[x_1, \ldots, x_r]$ with the property that, if $G_i := \langle x_i, \ldots, x_r \rangle$ for $1 \leq i \leq r+1$, then $G_{i+1} \lhd G_i$ for $1 \leq i \leq r$.

Polycyclic sequences, and the algorithms that use them and their associated polycyclic presentations as a framework for computing in polycyclic groups will form the subject of Chapter 8. Algorithms for finite solvable groups G defined using a polycyclic presentation of G generally perform faster than algorithms to carry out the same computations in a permutation representation of G. So, if we are given a finite solvable permutation group in which we wish to carry out nontrivial structural computations, then it is likely to be worthwhile to use the solvable BSGS algorithm to transfer to a polycyclic presentation of the group.

Another important feature is that the series $\{G_i\}$ of subnormal subgroups of G includes a series of normal subgroups of G with abelian factor groups as a subsequence. This can easily be refined, if required, to a series of normal subgroups with elementary abelian factors. As we shall see in Section 8.5, such series are used in structural computations in finite solvable groups.

We shall describe the method from the bottom up. We start with a procedure NORMALIZINGGENERATOR, which takes a BSGS (B, S) for a permutation group N and a permutation g that normalizes N as input, and extends (B, S) to a BSGS for $\langle g, N \rangle$. We shall prove in Proposition 4.9 that g permutes the orbits of N. So $\beta_1^{\langle N, g \rangle}$ is a union of m orbits of N, including the first basic orbit $\Delta^{(1)} = \beta_1^N$, where $m > 0$ is minimal such that g^m fixes $\Delta^{(1)}$.

Let $h := g^m u^{-1}$, where $u \in N$ satisfies $\beta_1^{g^m} = \beta_1^u$. Then $h \in \langle N, g \rangle_{\beta_1} = \langle N, g \rangle^{(2)}$, and $\langle N, g \rangle_{\beta_1} = \langle N, g^m \rangle_{\beta_1} = \langle N_{\beta_1}, h \rangle$. These considerations justify the procedure NORMALIZINGGENERATOR. In addition to updating the BSGS, a list U of the new strong generators appended to S is returned. It is not hard to show that, if T is a polycyclic sequence for N, then U cat T is a polycyclic sequence for $\langle N, g \rangle$.

The next procedure S-NORMALCLOSURE also takes a BSGS (B, S) for a permutation group N and a permutation w normalizing N as input. In addition, it takes generators X of a larger permutation group G as input, where it is known that $N \lhd G$, and that $w \in G \setminus N$. It starts to compute the normal closure $M := \langle N, w \rangle^G$ of $\langle N, w \rangle$ in G. There are two possible outcomes. If it turns out that M/N is abelian, then the BSGS is extended to a BSGS for M. In addition, a sequence U containing the new strong generators is returned and is ordered in such a way that, if T is a polycyclic sequence for N, then U cat T is a polycyclic sequence for M. If, on the other hand, M/N is not abelian, then a pair of G-conjugates u and v of w such that the commutator $u^{-1}v^{-1}uv \notin N$ is returned.

S-NORMALCLOSURE is reasonably self-explanatory. Notice that we make a copy of the original group N in line 1; this is for the purpose of testing

NORMALIZINGGENERATOR$(g, \sim B, \sim S, \sim \Delta^*)$

    Input: BSGS B, S, Δ^* for $N \leq \mathrm{Sym}(\Omega)$, $g \in \mathbf{N}_{\mathrm{Sym}(\Omega)}(N)$

    Output: B, S, Δ^* are updated to a BSGS for $\langle N, g \rangle$

                 List U of strong generators appended to S is returned

1   $U := []$; $h := g$; $i := 0$;

2   **while** $h \neq 1$

3   **do** $i := i + 1$;

4   **if** $i > k$ (∗ Add new base point ∗)

                then Let $\gamma \in \Omega$ with $\gamma^h \neq \gamma$;

5   APPEND$(\sim B, \gamma)$; $k := k+1$; $S^{(k)} := []$;

6   Let $m > 0$ be minimal with $\beta_i^{h^m} \in \Delta^{(i)}$;

7   Let $u \in H^{(i)}$ with $\beta_i^u = \beta_i^{h^m}$;

8   **if** $m > 1$

9   **then** (∗ Enlarge $S^{(i)}$ and $\Delta^{(i)}$ ∗)

10   APPEND$(\sim S^{(i)}, h)$; $H^{(i)} := \langle S^{(i)} \rangle$; $\Delta^{(i)} := \beta_i^{H^{(i)}}$;

11   APPEND$(\sim U, h)$;

12         $h := h^m u^{-1}$;

13 **return** U;

membership in N and, in an implementation, we would also need to remember the original input versions of B, S, Δ^*. The list C_1 initialized in line 2 consists of the conjugates of w to be processed, whereas C_2 consists of those that have already been processed. Note also that at line 5, a list $[g, h]$ of two group elements is returned, *not* a commutator!

S-NORMALCLOSURE$(w, \sim B, \sim S, \sim \Delta^*, X)$

    Input: BSGS B, S, Δ^* for $N \lhd G \leq \mathrm{Sym}(\Omega)$, $\langle X \rangle = G$, $w \in G \setminus N$

    Output: Either `false`, $[u, v]$, where u, v are G-conjugates of w with

                 $u^{-1}v^{-1}uv \notin N$; or `true`, U where B, S, Δ^* have been updated

                 to generate $M := \langle N, w \rangle^G$ with M/N abelian, and U is a list

                 of the strong generators that were appended to S.

1   $N := \langle S \rangle$;

2   $C_1 := [w]$; $C_2 := []$; $U := []$;

3   **for** $g \in C_1$ **do if** $g \notin \langle S \rangle$

4   **then for** $h \in C_2$ **do if** $g^{-1}h^{-1}gh \notin N$

5   **then return** `false`, $[g, h]$;

6            $V :=$ NORMALIZINGGENERATOR$(g, \sim B, \sim S, \sim \Delta^*)$;

7            $U := V$ cat U;

8   APPEND$(\sim C_2, g)$;

9   **for** $x \in X$ **do** APPEND$(\sim C_1, x^{-1}gx)$;

10 **return** `true`, U;

We turn now to the solvable BSGS algorithm SOLVABLEBSGS. The input consists of a generating set S and a possibly empty partial base B for a permutation group G.

Throughout the procedure, we maintain a BSGS (B, S) for a normal subgroup N of G. We attempt to use S-NORMALCLOSURE to update the current N to $M := \langle N, w \rangle^G$ with M/N is abelian. We try this with w equal to one of the original generators of G. If this fails, then S-NORMALCLOSURE returns $u, v \in G$ such that $w' := u^{-1}v^{-1}uv \notin N$, and we try again with w replaced by w'. We now have $w \in [G, G] = G^{[1]}$ and, after m successive failures of S-NORMALCLOSURE, we have $w \in G^{[m]}$ on the next call of S-NORMALCLOSURE.

By a result proved by Dixon in [Dix68], the derived length of a solvable permutation group of degree n is at most $5 \log_3 n/2$ so, if we get more than this number of consecutive failures of S-NORMALCLOSURE, then we can conclude that the input group G is not solvable and give up. On the other hand, if G is solvable, then we must eventually succeed in enlarging N, and so the procedure will succeed. In this event, a polycyclic sequence for G is returned, which will be a reordering of the strong generating set S of G that has been computed. Note that we have abbreviated S-NORMALCLOSURE to S-NCL at line 8.

SOLVABLEBSGS($\sim B, \sim S$)

    Input: Generators S of G and a partial base B for G
    Output: Either `false`, if G is not solvable or
              `true`, T where T is a polycyclic sequence for G.
              B and S are updated to form a BSGS for G but, unlike in
              SCHREIERSIMS, S is completely recomputed
1   $X := S$; $S := []$; $T := []$;
2 **for** $x \in X$
3   **do while** $x \notin \langle S \rangle$
4   **do** $c := 0$; $w := x$; $a := $ `false`;
5   **while** $a = $ `false`
6   **do** $c := c + 1$;
7   **if** $c > 5 \log_3 n/2$ **then return** `false`;
8                        $a, U := $ S-NCL$(w, \sim B, \sim S, \sim \Delta^*)$;
9   **if** $a = $ `false`
10   **then** $w := U[1]^{-1}U[2]^{-1}U[1]U[2]$;
11                $T := U$ cat T;
12 **return** `true`, T;

Let $T := [x_1, \ldots, x_r]$ be the polycyclic sequence returned by SOLVABLEBSGS, and let $r_i := |G_i/G_{i+1}|$ for $1 \le i \le r$, where the G_i are as defined earlier. If T is to be used in the algorithms to be described in Chapter 8, then we will also need to compute the relations in a polycyclic presentation on this generating set. As we shall see in Chapter 8, to do this it is sufficient to be able to write any element of $g \in G$ as a normal form word $x_1^{e_1} \cdots x_r^{e_r}$ where, for $1 \leq i \leq r$, we have $0 \leq e_i < r_i$. This is not difficult. The exponents e_i are computed and returned by the outline function NORMALFORMEXPONENTS.

NORMALFORMEXPONENTS(g, T, B, S, Δ^*)

1 $E := [];$
2 **for** $i \in [1 .. r]$
   (* At this stage $g \in G_i$ *)
3   **do** Use (B, S, Δ^*) to write g as a word w in $(S \cup S^{-1})^*$;
4   Let e be the total exponent count of x_i in w;
5   APPEND$(\sim E, e)$; $g := x_i^{-e} g$;
6 **return** E;

For this procedure to work as claimed, it is essential that the word computed for $g \in G_i$ at line 3 should involve the generators $x_j^{\pm 1}$ only for $j \geq i$. Provided that the data structure Δ^* is updated from its current value whenever S is extended during SOLVABLEBSGS rather than being completely recomputed each time, this property will hold automatically. However, extra care would need to be taken if extra generators had been adjoined to S in order to achieve the shallow Schreier trees described in Subsection 4.4.4.

4.4.7 Change of base

As we shall see in the following sections of this chapter, virtually all algorithms for computing properties of a permutation group G require a BSGS for G to be known, and in many cases they need, or function more efficiently with, a base for G that might be different from the one found by the Schreier-Sims algorithm. For example, if we want to compute the pointwise stabilizer in G of some points $\alpha_1, \ldots, \alpha_r \in [1 .. n]$, then the best way to proceed is to change the base of G to one beginning with the sequence $[\alpha_1, \ldots, \alpha_r]$, and then the required stabilizer is just $G^{(r+1)}$.

The fundamental procedure BASESWAP for carrying out base changes interchanges two adjacent points β_i and β_{i+1} in the existing base B. This is due to Sims [Sim71b, Sim71a].

To prove correctness of BASESWAP, we need to show that the new S at the end of the procedure is a strong generating set relative to the new base. The only member of the stabilizer-chain that has changed is $G^{(i+1)}$, which needs to be changed to $H := G^{(i)}_{\beta_{i+1}}$, so it is enough to show that the set T, which is adjoined to S in line 10 of the procedure, generates H. Initially $T = S^{(i+2)}$, and the only elements appended to T have the form yx, where $y, x \in G^{(i)}$ and yx fixes β_{i+1}, so we always have $T \subseteq H$.

By the orbit-stabilizer theorem, we have

$$|G^{(i)}| = |\Delta^{(i)}| |\Delta^{(i+1)}| |G^{(i+2)}| = |\beta_{i+1}^{G^{(i)}}| |H|$$

BASESWAP($\sim B, \sim S, \sim \Delta^*, i$)

Input: BSGS (B, S) for $G \leq \text{Sym}(n)$, $i \in [1 .. |B|-1]$

1    $s := |\Delta^{(i)}||\Delta^{(i+1)}|/|\beta_{i+1}^{G^{(i)}}|$;

     ($*$ s is the size of $|\Delta^{(i+1)}|$ after the base change $*$)

2    $T := S^{(i+2)}$; $\Gamma := \Delta^{(i)} \setminus \{\beta_i, \beta_{i+1}\}$;

3   **while** $|\beta_{i+1}^{\langle T \rangle}| \neq s$

4   **do** Choose $\gamma \in \Gamma$, $x \in U^{(i)}$ such that $\gamma = \beta_i^x$;

5   **if** $\beta_{i+1}^{x^{-1}} \notin \Delta^{(i+1)}$

6   **then** $\Gamma := \Gamma \setminus \gamma^{\langle T \rangle}$;

7   **else** Find $y \in G^{(i+1)}$ with $\beta_{i+1}^y = \beta_{i+1}^{x^{-1}}$;

8   **if** $\beta_i^{yx} \notin \beta_i^{\langle T \rangle}$

9   **then** $T := T \cup \{yx\}$; $\Gamma := \Gamma \setminus \beta_i^{\langle T \rangle}$;

10   $S := S \cup T$;

11   Swap β_i and β_{i+1};

12   Remove redundant generators from S;

13   Recalculate $\Delta^{(i)}$, $\Delta^{(i+1)}$, Schreier vectors;

so $|H| = s|G^{(i+2)}|$, where s is as defined in line 1 of the procedure. Hence $s = |\beta_{i+1}^H|$. Since $G^{(i+1)} \subseteq \langle T \rangle$, and we do not exit the main 'While' loop at line 3 until $|\beta_{i+1}^{\langle T \rangle}| = s$, we must have $\langle T \rangle = H$ when we exit the 'While' loop, so the question is whether the elements appended to T during the 'While' loop are enough to generate H.

A general element of $G^{(i)}$ can be written as yx for $y \in G^{(i+1)}$ (that is, the original $G^{(i+1)} = G_{\beta_i}^{(i)}$), and $x \in U^{(i)}$. For a given $x \in U^{(i)}$, there exists such an element $yx \in H$ if and only if there exists $y \in G^{(i+1)}$ with $\beta_{i+1}^{yx} = \beta_{i+1}$, which is the case if and only if $\beta_{i+1}^{x^{-1}} \in \beta_{i+1}^{G^{(i+1)}} = \Delta^{(i+1)}$. This explains the subdivision in the 'If' statement at line 5.

If $y_1 x$ and $y_2 x$ are two elements of H of this form, then $(y_1 x)(y_2 x)^{-1} = y_1 y_2^{-1} \in G^{(i+2)} \subseteq \langle T \rangle$, and so, for a fixed x, we only need to include one such element in T, which is what we do in the procedure. Furthermore, we do not need to adjoin any element yx for which $\beta_i^{yx} = \beta_i^t$ for some t that is already in $\langle T \rangle$; this justifies the removal of the elements of $\beta_i^{\langle T \rangle}$ from Γ in line 9.

On the other hand, if there is no element $yx \in H$ for the given x, then neither can there be such an element of the form yxt for any t already in $\langle T \rangle$; this justifies us in removing $\gamma^{\langle T \rangle}$ from Γ in line 6.

A variation on this method of swapping two base points, which is discussed in [Ser03], is to produce random elements of the subgroup H by calculating a Schreier vector v for $\beta_{i+1}^{G^{(i)}}$, and then applying the algorithm RANDOMSTAB($\beta_{i+1}, v, S^{(i)}$) repeatedly. We can use the same halting condition $|\beta_{i+1}^{\langle T \rangle}| = s$ as before.

A simpler method of changing bases, also noted by Sims in [Sim71b], is simply to conjugate by some $g \in G$, thereby changing the base from $[\beta_1, \ldots, \beta_k]$ to $[\beta_1^g, \ldots, \beta_k^g]$. To make an arbitrary base change, we can proceed as follows. Suppose that we want to replace β_1 to α for some α. If $\alpha \in \Delta^{(1)}$, then $\beta_1^g = \alpha$ for some $g \in G$, and we can do it by conjugating by g^{-1}. Otherwise, we can suppose, recursively, that we can replace the second base point β_2 by α (or, if $k = 1$, simply introduce α as a new redundant base point β_2), and then we apply BASESWAP($\sim S, \sim B, 1$).

In [CF92], Cooperman and Finkelstein describe an algorithm for a cyclic base change:

$$(b_1, \ldots, b_k, c) \to (c, b_1, \ldots, b_k).$$

It has been observed by several authors that, if an initial segment of the new base is very different from that of the original base, then it can be better to reconstruct the BSGS data structures from scratch, using RANDOMSCHREIER. In [CFS90], Cooperman, Finkelstein, and Sarawagi prove that a random base change algorithm can be described, which has probability $1 - \frac{1}{|G|}$ of constructing a 'short' Schreier vector data structure relative to a given base B after processing at most $20 \log_2 |G|$ random elements of G.

A problem that needs to be addressed when base changes are being carried out is how to keep track of the SLP elements that express the strong generators as words in the original group generators. If there is enough space available, then the simplest possibility is always to keep a copy of the BSGS calculated originally by SCHREIERSIMS or by RANDOMSCHREIER.

Exercises

1. Assuming the existence of a BSGS (B, S) for G and the data structure Δ^* as described above, write a function which, given a sequence C of k distinct points of Ω, either finds $g \in G$ with $B^g = C$, or reports that no such g exists.

2. Let $G \le \mathrm{Sym}(\Omega)$, Δ an orbit of G with $|\Delta| = m$, $\Gamma \subseteq \Delta$ with $|\Gamma| = l$, and let $g \in G$ be random, with uniform distribution on G. Prove that the expected value of $|\Gamma^g \setminus \Gamma|$ is $l(m - l)/m$.

3. Suppose that $g \in G$ is random with uniform distribution on G, H is a subgroup of G, and T is right transversal of H in G. Denote the element of $T \cap Hg$ by $\overline{g}$. Show that $g\overline{g}^{-1}$ is a uniformly distributed random element of H.

4. Suppose that (B, S) is not a BSGS for G, and let $g \in G$ be random with uniform distribution on G. Show that STRIP returns a nonidentity element with probability at least $1/2$.

5. Prove the claim that, if T is a polycyclic sequence for the input group N to NORMALIZINGGENERATOR, and U is the sequence returned, then U cat T is a polycyclic sequence for $\langle N, g \rangle$.

4.5   Homomorphisms from permutation groups

In this section, we study computations involving homomorphisms $\varphi : G \to H$, in the case when G is a finite permutation group. The reader should first review the general discussion of computation with group homomorphisms in Section 3.4.

Assuming that we have used SCHREIERSIMS or RANDOMSCHREIER (and possibly also SHALLOWORBITSV) to compute a BSGS (B, S) for G, and that we have taken care to store SLP elements that express the strong generators in S in terms of the original generators X of G, we can define group homomorphisms $\varphi : G \to H$ by specifying the images $\varphi(x)$ for $x \in X$. The images $\varphi(x)$ for $x \in S \setminus X$ can then be computed (and stored) by using their SLP expressions, as described above in Section 3.1.3. Now, the algorithm STRIP can be used to express an arbitrary group element $g \in G$ as a word in the strong generators, thereby enabling $\varphi(g)$ to be computed.

If membership testing and rewriting algorithms are available for subgroups of H, then we can efficiently carry out all of the desirable calculations involving homomorphisms that were discussed in Subsection 3.4.2. If H is a permutation group and we are using IMAGEKERNEL to compute the orders of the image and the kernel, then we can safely use RANDOMSCHREIER without verification for solving the rewriting and membership problem in $\operatorname{im}(\varphi)$, because IMAGE-KERNEL does not terminate until it has checked that $|\ker(\varphi)| |\operatorname{im}(\varphi)| = |G|$, which provides automatic verification that the order $|\operatorname{im}(\varphi)|$ is correct. (If RANDOMSCHREIER returns an incorrect answer, then it will always report that the group is smaller than it really is, never larger.)

In addition, if we can compute a presentation of G on S, then we can test whether the images $\varphi(x)$ for $x \in X$ really do define a homomorphism $\varphi : G \to H$. Finding such presentations will be handled in detail later, in Chapter 6.

There are, however, two specific and frequently occurring situations in which we can carry out some of these operations more efficiently than we can by the general methods. These are the induced action of G on an orbit of G^Ω (or, more generally, on a nonempty union of orbits), and the induced action of G on a block system preserved by G^Ω. Furthermore, there is an alternative approach to general group homomorphism computations for the case when H is a permutation group, which is probably no faster than the general method, but is interesting enough to merit consideration.

4.5.1   The induced action on a union of orbits

Computing with the induced orbit action $\varphi : G \to G^\Psi$ for a nonempty union Ψ of orbits of G^Ω is straightforward. We first rename the elements of Ψ as $1, 2, 3, \ldots$, so that $\varphi \leq \operatorname{Sym}(|\Psi|)$. Calculating $\varphi(g)$ for $g \in G$ is then just a matter of applying the renumbering to the action of g on Ψ.

We then change base so as to have as many base points as possible in Ψ. More precisely, we change to a base $[\beta_1, \ldots, \beta_k]$ in which, for some j, $\beta_i \in \Psi$ if and only if $1 \leq i \leq j$, and the $(j+1)$-th basic stabilizer $G^{(j+1)}$ fixes all points of Ψ. We then have $G^{(j+1)} = \ker(\varphi)$. The images $\varphi(g)$ of the strong generators of G that do not lie in $G^{(j+1)}$ form a strong generating set for $\operatorname{im}(\varphi)$ with respect to the base $[\varphi(\beta_1), \ldots, \varphi(\beta_j)]$. We can therefore test permutations for membership of $\operatorname{im}(\varphi)$. To compute an inverse image $\varphi^{-1}(h)$ for $h \in \operatorname{im}(\varphi)$, we can use STRIP in $\operatorname{im}(\varphi)$ to express h as a word in the strong generators of $\operatorname{im}(\varphi)$, but since these strong generators are just $\varphi(g)$ for $g \in S$, the same word defined over S will evaluate to $g \in G$ with $\varphi(g) = h$. We can therefore carry out all of the desired types of calculations with φ.

4.5.2   The induced action on a block system

Let us turn now to the induced action of a transitive group G^Ω on a block system Γ preserved by G^Ω, where $\Omega = \{1 \ldots n\}$. We assume that the block system is defined by an associated partition function p, for which $p[\alpha] \in \Omega$ is the representative point in Ω of the block containing α, so $p[p[\alpha]] = p[\alpha]$ for all $\alpha \in \Omega$. The function MINIMALBLOCK returns such a p.

We first want to number the blocks $1, \ldots, m$, where m is the number of blocks. This is accomplished by the following function, which returns m and lists b, ρ, where for, $\alpha \in \Omega$, $b[\alpha]$ is the number of the block containing α and, for $i \in \{1 \ldots m\}$, $\rho(i)$ is the representative in Ω of block number i.

NUMBERBLOCKS(p)

> **Input**: Partition function p as above
> **Output**: m, b, ρ as described above

1    $m := 0$;
2   **for** $i \in [1 \ldots n]$ **do if** $p[i] = i$
3   **then** $m := m + 1$; $b[i] := m$; $\rho[m] := i$;
4   **for** $i \in [1 \ldots n]$ **do** $b[i] := b[p[i]]$;
5   **return** m, b, ρ;

The induced action $\varphi : G \to G^\Gamma$ can now be calculated easily by the rule $i^{\varphi(g)} = b[\rho[i]^g]$ for $i \in \{1 \ldots m\}$ and $g \in G$.

PROPOSITION 4.6   *Let G^Ω preserve block system Γ, let $H \leq G$, $\alpha \in \Omega$, and let Δ be the block of Γ containing α. For each $\beta \in \alpha^H$, let u_β be an element of H with $\alpha^{u_\beta} = \beta$. Then the stabilizer H_Δ of Δ in H is generated by H_α together with $\{u_\beta \mid \beta \in \alpha^H \cap \Delta\}$.*

PROOF   Each element of H_α and each u_β for $\beta \in \alpha^H \cap \Delta$ maps a point in Δ to a point in Δ and hence, by the definition of a block system, it must lie in H_Δ. Conversely, if $g \in H_\Delta$ and $\alpha^g = \beta$, then $g = gu_\beta^{-1}u_\beta$ with $gu_\beta^{-1} \in H_\alpha$.

∎

The above result justifies the function BLOCKSTABILIZER for computing the stabilizer of the block containing α in the subgroup $H = \langle Y \rangle$ of G. It makes use of the array b returned by NUMBERBLOCKS. We start by changing base of H to have α as the first point. This enables generators of the stabilizer H_α and the elements u_β to be computed by the standard BSGS functions.

BLOCKSTABILIZER(Y, α, b)

    Input: Generators Y of $H \leq G$, $\alpha \in \Omega$, array b for block action
    Output: Generators of H_Δ for block Δ containing α
1   Change base of H to make α the first base point;
2   Let Z be the set of generators of H_α;
3   **for** $\beta \in \alpha^H$ **do if** $b[\beta] = b[\alpha]$ **and** $\beta \notin \alpha^{\langle Z \rangle}$
4   **then** APPEND$(\sim Z, u_\beta)$, where $u_\beta \in H$ with $\alpha^{u_\beta} = \beta$;
5   **return** Z;

The function BLOCKIMAGEKERNEL calculates a BSGS (B_Γ, S_Γ) for $\text{im}(\varphi)$, together with $\ker(\varphi)$. It uses a BSGS (B, S) for G.

BLOCKIMAGEKERNEL(G, b)

    Input: array b for block action $\varphi : G \to G^\Gamma$
    Output: B_Γ, S_Γ, $\ker(\varphi)$
1    $X :=$ initial generators of G;
2    $B_\Gamma := []$; $S_\Gamma := [\varphi(x) \mid x \in X, \varphi(x) \neq 1]$;
3    $i := 1$; Choose $\alpha_i \in \Omega$;
4   **while true**
5   **do** APPEND$(\sim B_\Gamma, b[\alpha_i])$;
6   Change BSGS (B, S) of G to make α_i the i-th base point;
7            $Z :=$ BLOCKSTABILIZER$(S^{(i)}, \alpha_i, b)$;
8            $\overline{Z} := [\varphi(x) \mid x \in Z, \varphi(x) \neq 1]$;
9   **if** $\overline{Z}$ is empty
10   **then return** $B_\Gamma, S_\Gamma, \langle Z \rangle$;
11   **else** $S_\Gamma := S_\Gamma$ cat $\overline{Z}$; $i := i+1$;
12   Choose $\alpha_i \in \Omega$ with $b[\alpha_i]$ not fixed by all $\varphi(x) \in \overline{Z}$;

As was the case in IMAGEKERNEL we would also need to keep a record of which $g \in G$ were used to define the strong generators $\varphi(g)$ of G^Γ, since that information is needed to compute inverse images under φ.

4.5.3 Homomorphisms between permutation groups

We now turn to a method for computing with $\varphi : G \to H$ when both G and H are finite permutation groups, and when $\varphi(x)$ is known for x in a generating set X of G. This is an alternative to the general methods described in Subsection 3.4.2.

Let $\varphi : G \to H$ be a homomorphism, where $G \leq \mathrm{Sym}(\Omega)$ and $H \leq \mathrm{Sym}(\Psi)$ with $\Omega \cap \Psi = \phi$. Let

$$\hat{G} := \{ (g, \varphi(g)) \mid g \in G \} \subseteq \mathrm{Sym}(\Omega) \times \mathrm{Sym}(\Psi).$$

The fact that φ is a homomorphism implies immediately that $\hat{G}$ is a subgroup of $\mathrm{Sym}(\Omega) \times \mathrm{Sym}(\Psi)$ with $\hat{G} \cong G$, and clearly $\hat{G}$ is generated by $\hat{X} := \{ (x, \varphi(x)) \mid x \in X \}$, where X generates G.

Of course, in practice, we are likely to be given $\Omega = \{1..n\}$ and $\Psi = \{1..m\}$, for some m, n, which are not disjoint, but we can easily renumber the points of Ψ to $\{n+1..n+m\}$, and then we can construct the set $\hat{X}$ explicitly in its natural action on $\{1..n+m\}$. The fact that $\hat{X}$ generates $\hat{G}$, which clearly has the same order as G, and that a base of G^Ω is also a base of $\hat{G}^{\Omega \cup \Psi}$ means that we can use the known base and known order versions of RANDOMSCHREIER to rapidly compute a BSGS for $\hat{G}$, assuming that a BSGS is known for G.

The homomorphism φ can now be identified with the induced action of $\hat{G}$ on the orbit Ψ of $\hat{G}^{\Omega \cup \Psi}$, and so we can use the methods described above in Subsection 4.5.1 to perform any desired computations with φ.

We may also wish to check that the map $\varphi : X \to H$ really does extend to a homomorphism from G to H. It turns out that this is the case if and only if a base for G^Ω is a base for the subgroup of $\hat{G}^{\Omega \cup \Psi}$ generated by $\hat{X}$; see the exercise below. We can check this condition using SCHREIERSIMS, but unfortunately we can no longer use the known base and known order versions. We could, however, use a version of SCHREIERSIMS that aborted as soon as it discovered that the initial base for G^Ω was not a base for $\langle \hat{X} \rangle$.

Exercise

1. Let $G = \langle X \rangle \leq \mathrm{Sym}(\Omega)$, $H \leq \mathrm{Sym}(\Psi)$ with $\Omega \cap \Psi = \phi$, and let $\varphi : X \to H$ be any map. Prove that φ extends to a homomorphism from G to H if and only if the subgroup $H := \langle (x, \varphi(x)) \mid x \in X \rangle$ of $\mathrm{Sym}(\Omega) \times \mathrm{Sym}(\Psi)$ satisfies $H_{(\Omega)} = 1$. (*Hint*: Use Theorem 2.52.)

4.6 Backtrack searches

The algorithms to be described in this section potentially involve searching through all of the elements of a permutation group G of degree n, and so their complexity is at least $O(|G|)$ in the worst case, which is worse than polynomial in n in general. It is crucial in designing such algorithms to find methods for skipping large numbers of the group elements during the search; we do this using a technique known as *pruning the search tree*. In favourable cases, this can lead to implementations of these algorithms running very fast, particularly on small examples. So, for example, in the early days of CGT, when typical examples under consideration had much smaller degrees than they do today, programs for computing centralizers of elements and intersections of two subgroups in permutation groups acquired the reputation of being fast.

Unfortunately, now that we are accustomed to dealing with groups of larger degree, this worse-than-polynomial behaviour has become noticeable, and algorithms for computing the centralizer of an element of order 2, for example, can be very slow. In fact the exponential (or worse) complexity can mean that a relatively small increase in degree could lead to a doubling of process times, and so for very large degrees these computations can become completely impractical. For this reason, considerable effort has been devoted to finding algorithms that are polynomial of as small degree as possible in n, for as many problems as possible. Unfortunately, there are some problems, including centralizers and normalizers of elements and subgroups, stabilizers of subsets of Ω, and intersections of subgroups, for which backtrack searching appears to be the only possible approach. Since these are important problems, we need to devote our efforts to making these searches run as efficiently as possible.

The reader who is interested in seeing a summary of which calculations in permutation groups are currently known to be possible in polynomial time could consult the survey article by Luks [Luk93]. Alternatively, [Ser03] contains complete information on complexity issues, including detailed proofs.

Throughout this section, we shall assume that a BSGS (B, S) with $B = [\beta_1, \ldots, \beta_k]$ is known for the permutation group G, and we shall use all of the notation that was introduced in Subsection 4.4.1. For $0 \leq l \leq k$, we shall denote the initial segment $[\beta_1, \ldots, \beta_l]$ of B by $B(l)$.

It is convenient to introduce an ordering $\prec$ on $\Omega = [1 \mathinner{.\,.} n]$ in which the base elements come first, and in order. That is, $\beta_i \prec \beta_j$ for $i < j$, and $\beta_i \prec \alpha$ if $\alpha \notin B$. By the definition of a base, an element $g \in G$ is uniquely determined by its base image $B^g = [\beta_1^g, \ldots, \beta_k^g]$, and so we can order the elements of G by their base images. That is, for $g, h \in G$, we define $g \prec h$ if B^g precedes B^h in the lexicographical ordering induced by $\prec$. To be precise, $g \prec h$ if and only if, for some l with $1 \leq l \leq k$, we have $\beta_i^g = \beta_i^h$ for $1 \leq i < l$ and $\beta_l^g \prec \beta_l^h$.

Notice that the definition of $\prec$ ensures that the identity element of G comes first and, for any l with $1 \leq l \leq k$, the elements of the l-th basic stabilizer $G^{(l)}$ precede those of $G \setminus G^{(l)}$.

Recall from Subsection 4.4.1 that each $g \in G$ has a unique representation as $g = u_k u_{k-1} \cdots u_1$ with $u_i \in U^{(i)}$, where $U^{(i)}$ is a right transversal of $G^{(i+1)}$ in $G^{(i)}$ for $1 \leq i \leq k$. The algorithms to be discussed in this section require us to consider representations of large numbers of group elements in this form.

The performance of implementations of these algorithms is highly dependent on a sensible choice of data structure for this representation of g. If there is enough space available, then the elements of $U^{(i)}$ should be precomputed rather than recomputed using Schreier vectors each time they are needed. The elements of $U^{(i)}$ for larger values of i are needed much more frequently than those for smaller values, so a sensible default decision is to precompute the $U^{(i)}$ for as many i as there is space available, but starting with $U^{(k)}$. It is also important to make sensible decisions as to when to keep permutations g as a permutation word, and when to multiply g out. A possible strategy here is to multiply g out only when it is to be adjoined as a new generator of a subgroup being computed, or if it needs to be returned by a program. But in some cases it may be faster to multiply out the subwords $u_l u_{l-1} \cdots u_1$, for smaller values of l, since these are changed relatively infrequently during the search algorithms. In our displayed code, we shall denote $u_l u_{l-1} \cdots u_1$ by $g[l]$ for $1 \le l \le k$, without any assumptions about whether the elements u_i or the $g[l]$ are stored as permutation words or as complete permutations. Note that the images $B(l)^g$ of the first l base points are determined by $g[l]$.

4.6.1   Searching through the elements of a group

We start by presenting a procedure PRINTELEMENTS that merely prints out each group element. This can be done simply by running through all possibilities for $u_l u_{l-1} \cdots u_1$ in turn. Although it complicates the code slightly, we shall see later that there are significant advantages in generating the group elements in increasing order of base images; that is, in the order $\prec$ of G defined above. In particular, it is important that, for each l, the elements of $G^{(l)}$ come before those of $G \setminus G^{(l)}$. This same basic method of looping through the group elements is used for subsequent, more ambitious, procedures.

PRINTELEMENTS(G)

    Input: Permutation group G with BSGS (B, S, Δ^*)

    ($*$ $g[l]$ will always denote the element $u_l u_{l-1} \cdots u_1$ of G $*$)

1   $l := 1$; $c[l] := 1$; $\Lambda[l] := $SORT$(\Delta^{(l)}, \prec)$; $u_l := 1_G$;

2 **while true**

3   **do while** $l < k$

4   **do** $l := l+1$; $\Lambda[l] := $ SORT$((\Delta^{(l)})^{g[l-1]}, \prec)$;

5                     $c[l] := 1$; $\gamma := \Lambda[l][c[l]]^{g[l-1]^{-1}}$; $u_l := u_\gamma^{(l)}$;

6   PRINT $g[l]$;

7   **while** $l > 0$ **and** $c[l] = |\Delta^{(l)}|$ **do** $l := l-1$;

8   **if** $l = 0$ **then return**;

9          $c[l] := c[l]+1$; $\gamma := \Lambda[l][c[l]]^{g[l-1]^{-1}}$; $u_l := u_\gamma^{(l)}$;

Notice that $\Lambda[l]$, as defined on lines 1 and 4 of the procedure, is set equal to the image of the l-th basic orbit $\Delta(l)$ under $g[l-1]$, and is sorted into increasing order under $\prec$. This enables us to find u_l at line 9 that defines the next group element under the ordering $\prec$; recall from Subsection 4.4.1 that, for $\gamma \in \Delta(l)$, $u_\gamma^{(l)}$ denotes the element of $U^{(l)}$ which maps β_l to γ.

Example 4.4

Let

$$G := \langle (1,2,3,4), (2,4), (5,6) \rangle < \mathrm{Sym}(6)$$

(so $G \cong D_8 \times C_2$ has order 16), $B = [1,2,5]$, $U^{(1)} = \{ (), (1,2,3,4), (1,3)(2,4), (1,4,3,2) \}$, $U^{(2)} = \{ (), (2,4) \}$, $U^{(3)} = \{ (), (5,6) \}$. Then the permutations (in vector format) are output in the order

[123456], [123465], [143256], [143265], [214356], [214365], [234156], [234165],

[321456], [321465], [341256], [341265], [412356], [412365], [432156], [432165].

⬜

The diagram below depicts the run of PRINTELEMENTS on this example. The tree is traversed during the run in order of a depth-first search, starting at vertex $(1,1)$ (that is, vertex 1 at level 1).

At the top of each iteration of the main 'While' loop at line 2, we are at one of the vertices of the diagram, where the level is specified by l. The current group element $g[l] = u_l \cdots u_1$ is equal to $u_{\gamma_l}^{(l)} \cdots u_{\gamma_1}^{(1)}$, where the γ_i are the labels on the edges joining the top vertex to the current vertex. At that point in the procedure, u_i is defined only for $1 \leq i \leq l$, but this is enough to specify uniquely the images $B(l)^g$ of the first l base elements under any group element $g = u_k \cdots u_1$ that has these values of u_i for $1 \leq i \leq l$.

For example, if the current vertex is $(2,5)$ (that is, vertex number 5 on level 2), then $u_1 = u_3^{(1)} = (1,3)(2,4)$, $u_2 = u_4^{(2)} = (2,4)$, and u_3 is undefined. But, since $u_3 \in G^{(3)} = G_{12}$, we know that $1^g = 1^{u_1} = 3$, and $2^g = 2^{u_2 u_1} = 2$ for any group element g represented by a vertex below $(2,5)$ in the tree. The vertices at the bottom layer correspond to the group elements themselves.

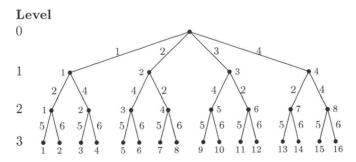

An important special case is when G is the full symmetric group $\mathrm{Sym}(n)$. This occurs, for example, when computing automorphism groups of combinatorial structures, which will be discussed briefly in Subsection 4.6.4, below.

SYMMETRICGROUPELEMENTS(n)

    Input: $n \in \mathbb{N}$
1   **for** $i \in [1 .. n]$ **do** $p[i] := i$;
2   **while** true
3   **do** PRINT $p \in \mathrm{Sym}(n)$;
4         $i := n - 1$;
5   **while** $i > 0$ **and** $p[i] > p[i{+}1]$
6   **do** $i := i{-}1$;
7   **if** $i = 0$ **then break**;
8         $j := n$;
9   **while** $p[i] > p[j]$
10   **do** $j := j{-}1$;
11         $s := p[i]$; $p[i] := p[j]$; $p[j] := s$;
12         $i := i{+}1$; $j := n$;
13   **while** $i < j$
14   **do** $s := p[i]$; $p[i] := p[j]$; $p[j] := s$;
15            $i := i{+}1$; $j := j{-}1$;

Here, we do not need to store or use the transversals $U^{(i)}$. We would like to run through the complete set of permutations of $[1 .. n]$ in lexicographical order, starting with the identity $[1, 2, \ldots, n]$ and ending with the order reversing permutation $[n, n-1, \ldots, 1]$. This is the same as the order defined by base images with respect to the base $[1, \ldots, n-1]$ of G. The problem is to find the permutation immediately following the current permutation $g \in \mathrm{Sym}(n)$ under this ordering. This can be done efficiently as follows.

Look at n^g, $(n-1)^g$, etc., until the sequence stops increasing, at i^g, say. For example, for $g = [6147532] \in \mathrm{Sym}(7)$, the images $2, 3, 5, 7, 4$ stop increasing at $i = 3$, where $i^g = 4$. We then look again at n^g, $(n-1)^g$, until we reach a number j^g larger than i^g. In the example, we stop at $j = 5$, where $5^g = 5 > i^g$. To compute the permutation g' immediately following g in the lexicographical ordering, we replace i^g by j^g, and then put the images $k^{g'}$ for $k > i$ into increasing order. So $g' = [6152347]$ in the example. An efficient way to achieve this is first to interchange i^g and j^g, which will have the effect of putting k^g for $k > i$ into decreasing order, and then to reverse that sequence. This is carried out in the procedure SYMMETRICGROUPELEMENTS. The permutation g is stored in the array p, and the variable s is used for swapping the values of $p[i]$ and $p[j]$.

Of course $|\mathrm{Sym}(n)| = n!$ grows very rapidly with n, and it would be unusual to attempt to search through the whole of $\mathrm{Sym}(n)$ for n larger than about 12 or 13. As we shall see in the next section, in most applications we try to avoid searching through all of the elements of a permutation group G, by making larger jumps forwards during the search. Making that possible would necessitate modifications to the above procedure.

4.6.2   Pruning the tree

Usually we are looking for a subset of G consisting of elements that satisfy some specific property P. Often, the group will be much larger than this subset, so it is essential to find methods of avoiding having to consider each group element individually. Fortunately, it is often possible to prove that an element $g \in G$ cannot satisfy P from a knowledge only of α^g for a few values of α. We can use that information by choosing the α in question to be base points.

More precisely, for l with $1 \leq l \leq k$, let us suppose that we have test functions $\text{TEST}(g, l)$ available, such that if $\text{TEST}(g, l)$ returns $\texttt{false}$, then there is no element h of G satisfying P for which $\beta_i^g = \beta_i^h$ for $1 \leq i \leq l$. In other words, TEST can rule elements out just by considering the first l base images. If we are at a vertex v at level l of the search tree, then β_i^g is defined at v for $1 \leq i \leq l$, and so we can apply $\text{TEST}(g, l)$. If it returns $\texttt{false}$, then we can immediately proceed to the next vertex at level l, and we do not need to consider any of the vertices below v. So we have pruned the tree below v.

We shall call the sequence $B(l)^g = [\beta_1^g, \ldots, \beta_l^g]$, for some l with $1 \leq l \leq k$, a *partial base image*. Then $\text{TEST}(g, l)$ must work using only a partial base image for g. We stress that TEST is a negative test only, and is used for ruling out candidates. If TEST returns $\texttt{true}$, then it does not imply that there exists g with these partial base images that satisfies P; it provides no information on that question.

The next procedure GENERALSEARCH assumes that we have a property P that can be evaluated on group elements $g \in G$, together with such tests $\text{TEST}(g, l)$ which are capable of returning $\texttt{false}$ from a knowledge of $B(l)^g$ only. Of course, in practice, such tests might only be available for some values of l, in which case we would make $\text{TEST}(g, l)$ always return $\texttt{true}$ for other values. Once again, the procedure as written simply prints the elements that satisfy the property. In some applications, we might only want a single element satisfying P, in which case we could modify the procedure to return such an element and stop as soon as it found one.

To take a rather artificial example, suppose, in Example 4.4, that we only want to print out just those elements $g \in G$ for which $1^g = 1$ or 3, and $2^g = 2$. Then $\text{TEST}(g, 1)$ can test for the first of these conditions, and $\text{TEST}(g, 2)$ for the second. The vertices of the tree will be traversed in the order shown in the table below, with the action specified.

Notice that if, instead, we were looking for an element $g \in G$ with $4^g = 1$ or 2, and $3^g = 3$ then, with the base $[1, 2, 5]$ for G, knowledge of partial base images would not be helpful in pruning the search tree. The solution, of course, would be to change the base of G to make $[4, 3]$ an initial segment of the base; $B = [4, 3, 5]$ would be a suitable base. The general point to be made here is that before embarking on any particular search, the base points of G should be chosen, in turn, to allow as many negative tests $\text{TEST}(g, l)$ as possible.

GENERALSEARCH(G)

> **Input**: Permutation group G with BSGS, property P, tests $\text{TEST}(g, l)$
> ($\ast$ $g[l]$ will always denote the element $u_l u_{l-1} \cdots u_1$ of G $\ast$)

1    $l := 1$; $c[l] := 1$; $\Lambda[l] := \text{SORT}(\Delta^{(1)}, \prec)$; $u_l := 1_G$;

2   **while true**

3   **do** ($\ast$ Test if base images are possible at current level $\ast$)

4   **while** $l < k$ **and** $\text{TEST}(g, l)$

5   **do** $l := l+1$; $\Lambda[l] := \text{SORT}((\Delta^{(l)})^{g[l-1]}, \prec)$;

6                  $c[l] := 1$; $\gamma := \Lambda[l][c[l]]^{g[l-1]^{-1}}$; $u_l := u_\gamma^{(l)}$;

7   **if** $l = k$ **and** $\text{TEST}(g, k)$ **and** $P(g[k])$ **then** PRINT $g[k]$;

8   **while** $l > 0$ **and** $c[l] = |\Delta^{(l)}|$ **do** $l := l-1$;

9   **if** $l = 0$ **then return**;

10         $c[l] := c[l]+1$; $\gamma := \Lambda[l][c[l]]^{g[l-1]^{-1}}$; $u_l := u_\gamma^{(l)}$;

Vertex	Action
(1,1)	$\text{TEST}(g, 1) = \texttt{true}$
(2,1)	$\text{TEST}(g, 2) = \texttt{true}$
(3,1)	$P(g) = \texttt{true}$, print $g = [123456]$
(3,2)	$P(g) = \texttt{true}$, print $g = [123465]$
(2,2)	$\text{TEST}(g, 2) = \texttt{false}$
(1,2)	$\text{TEST}(g, 1) = \texttt{false}$
(1,3)	$\text{TEST}(g, 1) = \texttt{true}$
(2,5)	$\text{TEST}(g, 2) = \texttt{true}$
(3,9)	$P(g) = \texttt{true}$, print $g = [321456]$
(3,10)	$P(g) = \texttt{true}$, print $g = [321465]$
(2,6)	$\text{TEST}(g, 2) = \texttt{false}$
(1,4)	$\text{TEST}(g, 1) = \texttt{false}$

4.6.3   Searching for subgroups and coset representatives

In many situations, the subset of elements in the group G for which we are searching forms a subgroup S of G. For example, we might want to find a Sylow p-subgroup for some prime p, or the centralizer or normalizer of a given subgroup. In that case, our aim is to find generators of S rather than each individual element of S.

At any stage during the course of the search of the tree, we will have found a subgroup K of S. Each time we find a new element g in K, we replace the current value of K by $\langle K, g \rangle$. The order of search ensures that generators of $K^{(i)}$ will be found before elements of $G \setminus G^{(i)}$ are considered for $2 \leq i \leq k$, and so the generating set of K computed will automatically be a strong generating set.

Since, for $k_1, k_2 \in K$ and $g \in G$, we have $\langle K, g \rangle = \langle K, k_1 g k_2 \rangle$, we need only adjoin a single element of the double coset KgK as a new generator. So we can restrict the number of elements that we need to consider during the search by rejecting all elements $g \in G$ that can be proved not to be minimal in their double coset KgK under the ordering $\prec$.

Since, at a node at level l in the search tree, we only know the partial base image $B(l)^g$ of the element g under consideration, we do not usually have enough information to decide conclusively whether g is minimal in its double coset. However, some useful tests for nonminimality of g are described in Sims' papers [Sim71b, Sim71a]. See also Leon's paper on automorphism groups of combinatorial objects, which contains pseudocode, and Section 9.1.1 of [Ser03]. The proof of the following result is straightforward, and is left to the reader.

PROPOSITION 4.7   *Let* $K < G \leq \mathrm{Sym}(\Omega)$, *let* $B := [\beta_1, \ldots, \beta_k]$ *be a base of* G, *and let* $\Delta^{(j)}$ *and* $\Delta_K^{(j)}$ *($1 \leq j \leq k$) be the basic orbits of* G *and* K *with respect to* B. *Let* $g \in G$, *let* $\gamma_j := \beta_j^g$ *for* $1 \leq j \leq k$, *and let* $\prec$ *be the ordering of* Ω *and of* G *defined by the base* B, *as described above. Then*

(i) g *is the* $\prec$-*least element of its coset* gK *if and only if* γ_j *is the* $\prec$-*least element of its orbit under* $K_{\gamma_1, \ldots, \gamma_{j-1}}$ *for* $1 \leq j \leq k$;

(ii) g *is the* $\prec$-*least element of its coset* Kg *if and only if* γ_j *is the* $\prec$-*least element of* $(\Delta_K^{(j)})^g$ *for* $1 \leq j \leq k$.

At a vertex at level l in the search tree, we can apply (i) of this result by checking whether γ_l is the least element of its $K_{\gamma_1, \ldots, \gamma_{l-1}}$ orbit. To do this, we need to continually change the base B_K of K to ensure that, at a vertex at level l, B_K has initial segment equal to the current partial base image $B(l)^g = [\gamma_1, \ldots, \gamma_l]$. This enables us to compute the orbits of the basic stabilizer $K_{\gamma_1, \ldots, \gamma_{l-1}}$ of K.

Since changing base carries with it a nontrivial cost, it may, on occasion, be faster to carry out the base changes and associated minimality test based on condition (i) only at the higher levels of the search tree, even if this means traversing more nodes of the tree at the lower levels. Note, however, that each individual base change involves only a single application of BaseSwap. In the code for SubgroupSearch below, we have applied this minimality test at all levels.

Condition (ii) or Proposition 4.7 can be applied most readily at level l of the search tree if $\beta_l \in \Delta_K^{(j)}$ for some $j < l$, because then we have $\gamma_l = \beta_l^g \in (\Delta_K^{(j)})^g$. So, if g is least in its coset Kg, then $\gamma_l \succ \gamma_j$, which is a testable condition. In practice, and in SubgroupSearch, we test the condition

$$\gamma_l \succ \mathrm{Max}\{\, \beta_j^g \mid 1 \leq j < l, \ \beta_l \in \Delta_K^{(j)} \,\}.$$

The following corollary of condition (ii) yields another testable condition.

COROLLARY 4.8   *With the notation of Proposition 4.7, if g is the $\prec$-least element of its coset Kg, then γ_l cannot be among the $|\Delta_K^{(l)}| - 1$ $\prec$-largest elements of its orbit under $G_{\gamma_1,\ldots,\gamma_{l-1}}$.*

PROOF   If g is the $\prec$-least element of Kg then, by Proposition 4.7, γ_l is the $\prec$-least element of $(\Delta_K^{(l)})^g = \{\,\beta_l^{hg} \mid h \in K_{\beta_1,\ldots,\beta_{l-1}}\,\}$. Since $hg = g(g^{-1}hg)$ with $g^{-1}hg \in G_{\gamma_1,\ldots,\gamma_{l-1}}$, all of the elements in $(\Delta_K^{(l)})^g$ lie in the same $G_{\gamma_1,\ldots,\gamma_{l-1}}$ orbit. Hence there must be at least $|\Delta_K^{(l)}| - 1$ elements of this orbit that are greater than γ_l under $\prec$.   ∎

In fact the orbit of γ_l under $G_{\gamma_1,\ldots,\gamma_{l-1}}$ has already been calculated and sorted as $\Lambda[l]$ in our code, so we can readily apply this test.

The function SUBGROUPSEARCH is rather complicated and will require careful study by the reader. The second input parameter, H, is intended to be a known subgroup of the sought subgroup S, and is used to initialize K. For example, when computing the normalizer in G of a subgroup L of G, we could put $H = L$, and when computing the centralizer in G of $g \in G$, we could put $H = \langle g \rangle$. If there is no suitable subgroup, then we put $H = \langle 1_G \rangle$.

The variable f in SUBGROUPSEARCH is equal to the smallest j for which $\beta_j^g \neq \beta_j$. The reason for keeping track of f is that, after finding a new element $g \in K$, we can reset the current level l in the search to f (line 22). The search starts with $l = f = k$ rather than with $l = 1$ as in GENERALSEARCH, and so the initializations are slightly different from those in GENERALSEARCH.

Since there are two different bases B and B_K of K in use, the reader must be particularly careful to be aware which is being used at which point in the code. The basic orbits $\Delta_K^{(j)}$ of K (lines 3, 7, 14, 15, 20 and 28) are always with respect to the base $B = [\beta_1, \ldots, \beta_k]$ of G, which is also the initial base of K. The basic stabilizers $K^{(j)}$ of K (lines 4, 12, 21 and 27) are always with respect to B_K.

Although it is not actually stated in the code, the orbit representatives $R[l]$ computed at lines 4, 12, 21 and 27 are chosen to be the $\prec$-least elements of these orbits. They are used in applying the test based on condition (i) of Proposition 4.7 at lines 10 and 17.

The points $\mu[l]$ and $\nu[l]$ of Ω defined at lines 7, 14, 15, and 28 are used to apply the tests described above that use condition (ii) of Proposition 4.7 and Corollary 4.8. The tests are also applied at lines 10 and 17.

In general, when searching for a subgroup K of G with SUBGROUPSEARCH, the larger K is, the faster the search will run. This is because the larger K is, the fewer orbits it and its basic stabilizers $K^{(i)}$ will have, which means fewer orbit representatives, and fewer images to be considered in the search. The most difficult subgroups to find are the very small ones.

As in GENERALSEARCH, we will need to make a choice of a base for G that allows use of the most effective TEST(g, l) for that particular problem. We

SUBGROUPSEARCH(G, H)

    Input: Permutation group G with BSGS, property P, tests TEST(g, l),
        subgroup $H \leq G$ of some elements satisfying P
    Output: Subgroup K of all elements of G satisfying P
    ($\ast$ $g[l]$ will always denote the element $u_l u_{l-1} \cdots u_1$ of G $\ast$)

1   $K := H$; $\ f := k$; $\ l := k$;
2   Let $B := [\beta_1, \ldots, \beta_k]$ be the base of G, and $B_K := B$ the base of K;
3   Calculate basic orbits $\Delta_K^{(l)}$ $(1 \leq l \leq k)$ of K;
4   Calculate orbit representatives $R[f]$ of $K^{(f)}$;
    ($\ast$ To avoid getting 1_G, remove β_f itself $\ast$)
5   Remove β_f from $R[f]$;
6   **for** $i \in [1 .. k]$ **do** $c[i] := 1$; $\ u_i := 1_G$; $\ \Lambda[i] := $SORT$(\Delta^{(i)}, \prec)$;
7   $\mu[l] := 0$; $\ \nu[l] := \Lambda[l][|\Delta^{(l)}| + 2 - |\Delta_K^{(l)}|]$;
8   **while true**
9   **do** ($\ast$ Test if base images are possible at current level $\ast$)
10   **while** $l < k$ **and** $\beta_l^{g[l]} \in R[l]$ **and** $\beta_l^{g[l]} \succ \mu[l]$ **and** $\beta_l^{g[l]} \prec \nu[l]$
               and TEST(g, l)
11   **do** Change l-th base point of K in base B_K to $\beta_l^{g[l]}$;
12   Calculate orbit reps. $R[l+1]$ of $K^{(l+1)}$ w.r.t. B_K;
13                   $l := l+1$; $\ \Lambda[l] := $SORT$((\Delta^{(l)})^{g[l-1]}, \prec)$;
14                   $\mu[l] := MAX\{\ \beta_j^{g[j]} \mid 1 \leq j < l, \ \beta_l \in \Delta_K^{(j)}\ \}$;
15                   $\nu[l] := \Lambda[l][|\Delta^{(l)}| + 2 - |\Delta_K^{(l)}|]$;
16                   $c[l] := 1$; $\ \gamma := \Lambda[l][c[l]]^{g[l-1]^{-1}}$; $\ u_l := u_\gamma^{(l)}$;
17   **if** $l = k$ **and** $\beta_l^{g[l]} \in R[l]$ **and** $\beta_l^{g[l]} \succ \mu[l]$ **and** $\beta_l^{g[l]} \prec \nu[l]$
               and TEST(g, l) **and** $P(g[l])$
18   **then** $K := \langle K, g[l] \rangle$;
19   Change base B_K of K back to $B = [\beta_1, \ldots, \beta_k]$;
20   Calculate basic orbits $\Delta_K^{(l)}$ $(1 \leq l \leq k)$ of K;
21   Recalculate orbit reps. $R[f]$ of $K^{(f)}$ w.r.t. B_K;
22               $l := f$;
23   **while** $l > 0$ **and** $c[l] = |\Delta^{(l)}|$ **do** $l := l-1$;
24   **if** $l = 0$ **then return** K;
25   **if** $l < f$
26   **then** $f := l$; $\ c[l] := 1$;
27   Calculate orbit reps. $R[f]$ of $K^{(f)}$ w.r.t. B_K;
28               $\mu[l] := 0$; $\ \nu[l] := \Lambda[l][|\Delta^{(l)}| + 2 - |\Delta_K^{(l)}|]$;
29        $c[l] := c[l]+1$; $\ \gamma := \Lambda[l][c[l]]^{g[l-1]^{-1}}$; $\ u_l := u_\gamma^{(l)}$;

shall see some specific examples of how this is done later.

    Typically, for each subgroup-searching problem, there is a corresponding problem in which we are looking for a single element that lies in a particular coset of an associated subgroup. For example, corresponding to the problem of finding the centralizer $\mathbf{C}_G(g)$ of an element $g \in G$, we have the problem of testing two elements h and g for conjugacy in G and, when they are conjugate, finding an element $x \in G$ with $x^{-1}hx = g$. If this holds for some $x \in G$, then it holds for $y \in G$ if and only if $y \in x\mathbf{C}_G(g)$, which is why we refer to it as a coset problem. Similarly, to the problem of computing the normalizer of a subgroup, there corresponds the coset problem of testing two subgroups for conjugacy, and finding a conjugating element.

Similar partial base image tests will often apply to the subgroup problem and to the corresponding coset problem. For example, as we shall prove later, elements in the normalizer of a subgroup H of G must permute the orbits of H, whereas an element conjugating subgroup K to subgroup H must map the orbits of K to those of H. Tests can be devised that check when these conditions cannot hold for the current partial base image.

When attempting to solve a coset problem, it can prove efficient to solve the corresponding subgroup problem first. If, for example, we are testing elements h and g for conjugacy in G, then we could compute $K = \mathbf{C}_G(g)$ first. This allows the orbits of the basic stabilizers $K^{(i)}$ of K to be used to limit the search for a conjugating element, in essentially the same way that they were used in SUBGROUPSEARCH in the test based on condition (i) of Proposition 4.7. This claim that we should solve the subgroup problem first is somewhat controversial, because we might of course be lucky and find a conjugating element much more quickly than it would take to compute the centralizer.

But in the author's experience, it is very rare for solving the subgroup problem first to do more than increase the total time taken for the coset problem by a very small amount, whereas there are some cases where it can avoid disasters. The author recalls one particular example, in which G was a direct product $H \times K$ of a small group H and a very large group K, and the problem was to find a conjugating element for two elements h, g that were both in H. The large centralizer of g was computed very quickly, which then allowed the conjugating element to be found almost instantly. But without this precomputation of the centralizer, the search for a conjugating element somehow managed to get lost inside of K, and took many hours!

Exercise

1. Prove Proposition 4.7.

4.6.4 Automorphism groups of combinatorial structures and partitions

The same method as that of SUBGROUPSEARCH can be used to compute automorphism groups of combinatorial structures, such as finite directed or undirected graphs, block designs and error-correcting codes. The corresponding coset problem is to test two such objects for isomorphism. There is a good introductory treatment of this topic by Leon in [Leo84]. Later, in Subsection 4.8.2, we shall discuss the opposite problem of using permutation groups to construct combinatorial objects.

The group G involved when computing $\text{Aut}(\Gamma)$ for a combinatorial object Γ is usually $\text{Sym}(\Omega)$ for some underlying set Ω of Γ. The base for G is then an ordered sequence $[\alpha_1, \ldots, \alpha_{n-1}]$ of $n-1$ points of Ω, where $|\Omega| = n$, but we choose this ordering to allow as many negative tests based on partial base images as possible. We can then identify $\Omega = [\alpha_1, \ldots, \alpha_n]$ with $[1 .. n]$, and use the procedure SYMMETRICGROUPELEMENTS described in Subsection 4.6.1 as the underlying looping mechanism. But it must be modified to allow moving on to the next base image at an arbitrary level l in the search tree, as in SUBGROUPSEARCH. We leave this modification as an exercise to the reader (Exercise 2 below).

As an example of a combinatorial structure, an (undirected) graph $\Gamma = (\Omega, E)$ is a set Ω of vertices, together with a set E of edges, which can be thought of as a subset of the set $\Omega^{\{2\}}$ of unordered pairs of elements of Ω. An automorphism of the graph is a permutation $\alpha \in \text{Sym}(\Omega)$ such that $E^\alpha = E$. There is a particularly well-developed and successful program available for computing automorphism groups of graphs and testing two graphs for isomorphism. This was written and is maintained by Brendan D. McKay, and is known as 'Nauty'; the name comes from the phrase "No AUTomorphisms, Yes?" See, for example, [McKay81] for a description of the ideas used in the algorithms involved, or the Nauty Web page for up-to-date information.

A graph can be made into a metric space with distance function d, in which an edge $\{\alpha, \beta\}$ is a path of length 1 connecting α to β. Then, for $\alpha, \beta \in \Omega$, $d(\alpha, \beta) = r$ means that there exist $\alpha = \alpha_0, \alpha_1, \ldots, \alpha_r = \beta$ with each $\{\alpha_{i-1}, \alpha_i\} \in E$, and no shorter path joining α to β.

A simple way of pruning the search tree in the computation of $\text{Aut}(\Gamma)$ is to use the fact that $g \in \text{Aut}(\Gamma)$ implies $d(\alpha, \beta) = a(\alpha^g, \beta^g)$ for any $\alpha, \beta \in \Omega$.

For example, suppose that $\Omega = \{1 .. 5\}$, and Γ is a pentagon, with edges $\{1, 2\}$, $\{2, 3\}$, $\{3, 4\}, \{4,5\}$, $\{5, 1\}$. Since 2 and 5 are joined to 1, we might choose the base to be $[1, 2, 5, 4, 3]$ rather than $[1, 2, 3, 4, 5]$, but that is not really critical in such a small example. Let $g \in \text{Aut}(\Gamma)$. If $1^g = 1$ and $2^g = 2$, then, since 2 and 5 are the only vertices at distance 1 from 1, we get $5^g = 5$, and similarly we deduce $3^g = 3$ and $4^g = 4$, so the search can proceed at level 2 to $2^g = 5$, which implies $5^g = 2$, $3^g = 4$, $4^g = 3$. A full test for $E^g = E$ reveals that $g_1 := (2, 5)(3, 4)$ really is in $\text{Aut}(\Gamma)$, so we set set $K = \langle g_1 \rangle$. Since $2^g = 3$ and $2^g = 4$ do not preserve distance from 1 in Γ, we backtrack to level 1, and move on to $1^g = 2$. Trying first $2^g = 1$, the edge distance 1 constraint forces $5^g = 3$, $4^g = 4$, and we find that $g_2 = (1, 2)(3, 5) \in \text{Aut}(\Gamma)$. Now $K := \langle g_1, g_2 \rangle$ has order 10 and is transitive on Ω, so the search completes immediately, with $\text{Aut}(\Gamma) = K$.

In this example, we discovered quickly that $g \in \text{Aut}(\Gamma)$ and $1^g = 1, 2^g = 2$ implies $g = 1$, so $[1, 2]$ is a base for $\text{Aut}(\Gamma)$. Generally, during the initial choice of base of an automorphism group calculation, we will look for some, preferably short, initial sequence $[\alpha_1, \ldots, \alpha_l]$ of our ordering of Ω, which will be a base for $\mathrm{Aut}(\Gamma)$. We can then use this fact during the search to deduce α^g for the remaining points α from $\alpha_1^g, \ldots, \alpha_l^g$.

A structure that arises naturally in many different combinatorial and group-theoretical searching algorithms is an *ordered partition* of the underlying set Ω. This consists of an ordered sequence $\mathcal{P} = [\Omega_1, \ldots, \Omega_r]$ of disjoint subsets of Ω of which the union is the whole of Ω. The individual subsets are treated as unordered sets in such a structure. An automorphism of $\mathcal{P}$ is a permutation that preserves the ordered structure, and the full automorphism group is clearly equal to the direct product of the groups $\mathrm{Sym}(\Omega_i)$.

There is an extensive theoretical account of the use of ordered partitions in backtrack search algorithms by J.S. Leon in [Leo91]. The topic is also covered in Section 9.2 of [Ser03]. Leon has written some efficient code for many of these algorithms, which is available from MAGMA and GAP. This includes basic procedures for computing the stabilizer of an ordered partition in a given permutation group G^Ω, and the associated coset problem of deciding whether one such partition can be mapped to another.

Notice that the problem of finding the setwise stabilizer G_S of a subset S of Ω in G is a special case of the partition stabilizing problem. In general, this is significantly more difficult than computing the pointwise stabilizer of S which, as we saw in Subsection 4.4.7, can be done simply by changing the base of G. To compute G_S, we choose our base $[\beta_1, \ldots, \beta_k]$ of G such that $\beta_i \in S$ for $1 \le i \le j$ only, with j as large as possible. We then use the fact that, for any $g \in G_S$, $\beta_i^g \in S$ for $1 \le i \le j$, and $\beta_i^g \notin S$ for $j+1 \le i \le k$ in our TEST functions.

In the calculation of $\mathrm{Aut}(\Gamma)$ for a graph $\Gamma = (\Omega, E)$, there are several useful ordered partitions of Ω. For example, for $\alpha \in \Omega$ and $r \in \mathbb{N}$, let $v(\alpha, r) := |\{\beta \in \Omega \mid d(\alpha, \beta) = r\}|$ be the number of vertices at distance r from α and, for $s > 0$, let $\Omega(r)_s := \{\alpha \in \Omega \mid v(\alpha, r) = s\}$. Then, for each $r = 1, 2, 3, \ldots$, the sets $\Omega(r)_s$ define an ordered partition of Ω, and each of these ordered partitions must be fixed by any $g \in \mathrm{Aut}(\Gamma)$. This information can be used to devise partial base image tests $\mathrm{TEST}(g, l)$. When searching through the basic stabilizer $\mathrm{Sym}(\Omega)_{\alpha_1}$ of the first base point 1, further partitions are stabilized, some of which are *refinements* of those stabilized by all of $\mathrm{Aut}(\Gamma)$. These ideas can be used to find a base for $\mathrm{Aut}(\Gamma)$. After considering a succession of basic stabilizers of $[\alpha_1, \ldots, \alpha_l]$ in $\mathrm{Sym}(\Omega)$ and the associated partition refinements, we hope eventually to find that the finest ordered partition, in which each subset is a singleton, must be preserved, which is of course equivalent to $[\alpha_1, \ldots, \alpha_l]$ being a base for $\mathrm{Aut}(\Gamma)$.

Exercises

1. It was claimed that testing two combinatorial objects for isomorphism is a coset problem. What coset of what subgroup, and in what group?

2. Write a general version of SUBGROUPSEARCH for which the group G is $\text{Sym}(n)$, using SYMMETRICGROUPELEMENTS as the underlying method for looping through the group elements.

3. Consider the so-called 'Petersen Graph' Γ below. Show that vertices can be labelled with unordered pairs $\{x, y\}$, $x, y \in \{1..5\}$, in such a way that two vertices are joined by an edge in Γ if and only if their labels are disjoint sets. Deduce that $\text{Aut}(\Gamma)$ has a subgroup isomorphic to $\text{Sym}(5)$ and hence that $|\text{Aut}(\Gamma)| \geq 120$. Find a base for $\text{Aut}(\Gamma)$ consisting of three vertices of Γ and deduce that $|\text{Aut}(\Gamma)| \leq 120$ and hence that $\text{Aut}(\Gamma) \cong \text{Sym}(5)$.

4.6.5 Normalizers and centralizers

Computing the normalizer $\mathbf{N}_G(H)$, where H and G are subgroups of $\text{Sym}(\Omega)$, together with the corresponding coset problem of testing two subgroups of $\text{Sym}(\Omega)$ for conjugacy in G, are among the most difficult problems in CGT and there are examples of degree less than 100, which will defeat even the best available implementations at the present time. Usually H is a subgroup of G, but the algorithms tend not to make use of that fact. Of course, if $H \leq G$, then we can initialize K to H in SUBGROUPSEARCH, but in general we could only initialize K to $H \cap G$. All sensible implementations will certainly use the following well-known result.

PROPOSITION 4.9   *Let $H, K, G \leq \text{Sym}(\Omega)$ and let $g \in G$ with $H^g = K$. Then g maps each orbit of H^Ω to an orbit of K^Ω. In particular, $\mathbf{N}_G(H)$ permutes the orbits of H^Ω.*

PROOF   We have to show that if α and β are in the same orbit of H^Ω, then α^g and β^g are in the same orbit of K^Ω. So, suppose there exists $h \in H$ with $\alpha^h = \beta$. Let $k := g^{-1}hg \in K$. Then $\alpha^{gk} = \alpha^{hg} = \beta^g$, which proves the result. ∎

So $\mathbf{N}_G(H)$ permutes the orbits of H^Ω and, for an initial segment $B(l-1)$ of our base B of G and $g \in \mathbf{N}_G(H)$, g maps the orbits of $H_{B(l-1)}$ to the orbits of $H_{C(l-1)}$, where $C(l-1) = B(l-1)^g$. To make use of this property in SUBGROUPSEARCH for $l > 1$, we require that the bases of both H and the currently known subgroup of $\mathbf{N}_G(H)$ have the initial segment $C(l-1)$; we must therefore carry out frequent base changes while running the procedure. To make the resulting partial base image tests as effective as possible, the initial base of G should be chosen such that as many base points as possible lie in the same orbit of H^Ω. An implementation of the normalizer and conjugacy testing algorithms that uses orbital information of this type is described by Butler in [But83]

With H, we can associate an ordered partition $\mathcal{P}_H$ of Ω as follows. Let $d_1, \ldots, d_r$ be the distinct sizes of the orbits of H^Ω, and let

$$\Omega_i := \{\, \alpha \in \Omega \mid |\alpha^H| = d_i \,\}$$

for $1 \le i \le r$. Then Proposition 4.9 implies that $\mathbf{N}_G(H)$ preserves $\mathcal{P}_H$ and, if $H^g = K$, then g maps $\mathcal{P}_H$ to $\mathcal{P}_K$. Of course, this fact is only useful computationally if $r > 1$. But the author has observed that, provided that $r > 1$, it is often significantly faster with current implementations to compute $L := \mathbf{N}_G(\mathcal{P}_H)$ first and then to compute $\mathbf{N}_G(H)$ as $\mathbf{N}_L(H)$, than it is to compute $\mathbf{N}_G(H)$ directly. Similarly, to test H and K for conjugacy in G, first look for $g_1 \in G$ with $\mathcal{P}_H^{g_1} = \mathcal{P}_K$ and, if g_1 exists, then look for $g_2 \in \mathbf{N}_G(\mathcal{P}_K)$ with $H^{g_1 g_2} = K$. Theoretically it is not surprising that this is a good policy, because the speed of a backtrack search is heavily dependent on the index of the sought subgroup in the whole group, and so splitting the problem into two parts with smaller associated indices is likely to be a good policy.

There are some situations in which none of the above methods provides any useful information at all. The most extreme case is when H^Ω is regular (see Definition 2.24). In that and similar cases, it is possible to make use of the fact that elements of $\mathbf{N}_G(H)$ induce automorphisms of H by conjugation. For example, if H is a cyclic group with H^Ω regular, then the automorphism induced by g, and hence g^Ω, is uniquely determined once β_1^g and β_2^g are known, where $\beta_1^h = \beta_2$ for a generator h of H. These properties were used in an implementation described by Holt in [Hol91]

Another property of normalizers that can be used effectively is that $\mathbf{N}_G(H)$ permutes the orbital graphs of H^Ω; that is, the orbits of H on the set $\Omega^{(2)}$ of ordered pairs of elements of Ω. This idea was originally proposed by Sims, and used in an implementation in CAYLEY in the case when H^Ω is transitive. It is also used in the GAP implementation described by Theissen in [The97].

For the calculation of centralizers of elements and of subgroups, a much stronger result is available.

PROPOSITION 4.10 *Let $H, G \le \mathrm{Sym}(\Omega)$ and let $g \in \mathbf{C}_G(H)$. Suppose that Δ_1 and Δ_2 are (not necessarily distinct) orbits of H^Ω and that $\Delta_1^g = \Delta_2$. Then the action of g on the points in Δ_1 is uniquely determined by H^Ω and the action of g on any single point $\alpha \in \Delta_1$.*

PROOF   Suppose $\alpha^g = \beta \in \Delta_2$, and let $\gamma \in \Delta_1$. Then there exists $h \in H$ with $\alpha^h = \gamma$, and then $\gamma^g = \alpha^{hg} = \alpha^{gh} = \beta^h$.   ∎

Similarly, it can be shown that, if $h_1^g = h_2$ with $h_1, h_2, g \in \text{Sym}(\Omega)$, then the complete action of g on any orbit of $\langle h_1 \rangle$ is uniquely determined by h_1 and the action of g on a single point in that orbit. As is the case for normalizer calculations, we should choose as many base points of G as possible in the same orbit of H or $\langle h \rangle$. Typically, we can then deduce several base images β_{i+j}^g from a single base image β_i^g, and in many examples the calculation of centralizers is extremely fast.

Example 4.5

Let us compute the centralizer in $G := \text{Alt}(8)$ of $h := (1,2,3)(4,5,6)$. The obvious base $B := [1,2,3,4,5,6]$ of G satisfies the properties of the preceding paragraph. We can initialize the centralizer K of h in G to be $\langle h \rangle$. If $[1,2,3,4]^g = [1,2,3,4]$, then $g = 1$, so the search will quickly jump back to $l = 4$ and consider g with $[1,2,3,4]^g = [1,2,3,5]$. Now $g \in \mathbf{C}_G(h)$ implies that $5^g = 6$ and $6^g = 4$, and we find $g_1 := (4,5,6) \in \mathbf{C}_G(h)$ and adjoin g_1 to K. Now $K^{(4)} = K_{[1,2,3]}$ has orbits $\{4,5,6\}$, $\{7\}$ and $\{8\}$ on $\Omega \setminus \{1,2,3\}$. We have already considered $g \in K^{(4)}$ with $4^g = 5 \in \{4,5,6\}$, and $4^g = 7$ and $4^g = 8$ are not possible, because $g \in \mathbf{C}_G(h)$ cannot map an orbit of h of length 3 to one of length 1. So we backtrack to $l = 3$.

Then $1^g = 1$ implies $2^g = 2$ and $3^g = 3$, so we immediately backtrack to $l = 1$. Since $1, 2,$ and 3 are already in 1^K, we consider $1^g = 4$, which implies $2^g = 5$ and $3^g = 6$. Now since g permutes the orbits of h, it must map the orbit $\{4,5,6\}$ to $\{1,2,3\}$, so we try $4^g = 1$, which implies $5^g = 2$ and $6^g = 3$. Now $g \in G$ is uniquely determined as $g_2 := (1,4)(2,5)(3,6)(7,8)$ and we find that $g_2 \in \mathbf{C}_G(h)$, so we adjoin g_2 to K. Now K^Ω has orbits $\{1,2,3,4,5,6\}$ and $\{7,8\}$. But $g \in \mathbf{C}_G(h)$ cannot map 1 to 7 or 8, so the search finishes with $K = \mathbf{C}_G(h) = \langle h, g_1, g_2 \rangle$, which has order 18.   ☐

The theoretical complexity of this algorithm for centralizer calculation is exponential in n in general; the worst case in this respect is when $|H| = 2$. There are some special situations in which polynomial algorithms for $\mathbf{C}_G(H)$ can be devised. For example, if H^Ω is transitive, then the search will complete in polynomial time, because $g \in \mathbf{C}_G(H)$ is uniquely determined by its action on a single point of Ω.

An important special case is the calculation of the *centre* $\mathbf{Z}(G) = \mathbf{C}_G(G)$ of G. For that calculation, we can modify the basic search algorithm to make it run in polynomial time as follows. Let the orbits of G be $\Omega_1, \ldots, \Omega_r$. Define $G(0) := G$ and, inductively, for $i > 0$, define $G(i)$ to be the complete inverse image of the centralizer of G^{Ω_i} in $\mathbf{Z}(G(i-1)^{\Omega_i})$. Then $\mathbf{Z}(G) = G(r)$ (see Exercise 1 below). Since each G^{Ω_i} is transitive, each $G(i)$ can be computed from $G(i-1)$ in polynomial time, and hence so can $\mathbf{Z}(G)$.

More generally, the centralizer in G of any normal subgroup H of G can be found in polynomial (and even in nearly-linear) time; see [Ser03, Sec. 6.1.4].

The calculation of $\mathbf{C}_G(H)$ when $G = \mathrm{Sym}(\Omega)$ can also be accomplished in polynomial time; see Exercise 2 below or [Ser03, Sec. 6.1.2].

The following corollary of Proposition 4.10 is often useful, and will be applied in Section 4.7.

COROLLARY 4.11   Let G act on Ω, let $H \leq G$, and suppose that H^Ω is transitive and abelian. Then H^Ω is regular and self-centralizing in G^Ω.

PROOF   Let $g \in G$ such that g^Ω centralizes H^Ω. Choose any $\alpha \in \Omega$ and let $\alpha^g = \beta$. By transitivity of H^Ω, there is $h \in H$ with $\alpha^h = \beta$. Since H is abelian, h^Ω also centralizes H^Ω, and then Proposition 4.10 tells us that $g^\Omega = h^\Omega$, so H^Ω is self-centralizing in G^Ω. This also tells us that h^Ω is the only element of H^Ω with $\alpha^h = \beta$, and in particular, choosing $\alpha = \beta$, we get $H_\alpha = \{1\}$, so H^Ω is regular.   ∎

Exercises

1. Prove that $\mathbf{Z}(G) = G(r)$ in the above discussion of the computation of $\mathbf{Z}(G)$.

2. Prove that $\mathbf{C}_G(H)$ can be calculated in polynomial time when $H \leq G = \mathrm{Sym}(\Omega)$. (*Hint*: This can be done by computing $\mathbf{C}_{\mathrm{Sym}(\Omega_i)}(H^{\Omega_i})$ and testing the groups H^{Ω_i} for conjugacy in G, where Ω_i are the orbits of H^Ω.)

4.6.6   Intersections of subgroups

Suppose that $G, H \leq K \leq \mathrm{Sym}(\Omega)$, and we want to compute the intersection $G \cap H$. Let $B = [\beta_1, \ldots, \beta_k]$ be a base for K and, as before, let $B(l)$ be the initial segment $[\beta_1, \ldots, \beta_l]$. Since elements of K are uniquely determined by their base images, an element $g \in G$ lies in H if and only if there exists $h \in H$ with $B^g = B^h$, in which case $g = h$.

We proceed by running SUBGROUPSEARCH on G. At any level l in the search tree, we are considering a candidate $g \in G$ for which $B(l-1)^g$ is known, and we also maintain an element $h \in H$ with $B(l-1)^h = B(l-1)^g$. Initially, $g = h = 1$. Suppose that we are at a vertex at level l in the search tree, and we are considering g with $\beta_l^g = \gamma$. If $h' \in H$ with $B(l-1)^{h'} = B(l-1)^g$ and $\beta_l^{h'} = \gamma$, then $h' = h_1 h$ for some $h_1 \in H^{(l)} = H_{B(l-1)}$, where h is our stored element of H with $B(l-1)^h = B(l-1)^g$. Hence $\beta_l^{gh^{-1}} = \beta_l^{h_1}$ and so $\beta_l^{gh^{-1}}$ must be in the orbit of β_l under $H^{(l)}$. But this orbit is just the l-th basic orbit of H, so this condition can be tested immediately. So our function TEST(g, l) will return `false` if $\beta_l^{gh^{-1}}$ is not in the orbit. Otherwise, $\text{Test}(g, l)$ will return `true` and, if $l < k$, then h_1 with $\beta_l^{gh^{-1}} = \beta_l^{h_1}$ is computed and the memorized value of h replaced by $h_1 h$. If $\text{Test}(g, l)$ returns `true` with $l = k$, then we know that there exists $h \in H$ with $B^h = B^g$ and so $g \in G \cap H$, and we can adjoin g to the intersection.

Example 4.6

Let $K := \text{Alt}(8)$ with base $[1, 2, 3, 4, 5, 6]$. Let

$$G := \langle (1, 2, 3), (4, 5, 6), (1, 4)(2, 5)(3, 6)(7, 8) \rangle$$

be the subgroup of order 18 computed in Example 4.5. We saw that the basic orbits of G are $\{1, 2, 3, 4, 5, 6\}$, $\{2\}$, $\{3\}$, $\{4, 5, 6\}$, $\{5\}$, and $\{6\}$. Let

$$H := \langle (1, 6)(2, 4)(3, 5)(7, 8), (1, 2)(3, 7)(4, 6)(5, 8), (2, 3, 7)(4, 5, 8) \rangle.$$

Then $|H| = 24$, H^Ω is transitive and $H^{(2)}$ is generated by $(2, 3, 7)(4, 5, 8)$, so the basic orbits of H are $\{1 .. 8\}$, $\{2, 3, 7\}$, $\{3\}$, $\{4\}$, $\{5\}$, and $\{6\}$.

In the search through G, at $l = 4$ (the first nontrivial level of G), we find that 5 and 6 are not in the 4th basic orbit of H, so we backtrack to $l = 1$. We first try $1^g = 2$. Then $\text{Test}(g, 1)$ returns `true` because 2 is in the first basic orbit of H, and stores $h = (1, 2)(3, 7)(4, 6)(5, 8)$ (for example), with $1^g = 2$. Moving down to level 2 in the tree, we must have $2^g = 3$, and then $2^{gh^{-1}} = 7$ is in the second basic orbit of H, so $\text{Test}(g, 2)$ returns true, and we take $h_1 = (2, 7, 3)(4, 8, 5)$ and replace h by $h_1 h = (1, 2, 3)(4, 5, 6)$. Now we move down to level 3, where we must have $3^g = 1$, so $3^{gh^{-1}} = 3$ and $\text{Test}(g, 3)$ returns `true` with h unchanged. Moving down to level 4, we have three choices for 4^g. With $4^g = 4$, we get $4^{gh^{-1}} = 6$, which is not in the fourth basic orbit $\{4\}$ of H, so $\text{Test}(g, 4)$ returns `false`. With $4^g = 5$, however, we get $4^{gh^{-1}} = 4$, so $\text{Test}(g, 4)$ returns `true` with h unchanged. Now we must have $5^g = 6$ and $6^g = 4$, and $\text{Test}(g, l)$ returns `true` for $l = 5$ and 6 and so we have $g_1 := g = (1, 2, 3)(4, 5, 6)$ in $G \cap H$.

The search then backtracks to level 1. Since we have $1, 2, 3$ already in $1^{G \cap H}$, we next try $1^g = 4$, which leads to $\text{Test}(g, 1)$ returning `true` with $h = (1, 4)(2, 6)(3, 8)(5, 7)$, for example. At level 2, we must have $2^g = 5$, leading to $2^{gh^{-1}} = 7$, and $\text{Test}(g, 2)$ returns `true` with h replaced by $h_1 h = (1, 4, 3, 6, 2, 5)(7, 8)$. At level 3, we must have $3^g = 6$, and then $3^{gh^{-1}} = 3$ so $\text{Test}(g, 3)$ returns `true` with h unchanged. At level 4, $\text{Test}(g, 4)$ returns `true` only for $4^g = 3$, with h unchanged, and this forces $5^g = 1$, $6^g = 2$ at levels 5 and 6, but $\text{Test}(g, l)$ returns `true` both times. The only element of G with these base images is $g_2 := g = (1, 4, 3, 6, 2, 5)(7, 8)$, so this is adjoined to the generating set of $G \cap H$. The search then completes. In fact $g_1 = g_2^4$, so g_1 is redundant, and $G \cap H$ is cyclic of order 6. ∎

There is a (nearly-linear) polynomial-time algorithm for computing the intersection $G \cap H$ when $G \leq \mathbf{N}(H)$; see [Ser03, Sec. 6.1.1].

Exercises

1. Show that $|H| = 24$ in the example above.

The following exercises are based on Proposition 4.3 of [Luk93]. They show that the problems of finding setwise stabilizers, centralizers of elements, and intersections of subgroups are *polynomially equivalent*. That is, if any one of them is solvable in polynomial time, then so are all three.

2. Let $G \le \text{Sym}(\Omega)$ and $\Delta \subseteq \Omega$. Let $\hat{\Omega} := \Omega_1 \cup \Omega_2$ be the union of two disjoint copies of Ω where, for $\alpha \in \Omega$, the corresponding points in Ω_i will be denoted by α_i for $i = 1, 2$. Let G act diagonally on $\hat{\Omega}$. (That is, if $\alpha^g = \beta$ for $\alpha, \beta \in \Omega$, then $\alpha_i^g = \beta_i$ for $i = 1, 2$.) Define $x \in \text{Sym}(\hat{\Omega})$ by

   (i) $\alpha_i^x = \alpha_i$ for $i = 1, 2$ and $\alpha \notin \Delta$;

   (ii) $\alpha_1^x = \alpha_2$, $\alpha_2^x = \alpha_1$ for $\alpha \in \Delta$.

   Prove that $G_\Delta = G \cap G^x = \mathbf{C}_G(x)$ (where we have identified G with $G^{\hat{\Omega}}$). Hence, if either the intersection or the centralizer problem is solvable in polynomial time, then so is the setwise stabilizer problem.

3. Let $G, H \le \text{Sym}(\Omega)$ and let $G \times H$ act on $\Omega \times \Omega$ in the natural way. Let $\Delta = \{(\alpha, \alpha) \mid \alpha \in \Omega\}$. Prove that $\{(g, g) \mid g \in G \cap H\} = (G \times H)_\Delta$. Hence, if the setwise stabilizer problem is solvable in polynomial time, then so is the intersection problem.

4. Let $G \le \text{Sym}(\Omega)$ and $x \in \text{Sym}(\Omega)$, and let G and x act diagonally on $\Omega \times \Omega$. Prove that $\{(g, g) \mid g \in \mathbf{C}_G(x)\} = G_\Delta$ with $\Delta = \{(\alpha, \alpha^x) \mid \alpha \in \Omega\}$. Hence, if the setwise stabilizer problem is solvable in polynomial time, then so is the centralizer problem.

4.6.7   Transversals and actions on cosets

In this subsection, we discuss the problem of computing a transversal of a subgroup H of a permutation group G, and the related problem of computing the action of G by right multiplication on the set of right cosets of H in G, as defined in Example 2.4.

We shall describe two different methods of computing transversals. The first of these generates the transversal elements one at a time, with very little storage overheads. It is not well adapted for finding the coset representative of a given group element, however, so it is not so useful for computing the action of G on the cosets. The second method has much larger storage overheads, but is well adapted for locating coset representatives of group elements. Both methods involve a backtrack search through the elements of G.

The first method naturally computes left coset rather than right coset representatives, but this in itself is not significant, because we can move from a left transversal to a right transversal, if required, simply by inverting all elements. As the representative of the coset gH, we choose the least element in the coset gH under the ordering $\prec$ defined at the beginning of this section.

We use Proposition 4.7 (i) to test whether g is the required representative. We do this in the same way as in SUBGROUPSEARCH: at a vertex at level l in the search tree, TEST(g, l) checks that $\gamma_l = \beta_l^g$ is the least element of its $H_{\gamma_1, \ldots, \gamma_{l-1}}$ orbit. This requires us to carry out base changes to make $B(l)^g$ the initial segment of the base B_H of H that we are using. For a vertex representing a group element g at the bottom layer k of the tree, we will have carried out this test for all l with $1 \leq l \leq k$, and hence, by Proposition 4.7, we will know that g is the required coset representative.

Example 4.7

Let $G := \mathrm{Alt}(5)$ and $H := \langle (1, 2, 3), (2, 3)(4, 5) \rangle$, with $B := [1, 2, 3]$. So $|H| = 6$ and $|G : H| = 10$. Since $H^{(3)} = 1$, the three elements 1, $(3, 4, 5)$, $(3, 5, 4)$ are all appended to the (initially empty) transversal T. Moving up to level 2, with $1^g = 1$, the base image $2^g = 3$ is rejected, because that would result in $2^{gh} = 2 \prec 3$ with $h = (2, 3)(4, 5) \in H$. (In this example, the ordering of Ω that we are using is the natural one because the base is $[1, 2, 3, 4]$.) With $2^g = 4$, TEST$(g, 2)$ returns true, so we descend to level 3, and change base of H from $[1, 2]$ to $[1, 4]$. Since $H_{[1,4]} = 1$, all three elements $(2, 4, 5)$, $(2, 4)(3, 5)$, $(2, 4, 3)$ are appended to T. Now $2^g = 5$ is rejected at level 2, because $2^{gh} = 4 \prec 5$ with $h = (2, 3)(4, 5)$, so we go up to level 1.

The base images $1^g = 2$ and $1^g = 3$ are rejected, because they would result in $1^{gh} = 1$ for $h = (1, 3, 2)$ and $h = (1, 2, 3)$, respectively. So we move on to $1^g = 4$, which is not rejected, change the first base point of H to 4, and go down to level 2. Now $2^g = 1$ passes the test, as do all three of its descendants, so $(1, 4, 5, 3, 2)$, $(1, 4, 3, 5, 2)$, $(1, 4, 2)$ are appended to T. Base images $2^g = 2$ and $2^g = 3$ are rejected, since then $2^{gh} = 1$ with $h = (1, 3, 2)$ or $(1, 2, 3) \in H_4$. So we move on to $2^g = 5$, which is not rejected, because $\{5\}$ is an orbit of H_4. We then change base of H to make $[4, 5]$ an initial segment, yielding the base $[4, 5, 1]$. Now base image $3^g = 1$ is accepted, yielding the final transversal element $(1, 4, 2, 5, 3)$. The complete left transversal is

$$\{ 1_G, (3, 4, 5), (3, 5, 4), (2, 4, 5), (2, 4)(3, 5), (2, 4, 3), (1, 4, 5, 3, 2),$$
$$(1, 4, 3, 5, 2), (1, 4, 2), (1, 4, 2, 5, 3) \}.$$

□

It is possible to find the coset representative of a given element of G, but this involves base changes to H, so we would not want to have to do this for large numbers of elements. For example, to find the coset representative of $g = (1, 5, 3)$, choose $h \in H$ to minimize 1^{gh} and replace g by gh. So we could take $h = (2, 3)(4, 5)$ and replace g by $gh = (1, 4, 5, 2, 3)$. Now change first base point of H to 4 and choose $h \in H_4$ to minimize 2^{gh}. So we take $h = (1, 2, 3)$, and replace g by $gh = (1, 4, 5, 3, 2)$. Now, make the second base point of H equal to $2^g = 1$, and we have $H_{4,1} = 1$, so g must now be in the transversal.

We turn now to the second method, in which we compute right coset representatives of H in G. Unlike the first method, this does not quite follow the pattern of our standard backtrack search through G. The procedure takes place in stages corresponding to the base points, which happen in reverse order k, $k-1$, ..., 1. Initially, we define $\overline{T}^{(k+1)} := \{1_G\}$, and then, in stage l, we compute coset representatives $\overline{T}^{(l)}$, such that

$$T^{(l)} := \bigcup_{l \leq i \leq k+1} \overline{T}^{(i)}$$

is a right transversal of $H^{(l)}$ in $G^{(l)}$. The required transversal is then $T^{(1)}$.

As usual, let $U^{(l)}$ be a right transversal of $G^{(l+1)}$ in $G^{(l)}$. Then an element of $\overline{T}^{(l)} = T^{(l)} \setminus T^{(l+1)}$ will have the form hu for $h \in G^{(l+1)}$ and $u \in U^{(l)} \setminus \{1\}$, and we can clearly assume that h is in $T^{(l+1)}$, which we have already computed. So the method is to consider all $u \in U^{(l)} \setminus \{1\}$ in turn, in order of increasing base image β_l^u and, for each such u, we consider tu for all $t \in T^{(l+1)}$. We then test whether $tu \in Hg$ for any g that is already in $T^{(l)}$ and, if not, then we adjoin tu to $\overline{T}^{(l)}$ and (by definition) to $T^{(l)}$.

Suppose that $tu \in Hg$ with $g \in T^{(l)}$. Then $g = t'u'$ for some $t' \in T^{(l+1)}$ and $u' \in U^{(l)}$. We cannot have $u = u'$, because in that case t and t' would be in distinct cosets of H in G by assumption. So, since we are considering $u \in U^{(l)} \setminus \{1\}$ in order of increasing β_l^u, we must have $\beta_l^{t'u'} = \beta_l^{u'} \prec \beta_l^u$. But $t'u' = htu$ for some $h \in H^{(l)}$, so $\gamma^{tu} \prec \beta_l^u$ for some γ $(= \beta_l^h)$ in the l-th basic orbit of H. Conversely, if $\gamma^{tu} \prec \beta_l^u$ for some such γ, then $tu \in Hg$ for some $g \in T^{(l)}$. This condition is readily testable from a knowledge of the original basic orbits of H, and no base changes are needed. We can stop as soon as $|T^{(l)}| = |G^{(l)}:H^{(l)}|$.

This method is generally a little faster than the first method, because base changes to H are not necessary. But it has considerably larger memory requirements. In order to compute $\overline{T}^{(l)}$ we need first to store $T^{(l+1)}$, and so $T^{(2)}$ will need to be stored in order to complete the computation. If the index is too large for this to be practical, and we wish to consider the coset representatives one at a time, then the first method should be used. However, we would not normally store elements of $T^{(l)}$ as complete permutations, unless they were all explicitly required. They can be more economically stored as permutation words in the transversals $U^{(l)}$. Further space can be saved by the use of a tree structure, which we shall describe below.

Let us carry out this computation on Example 4.7. We choose:

$$U^{(1)} := \{\, 1_G,\ (1,2,3),\ (1,3,4),\ (1,4,5),\ (1,5,2)\,\},$$

$$U^{(2)} := \{\, 1_G,\ (2,3,4),\ (2,4,5),\ (2,5,3)\,\},$$

$$U^{(3)} := \{\, 1_G,\ (3,4,5),\ (3,5,4)\,\}.$$

We start with $\overline{T}^{(4)} = \{1\}$ and then $H^{(3)} = 1$ implies that $\overline{T}^{(3)} = \{(3,4,5),$ $(3,5,4)\}$. Now moving on to Stage 2, we first consider $g := t(2,3,4)$ for $t \in T^{(3)}$. The second basic orbit of H is $\{2,3\}$, so g will be put into $\overline{T}^{(2)}$ if and only if $3^g \geq 3$, which is the case for $t = 1$ and $t = (3,5,4)$, but not for $t = (3,4,5)$. So we put $(2,3,4)$ and $(2,3,5)$ into $\overline{T}^{(2)}$. Next we consider $t(2,4,5)$ which is put into $\overline{T}^{(2)}$ only for $t = (3,4,5)$, when $g = (2,4)(3,5)$.

We now have $|T^{(2)}| = 6 = |G^{(2)}:H^{(2)}|$, so we move on to Stage 1. First we consider $g := t(1,2,3)$ for $t \in T^{(2)}$. The first basic orbit of H is $\{1,2,3\}$, so g will be put into $\overline{T}^{(1)}$ if and only if $2^g \geq 2$ and $3^g \geq 2$. This occurs for $t = (3,4,5), (3,5,4)$ and $(2,4)(3,5)$, so the corresponding elements $(1,2,3,4,5)$, $(1,2,3,5,4)$ and $(1,2,4,3,5)$ are put into $\overline{T}^{(1)}$. Finally $g = t(1,3,4)$ goes into $\overline{T}^{(1)}$ only for $t = (2,3,5)$, when $g = (1,3,5,2,4)$, and $T^{(1)}$ then contains 10 elements and is complete. The resulting right transversal is

$$\{1, (3,4,5), (3,5,4), (2,3,4), (2,3,5), (2,4)(3,5), (1,2,3,4,5),$$

$$(1,2,3,5,4), (1,2,4,3,5), (1,3,5,2,4)\}.$$

With this method, we can also identify the coset representative of a given element of G without having to change base of H. Let us demonstrate how to do this in the example on $g = (1,4,2,3,5)$. We first find the element γ in the first basic orbit $\{1,2,3\}$ of H such that γ^g is as small as possible. So $\gamma = 2$, and we replace g by hg where $h \in H$ with $1^h = 2$; we can take $h = (1,2,3)$ and replace g by $(1,3,4,2,5)$. Now we know that the required representative is tu, where $t \in T^{(2)}$ and $u \in U^{(1)}$ with $1^u = 3$; that is, $u = (1,3,4)$, and we have to find the coset representative in $T^{(2)}$ of $g_2 := gu^{-1} = (2,5,4)$. For this, we look for γ in the second basic orbit $\{2,3\}$ of H with γ^{g_2} minimal, so $\gamma = 3$ and we replace g_2 by hg_2 with $h = (2,3)(4,5)$; that is, by $(2,3,5)$. Since $H^{(3)}$ is trivial, we now have $g_2 \in T^{(2)}$, so the required coset representative of $(1,4,2,3,5)$ is $g_2u = (1,3,5,2,4)$.

We can save some space by storing the coset representatives in a tree structure, as in the diagram below for the above example. This is similar to the diagram in Subsection 4.6.1 above rotated through 90 degrees, but we have only shown the branches of the tree that lead to transversal elements. These are numbered on the right of the diagram, and occur in the same order as the list above. So, for example, by reading the edge labels of its branch, we see that element number 8 of T is $u_5^{(3)} u_2^{(2)} u_2^{(1)} = (3,5,4)(1,2,3) = (1,2,3,5,4)$.

Storing this data structure requires two machine words for each node of the tree: a pointer to the next node at the same level, and a pointer to the first child at the next level down, or the number of the coset in the case of a leaf. This is typically significantly less than the $|B| |G:H|$ words required to store the coset representatives as individual permutation words. The tree structure has the added advantage that we can identify the coset of an arbitrary group element by tracing its path through the tree, and then we immediately know

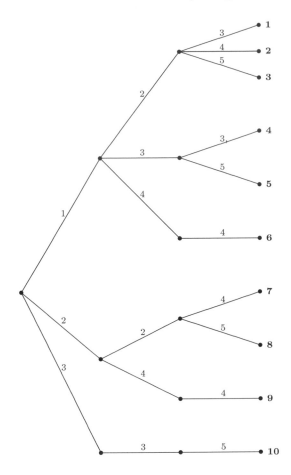

the number of coset when we reach its leaf. If we were just using permutation words, then we would need to use a hashing algorithm to attach numbers to transversal elements, which would require yet more space.

Now let Ω be the set of right cosets of H in G, and let $\varphi : G \to \mathrm{Sym}(\Omega)$ be the action of G by right multiplication on Ω, as defined in Example 2.4. Since we can compute the coset representative of any element in G, it is straightforward to compute $\varphi(g)$ for any $g \in G$; we just need to find the coset representative in T of tg for each $t \in T$. If we need to calculate $\varphi(g)$ for many $g \in G$, then there are two possible methods. We could simply do it for each g, as just described. Or we could calculate $\varphi(g)$ in this manner just for the generators of G, and then use the general method for group homomorphisms described in Section 3.4 and Section 4.5 to compute $\varphi(g)$ for subsequent $g \in G$. It turns out that the first method is faster if $|G:H|$ is small, particularly if $\mathrm{im}(\varphi)$ is much smaller than G itself, but the second method can be faster if $|G:H|$ is very large, so this choice of methods can be a difficult one.

4.6.8 Finding double coset representatives

DEFINITION 4.12 If H and K are subgroups of a group G, then a *double coset* of H and K in G is defined to be a subset of G of the form $HgK := \{\, hgk \mid h \in H,\, k \in K \,\}$ for some fixed $g \in G$.

As with single cosets, it is straightforward to prove that two double cosets of H, K in G are either equal or disjoint, and so the double cosets partition G, and we can define a set of double coset representatives of H, K in G to be a subset $\{g_1, \ldots, g_t\}$ of G such that G is the disjoint union of the double cosets Hg_iK $(1 \le i \le t)$.

There are a number of algorithms in CGT that require the computation of sets of double coset representatives; we shall see one later, in Section 9.1. Unfortunately, no really satisfactory algorithm for solving this problem has been found to date.

The straightforward approach is as follows. First find representatives of the set Ω of right coset representatives of H in G, together with the 'coset action' homomorphism $\varphi : G \to \mathrm{Sym}(\Omega)$, as described in the preceding subsection. Then $\{g_1, \ldots, g_t\}$ is a set of double coset representatives of H, K in G, where $Hg_1, \ldots, Hg_t$ are representatives of the orbits of the action of K on Ω defined by φ. (We leave the proof of that statement to the reader.) So we can use the standard orbits procedures described in Section 4.1 to find them.

The problem, of course, with this method is that it involves calculation of the set of coset representatives of H in G, which is typically much larger than the required set of double coset representatives. One simple possibility for improved performance is to notice that, if K is larger than H, then it is quicker to compute the transversal and coset action homomorphism using K rather than H. This will enable us to find representatives $\{g_1, \ldots, g_t\}$ of the double cosets KgH, and then $\{g_1^{-1}, \ldots, g_t^{-1}\}$ is a set of representatives of the double cosets HgK.

If H is not a maximal subgroup of G, then the following divide-and-conquer approach due originally to Laue for solvable groups [Lau82], and implemented for general finite groups in GAP by Hulpke can be used. Let X be a subgroup of G with $H < X < G$. First construct the coset action homomorphism of G on the set Ω of right cosets of X in G, and use this to find orbit representatives $Xg_1, \ldots, Xg_t$ of the action of K on Ω, together with the corresponding stabilizers $K_1, \ldots, K_t$ of Xg_i in K. Let $Hx_1, \ldots Hx_s$ be the right cosets of H in X and, for $1 \le i \le t$, compute orbit representatives of the action of K_i by multiplication on the right on the set $\{\, Hx_1g_i, \ldots Hx_sg_i \,\}$. If these orbit representatives are $Hx_{i_1}g_i, \ldots Hx_{i_{u_i}}g_i$, then the required double cosets representatives of H and K in G are the elements $x_{i_k}g_i$ for $1 \le i \le t$ and $1 \le k \le u_i$.

More generally, a chain of subgroups $H < X_1 < \cdots < X_k < G$ can be used in this manner to reduce the problem to a union of smaller problems.

We also mention a method due to G. Butler for finding a canonical representative of a given double coset, which is described in [But84]. This is, however, known to be an NP-hard problem in general; see [Luk93].

Exercises

1. Show that two double cosets of $H, K \leq G$ are either equal or disjoint.

2. Prove that the stabilizer in K of the coset Hg in the action of K by right multiplication on the right cosets of H in G is equal to $K \cap g^{-1}Hg$. Deduce that $|HgK| = |H||K|/|K \cap g^{-1}Hg|$.

4.7   Sylow subgroups, p-cores, and the solvable radical

Finding the Sylow subgroups of a given finite group is one of the most basic problems in CGT, and algorithms for solving this and related problems are among the most frequently used in practice. It is also unusually interesting as one of the few problems in finite permutation groups that are known to be solvable in polynomial time, but only by algorithms that rely on the classification of finite simple groups and on nontrivial properties of the finite simple groups. These algorithms were devised by Kantor and described in [Kan85] and [Kan90]. They have not yet been fully implemented, and there remain significant impediments to an implementation that could efficiently handle groups having exceptional groups of Lie type as composition factors.

The current implementations in GAP and MAGMA still rely partly on backtrack searches, so they are not polynomial-time algorithms. The backtrack searches arise in certain centralizer computations, which fortunately are very fast in practice much of the time. Much of the work in developing these methods was by Butler and Cannon, and is described in [BC91] and various earlier papers, such as [BC89]. Various later ideas for cutting down the number of situations where one needs to resort to backtracks searches, as well as the size of the groups involved in these searches are discussed by P.M. Neumann in [Neu87] and by Cannon, Cox, and Holt in [CCH97b], and we shall describe some of these techniques in this section. Current implementations seem to run reasonably quickly on moderately large examples, but the existence of the backtrack searches means that there remains a danger of encountering unsatisfactory behaviour whenever we try to compute with larger examples than previously.

Other subgroups that we should like to be able to compute efficiently include the *p-core* $\mathbf{O}_p(G)$, for a prime p, which is defined to be the largest normal subgroup of G of order a power of p, and the *solvable radical* $\mathbf{O}_\infty(G)$, which is defined to be the largest normal solvable subgroup of G. Although the p-core was at one time computed as the intersection of the Sylow p-subgroups of G, this turned out not to be a good approach, since it involves backtracks in both the Sylow subgroup and intersection phases.

It is shown, for example, by Luks in [Luk93], that both $\mathbf{O}_p(G)$ and $\mathbf{O}_\infty(G)$ are computable in polynomial time, and nearly-linear-time algorithms are described in Section 6.3.1 of [Ser03]. Fast practical algorithms are also available for computing these subgroups. The approach that we shall describe here for $\mathbf{O}_p(G)$ is due to Unger [Ung02].

We have already (in Subsection 3.1.1) remarked on the unfortunate fact that we cannot in general find a faithful permutation representation of a quotient group G/N of a permutation group G of degree comparable to that of G. However, it turns out that we can find such a representation for $G/\mathbf{O}_p(G)$ and for $G/\mathbf{O}_\infty(G)$. This result was first proved by Easdown and Praeger in [EP88] and is also proved by Holt in [Hol97], where it is shown that the homomorphisms involved can be computed explicitly. We shall include a proof of it here. It enables us to compute $\mathbf{O}_\infty(G)$ by first finding some nontrivial $\mathbf{O}_p(G)$ and then solving the problem recursively in $G/\mathbf{O}_p(G)$.

4.7.1 Reductions involving intransitivity and imprimitivity

We can reduce the difficulty of many computational problems in finite permutation groups by a divide-and-conquer technique using actions on orbits, in the case of intransitive groups, and actions on blocks, in the case of imprimitive groups.

Some of these reductions are based on the idea of solving the problem, in turn, in two proper quotients G/M and G/N of G, where $M \cap N = 1$. More precisely, we have the following results for Sylow subgroups and $\mathbf{O}_p(G)$.

LEMMA 4.13 *Let G be a finite group having normal subgroups M and N with $M \cap N = 1$ and let $\mu : G \to G/M$ and $\nu : G \to G/N$ be the natural epimorphisms. Let p be a prime.*

(i) *Let $Q/M \in \mathrm{Syl}_p(G/M)$ and $Q := \mu^{-1}(Q/M)$, and let $R/N \in \mathrm{Syl}_p(\nu_Q(Q))$ and $P := \nu_Q^{-1}(R/N)$. Then $P \in \mathrm{Syl}_p(G)$.*

(ii) *Let $Q/M = \mathbf{O}_p(G/M)$ and $Q := \mu^{-1}(Q/M)$, and let $R/N = \mathbf{O}_p(\nu_Q(Q))$ and $P := \nu_Q^{-1}(R/N)$. Then $P = \mathbf{O}_p(G)$.*

PROOF (i) Clearly Q contains a Sylow p-subgroup of G and, since $M \cap N = 1$, $Q \cap N$ is a p-group. Similarly, R contains a Sylow p-subgroup of Q and hence of G. Now $R = PN$, where $PN/N \cong P/(P \cap N)$ is a p-group. Furthermore $P \cap N = Q \cap N$, so $P \cap N$ and hence P is a p-group, and the result follows.

(ii) As in (i), $P \cap N = Q \cap N$ is a p-group, so $P = \mathbf{O}_p(Q)$. But $\mathbf{O}_p(G) \leq Q$, so $\mathbf{O}_p(G) = \mathbf{O}_p(Q) = P$. ∎

There are two specific situations in which we will apply this lemma. The first is when G^Ω is intransitive, in which case we can write $\Omega = \Omega_1 \cup \Omega_2$ where G fixes the nonempty and nonintersecting sets Ω_1 and Ω_2. As we saw in Subsection 4.5.1, we can compute easily and efficiently with the induced action homomorphisms $\mu : G \to G^{\Omega_1}$ and $\nu : G \to G^{\Omega_2}$. Let $M := \ker(\mu)$ and $N := \ker(\nu)$. Then, since G is a permutation group on $\Omega = \Omega_1 \cup \Omega_2$, we have $M \cap N = 1$.

We can then recursively call our algorithm for computing $\mathbf{O}_p(G)$ or a Sylow p-subgroup of G on $G^{\Omega_1} \cong G/M$ and $Q^{\Omega_2} \cong QN/N$, which both have smaller permutation degree than G. Then we can use the recipe in Lemma 4.13 to construct $\mathbf{O}_p(G)$ or a Sylow p-subgroup of G.

The second situation is when G^Ω is imprimitive and has two distinct minimal block systems Σ_1 and Σ_2, say. Of course, to apply this reduction, we first need to apply the algorithms described in Section 4.3 to find all such minimal block systems. Here, we let $\mu : G \to G^{\Sigma_1}$ and $\nu : G \to G^{\Sigma_2}$ be the induced action homomorphisms. We saw in Subsection 4.5.2 that we can compute effectively with these homomorphisms. Again, we must have $M \cap N = 1$, because otherwise the orbits of $M \cap N$ would form a block system with blocks strictly contained in the blocks of Σ_1 and Σ_2, thereby contradicting their minimality. So again we can solve our problems by using recursive calls to smaller examples.

In the case of a unique minimal block system Σ, we can still use recursion on G^Σ, which enables us to assume that G^Σ is a p-group, but we cannot make any immediate further reductions. (A p-group is, by definition, a group of order p^n for some $n \geq 0$.)

In the following two subsections, we shall consider in more detail the cases of computing Sylow p-subgroups and $\mathbf{O}_p(G)$. It will turn out that, for $\mathbf{O}_p(G)$, the unique minimal block system obstacle does not arise.

4.7.2 Computing Sylow subgroups

As explained in the previous subsection, when computing a Sylow p-subgroup of G, we can assume, by recursion and part (i) of Lemma 4.13, that G is transitive, and is either primitive, or is imprimitive with a unique minimal block system Σ such that G^Σ is a p-group. We can also assume that $|\Omega|$ is divisible by p, since otherwise we just compute a Sylow p-subgroup of the smaller group G_α for some $\alpha \in \Omega$.

There are other possible reductions we could make. One approach, which is advocated in [CCH97b], is to solve the Sylow subgroup finding algorithm at the same time as solving the problem of finding a conjugating element for two Sylow subgroups P and Q of G. This allows us to make further reductions in the imprimitive case, by using mutual recursion between the two problems.

Another possibility, which we shall mention again in a later chapter, is to compute $\mathbf{O}_\infty(G)$ and, if it is nontrivial, recursively compute $R/\mathbf{O}_\infty(G) \in$ $\mathrm{Syl}_p(G/\mathbf{O}_\infty(G))$, and then find a Sylow p-subgroup of the solvable group R, possibly using a PC-representation.

But however many recursions we make, if our group is primitive and non-abelian simple, then we can make no more of these types of reductions. In the polynomial-time algorithm of Kantor, we would first need to identify the isomorphism type of the simple group. This is not difficult and, by Proposition 10.8, it can nearly always be done just from a knowledge of $|G|$. In the few ambiguous cases, there are simple tests to distinguish between the two possibilities. We then need to set up an isomorphism between G and the natural representation of the isomorphic simple group. For example, if $G \cong \mathrm{PSL}(n, q)$, then we would define an epimorphism from the natural representation of $\mathrm{SL}(n, q)$ as a group of $d \times d$ matrices over $\mathbb{F}_q$ to G. We then simply 'write down' a Sylow subgroup in the natural representation, and compute a Sylow subgroup of G by using the isomorphism.

There is one situation in which we do this without any further work at all, namely when G is the alternating or symmetric group on Ω, when it is straightforward to simply write down a Sylow p-subgroup, so we should certainly do that.

Otherwise, in the absence of the relatively large amount of machinery required to implement the Kantor approach, which we have only hinted at here, we are forced to fall back on backtrack algorithms. The most successful approach to this has been the method of Butler and Cannon, which we shall describe here. The backtrack part of the method occurs in the calculation of centralizers of group elements. Other possible approaches use normalizer calculations, but they are typically significantly slower than centralizer calculations.

The basic aim of the Butler-Cannon method is to find an element $g \in G$ of order p such that $C := \mathbf{C}_G(g)$ contains a Sylow p-subgroup of G. Once we have done that, and calculated C, then we apply recursion to C and thereby assume that $C = G$. Now g has order p, so the orbits of g have size 1 or p. Since we are assuming that G^Ω is transitive, and, by Proposition 4.9, G permutes the orbits of g, all of the orbits of g must have order p, and they form a minimal block system Σ of G^Ω. We are also assuming that G^Σ is a p-group. We claim that G is p-group, and so in fact we are done.

To prove this, we must show that the kernel K of the action of G on Σ is a p-group. Let Ω_i $(1 \le i \le r)$ be the orbits of g and, for $1 \le i \le r$, let g_i be the permutation of Ω that acts like g on Ω_i, and fixes all points of Ω_j for $j \ne i$. Then the elements g_i generate an elementary abelian p-group Q of order p^r, which contains g. We claim that $K \le Q$. To see this, let $h \in K$. Since $K \le C$, h^{Ω_i} centralizes $g_i^{\Omega_i} = g^{\Omega_i}$ for each i, and so, by Corollary 4.11, $h^{\Omega_i} = (g_i^{m_i})^{\Omega_i}$ for some m_i with $0 \le m_i < p$. Then $h = \prod_{i=1}^r g_i^{m_i}$, which proves the claim.

There remains the problem of finding a suitable element g. We start by

SYLOW(G, p)

    Input: A permutation group G and a prime p

    Output: Generators of $P \in \mathrm{Syl}_p(G)$

1 **if** G is a p-group **then return** G;

2 **if** G^Ω is intransitive **then** recurse on orbit actions;

3 **if** p does not divide $|\Omega|$ **then** recurse on G_α;

4 **if** G^Ω is imprimitive with two minimal block systems

5   **then** recurse on block actions;

6 **if** G^Ω is imprimitive with unique minimal block system Σ **and**

                 G^Σ is not a p-group

7   **then** recurse on G^Σ;

8 **if** $G = \mathrm{Sym}(\Omega)$ or $\mathrm{Alt}(\Omega)$ **then return** a Sylow p-subgroup;

9 Find an element g of G of order p;

10 **if** p^2 does not divide $|G|$ **then return** $\langle g \rangle$;

11 Compute $C := \mathbf{C}_G(g)$;

12 **while** C does not contain a Sylow p-subgroup of G

13   **do** Find $S \in \mathrm{Syl}_p(C)$ and compute $Z := \mathbf{Z}(S)$;

14   **for** elements $h \in Z$ with $|h| = p$ **do** compute $C_h := \mathbf{C}_G(h)$

15   until we find an h such that p divides $|C_h/S|$;

16   Replace C by C_h;

17 Recurse on C;

looking for any element g of order p. To do this, we choose random elements h of G until we find one whose order pm is multiple of p, and take $g := h^m$. In many groups this works very quickly. There are a few examples in which it does not; for example, there are groups of order $p(p-1)$ in which only $p-1$ elements have order a power of p, which is a small proportion when p is large. In this, and other similar examples, a policy of sometimes choosing h to be a commutator of random elements already considered will often find the required element quickly.

Having found g of order p we must then compute $C := \mathbf{C}_G(g)$. If C contains a Sylow p-subgroup of G then we are done. Otherwise, we still compute $S \in \mathrm{Syl}_p(C)$. Now, by Sylow's theorem, S is contained in some $P \in \mathrm{Syl}_p(G)$, and P has nontrivial centre, so there exists $h \in \mathbf{Z}(P)$ with h of order p. Since $h \in C$, we must have $h \in S$, because otherwise S would be properly contained in $P \cap C$, contradicting $S \in \mathrm{Syl}_p(C)$. So we know that there exists an element g of order p in $\mathbf{Z}(S)$ having the required property that $\mathbf{C}_G(g)$ contains a Sylow p-subgroup of G.

So we compute $\mathbf{Z}(S)$, which is not difficult, and then search through the elements g of order p (see Exercise 3 below) in $\mathbf{Z}(S)$ looking for the required g. This necessitates computing $\mathbf{C}_G(g)$ for each such g, and is potentially the slowest part of the algorithm because, in a bad case, there could be many such elements g, with only a small proportion of them having the required centralizer. In practice, if we find any g for which the Sylow p-subgroup of $\mathbf{C}_G(g)$ is larger than S, then we replace S with this new Sylow p-subgroup. SYLOW is a summary of the complete Sylow p-subgroup algorithm.

4.7.3   A result on quotient groups of permutation groups

In this subsection, we shall prove the following result, which was probably first proved by Easdown and Praeger in [EP88]. The specific classes of groups to which we shall apply the result are the classes of all abelian groups, and of all elementary abelian p-groups for some prime p.

THEOREM 4.14   *Let G be a finite permutation group on Ω of degree n. Let $\mathcal{A}$ be a class of finite abelian groups closed under subgroup, quotient, and direct product, and let N be a normal $\mathcal{A}$-subgroup of G. Then there exists a permutation representation φ of G of degree at most n, such that the kernel K of φ is an $\mathcal{A}$-group that contains N.*

PROOF   We may clearly assume that $N \neq 1$. Let $S := \mathrm{Sym}(\Omega)$, let Σ be the set of orbits of N on Ω, and let Γ be a subset of Ω containing one representative from each orbit $\Delta \in \Sigma$. Let L be the subgroup of S consisting of all $g \in S$ such that g fixes each $\Delta \in \Sigma$ and, for all $\Delta \in \Sigma$, there exists $h \in N$ with $g^\Delta = h^\Delta$. Then L is isomorphic to the direct product of the induced action groups N^Δ for $\Delta \in \Sigma$, and so, by the assumptions on the class $\mathcal{A}$, we have $L \in \mathcal{A}$. It is easy to check that L is normalized by G. Let $E := GL$, and let $K := G \cap L$. Then clearly $N \leq K$, $K \in \mathcal{A}$ (since $\mathcal{A}$ is subgroup closed), and $E/L = GL/L \cong G/K$.

Let $D := E_\Gamma$ be the setwise stabilizer of Γ in E. We claim that $E = DL$. Let $e \in E$. To establish the claim, we shall define an element $f \in L$ such that $ef \in D$. To do this, we need to define f^Δ for each $\Delta \in \Sigma$. Since G permutes the set Σ, we have $\Delta^{e^{-1}} \in \Sigma$ and there is a unique point $\gamma \in \Gamma \cap \Delta^{e^{-1}}$. Then $\gamma^e \in \Delta$, and so there exists $h \in N$ with γ^{eh} equal to the unique point in $\Gamma \cap \Delta$. We define f^Δ to be equal to h^Δ. Then, by definition of L, we have $f \in L$, and by choice of h, we have $\gamma^{ef} \in \Gamma$ for all $\gamma \in \Gamma$. Thus $ef \in D$, and $E = DL$ as required. Since L is abelian, by Corollary 4.11, any element in L that fixes a point of one of its orbits fixes the whole orbit pointwise. But $L \cap D$ must fix every point in Γ, and hence $L \cap D = 1$, and D is a complement of L in E.

Now we have $G/K \cong E/L \cong DL/L \cong D$, and since D is a permutation group of degree n, we can define the required permutation representation φ of G by $\varphi(g) = gf \in D$, where f is as defined above for $e = g$. This completes the proof.   ∎

For computational applications, it is important to observe that the proof allows us easily to compute $\varphi(g) = gf$ for $g \in G$, and hence we can compute effectively with the homomorphism φ.

COROLLARY 4.15 *Let G be a finite permutation group on Ω of degree n. Then there exist faithful permutation representations of degree at most n of $G/\mathbf{O}_p(G)$, for any prime p, and of $G/\mathbf{O}_\infty(G)$.*

PROOF The proof is by induction on $|G|$. The results are all clear if $\mathbf{O}_\infty(G) = 1$, so assume not, and let N be a minimal normal subgroup of G contained in $\mathbf{O}_p(G)$ or in $\mathbf{O}_\infty(G)$, as appropriate. Since $\mathbf{O}_\infty(G)$ is solvable, its composition factors are all cyclic groups of prime order, and it follows from Corollary 2.32 that N is an elementary abelian p-group. So, by Theorem 4.14, there is an abelian group K containing N such that G/K has a permutation representation of degree at most n, and the result follows by induction applied to G/K. $\blacksquare$

Provided that we can compute $\mathbf{O}_p(G)$ for primes p, which we shall be explaining how to do in the next subsection, then the representations in Corollary 4.15 can again be computed explicitly. For the representation of $G/\mathbf{O}_p(G)$, we could, for example, first compute $\mathbf{O}_p(G)$, then $N := \mathbf{Z}(\mathbf{O}_p(G))$, then use Theorem 4.14 to find a faithful permutation representation φ_1 of G/K_1 where $N \leq K_1 \leq \mathbf{O}_p(G)$. Then, if $K_1 \neq \mathbf{O}_p(G)$, we repeat the process on $\mathrm{im}(\varphi_1)$ to get a faithful permutation representation of G/K_2 where $K_1 < K_2 \leq \mathbf{O}_p(G)$, and carry on like this until we have the required representation φ of $G/\mathbf{O}_p(G)$. (Notice that it is neither necessary nor desirable to choose N to be elementary abelian like we did in the proof of the corollary.)

To get the faithful representation of $G/\mathbf{O}_\infty(G)$, we first find a prime p with $\mathbf{O}_p(G) \neq 1$, then find a faithful permutation representation φ_1 of $G/\mathbf{O}_p(G)$, as described above. Then, if $\mathbf{O}_p(G) \neq \mathbf{O}_\infty(G)$, we repeat the process on $\mathrm{im}(\varphi_1)$, and carry on until like this until we have found the required representation.

4.7.4 Computing the p-core

The method that we shall describe here for $\mathbf{O}_p(G)$ is due to Unger [Ung02]. As was the case for Sylow p-subgroups, we can reduce, by recursion and part (ii) of Lemma 4.13, to the situation where G is transitive, and is either primitive, or is imprimitive with a unique minimal block system Σ such that G^Σ is a p-group.

Let us first deal with the unique minimal block system case. Let K be the kernel of the action of G on Σ. If $K = 1$, then G is a p-group and $G = \mathbf{O}_p(G)$, so assume not. We recursively compute $Q := \mathbf{O}_p(K)$. Suppose that $Q = 1$. Then we claim that $\mathbf{O}_p(G) = 1$. If not, then let N be a minimal normal subgroup of G contained in $\mathbf{O}_p(G)$. Since the composition factors of any p-group are cyclic of order p, it follows from Corollary 2.32 that N is an elementary abelian p-group. Since $Q = 1$, we have $K \cap N = 1$, and hence $[K, N] \leq K \cap N = 1$, so $K \leq \mathbf{C}_G(N)$. Let Δ be an orbit of N^Ω. Then, by Proposition 2.29, either $\Delta = \Omega$, or Δ is a block of imprimitivity of G^Ω. Since

Σ is the unique minimal block system of G^Ω, it is true in either case that Δ is a union of blocks of Σ. So K fixes the set Δ, and then Corollary 4.11 implies that $K^\Delta \leq N^\Delta$ and hence K^Δ is a p-group. But this is true for all orbits Δ of K, so K is a p-group, contradicting $Q = 1$. Hence $\mathbf{O}_p(G) = 1$ as claimed, and so we may assume that $Q = \mathbf{O}_p(K) \neq 1$.

Now we compute $N := \mathbf{Z}(Q)$ and use Theorem 4.14 to construct a epimorphism $\varphi : G \to H$ for which $M := \ker(\varphi)$ is a p-group containing N. Then we can apply recursion to compute $\mathbf{O}_p(H)$ and hence $\mathbf{O}_p(G) := \varphi^{-1}(\mathbf{O}_p(H))$.

It remains to deal with the case in which G^Ω is primitive. In that case, if $\mathbf{O}_p(G) \neq 1$, then let N be a minimal normal subgroup of G contained in $\mathbf{O}_p(G)$. As above, N is elementary abelian, and N^Ω is transitive by Proposition 2.29 and regular by Corollary 4.11. So it is an elementary abelian regular normal subgroup, which is often abbreviated to EARNS, of G.

If $N < \mathbf{O}_p(G)$, then $\mathbf{Z}(\mathbf{O}_p(G)) \cap N \neq 1$, but by Corollary 4.11 again, we cannot have $N \leq \mathbf{Z}(\mathbf{O}_p(G))$, so $1 < \mathbf{Z}(\mathbf{O}_p(G)) \cap N < N$, contradicting the minimality of N as a normal Subgroup of G. Hence $N = \mathbf{O}_p(G)$.

So the problem reduces to deciding whether a primitive group has an EARNS. The method that we shall describe for this comes originally from P.M. Neumann [Neu87]. If the degree n of G is not equal to a prime power p^d, then clearly G has no EARNS, so we may assume that $n = p^d$.

First we need to recall some basic properties of Frobenius groups.

DEFINITION 4.16 A *Frobenius group* is a transitive permutation group on a set Ω such that $G_\alpha \neq 1$, but $G_{\alpha\beta} = 1$ for all distinct $\alpha, \beta \in \Omega$.

Let G^Ω be a finite Frobenius group of degree n. By a simple counting argument, we find that there are exactly $n-1$ elements of G that fix no point of Ω. The fundamental result of Frobenius, dating from 1901, is that these $n-1$ elements, together with the identity, form a normal subgroup K of G, known as the *Frobenius kernel* of G. The proof uses character theory, and no proof is currently known that does not use character theory. The stabilizer G_α is known as the *Frobenius complement*. Various properties of the Frobenius complement can be shown without too much difficulty. For example, the centre is nontrivial, the Sylow p-subgroups are cyclic for p odd, and cyclic or generalized quaternion for p even, and A_5 is the only possible nonabelian composition factor of a Frobenius complement. A much deeper result, due to J.G. Thompson (1959), is that the Frobenius kernel is nilpotent. For proofs of all of these properties see, for example, [Pas68].

Returning to our EARNS algorithm, we first check whether $G_{\alpha\beta} = 1$ for all distinct $\alpha, \beta \in \Omega$. If so, then either $G_\alpha = 1$, in which case G^Ω primitive implies that $|G| = p$, so G itself is the EARNS, or G is a Frobenius group. In that case, we compute $Z = \mathbf{Z}(G_\alpha)$ for some α, which we know is nontrivial, choose $z \in Z \setminus \{1\}$, choose $h \in G \setminus G_\alpha$, and let $g := [h, z]$. Then $g \neq 1$ is in the Frobenius kernel of G (exercise) and, since G^Ω is primitive, we can compute the full kernel, which is the required EARNS, as the normal closure $\langle g^G \rangle$.

So we assume that $G_{\alpha\beta} \neq 1$ for some distinct $\alpha, \beta \in \Omega$. We fix α and then choose β from the orbit of G_α on $\Omega \setminus \{\alpha\}$ of the smallest size. This gives the largest possible $|G_{\alpha\beta}|$, so clearly $G_{\alpha\beta} \neq 1$ with this choice. Let Δ be the set of fixed points of $G_{\alpha\beta}$; that is, $\Delta := \{ \gamma \in \Omega \mid \gamma^g = \gamma \ \forall g \in G_{\alpha\beta} \}$.

Suppose that G does have an EARNS N. Then, for $\gamma \in \Delta$, there is a (unique) $g \in N$ with $\alpha^g = \gamma$. For $h \in G_{\alpha\beta}$, we have $g^{-1}hg \in G_\gamma$, so $[h, g] \in N \cap G_\gamma = 1$, and hence $g \in C := \mathbf{C}_G(G_{\alpha\beta})$. In particular, $N \cap C$ acts regularly on Δ, so if $|\Delta|$ is not a power of p, then G has no EARNS, and we are done.

Otherwise, we compute C. This is typically an easy centralizer computation, but it can be done in polynomial (indeed, in almost linear) time; see Section 6.2.3 of [Ser03] for details. If C does not act transitively on Δ, then again G has no EARNS, so assume that it does.

For any $\gamma \in \Delta \setminus \{\alpha\}$, we have $G_{\alpha\beta} \leq G_{\alpha\gamma}$ by definition of Δ, but the the choice of β forces $G_{\alpha\beta} = G_{\alpha\gamma}$, so $G_{\alpha\gamma}$ fixes all points of Δ, and C^Δ must be regular or a Frobenius group. In the latter case, $(N \cap C)^\Delta$ must be the Frobenius kernel. So we compute this kernel K^Δ as described above for the case when G^Ω was a Frobenius group, and then let K be the inverse image of K^Δ in C. (There is an extra complication here, because C^Δ might be an imprimitive Frobenius group, but we can choose further generators of the form $g := [h, z]$ for $z \in \mathbf{Z}(C_\alpha)$ and random $h \in C \setminus C_\alpha$ until we have generated the whole of K.) If C^Δ is regular, then we just let $K = C$.

In either case, we now have a subgroup $K \leq C$ with K^Δ regular which, if the EARNS N exists at all, satisfies $N \cap C \leq K$ and $(N \cap C)^\Delta = K^\Delta$. Next we compute the subgroup $P := \{ x \in C_{\alpha\beta} \mid x^p = 1 \}$; see Exercise 3 below. Choose any $x \in C$ with $|x| = p$ and $x \notin C_{\alpha\beta}$. (Any $1 \neq x \in N \cap C$ has this property, so if there is no such x, then G has no EARNS.)

Then $x^\Delta = g^\Delta$ for some $g \in N \cap C$, so $x^{-1}g \in P$. Hence we can find g as xh for some $h \in P$. For each such xh, we test whether $g := xh \in N$ by computing $\langle g^G \rangle$, and checking whether it is an EARNS. If we try all $h \in P$ and find no EARNS, then G has no EARNS. It can be shown (see exercises below) that if N exists, then $|P| \leq p^{d-1}$ (recall that $n = |\Omega| = p^d$), so there are not too many elements to try! As a refinement, if, on calculating P, we find that $|P| > p^{d-1}$, then there can be no EARNS, and we can stop immediately.

P-CORE is a summary of the complete p-core algorithm.

4.7.5   Computing the solvable radical

It is now straightforward to compute the solvable radical $\mathbf{O}_\infty(G)$ of G. We consider the primes p dividing $|G|$ and compute $\mathbf{O}_p(G)$ as described above. If $\mathbf{O}_p(G) = 1$ for all such p, then $\mathbf{O}_\infty(G) = 1$.

If we find a prime p with $\mathbf{O}_p(G) \neq 1$ then, by Corollary 4.15, we can compute an explicit epimorphism $\varphi : G \to H := G/\mathbf{O}_p(G)$. Now we can recursively compute $\mathbf{O}_\infty(H)$ and hence compute $\mathbf{O}_\infty(G) := \varphi^{-1}(\mathbf{O}_\infty(H))$.

P-CORE(G, p)

> **Input:** A permutation group G^Ω and a prime p
> **Output:** Generators of $\mathbf{O}_p(G)$
>
> 1   **if** G is a p-group **then return** G;
> 2   **if** G^Ω is intransitive **then** recurse on orbit actions;
> 3   **if** G^Ω is imprimitive with two minimal block systems
> 4   **then** recurse on block actions;
> 5   **if** G^Ω is imprimitive with unique minimal block system Σ
> 6   **then** Compute the kernel K of the action of G on Σ;
> 7   **if** $K = 1$ **then** recurse on G^Σ;
> 8   Compute $Q := \mathbf{O}_p(K)$;
> 9   **if** $Q = 1$ **then return** $\{\}$;
> 10   Compute $N := \mathbf{Z}(Q)$;
> 11   Find a permutation representation φ of G such that
> $M := \ker(\varphi)$ is a p-group with $N \le M$;
> 12   Recursively compute $L := \mathbf{O}_p(\mathrm{im}(\varphi))$;
> 13   **return** generators of $\varphi^{-1}(L)$;
> $(\ast$ Now G^Ω is primitive – we use Neumann's EARNS routine $\ast)$
> 14   **if** $n := |\Omega|$ is not a power p^d of p **then return** $\{\}$;
> 15   Choose $\alpha \in \Omega$;
> 16   **if** G^Ω is a Frobenius group
> 17   **then** Choose $1 \ne z \in \mathbf{Z}(G_\alpha)$ and $h \in G \setminus G_\alpha$;
> 18   **return** generators of $\langle [h, z]^G \rangle$;
> 19   Choose β in an orbit of G_α on $\Omega \setminus \{\alpha\}$ of minimal length;
> 20   Compute the fixed point set Δ of $G_{\alpha\beta}$;
> 21   **if** Δ is not a power of p **then return** $\{\}$;
> 22   Compute $C := \mathbf{C}_G(G_{\alpha\beta})$;
> 23   If C^Δ is not transitive **then return** $\{\}$;
> 24   Compute subgroup $R \le C$, where R^Δ is the regular normal subgroup
>   of the regular or Frobenius group C^Δ;
> 25   Compute $P := \{\, x \in C_{\alpha\beta} \mid x^p = 1 \,\}$;
> 26   **if** $|P| > p^{d-1}$ **then return** $\{\}$;
> 27   Find $x \in R \setminus C_{\alpha\beta}$ with $|x| = p$;
> 28   **for** $h \in P$
> 29   **do** Compute $N := \langle (xh)^G \rangle$;
> 30   **if** N is an EARNS of G **then return** generators of N;
> 31   **return** $\{\}$;

4.7.6   Nonabelian regular normal subgroups

In this final section, we shall take note of the fact that Neumann's EARNS algorithm can be used almost unchanged to test for the existence of, and find nonabelian regular normal subgroups in a primitive permutation group G^Ω. We shall be using this in Subsection 10.1.4 as part of an algorithm to find the socle of a general primitive permutation group.

By Thompson's result, a Frobenius kernel is nilpotent, and so the Frobenius kernel of a primitive group is elementary abelian, and hence a primitive Frobenius group cannot have a nonabelian regular normal subgroup.

As in the EARNS routine, we fix $\alpha \in \Omega$ and choose β from the orbit of G_α on $\Omega \setminus \{\alpha\}$ of the smallest size; then $G_{\alpha\beta} \neq 1$. Let Δ be the set of fixed points of $G_{\alpha\beta}$, and compute $C := \mathbf{C}_G(G_{\alpha\beta})$.

As in the EARNS case, for any regular normal subgroup N of G, $N \cap C$ acts regularly on Δ, so if C^Δ is not transitive, then there is no such N, and we are done. By choice of β, C^Δ must either be a Frobenius group or act regularly. In the first case we find $K \leq C$ such that K^Δ is the Frobenius kernel, and in the second case we let $K = C$.

Choose $g \in K$ of prime order p, with $x^\Delta \neq 1$ (if there is no such g, then there is no regular normal subgroup), and compute the subgroup P consisting of elements of order 1 or p in $C_{\alpha\beta}$. Then, as in the EARNS case, a regular normal subgroup of G is the normal closure of xh for some $h \in P$, and so we just try all possible h.

We shall see later, in Theorem 10.1.3, that a primitive group can have one or two nonabelian regular normal subgroups. If we find one such subgroup N, then we could either continue the search through $h \in P$ to look for a possible second subgroup, or use the fact that this second subgroup, if it exists, must be the centralizer in $\mathrm{Sym}(\Omega)$ of N.

It can be shown, using the O'Nan-Scott theorem (Theorem 10.1.3), that $|P| < |\Omega|$, and so this algorithm can be made to run in polynomial time. Seress points out in Remark 6.2.14 of [Ser03] that it is not almost linear, and he presents an alternative almost linear Monte Carlo algorithm due to Luks to find nonabelian regular normal subgroups.

Exercises

Assume that G is finite in these exercises.

1. Prove that $\mathbf{O}_p(G)$ is the intersection of the Sylow p-subgroups of G.

2. Prove that $\mathbf{O}_\infty(G) \neq 1$ if and only if $\mathbf{O}_p(G) \neq 1$ for some prime p.

3. Let $G = \langle X \rangle$ be a finite abelian p-group. Devise an algorithm for computing the subgroup P of G consisting of all elements of order 1 or p.

   (*Hint*: Let $X = \langle x_1, \ldots, x_r \rangle$ with $o_i \geq o_{i+1}$ for $1 \leq i < r$, where $o_i = |x_i|$. For each i, if $x_i^{o_i/p} \notin \langle x_j^{o_j/p} \mid 1 \leq j < i \rangle$, then include $x_i^{o_i/p}$ as a generator of P. Otherwise, replace x_i by $x_i g$ for suitable $g \in \langle x^j \mid 1 \leq j < i \rangle$ to reduce $|x_i|$.)

4. Work out how to write down generators of a Sylow p-subgroup of $\mathrm{Sym}(n)$ using the following steps.

(i) First write n in base p. That is, $n = \sum_{i=0}^{r} a_i p^i$ with $0 \le a_i < p$. Using group orders, show that a Sylow p-subgroup of $\mathrm{Sym}(n)$ is contained in the direct product of groups X_i $(1 \le i \le r)$, where each X_i is itself the direct product of a_i copies of $\mathrm{Sym}(p^i)$. This effectively reduces the problem to the case when n is a power p^i of p.

(ii) When $n = p^i$, show that a Sylow p-subgroup of $\mathrm{Sym}(n)$ is generated by a Sylow p-subgroup of $\mathrm{Sym}(p^{i-1})$ acting on the set $\{1..p^{i-1}\}$ together with the permutation $g \in \mathrm{Sym}(n)$ with $j^g = j + p^{i-1} \pmod{p^i}$ for $1 \le j \le p^i$.

(For $P_i \in \mathrm{Syl}_p(\mathrm{Sym}(p^i))$, we have $P_1 \cong C_p$ and, for $i > 1$, P_i is the wreath product $P_{i-1} \wr C_p$; see, for example, [Hal59, Sec. 5.9].)

5. Modify the method of Exercise 4 to write down $P \in \mathrm{Syl}_2(\mathrm{Alt}(n))$.

6. Show that there are exactly $n-1$ fixed-point-free elements in a Frobenius group of degree n.

7. In the description of the $\mathbf{O}_p(G)$ algorithm for Frobenius groups, prove the claim that the element g defined as $[h, z]$ is a nontrivial element in the Frobenius kernel.

8. Let H be any finite group and p a prime with $\mathbf{O}_p(H) = 1$. Let P be an abelian p-subgroup of H, and $C := \mathbf{C}_H(P)$. Prove that $|P| \le |H:C|$.

   (*Hint*: Use induction on $|H:C|$. Let Q be a conjugate of P in H that does not centralize P. Let $P^* := P \cap \mathbf{C}_H(Q)$, $C^* := \mathbf{C}_H(P^*)$, and deduce by induction that $|P^*| \le |H:C^*|$. Assume WLOG that $|P^*| \ge |Q \cap \mathbf{C}_H(P)|$, and deduce $|C^*:C| \ge |P:P^*|$.)

9. We saw earlier that, if a primitive group G^{Ω} has an EARNS N, then $N = \mathbf{O}_p(G)$ and hence $\mathbf{O}_p(G_\alpha) = 1$. Taking P to be the elements of order dividing p in $C_{\alpha\beta}$ with $C := \mathbf{C}_G(G_{\alpha\beta})$ as in P-CORE, apply the result in the previous exercise to $H = G_\alpha$ to prove that $|P| \le p^{d-1}$.

4.8 Applications

In this section we briefly review three applications of the methods discussed in the chapter. The first, to card shuffling, is primarily recreational, but has some relevance to computer networks. The second is to the construction of various types of combinatorial objects, including error-correcting codes, which are important outside of mathematics. The third is another application to computer networks.

4.8.1   Card shuffling

Consider a deck of $2n$ cards for some $n \in \mathbb{N}$. A perfect riffle shuffle is executed by dividing the deck into two equal piles, and then reassembling it by taking cards alternately from the two piles. There are two ways of doing this, the *out-shuffle*, in which we take the top card of the original deck first, and the *in-shuffle*, in which we take the top card of the original deck second when we re-assemble.

If we number the cards $1, \ldots, 2n$ from the top downwards in the original deck, then the permutations g_o and g_i of $[1 \mathinner{.\,.} 2n]$ defined by the out- and in-shuffles are given by:

$$(2i-1)^{g_o} = i, \ (2i)^{g_o} = n+i; \quad (2i-1)^{g_i} = n+i, \ (2i)^{g_i} = i.$$

for $1 \le i \le n$.

Persi Diaconis, who is one of the few mathematicians who can carry out perfect riffle shuffles accurately in practice, had been wondering for several years about the order and structure of the permutation group $\mathrm{Sh}(n) := \langle g_o, g_i \rangle \le \mathrm{Sym}(2n)$ generated by the two types of riffle shuffle when, in the early 1980s he got together with R.L. Graham and W.M. Kantor and investigated the problem seriously. Eventually they solved the problem completely [DGK83]. An implementation of the Schreier-Sims algorithm was used for experimental purposes, and some of the results obtained from these computations were used in the final proof. Donald E. Knuth was also involved in these computational experiments, and it was as a direct result of this application that he carried out his analysis of the Schreier-Sims algorithm in [Knu81], which was mentioned in Subsection 4.4.3.

It is straightforward to show that $\{ \{i, 2n-i+1\} \mid 1 \le i \le n \}$ is a system of imprimitivity preserved by $\mathrm{Sh}(n)$, and so $\mathrm{Sh}(n)$ is a subgroup of the wreath product $\mathrm{Sym}(2) \wr \mathrm{Sym}(n) = C_2 \wr \mathrm{Sym}(n)$ (see Subsection 2.2.5).

It turns out that, for most values of n, $\mathrm{Sh}(n)$ has index 1, 2 or 4 in $C_2 \wr \mathrm{Sym}(n)$, depending on the value of n mod 4. For example, in the case $n = 26$ of the standard deck of cards, we have $\mathrm{Sh}(26) = C_2 \wr \mathrm{Sym}(n)$. But there are some interesting exceptional cases. When n is a power of 2, $\mathrm{Sh}(n)$ is much smaller, and is isomorphic to $C_2 \wr C_n$. Furthermore $\mathrm{Sh}(6) \cong C_2^6 \rtimes \mathrm{PGL}(2,5)$ and, the most interesting exceptional case of all, which was partly what motivated Diaconis to want to crack the problem, is $\mathrm{Sh}(12) \cong C_2^{11} \rtimes M_{12}$. Here M_{12} is one of the five sporadic simple groups discovered in the 19th century by Mathieu.

This particular problem has applications to the interconnection of computers in networks for parallel processing, particularly in the case when n is a power of 2. See also the article of G. Kolata [Kol82] for further interesting related details, and that of Medvedoff and Morrison87 [MM87] for generalizations to the situation where the original deck is split into more than two piles.

4.8.2 Graphs, block designs, and error-correcting codes

In Subsection 4.6.4, we briefly discussed the use of backtrack searches in the computation of the automorphism groups of finite combinatorial objects and testing two such objects for isomorphism. It is also possible to go in the opposite direction; namely, start with a permutation group G on a finite set Ω and use it to construct combinatorial structures with vertex set Ω. These objects will generally contain G in their automorphism group, so it is a useful technique for finding examples with a relatively large automorphism group. 'Random' graphs or designs tend to have trivial or at most very small automorphism groups.

Later in the book, in Section 10.5, we shall briefly discuss the more fundamental problems of counting and enumerating *all* combinatorial structures of a given type.

The most straightforward construction of a combinatorial structure from a permutation group produces finite graphs. Usually we choose a transitive group G^Ω, fix $\alpha \in \Omega$, and let Δ be an orbit of the point stabilizer G_α other than $\{\alpha\}$. Then the graph with vertex set Ω and edge set $\{ (\alpha, \beta)^g \mid \beta \in \Delta, g \in G \}$ is easily seen to be regular of valency $|\Delta|$. (That is, each point α is the source of exactly $|\Delta|$ edges.) If G contains an element interchanging α and β for some $\beta \in \Delta$, then we will have $(\alpha, \beta) \in E$ if and only if $(\beta, \alpha) \in E$, and so the graph is undirected. Otherwise, it is directed. Its automorphism group will always contain G, but it is larger than G in many examples.

In general, a t-(n, k, λ) design on Ω where $|\Omega| = n$ is defined to be a set $\mathcal{B}$ of subsets of Ω of size k known as *blocks*, with the property that any t points in Ω are contained in exactly λ blocks. (Note: many authors prefer a more symmetrical definition in which Ω and $\mathcal{B}$ are sets with a given incidence relation between them.) Well-known examples with large t include the 5-$(12, 6, 1)$ and 5-$(24, 8, 1)$ designs, which have the Mathieu groups M_{12} and M_{24} as automorphism groups.

Error-correcting codes are used to correct errors introduced during transmission through noisy communication networks. So techniques for constructing such codes potentially have technological applications outside of mathematics. Pless [Ple81] provides an elementary treatment of coding theory, and Assmus and Key [AK92] describe the connection with designs, which we shall introduce here.

In general, the codewords to be transmitted will lie in $\mathbb{F}_q^n$, for some q and n. Since binary bits are universally used in practice, we might prefer to assume that $q = 2$, but there are some mathematically interesting codes for larger q. Only words from a fixed subset C of $\mathbb{F}_q^n$, the set of *codewords*, are transmitted. For $v, w \in \mathbb{F}_q^n$, we define the distance $d(v, w)$ between v and w to be the number of components in which v and w differ. The *minimum distance* d of C is defined to be the minimum of $d(v, w)$ for $v, w \in C$, with $v \neq w$. Suppose that codeword v is transmitted, but e of its components are changed during transmission, so a word v' is received with $d(v, v') = e$. Then, provided that $2e + 1 \leq d$, v will be the unique element of C with $d(v, v') \leq e$, and so the word v transmitted can be correctly reconstructed from the word v' received. In that case C is known as an *e-error-correcting code*.

There are various desirable properties of a good code. We would like e to be as large as possible but, for reasons of efficiency, we would also prefer $|C|/q^n$ to be large; unfortunately, these two requirements tend to conflict with each other! Another important property is the ease of decoding; that is, for given $v' \in \mathbb{F}_q^n$, we need to be able to locate the $v \in C$ with $d(v, v')$ minimal as quickly as possible.

A *linear code* C is one which is a subspace of $\mathbb{F}_q^n$. Most of the codes in widespread usage are linear, largely because, in favourable examples, decoding can be accomplished by means of a matrix multiplication. Their minimum distance is equal to the minimal weight (that is, the number of nonzero components) of a nonzero vector in C.

The technique that we shall discuss here is to start with a t-(n, k, λ) design on a set $\Omega = \{1 .. n\}$ with automorphism group containing permutation group G^Ω. Let $\mathcal{B}$ be the set of blocks in the design. Now choose a prime p and, for $B \in \mathcal{B}$, define v_B to be the characteristic function of B as a subset of Ω, regarded as a vector in F_p^n. Then let C be the subspace of F_p^n spanned by the vectors v_B with $B \in \mathcal{B}$. This construction results in some interesting codes. The two best-known examples are the *Golay codes* derived in this way from the 5-$(12, 6, 1)$ design with $p = 3$, and the 5-$(24, 8, 1)$ design with $p = 2$. They have minimum weights 6 and 8, respectively.

Codes of this kind can be looked for systematically in the following way. For details of specific examples, see [KMR03] and other recent papers by the same authors. As in the construction of graphs mentioned above, we start with a transitive (in fact, primitive) permutation group G^Ω with $|\Omega| = n$, fix $\alpha \in \Omega$ and let Δ be an orbit of G_α other than $\{\alpha\}$. (Or, more generally, we can let Δ be a union of such orbits.) It turns out that $\{\Delta^g \mid g \in G\}$ is a 1-$(n, |\Delta|, |\Delta|)$ design with automorphism group containing G. We can construct codes from these designs as described above for small primes p.

The MAGMA system is particularly well-suited for carrying out such computations since, in addition to the standard permutation group machinery, it contains functions for computing standard properties of designs and codes, such as their minimum weights and automorphism groups.

As a specific example that was considered by Key and Moori in [KM02], start with the second Janko group J_2. Its primitive permutation representations correspond to its maximal subgroups, of which there are nine conjugacy classes. (These maximal subgroups and corresponding permutation representations acting by right multiplication on their cosets can also be found quickly by means of standard function calls.) The two smallest degree primitive representations have degrees 100 and 280.

- A G_α-orbit Δ of length 36 in the first of these representations gives rise to a code of dimension 36 in $\mathbb{F}_2^{100}$ with minimum weight 16.

- A G_α-orbit Δ of length 108 in the second representation yields a code of dimension 14 in $\mathbb{F}_2^{280}$ with minimum weight 108.

An additional feature and advantage of this type of code is that the automorphism group can be used to speed up the decoding process; we omit the details.

4.8.3 Diameters of Cayley graphs

We recall from Definition 2.13 that the *Cayley graph* $\Gamma_X(G)$ of a group G with respect to a set X of generators of G is a directed labelled graph with vertex set $V := \{ v_g \mid g \in G \}$ in one-one correspondence with the set of elements of G and, for each $g \in G$ and $x \in X$, a directed edge labelled x from v_g to v_{gx} (which is regarded also as being an edge labelled x^{-1} from v_{gx} to v_g). The group G acts transitively as a group of automorphisms of $\Gamma_X(G)$ by multiplication on the left.

The *diameter* of a (connected) graph is defined to be the maximal distance between two of its vertices, where all edges have length 1. So the diameter of $\Gamma_X(G)$ is equal to $\max\{ l_X(g) \mid g \in G \}$, where $l_X(g)$ is the length of the shortest word in $(X \cup X^{-1})^*$ which represents g.

Finding the diameter of Cayley graphs of large finite groups, and the related problem of finding $l_X(g)$ for given $g \in G$, are among the most basic problems in CGT. Unfortunately, these problems represent a failure of CGT, in that no methods have been found for solving them that are significantly better than a brute-force depth first search through the vertices of $\Gamma_X(G)$ starting at v_{1_G}. The results by Even and Goldreich in [EG81] and by Jerrum in [Jer85] suggest that this is a fundamentally difficult problem. Some efforts at finding bounds can be found in the paper of Driscoll and Furst [DF87].

For example, one might have hoped that the monumental effort that has devoted to developing algorithms for permutation groups would have enabled us, by now, to compute the diameter of the Cayley graph of the Rubik cube group on its natural set of 6 generators. This is of course the same as the maximum number of quarter twists required to restore the cube to its pristine state starting from an arbitrary position, and is often referred to as the number of moves in *God's algorithm*. Unfortunately, this number is still unknown; the best-known lower and upper bounds for it are 24 and 42, but the answer is likely to be closer to 24 than to 42. The order of the group is $43\,252\,003\,274\,489\,856\,000 \approx 4.32 \times 10^{19}$.

As we saw in Section 4.4, in a permutation group we can easily express elements $g \in G$ as words in the strong generators, and hence in the original generators via straight-line programs, which is adequate for many computational applications. But this approach, when applied using the output of the basic Schreier-Sims algorithm, typically results in very long words for g in $(X \cup X^{-1})^*$, and so it is useless if we really want reasonable length words for g in the original generators.

The general problem of expressing elements in a permutation group as moderately short, but not necessarily minimal-length words in the original generators and their inverses is known as the *factorization problem* in permutation groups. In [Min98], Minkwitz describes some heuristic methods for finding reasonably short words in $(X \cup X^{-1})^*$ for the strong generators in a BSGS that has already been computed. These can then be used to produce moderately short words for arbitrary group elements in $(X \cup X^{-1})^*$, and thereby solve the factorization problem. Using this method, he was able to express arbitrary elements of the Rubik cube group as words of length at most 144 in its six natural generators. These ideas have been implemented in GAP by Hulpke, who reports reducing this number to about 100. The use of *genetic algorithms* has also been suggested as a novel means of attacking this problem, and we await future developments in that direction with interest!

4.8.4 Processor interconnection networks

We turn now to a successful application involving diameters of Cayley graphs and the factorization problem in permutation groups.

A network of processors for parallel computation can be regarded as an undirected graph in which the vertices are the processors, and two vertices are joined by an edge if there is a direct connection between the two processors represented. Let us assume that each processor is connected to the same number r of other processors; that is, that the graph is regular with valency r. The physical design of a large network is made more feasible if r is small, but computations are rendered more efficient if there are short paths connecting any two vertices (that is, the diameter of the graph should be as small as possible). Of course, these two requirements tend to conflict with each other!

In [SS92], Schibbell and Stafford observed that the underlying undirected graphs of Cayley graphs $\Gamma_X(G)$ of certain types of finite groups, particularly finite nonabelian simple groups, are very suitable for processor interconnection networks in that they have a smaller diameter for a given valency than other types of graphs that had been used for this purpose.

The principal problem to be solved in this application is calculating the shortest path between two given vertices v_g and v_h for $g, h \in G$ which, in the notation of the preceding subsection, is equivalent to finding $l_X(gh^{-1})$. When the network is in use for a large parallel computation, such shortest paths need to be computed frequently, and it is vital that the calculation of these paths should require minimal time and space resources, since otherwise they would impair the performance of the main computation.

In contrast to problems like finding the number of moves required to restore the Rubik cube, we are free to choose a convenient generating set for G. In [SS92], the authors achieved excellent results in terms of minimizing resources, by using a strong generating set in a low-degree permutation representation of G, and using the algorithm STRIP to calculate the paths. See also the more recent article by Heydemann [Hey97] for further references.

Chapter 5

Coset Enumeration

Most of the natural questions that one might ask about a group G that is defined by a finite presentation, such as "is G finite?" or even "is G trivial?" have been proved to be theoretically undecidable in general. In particular, the basic problem of deciding whether two words in the generators represent the same element of the group, which is known as the *word problem*, was proved to be undecidable by Novikov [Nov55], and then by Boone, with a simpler proof, in [Boo59]. See the survey article by Miller [Mil92] for a detailed discussion and survey of results of this type.

However, many of these problems are semidecidable, in the following sense: although we cannot decide whether G is trivial or finite, if G does happen to be trivial or finite, then it is possible to verify (that is, to prove) that fact.

Suppose that G is defined by a finite presentation, and H is a subgroup of G, which is specified by words in the generators of G that generate H. The procedure to be described in this chapter is known as *coset enumeration* and is one of the most fundamental methods in CGT.

Its aim is to attempt to verify that $|G : H|$ is finite. It will terminate only when $|G : H|$ is finite, and so it cannot correctly be described as an algorithm. When the index is infinite, it will run forever in theory, but in practice it will usually run out of space. Of course it may also run out of space when $|G : H|$ is finite but very large, or even when $|G : H|$ is small, but the presentation is a difficult one. So, we learn nothing when this happens. However, if the procedure does terminate, then it will have verified that $|G : H|$ is finite, and it will return the index, together with a complete *coset table*, which is equivalent to the permutation representation of the input group G in its action by right multiplication on the right cosets of H.

The first coset enumeration procedure was described by Todd and Coxeter in [TC36], and, for this reason, it is often known as the *Todd-Coxeter algorithm*. At that time, it was intended for computation by hand. The first machine implementation, which was probably also the first implementation of any procedure in group theory, was by B. Hazelgrove in 1953. For details of this and other early implementations, see the article by Leech [Lee63].

For a much more detailed treatment of both the theoretical and practical aspects of coset enumeration, the reader could consult [Sim94]. For an elementary overview of coset table methods and their variations, which nevertheless contains outlines of proofs, the article by Neubüser [Neu82] is to be recommended. In this book, we shall take an intermediate approach, but we have made extensive use of some of the pseudocode from [Sim94] in our own pseudocode.

Later in the chapter, we shall present some methods derived from coset enumeration, including computing presentations of subgroups of finite index in a finitely presented group, and finding all subgroups up to a specified finite index in a finitely presented group.

5.1   The basic procedure

Since even the simplest versions of coset enumeration seem rather complicated when we attempt to describe them precisely, we shall start with a simple example. The reader can then refer back to the example when reading the formal description.

Example 5.1

Let $G := \langle x, y \mid x^3, y^3, x^{-1}y^{-1}xy \rangle$, and $H := \langle x \rangle$. It is hopefully clear immediately to the reader that G is a direct product of two cyclic groups of order 3, and that $|G:H| = 3$.

We start by letting the number 1 represent the trivial coset H of H in G. Since $x \in H$, we have $Hx = H$, and so we write $1^x = 1$. We now define $1^y = 2$ and $1^{y^{-1}} = 3$. What this means is that we are using the integers 2 and 3 to represent the cosets Hy and Hy^{-1}, respectively. We then have $2^{y^{-1}} = 1$ and $3^y = 1$.

We now perform a process known as *scanning* relators under cosets. Since relators w represent the identity element of G, we must have $\alpha^w = \alpha$ for any coset α and any relator w. This is already true for 1^{x^3}. When we scan 1 under y^3, however, we find that $1^y = 2$, but 2^y is not yet defined, so the scan is incomplete. We can also start scanning from the end of the relator, and we find that $3^y = 1$, but $3^{y^{-1}}$ is undefined. At this stage, the fact that $1^{y^3} = 1$, enables us to make the *deduction* $2^y = 3$. This process is usually carried out using a table like the one below. The deduction is underlined.

$$
\begin{array}{ccccccc}
 & y & & y & & y & \\
1 & & 2 & & 3 & & 1 \\
 & & & & \underline{} & &
\end{array}
$$

We also have $3^{y^{-1}} = 2$. Scanning 1 and 2 under $x^{-1}y^{-1}xy$ yields the deductions $3^x = 3$ and $2^x = 2$:

$$
\begin{array}{ccccc}
x^{-1} & y^{-1} & x & y & \\
1 & 1 & 3 & 3 & 1 \\
 & & & \overline{} & \\
2 & 2 & 1 & 3 & 2 \\
 & \overline{} & & &
\end{array}
$$

The reader can verify that we now have $\alpha^w = \alpha$ for all $\alpha = 1, 2, 3$ and all relators $w \in R$. It is also true that α^x is defined for $\alpha = 1, 2, 3$ and all $w \in \Omega$. It is often helpful to write down all of these values in a table known as a *coset table*.

	x	x^{-1}	y	y^{-1}
1	1	1	2	3
2	2	2	3	1
3	3	3	1	2

Since all cosets now scan correctly under all relators, and all entries in the coset table are defined, the procedure terminates, and the index of H in G is equal to the number of cosets defined, which is 3. ▯

5.1.1 Coset tables and their properties

Now we are ready to embark upon a more formal description of the procedure. The input consists of a finitely presented group $G = \langle X \mid R \rangle$, and a finitely generated subgroup $H = \langle Y \rangle$. As in Section 2.4 of Chapter 2, we define $A := X \cup X^{-1}$. Here R is a set of defining relators for G, and is given as a set of words in A^*. The set Y of generators of H is also a subset of A^*. We shall assume that both R and Y consist of reduced words.

Throughout the procedure, we maintain a coset table $\mathcal{C}$ for G. Formally, we define a coset table $\mathcal{C}$ for a group G with finite generating set X to be a quintuple (τ, χ, p, n, M), where $M, n \in \mathbb{N}$ with $1 \leq n \leq M$, τ is a map from $[1 \mathinner{\ldotp\ldotp} n]$ to A^*, p is a map from $[1 \mathinner{\ldotp\ldotp} n]$ to $[1 \mathinner{\ldotp\ldotp} n]$ with $p(\alpha) \leq \alpha$ for all α, and χ is a partial map from $[1 \mathinner{\ldotp\ldotp} n] \times A$ to $[1 \mathinner{\ldotp\ldotp} n]$. However, we shall always write $\chi(\alpha, x)$ as α^x, and suppress all further reference to χ. For a given $\alpha \in [1 \mathinner{\ldotp\ldotp} n]$ and $x \in A$, α^x may or may not be defined. Throughout the remainder of this chapter $\mathcal{C} = (\tau, \chi, p, n, M)$ will be a coset table for a fixed group $G = \langle X \mid R \rangle$.

The number M represents the largest number of cosets that we allow. This is determined in practice by the amount of computer memory available, and we shall assume that it is fixed throughout the procedure. The other components, τ, χ, p, and n may change as the coset enumeration is executed. The procedures that we shall write down for performing the enumeration will usually take $\mathcal{C}$ as an argument, and will often alter one or more of these components.

We define the set Ω of *live cosets* to be $\{\alpha \in [1 \mathinner{\ldotp\ldotp} n] \mid p(\alpha) = \alpha\}$. We need p, because it sometimes turns out during the course of the enumeration that the cosets represented by two numbers α and β with $\alpha < \beta$ are equal; we record this occurrence by setting $p[\beta] := \alpha$, which has the effect of removing β from the set of live cosets. We shall always have $1 \in \Omega$. The number n represents the largest number that has been used so far for a live coset. The word $\tau(\alpha)$ is a coset representative of the coset corresponding to $\alpha \in \Omega$. We initialize an enumeration by setting $n := 1$, $p[1] := 1$, and $\tau(1) := \varepsilon$. The coset table is called *complete* if it has no undefined entries on the live cosets; that is, if α^x is defined for all $\alpha \in \Omega$ and $x \in A$.

In an implementation, the map χ will typically be stored as a simple rectangular array, as in the example above, with rows indexed by Ω and columns by A, and undefined entries will be set to 0. But there are other possibilities. If a very small percentage of the entries are defined, then it might use less memory to use some kind of sparse-matrix storage method. The coset representatives $\tau(\alpha)$ are not stored explicitly in the implementation, because they are not required for the mechanics of carrying out the computation, but they are important for the theoretical description.

We shall now state a number of properties that $\mathcal{C}$ will satisfy throughout the procedure. Of course, we shall need to prove that these properties are maintained by all changes that we make during the course of the procedure.

Property 1 $1 \in \Omega$ *and* $\tau(1) = \varepsilon$.

Property 2 $\alpha^x = \beta \iff \beta^{x^{-1}} = \alpha$.

Property 3   *If* $\alpha^x = \beta$, *then* $H\tau(\alpha)x = H\tau(\beta)$.

In other words, the action of A on Ω represents multiplication on the right when the elements of Ω are thought of as cosets of H in G.

For $\alpha \in \Omega$ and $w = x_1 x_2 \cdots x_r \in A^*$, let $\alpha_0 := \alpha$ and, for $1 \leq i \leq r$, let $\alpha_i := \alpha_{i-1}^{x_i}$ if this is defined; otherwise α_i is undefined. Then, we say that α^w is defined and equal to α_r whenever the α_i are all defined for $0 \leq i \leq r$.

Property 4   *For all* $\alpha \in \Omega$, $1^{\tau(\alpha)}$ *is defined and is equal to* α.

5.1.2   Defining and scanning

If some α^x is undefined, then the simplest way of remedying this is to adjoin a new element β to Ω and make that α^x. This is accomplished by the following procedure. The final statement of the procedure, which defines $\tau(\beta)$, is written as a comment, because it is not part of the implementation. The procedure starts by checking that we have space available for a new definition. If not, then it aborts.

For ease of notation, in this and other procedures taking a coset table $\sim \mathcal{C}$ as a modifiable input parameter, we write the components τ, χ, p, n, M of $\mathcal{C}$

DEFINE($\sim \mathcal{C}, \alpha, x$)

    Input: $\alpha \in \Omega$, $x \in A$, with α^x undefined.
1 **if** $n = M$ **then abort;**
2 $n := n+1$; $\beta := n$; $p[\beta] := \beta$;
3 $\alpha^x := \beta$; $\beta^{x^{-1}} := \alpha$;
4 (* Set $\tau(\beta)$ equal to $\tau(\alpha)x$. *)

simply as τ, χ, p, n, M, rather than as $\mathcal{C}.\tau$, etc.

It is clear that Properties 1 and 2 remain true after executing this procedure. There are only two new entries in the coset table, and we have defined $\tau(\beta)$ to ensure Property 3 holds for these entries. As for Property 4, we are assuming that $1^{\tau(\alpha)} = \alpha$, so $1^{\tau(\beta)} = \alpha^x$ is defined and equal to β. Incidentally, the fact that $1^{\tau(\alpha)} = \alpha$ implies that the word $\tau(\alpha)$ cannot end with x^{-1}, because otherwise α^x would be defined already, contrary to assumption. This ensures that $\tau(\beta)$ is a reduced word.

Now we turn to the *scanning* process. There are two situations under which we might scan an element $\alpha \in \Omega$ under a word $w \in A^*$. The first is when $\alpha = 1$ and $w \in Y$, and the second is when α is any element of Ω and $w \in R$. This is because $w \in R$ implies that $Hgw = Hg$ for any $g \in G$, whereas $w \in Y$ only implies that $Hw = H$.

To scan α under w, we first locate the largest prefix s of w for which α^s is defined. Of course, s might be empty. Let $w = sv$, let t be the longest suffix of v for which $\alpha^{t^{-1}}$ is defined, and let $v = ut$. If $t = v$, then we say that the scan completes, and if, in addition, $\alpha^s = \alpha^{t^{-1}}$, then we say that the scan completes correctly. (In fact, because of Property 2, it can only complete correctly when $s = w$, $t = \varepsilon$.) Before considering the unfortunate situation when the scan completes but not correctly, we consider another interesting situation, namely where the scan does not complete, but $|u| = 1$; that is, the word u consists of a single generator $x \in A$. In that case, if $\alpha^s = \beta$ and $\alpha^{t^{-1}} = \gamma$, then we can set $\beta^x = \gamma$ and $\gamma^{x^{-1}} = \beta$. These assignments are known as a *deduction*, and it is clear that deductions enable the scan to complete correctly. Formally, we have:

PROPOSITION 5.1   *Suppose that the assignments $\beta^x = \gamma$ and $\gamma^{x^{-1}} = \beta$ result from a deduction on scanning an element $\alpha \in \Omega$ under $w \in R$, or on scanning 1 under $w \in Y$. Then Properties 1, 2, 3, 4 remain true after making these deductions.*

PROOF   These are all clear except for Property 3. Using the notation of the preceding paragraph, we need to prove that $H\tau(\beta)x = H\tau(\gamma)$. Since $\alpha^s = \beta$ and $\gamma^t = \alpha$ before making the deduction, Property 3 implies that $H\tau(\alpha)s = H\tau(\beta)$ and $H\tau(\gamma)t = H\tau(\alpha)$. Combining these gives $H\tau(\beta)x =$

$H\tau(\gamma)tsx$. If $sxt \in R$, then $tsx =_G 1$, and we have $H\tau(\beta)x = H\tau(\gamma)$, as required. If $\alpha = 1$ and $w = sxt \in Y$, then $H\tau(\alpha) = H$, so

$$H\tau(\beta) = H\tau(\alpha)s = Hs = Ht^{-1}x^{-1} = H\tau(\gamma)tt^{-1}x^{-1} = H\tau(\gamma)x^{-1}$$

and again we have $H\tau(\beta)x = H\tau(\gamma)$.   ∎

It is time for another example.

Example 5.2

Let $G := \langle\, x,y \mid x^2, y^3, (xy)^3 \,\rangle$ and $H := \langle xy \rangle$. In practice, when there is a relator x^2, we know that $x =_G x^{-1}$, and we can save time and space by omitting the column for x^{-1} in the coset table, since its entries can be assumed to be the same as those for x. We shall do that here. Define $1^x = 2$. So $2^x = 1$. Scanning 1 under the subgroup generator xy yields the deduction $2^y = 1$, $1^{y^{-1}} = 2$. Define $1^y = 3$, $3^{y^{-1}} = 1$. Then scanning 1 under y^3 yields the deduction $3^y = 2$, $2^{y^{-1}} = 3$. Define $3^x = 4$, $4^x = 3$. Scanning 2 under $(xy)^3$ leads to the deduction $4^y = 4$:

$$\begin{array}{cccccc} x & y & x & y & x & y \\ 2 & 1 & 3 & 4 & 4 & 3 & 2 \end{array}$$

We now find that the coset table is complete (see below), and elements of Ω scan correctly under all relators, so we have $|G:H| = |\Omega| = 4$.

	x	y	y^{-1}
1	2	3	2
2	1	1	3
3	4	2	1
4	3	4	4

▯

We can now prove the correctness of the basic procedure, assuming that all of the required properties hold.

THEOREM 5.2   *Assume that:*

(i) *Properties 1 – 4 all hold;*

(ii) *the coset table is complete;*

(iii) *1 scans correctly under all $w \in Y$;*

(iv) *all $\alpha \in \Omega$ scan correctly under all $w \in R$.*

Then $|G : H| = |\Omega|$. Furthermore, for each $x \in A$, the map $\varphi(x) : \Omega \to \Omega$ defined by $\alpha \mapsto \alpha^x$ is a permutation of Ω, and φ extends to a homomorphism from G to $\mathrm{Sym}(\Omega)$ which is equivalent to the action of G on the cosets of H by right multiplication.

PROOF   Property 2 says that $\varphi(x)$ and $\varphi(x^{-1})$ are inverse maps for any $x \in A$, and so $\varphi(x)$ is a permutation. Assumption (iv) says that for any $w = x_1 \cdots x_r \in R$, we have $\alpha^{\varphi(x_1)\varphi(x_2)\cdots\varphi(x_r)} = \alpha$ for all $\alpha \in \Omega$, so $\varphi(x_1)\varphi(x_2)\cdots\varphi(x_r) = 1_{\mathrm{Sym}(\Omega)}$. Now Theorem 2.52 implies that φ extends to a group homomorphism.

To prove the equivalence of φ with the action of G on the set C of cosets of H, we define $\overline{\tau} : \Omega \to C$ by $\overline{\tau}(\alpha) = H\tau(\alpha)$. For any $v \in A^*$, assumption (ii) and Property 3 imply that $\overline{\tau}(1^v) = Hv$, so $\overline{\tau}$ is surjective. If $\overline{\tau}(\alpha) = \overline{\tau}(\beta)$, then $H\tau(\alpha) = H\tau(\beta)$, so $\tau(\alpha)\tau(\beta)^{-1} \in H$, and hence $\tau(\alpha)\tau(\beta)^{-1}$ is a product $w_1 \cdots w_r$ of elements $w_i \in Y \cup Y^{-1}$. By assumption (iii), we have $1^{w_i} = 1$ for all w_i and hence $1^{\tau(\alpha)\tau(\beta)^{-1}} = 1$ and, using Property 4, $\alpha = 1^{\tau(\alpha)} = 1^{\tau(\beta)} = \beta$. So $\overline{\tau}$ is a bijection. It now follows that $|G : H| = |\Omega|$ and Property 3 implies that $\overline{\tau}$ defines an equivalence between φ and the action of G on C by right multiplication.   ∎

The code for the procedure for scanning $\alpha \in \Omega$ under $w \in A^*$ is as follows. This is called only when either $w \in R$, or when $\alpha = 1$ and $w \in Y$.

```
SCAN(∼C, α, w)
       Input: α ∈ Ω, w ∈ A*.
       (∗ Scan forwards ∗)
 1     Let w = x₁x₂ ··· xᵣ, with xᵢ ∈ A;
 2     f := α;  i := 1;
 3     while i ≤ r and f^xᵢ is defined
 4          do f := f^xᵢ;  i := i+1;
 5     if i > r
 6        then if f ≠ α then COINCIDENCE(∼C, f, α);
 7             return ;
       (∗ Forward scan was incomplete. Scan backwards ∗)
 8     b := α;  j := r;
 9     while j ≥ i and b^{xᵢ⁻¹} is defined
10          do b := b^{xᵢ⁻¹};  j := j−1;
11     if j < i
12        then COINCIDENCE(∼C, f, b);
13     elseif j = i
14        then (∗ Deduction ∗) f^xᵢ := b;  b^{xᵢ⁻¹} := f;
15     (∗ Otherwise j > i – scan is incomplete and yields no information ∗).
```

5.1.3 Coincidences

The calls to the COINCIDENCE routine, which we have not yet explained, occur when a scan completes incorrectly. In the notation of the procedure SCAN, this occurs when, for some i with $1 \le i \le r+1$, we have $w = st$ with $s = x_1 x_2 \cdots x_{i-1}$, $t = x_i x_{i+1} \cdots x_r$, and $\beta := \alpha^s$ and $\gamma := \alpha^{t^{-1}}$ are defined but unequal. This means that β and γ represent the same coset of H in G. The proof of this fact, which is stated in the following proposition, is similar to that of Proposition 5.1 and will be left to the reader.

PROPOSITION 5.3 *Suppose that either $\alpha \in \Omega$ and $w \in W$, or $\alpha = 1$ and $w \in Y$. Suppose also that $w = st$, where $\beta := \alpha^s$ and $\gamma := \alpha^{t^{-1}}$ are both defined. Then $H\tau(\beta) = H\tau(\gamma)$.*

We call the eventuality just described a *coincidence* between β and γ. By Property 3, if β^x and γ^x are both defined for some $x \in A$, then Proposition 5.3 says that $H\tau(\beta) = H\tau(\gamma)$, and then Property 3 implies that $H\tau(\beta^x) = H\tau(\gamma^x)$. By induction on $|w|$, it follows that the same is true with x replaced by any $w \in A^*$.

Although α^w is not usually defined for all $\alpha \in \Omega$ and $w \in A^*$, we can still use Definition 2.26 and call an equivalence relation $\sim$ on Ω a *G-congruence* if $\alpha \sim \beta$ implies $\alpha^w \sim \beta^w$ whenever α^w and β^w are both defined. We can then define the *G*-congruence generated by $S \subseteq \Omega \times \Omega$ as in Definition 2.27.

In the coset enumeration procedure, as soon as a coincidence between two elements $\alpha, \beta \in \Omega$ is found, we immediately compute $\sim$, the right congruence generated by this coincidence. We shall see in the proof of Proposition 5.4 below that $\gamma \sim \delta$ implies $H\tau(\gamma) = H\tau(\delta)$, so we only want to keep one representative of each $\sim$-equivalence class in Ω. All other elements in the class are removed from Ω. It is also essential that we avoid losing information, and to do this, we need to transfer existing coset table entries between the deleted elements of Ω to the equivalent $\sim$-class representatives.

Let us first work through an example.

Example 5.3

Let $G := \langle x, y \mid x^2 y^2, x^3 y^5 \rangle$ and $H := 1$ (so Y is empty). Define $1^x = 2$, $2^x = 3$, $3^y = 4$ and deduce $4^y = 1$ by scanning 1 under $x^2 y^2$. Then define $3^x = 5$, $5^y = 6$ and deduce $6^y = 2$ by scanning 2 under $x^2 y^2$, deduce $2^y = 3$ by scanning 1 under $x^3 y^5$, and deduce $5^x = 6$ by scanning 3 under $x^2 y^2$, Now, when we scan 2 under $x^3 y^5$, we find that $2^{x^3 y^4} = 1$, but $2^{y^{-1}} = 6$, so we have an incorrect completed scan with the coincidence $6 \sim 1$. We therefore remove 6 from Ω. When we compare the elements 6^z with 1^z for $z \in A$, we find that $6^{y^{-1}} = 5$ but $1^{y^{-1}} = 4$, so we have another coincidence $5 \sim 4$, and we remove 5 from Ω. After this there are no further coincidences, and we end up with the complete coset table below, with all $\alpha \in \Omega$ scanning correctly under all

$w \in R$. So $|G| = 4$. Note that $x =_G y$.

	x	x^{-1}	y	y^{-1}
1	2	4	2	4
2	3	1	3	1
3	4	2	4	2
4	1	3	1	3

☐

Now we are ready to write down the coincidence procedure. This is very similar to the procedure MINIMALBLOCK that we studied in Subsection 4.3.3, but there are some important differences. There are some extra complications resulting from the facts that not all coset table entries α^x are defined, and that we wish to transfer defined entries on the eliminated cosets to their equivalence class representatives. Another difference is that when merging two equivalence classes, virtually all implementations of coset enumeration to date have chosen to use the smaller of the two (integer) representatives as the representative of the merged class, rather than the theoretically more efficient choice of the representative from the larger of the two classes. The main reason for this is to avoid having to store the array c that was used in MINIMALBLOCK for the size of the classes. In coset enumeration, space is much more critical than time and, in any case, the time difference resulting from the two possible methods appears not to be significant in practice. So we have have adopted that approach here.

We assume that, before each call of COINCIDENCE, for $\alpha \in [1 .. n]$, we have $p[\alpha] = \alpha$ if and only if $\alpha \in \Omega$. Proposition 5.4 will guarantee that this remains true after the call.

REP$(\kappa, \sim p)$

    Input: $\kappa \in \Omega$, array p
1    $\lambda := \kappa;$ $\rho := p[\lambda];$
2   **while** $\rho \neq \lambda$ **do** $\lambda := \rho;$ $\rho := p[\lambda];$
3    $\mu := \kappa;$ $\rho := p[\mu];$
4   **while** $\rho \neq \lambda$ **do** $p[\mu] := \lambda;$ $\mu := \rho;$ $\rho := p[\mu];$
5   **return** $\lambda;$

MERGE$(\kappa, \lambda, \sim p, \sim q, \sim l)$

    Input: $\kappa, \lambda \in \Omega$, p, q, l
    (∗ q is a queue of length l of elements to be deleted from Ω ∗)
1    $\varphi := $ REP$(\kappa, \sim p);$ $\psi := $ REP$(\lambda, \sim p);$
2   **if** $\varphi \neq \psi$
3   **then** $\mu := \min(\varphi, \psi);$ $\nu := \max(\varphi, \psi);$
4            $p[\nu] := \mu;$ $l := l+1;$ $q[l] := \nu;$

COINCIDENCE($\sim \mathcal{C}, \alpha, \beta$)

Input: A coincident pair $\alpha, \beta \in \Omega$ with $H\tau(\alpha) = H\tau(\beta)$

1    $l := 0$; $q := []$; MERGE($\alpha, \beta, \sim p, \sim q, \sim l$); $i := 1$;
2   **while** $i \leq l$
3   **do** $\gamma := q[i]$; $i := i+1$;
4            $\Omega := \Omega \setminus \{\gamma\}$;
5   **for** $x \in A$ **do if** γ^x is defined and equal to δ
6   **then** (∗ Remove the entry $\delta^{x^{-1}} = \gamma$ from the coset table ∗)
7   Undefine $\delta^{x^{-1}}$;
                    (∗ Queue new coincidences, and transfer definitions ∗)
8                    $\mu := \text{REP}(\gamma, \sim p)$; $\nu := \text{REP}(\delta, \sim p)$;
9   **if** μ^x is defined **then** MERGE($\nu, \mu^x, \sim p, \sim q, \sim l$);
10   **elseif** $\nu^{x^{-1}}$ is defined
11   **then** MERGE($\mu, \nu^{x^{-1}}, \sim p, \sim q, \sim l$);
12   **else** $\mu^x := \nu$; $\nu^{x^{-1}} := \mu$;

We now prove that this procedure does what we want it to do.

PROPOSITION 5.4   *After executing* COINCIDENCE($\sim \mathcal{C}, \alpha, \beta$), *only the least representatives of the equivalences classes of the the G-congruence $\sim$ generated by (α, β) remain in Ω and, for $\nu \in [1 .. n]$, $p[\nu] = \nu$ if and only if $\nu \in \Omega$. Furthermore, if for some $x \in A$ we have $\gamma^x = \delta$ before executing the procedure, then $\text{REP}(\gamma)^x = \text{REP}(\delta)$ afterwards. Finally, Property 1 to Property 4 all continue to hold after execution of the procedure.*

PROOF   We are assuming that, before the call, for $\nu \in [1 .. n]$, we have $p[\nu] = \nu$ if and only if $\nu \in \Omega$. The value of $p[\nu]$ is changed only in REP and MERGE, and it is only changed in REP when $p[\nu] \neq \nu$. If it is changed in MERGE then it is set equal to a value $\mu < \nu$. It follows by a straightforward induction argument that any changes to $p[\nu]$ in REP decrease its value further. Note also that any element ν returned by REP satisfies $p[\nu] = \nu$, and the ν for which $p[\nu]$ is changed in MERGE has just been returned by REP. It follows that, for each $\nu \in \Omega$, the initial value ν of $p[\nu]$ is changed at most once in MERGE, and at that stage it is added to the queue q. Since the entries in q are processed and removed from Ω during the main loop of the procedure, the procedure must halt, and we have proved the assertion that, at the end of the procedure, $p[\nu] = \nu$ if and only if $\nu \in \Omega$.

As in the proof of Theorem 4.3, we shall denote the equivalence relation on Ω defined by p throughout the procedure by $\sim$, and the final value of $\sim$ at the end of the procedure by $\sim_F$. The G-congruence generated by (α, β) will be denoted by $\equiv$.

We next show that $\kappa \sim \lambda$ always implies $\kappa \equiv \lambda$ and $H\tau(\kappa) = H\tau(\lambda)$. Note that $\sim$ is only changed during calls of MERGE, when two $\sim$-classes are merged. The first call of MERGE at line 1 merely makes $\alpha \sim \beta$, and we are assuming that $H\tau(\alpha) = H\tau(\beta)$. Before the other two calls, $\text{MERGE}(\nu, \mu^x)$ on line 9 and $\text{MERGE}(\mu, \nu^{x^{-1}})$ on line 11, we have $\mu = \text{REP}(\gamma)$ and $\nu = \text{REP}(\delta)$, where $\gamma^x = \delta$, so we can assume inductively that $\mu \sim \gamma$, $\nu \sim \delta$, $H\tau(\mu) = H\tau(\gamma)$ and $H\tau(\nu) = H\tau(\delta)$. Hence $\mu^x \sim \nu$, $\mu \sim \nu^{x^{-1}}$ and, by Property 3, $H\tau(\mu^x) = H\tau(\nu)$, and $H\tau(\nu^{x^{-1}}) = H\tau(\mu)$ Thus, on any call of $\text{MERGE}(\kappa, \lambda)$ during the COINCIDENCE procedure, we have $\kappa \equiv \lambda$ and $H\tau(\kappa) = H\tau(\lambda)$, and so the claim that $\kappa \sim \lambda$ implies $\kappa \equiv \lambda$, and $H\tau(\kappa) = H\tau(\lambda)$ remains true after the call. Hence it is always true.

Properties 1 and 4 are clearly unaffected by calls of COINCIDENCE. The only entries that are removed from the coset table are those of the form $\delta^{x^{-1}} = \gamma$ at line 7, where γ has been removed from Ω at line 4, and the only entries to be added are $\mu^x := \nu$; $\nu^{x^{-1}} := \mu$; together, at line 12, in situations where both were previously undefined. So Property 2 continues to hold. As we saw in the preceding paragraph, whenever we add these entries, we have $\mu^x \sim \nu$ and $H\tau(\mu^x) = H\tau(\nu)$, so Property 3 continues to hold.

At any stage, let us call two coset table entries $\alpha^x = \beta$ and $\overline{\alpha}^x = \overline{\beta}$ equivalent, if $\alpha \sim \overline{\alpha}$ and $\beta \sim \overline{\beta}$. Suppose we have an entry $\alpha^x = \beta$ in the coset table before the call of COINCIDENCE. We claim that at all times during the procedure, there is an equivalent entry in the coset table. After removing an element γ from Ω at line 4, any entries $\gamma^x = \delta$ (implicitly, since γ is no longer in Ω) and $\delta^{x^{-1}} = \gamma$ (at line 7) are removed from the coset table. We can assume that the claim is true immediately before these entries are removed, and we must show that it remains true afterwards.

On removal of the entries, we set $\mu := \text{REP}(\gamma)$ and $\nu := \text{REP}(\delta)$, and one of three things can happen. If μ^x is defined and equal to λ, say, then we call $\text{MERGE}(\nu, \lambda)$, making $\nu \sim \lambda$. But then the coset table entries $\gamma^x = \delta$ and $\mu^x = \lambda$ become equivalent, as do $\delta^{x^{-1}} = \gamma$ and $\lambda^{x^{-1}} = \mu$. The situation is similar if $\nu^{x^{-1}}$ is defined. Finally, if neither μ^x nor $\nu^{x^{-1}}$ is defined, then we add the entries $\mu^x = \nu$ and $\nu^{x^{-1}} = \mu$, and these are equivalent to $\gamma^x = \delta$ and $\delta^{x^{-1}} = \gamma$. Thus the claim is true.

At the end of the procedure, the only elements of Ω remaining are those with $\text{REP}(\nu) = p[\nu] = \nu$. So, for an initial coset table entry $\alpha^x = \beta$, the only possible equivalent entry at the end is $\text{REP}(\alpha)^x = \text{REP}(\beta)$.

Finally, we prove that $\kappa \equiv \lambda$ implies $\kappa \sim_F \lambda$ for all $\kappa, \lambda \in \Omega$. Since we have already proved the converse, by definition of the G-congruence $\equiv$, it suffices to prove that $\sim_F$ is a G-congruence. So, suppose that $\kappa \sim_F \lambda$ and that κ^w and λ^w are both defined for some $w \in A^*$. We have to prove that $\kappa^w \sim_F \lambda^w$ and, by induction, it suffices to do this for $w = x \in A$. (Since G is not necessarily finite, we cannot restrict to $x \in X$, as we could in MINIMALBLOCK!) As we have just seen, at the end of the procedure, there are coset table entries that are equivalent to the entries for κ^x and λ^x. Since these entries have the same source $\text{REP}(\kappa) = \text{REP}(\lambda)$, they must be the same entry $\gamma^x = \delta$, say, but then $\kappa^x \sim_F \delta \sim_F \lambda^x$, which proves the claim. This completes the proof. ∎

Example 5.4

This is the example $F(2,5)$, one of the *Fibonacci groups*, which have provided a much-studied source of examples in combinatorial group theory. See, for example, [Joh98]. Let

$$G := \langle\, a, b, c, d, e \mid ab = c,\ bc = d,\ cd = e,\ de = a,\ ea = b \,\rangle$$

and $H := \langle a \rangle$. In the computation, we of course rewrite the defining relations as relators abc^{-1}, etc. We have $1^a = 1$. Define $1^b = 2$, and deduce $1^c = 2$ by scanning 1 under abc^{-1}. Then define $2^c = 3$ and deduce $1^d = 3$ from scanning 1 under bcd^{-1}, $3^e = 1$ from scanning 1 under dea^{-1}, $3^b = 1$ from scanning 3 under eab^{-1}, $3^d = 2$ from scanning 3 under bcd^{-1}, $2^e = 2$ from scanning 2 under cde^{-1}, $3^a = 2$ from scanning 3 under dea^{-1}, $1^e = 3$ from scanning 1 under eab^{-1}.

Now, when we scan 1 under cde^{-1}, we find that $1^c = 2$, but $1^{ed^{-1}} = 1$, so we have an incorrect completed scan with the coincidence $2 \sim 1$. Since $2^e = 2$ but $1^e = 3$, we also get the coincidence $3 \sim 2$, and then the table is complete with $1^a = 1^b = 1^c = 1^d = 1^e = 1$. So we have a complete collapse, and we have shown that $G = \langle a \rangle$. This tells us that G is abelian and, using that information, it is not hard to use the relations to show directly that G has order 11, and that $b = a^4$, $c = a^5$, $d = a^9$, $e = a^3$. □

In the examples that we have worked through so far, we have not explained what strategy we have adopted to decide which definitions to make. When doing examples by hand, the overriding objective is to avoid having to process coincidences, and in fact we have carefully chosen our definitions so as to minimize the amount of coincidence checking that needs to be done. A variety of strategies can be adopted for this purpose, including trial and error.

When implementing coset enumeration on a computer, we of course need to specify precisely a policy for choice of definitions, which means that we cannot be quite so flexible as with a hand enumeration. Avoiding coincidence processing is still important, since it can reduce overall space requirements, but it is not quite so critical as with hand enumeration. We shall discuss the policies that have been most commonly used in more detail in the next section.

To complete our theoretical analysis of coset enumeration, we still need to prove that, if $|G : H|$ is finite, then the procedure will eventually halt. In fact, whether this is true depends on the strategy used for definitions. In [War74], J.N. Ward presents a simple example in which $|G : H|$ is finite, but in which certain apparently reasonable strategies for choosing definitions result in the enumeration never completing. The property that is required from our definition strategy is that all images α^x eventually get defined. We also require that 1 is scanned under all subgroup generators, and all α are scanned under all relators.

Property 5 *Let $\overline{\Omega}$ be the set of elements of Ω that are not eliminated in any call of* COINCIDENCE. *Then*

(i) *For each $\alpha \in \overline{\Omega}$ and each $x \in A$, α^x will at some stage be defined;*

(ii) *For each $w \in Y$, SCAN($\sim C, 1, w$) is called at some stage;*

(iii) *For each $\alpha \in \overline{\Omega}$ and each $w \in R$, SCAN($\sim C, \alpha, w$) is called at some stage.*

The ingenious idea of using an infinite coset table in the proof of the proposition below is taken from Theorem 2 of [Neu82]. There are other ways of proving this, which do not involve infinite coset tables, but they involve more technicalities.

THEOREM 5.5 *Assume that Property 1 to Property 5 hold throughout a coset enumeration of G with respect to the subgroup H, and that $|G:H|$ is finite. Then the coset enumeration procedure will eventually terminate.*

PROOF Let $\overline{\Omega}$ be as in Property 5. Let $\alpha \in \overline{\Omega}$ and $x \in A$. By Property 5, α^x is defined at some stage; say $\alpha^x = \beta$. Since α is not eliminated in any call of COINCIDENCE, by Proposition 5.4, we must have $p[\alpha] = \alpha$ at all times during any such call, and also $\alpha^x = p[\beta] \leq \beta$ at the end of the call. Since α^x can only decrease, it must eventually become stable. Thus $\overline{\Omega}$ must be infinite, because otherwise the coset table would complete with $\Omega = \overline{\Omega}$.

We can now define an infinite complete coset table, in which the rows are indexed by the elements $\alpha \in \overline{\Omega}$, and the entries α^x are their stable values. Property 5 ensures that the hypotheses of Theorem 5.2 applies to this infinite complete coset table, and we can conclude from Theorem 5.2 that the action of G by right multiplication on the right cosets of H has infinite degree. But this contradicts the assumption that $|G:H|$ is finite. ∎

Exercises

1. Use coset enumeration to show that the subgroup $H := \langle x, y^2, yxyx^{-1}y^{-1}, yx^2yx^{-3}y^{-1}, yx^4y^{-1}, yx^3yx^{-2}y^{-1} \rangle$ has index 5 in the free group G on $\{x, y\}$.

2. Show by coset enumeration with the trivial subgroup that the group $G := \langle x, y \mid x^2y^2, y^{-1}xyx^{-3} \rangle$ has order 8. (It is possible to do this without any coincidences occurring.) Show also that $G \cong Q_8$.

5.2 Strategies for coset enumeration

As was mentioned earlier, the main problem to be resolved when implementing the coset enumeration procedure is to specify a strategy for making definitions. There are two principal strategies that have been employed in the past, and the most recent implementations, such as the ACE3 [HR99] package by Havas and Ramsey, allow the user to select a combination of these two methods in any desired proportion, together with a selection of modifications that have been observed to perform well on some examples.

It is rare for an enumeration to take a very long time without using a correspondingly large amount of space, so we regard the space used, and specifically the maximum value of $|\Omega|$ during the course of the enumeration as the principal measure of how well that enumeration is performing. (The total number of cosets defined during the enumeration could also be used, and would provide an indication of the time requirements as well as the space requirements.) Roughly speaking, we regard a particular enumeration as performing well if the maximum size of $|\Omega|$ is not more than perhaps 25% larger than $|G\!:\!H|$.

Fortunately, for many of the examples for which coset enumeration has been used in earnest, at least one and often both of the two principal strategies performs satisfactorily. But one problem to which coset enumeration is often applied is trying to find presentations of a given group with the smallest possible number of defining relators, and for an application like that, where we are in some sense deliberately trying to make the problem as hard as possible, it can be crucial to find the appropriate strategy.

It seems to be a feature of coset enumeration that, for any single strategy, there are examples that will perform badly using that strategy, but will perform relatively well using some other strategy. It is also not hard to construct presentations, even of the trivial group, in which the maximum value of $|\Omega|$ is arbitrarily large; this is an inevitable consequence of the fact that deciding whether $|G\!:\!H|$ is finite is an unsolvable problem in general.

We shall begin by describing the two principal strategies in detail, and then discuss some other ideas and modifications that have been used.

5.2.1 The relator-based method

This is probably the easier of the two methods to describe and implement, and it has often been called the HLT method, after Hazelgrove, Leech [Lee63] and Trotter [Tro64], who have all contributed to its development. There is a detailed description of an implementation in [CDHW73].

The idea is that whenever we call SCAN and the scan is incomplete, then we make new definitions to enable the scan to complete; that is, we fill in the gaps in the scan of the relator or subgroup generator. This is why we call it the relator-based method. To carry this out, we first need to write a modified version of SCAN, which makes the required definitions. Note that the 'While' loop on line 3 of this procedure is exited as soon as the scan completes, when a **return** statement is executed.

SCANANDFILL($\sim\mathcal{C}, \alpha, w$)

    Input: $\alpha \in \Omega$, $w \in A^*$, with α^x undefined.

1   Let $w = x_1 x_2 \cdots x_r$, with $x_i \in A$;

2   $f := \alpha$; $i := 1$; $b := \alpha$; $j := r$;

3   **while** true

4   **do** (* Scan forwards *)

5   **while** $i \le r$ **and** f^{x_i} is defined

6   **do** $f := f^{x_i}$; $i := i+1$;

7   **if** $i > r$

8   **then if** $f \neq \alpha$ **then** COINCIDENCE($\sim\mathcal{C}, f, \alpha$);

9   **return** ;

            (* Forward scan was incomplete. Scan backwards *)

10   **while** $j \ge i$ **and** $b^{x_i^{-1}}$ is defined

11   **do** $b := b^{x_i^{-1}}$; $j := j-1$;

12   **if** $j < i$

13   **then** COINCIDENCE($\sim\mathcal{C}, f, b$);

14   **elseif** $j = i$

15   **then** (* Deduction *) $f^{x_i} := b$; $b^{x_i^{-1}} := f$;

16   **return** ;

            (* Otherwise $j > i$ and the scan is incomplete.

                Make a new definition and continue with scan. *)

17   **else** DEFINE($\sim\mathcal{C}, f, x_i$);

In an implementation, whenever we make a call to DEFINE, as we did here, we would need to check whether DEFINE aborted due to lack of space and, if so, take some appropriate action. For example, we might give up, or we might attempt to reuse any space available from cosets that have been eliminated by calls of COINCIDENCE. We shall discuss these possibilities later.

It is now easy for us to write down a function for relator-based coset enumeration, which returns the coset table completed. The array p used by COINCIDENCE is global to this procedure. Of course, this function can only complete if $|G : H|$ is finite. In this and subsequent functions in this subsection, we have described a loop over all elements of Ω by using the statement

$$\textbf{for } \alpha \in \Omega \textbf{ do}$$

However, recalling that n is the largest element of Ω that has been defined so far, and elements $\alpha \in \Omega$ are recognized by the fact that $p[\alpha] = \alpha$, this should really be coded as

$$\textbf{for } \alpha \in [1 \mathbin{..} n] \textbf{ do if } p[\alpha] = \alpha \textbf{ then}$$

COSETENUMERATIONR($\langle X|R\rangle, Y$)

    Input: A finite group presentation $\langle X|R\rangle$ and a finite subset Y of A^*
1   Initialize a coset table C for $\langle X|R\rangle$;
2   $p[1] := 1$; $n := 1$;
3   **for** $w \in Y$ **do** SCANANDFILL($\sim C, 1, w$);
4   **for** $\alpha \in \Omega$
5   **do for** $w \in R$
6   **do** SCANANDFILL($\sim C, \alpha, w$);
                (* If α was eliminated during the scan then break *)
7   **if** $p[\alpha] < \alpha$ **then break**;
8   **if** $p[\alpha] < \alpha$ **then continue**;
9   **for** $x \in A$
10   **do if** α^x is undefined **then** DEFINE($\sim C, \alpha, x$);
11 **return** C;

It is clear that the conditions of Property 5 hold for this function, and so the results of the last section ensure that it will perform as required. Note that the lines 9,10 ensure that Property 5(i) holds. This line has not always been included in past implementations, with the consequence that the enumeration would not terminate even when $|G:H|$ was finite on certain examples, such as those described by Ward in [War74]; see Exercise 3 below.

The above function is not always optimal in terms of the maximal value of $|\Omega|$ attained, but it tends to perform satisfactorily in most straightforward examples and, provided there is enough memory available, it often completes faster than other methods that are more careful about minimizing memory usage. In some examples in which it does run out of memory, a modification known as HLT+Lookahead, which is described in [CDHW73], enables it to complete. The idea is that whenever a call of DEFINE aborts due to lack of space available for further definitions, the current value of α is remembered, and the procedure below is executed.

LOOKAHEAD($\sim C, R$)

1 **for** $\beta \in \Omega, w \in R$
2   **do** SCAN($\sim C, \beta, w$);
3   **if** $p[\beta] < \beta$ **then continue** β;

This performs a complete scan of all cosets under all relators, but without making new definitions. The idea is that space may be recovered as a result of cosets being eliminated from Ω in calls of COINCIDENCE. If any space is recovered, then we will need to renumber cosets so as to make the set Ω of active cosets equal to $[1..n]$ with $n < M$. We shall describe this renumbering process in more detail in Subsection 5.2.3, below. The main function can then be resumed, starting again from the remembered value of α. (But of course the implementation will need to be able to deal with the case in which this α has been eliminated during LOOKAHEAD!)

5.2.2 The coset table-based method

This has often been called the *Felsch* method, and was first described in [Fel61]. The policy here is to define α^x where α^x is the first undefined entry in the coset table. By this, we mean we choose the first $\alpha \in \Omega$ for which α^x is undefined for some $x \in A$, and then choose the first $x \in A$, with respect to some fixed ordering of A, for which α^x is undefined. Notice that this means that we are defining elements 1^w for $w \in A^*$ in order of increasing w in the shortlex ordering (Definition 2.60) of A^* with respect to the ordering of A that we are using. In practice, we number the elements of A before starting the enumeration, and this provides our ordering of A. Occasionally examples are encountered in which this ordering drastically affects the performance of the enumeration, so the implementation should allow the user the possibility of setting this ordering.

The usual practice with the Felsch strategy is to start by calling SCANAND-FILL$(\sim C, 1, w)$ for all $w \in Y$, just as we did in HLT. After doing that, the policy is to derive all possible consequences of each definition before making any further definitions. The easiest way to do that is simply to call the procedure LOOKAHEAD after each definition, but in practice that would be very inefficient. A better method is to store all pairs (α, x) for which a value is assigned to α^x during a definition or deduction on a *deduction stack* and then, when processing the entries in the stack, we only call SCAN$(\sim C, \alpha, w)$ on those α and w for which new information could conceivably be deduced from the definition or deduction currently being processed. To accomplish this, the code of the procedures DEFINE, SCAN, and COINCIDENCE are first modified such that each time the value of α^x is defined or altered for any α and x, the pair (α, x) is pushed onto this deduction stack. However, since such alterations are always made in inverse pairs $\alpha^x := \beta$ and $\beta^{x^{-1}} := \alpha$, we only need save the first alteration of such a pair.

Consequences of a particular new assignment $\alpha^x := \beta$ can only occur for a call of SCAN$(\sim C, \gamma, w)$ for which $w \in R$ and $w = uxv$ with either $\gamma^u = \alpha$ or $\gamma^{v^{-1}} = \beta$. The new information $\alpha^x = \beta$ can then be used to extend this scan, which might then result in a new deduction or coincidence. We can obtain the same new information by calling SCAN$(\sim C, \alpha, xvu)$ and SCAN$(\sim C, \beta, x^{-1}u^{-1}v^{-1})$ rather than SCAN$(\sim C, \gamma, w)$.

To implement this idea, it is convenient to start by computing the set R^c of all cyclic conjugates of all elements of $R \cup R^{-1}$, and then, for each $x \in A$, to compute the set R_x^c of all elements of R^c having x as their first symbol. (And before doing that, we should replace the elements of R by cyclically reduced conjugates if necessary.) We can then look for consequences of the new assignment $\alpha^x := \beta$ by calling SCAN$(\sim C, \alpha, w)$ for all $w \in R_x^c$ and SCAN$(\sim C, \beta, w)$ for all $w \in R_{x^{-1}}^c$.

In implementations, a fixed amount of space is generally reserved for the deduction stack. If this space is exceeded, then rather than processing individual deductions as just described, we call LOOKAHEAD, which means that all consequences of all deductions are automatically found. This has been found not to be unduly time-consuming, because usually the deduction stack does not grow very large and, when it does, it means that there is a large number of deductions to process, so we might as well call LOOKAHEAD.

We can now write down the code for coset table-based coset enumeration, but we write the code for processing deductions as a separate procedure.

PROCESSDEDUCTIONS($\sim \mathcal{C}, \{R_x^c\}$)

```
1   while deduction stack is not empty
2       do if deduction stack is full
3           then LOOKAHEAD(∼C, R);
4               Empty deduction stack;
5           else
6               Pop next pair (α, x) from deduction stack;
7                   if p[α] = α then for w ∈ R_x^c
8                       do SCAN(∼C, α, w);
9                           if p[α] < α then break;
10              β := α^x;
11              if p[β] = β then for w ∈ R_{x^{-1}}^c
12                  do SCAN(∼C, β, w);
13                      if p[β] < β then break;
```

COSETENUMERATIONC($\langle X|R \rangle, Y$)

Input: A finite group presentation $\langle X|R \rangle$ and a finite subset Y of A^*

```
1   Replace all elements of R by their cyclic reductions;
2   Compute the set R^c of all cyclic conjugates of all elements of R ∪ R^{-1};
3   for x ∈ A,
4       do compute set R_x^c of elements of R^c having first letter x;
5   Initialize a coset table C for ⟨ X | R ⟩;
6   p[1] := 1;  n := 1;
7   for w ∈ Y do SCANANDFILL(∼C, 1, w);
8   PROCESSDEDUCTIONS(∼C, {R_x^c});
    (∗ Make definitions using first undefined element of coset table ∗)
9   for α ∈ Ω,  x ∈ A do if α^x is undefined
10      then DEFINE(∼C, α, x);  PROCESSDEDUCTIONS(∼C, {R_x^c});
11  return C;
```

To prove that this procedure works as claimed, we need to verify that the conditions of Property 5 are satisfied. The first two conditions of this property are clearly satisfied. For the third condition, let $\alpha \in \Omega$, $w \in R$, and suppose that w begins with x. Then α^x will certainly get defined at some stage, and then (α, x) (or its inverse assignment) will be put on the deduction stack. Then, provided that α remains within Ω, SCAN($\sim \mathcal{C}, \alpha, w$) will be called from the next call of PROCESSDEDUCTIONS.

5.2.3 Compression and standardization

The procedure DEFINE always uses $n + 1$ as the next coset number, where n is the largest number that has been used to date. Since cosets get eliminated during calls of COINCIDENCE, this means that Ω will not always equal $[1 .. |\Omega|]$. Normally, at the end of an enumeration, and also possibly during the enumeration if we run out of space to define new cosets (see Subsection 5.2.5 below), we will want to redefine the coset table to make $\Omega = [1 .. n]$, where $n = |\Omega|$. The following procedure does exactly that.

COMPRESS($\sim\mathcal{C}$)

```
1   γ := 0;
2   for α ∈ [1 .. n] do if p[α] = α
3       then γ := γ + 1;
4           if γ ≠ α
5               then (* Replace α by γ in coset table *)
6                   for x ∈ A
7                       do β := αˣ;
8                           if β = α then β := γ;
9                               γˣ := β;  βˣ⁻¹ := γ;
10  n := γ;
11  For α ∈ [1 .. n] do p[α] := α;
```

In some situations, it is desirable at the end of an enumeration to permute the cosets so that they occur in some sort of standard order. This could be used, for example, to test for the equivalence of the permutation actions of G defined by two coset tables. It is also convenient, in terms of simplifying code, to standardize the coset table before using it as input to the subgroup presentation algorithm that will be described in Subsection 5.3.1 below.

The code below reorders the elements of Ω such that, if we scan the coset table first by elements of Ω and then by elements of A, then the cosets occur in ascending numerical order. This is the order in which they would occur with a coset-based enumeration that completed with no coincidences. We shall call the resulting coset table *standardized*. We assume that COMPRESS has been called before STANDARDIZE; that is, that $\Omega = [1 .. n]$.

SWITCH($\sim\mathcal{C}, \beta, \gamma$)

```
    (* Switch the elements β, γ of Ω in C *)
1   for x ∈ X
2       do z := γˣ;  γˣ := βˣ;  βˣ := z;
3           for α ∈ Ω
4               do if αˣ = β then αˣ := γ;
5                   elseif αˣ = γ then αˣ := β;
```

STANDARDIZE($\sim\mathcal{C}$)

    Input: A complete coset table $\sim\mathcal{C}$ on $\Omega = [1 \mathinner{.\,.} n]$

1 $\gamma := 2$;

    ($*$ γ is the next entry that we want to encounter in the coset table $*$)

2 **for** $\alpha \in [1 \mathinner{.\,.} n]$, $x \in A$

3   **do** $\beta := \alpha^x$;

4   **if** $\beta \geq \gamma$

5   **then if** $\beta > \gamma$ **then** SWITCH($\sim\mathcal{C}, \gamma, \beta$);

6                 $\gamma := \gamma + 1$;

7   **if** $\gamma = n$ **then return** ;

As an example, consider the following coset table that was computed in Example 5.3:

	x	x^{-1}	y	y^{-1}
1	2	4	2	4
2	3	1	3	1
3	4	2	4	2
4	1	3	1	3

When we work through the above procedure, we start with $\alpha = 1$, $\gamma = 2$ and find $\alpha^x = 2 = \gamma$, which is correct for a standardized table. So γ is increased to 3. We then look at $\alpha^{x^{-1}} = 4$, which is not what we want in a standardized table. We want the next new entry to be $\gamma = 3$, so we call SWITCH(3,4). This interchanges 3 and 4 in the coset table, and the modified table is:

	x	x^{-1}	y	y^{-1}
1	2	3	2	3
2	4	1	4	1
3	1	4	1	4
4	3	2	3	2

The table is now in fact standardized. This is because the entries $1, 2, \ldots n-1$ appear in the correct order in the table, and so n must also occur in the correct place. Thus, in the procedure, we can interrupt the loop and return on observing that the incremented value of γ is n.

5.2.4 Recent developments and examples

Although both COSETENUMERATIONR and COSETENUMERATIONC will work adequately on most straightforward examples in which $|G : H|$ is relatively small compared with the maximum number M of possible cosets, there are examples in which one performs significantly better than the other.

    In its present form, and for large indices $|G : H|$, COSETENUMERATIONR tends to be less good when memory is in short supply, even when used in conjunction with LOOKAHEAD. A possibility proposed by Sims in [Sim94] is to use the deduction stack and PROCESSDEDUCTIONS in combination with COSETENUMERATIONR.

COSETENUMERATIONC in its pure form can struggle when the presentation has very long relators, because there is a danger that consequences of such long relators are not derived for a very long time. One simple method that has been found to improve the performance of COSETENUMERATIONC significantly is the rather strange-sounding technique of including the group relators among the subgroup generators; that is, by replacing the original Y by $Y \cup R$. This has the effect of ensuring that at least SCAN$(\sim\mathcal{C}, 1, w)$ is called immediately for all $w \in R$.

We shall now describe the performance of these procedures on a selection of examples. We take H to be the trivial subgroup in each case.

First a straightforward enumeration: M_{12}, of order 95040, has a presentation

$$\langle\, a, c, d \mid a^{11}, c^2, d^2, (ac)^3, (ad)^3, (cd)^{10}, a^2(cd)^2a(cd)^{-2}\,\rangle.$$

This completes within a second on a PC using any strategy. It is very slightly faster using the relator-based method (HLT+LookAhead), but defines a maximum of 125293 and a total of 315238 cosets. With the coset table-based method (Felsch), it uses a maximum of 95040, which is optimal, and a total of 97060 cosets.

The remaining examples have been selected for their difficulty. Fortunately, in many applications, the straightforward behaviour exhibited by the above example M_{12} is more typical.

The following deficiency-zero presentations of $G = \mathrm{SL}(2, p)$ for p prime are defined by Campbell and Robertson in [CR80]:

$$\langle\, x, y \mid x^2(xy)^{-3}, (xy^4xy^{(p+1)/2})^2 y^p x^{2k}\,\rangle,$$

where k is the integral part of $p/3$. These are very difficult examples relative to the order of the group. The long relator has the effect that they perform better with HLT+LookAhead. For example, with $p = 13$, $|G| = 2184$, the Felsch method defined over 3 million cosets and took about 5 times longer than HLT+LookAhead, which completed using a maximum of 2 million cosets. For $p = 17$, $|G| = 4896$, HLT+LookAhead completed with a maximum of 10 million cosets, whereas Felsch needed more than 40 million.

The group

$$\langle\, a, b, c \mid b^2c^{-1}bc, a^2b^{-1}ab, ca^{-1}b^{-1}cabc\,\rangle$$

is the first of the 14 groups defined in [HNO95]. It has order 6561. It needs just over 14000 cosets to complete with Felsch, and more than 86000 to complete with HLT+Lookahead, but this decreases to about 40000 if cyclic conjugates of relators are included.

The eighth of these examples, also of order 6561, is

$$\langle\, a, b, c \mid a^2c^{-1}ac, acb^2a^{-1}c^{-1}b, a^{-1}b^{-1}abc^3\,\rangle.$$

This is much harder. It needed 30 million cosets with the Felsch method, and failed to complete with 100 million cosets using HLT+Lookahead.

The most versatile and carefully implemented coset enumeration package at the present time is almost certainly ACE3 [HR99] by Havas and Ramsay, and includes virtually all possible variations that have ever been suggested as user options. There are interfaces to this implementation from both GAP and MAGMA. The default mode of running is to alternate between the definition strategies of COSETENUMERATIONR and COSETENUMERATIONC, in the sense that m cosets are defined using the relator-based method and then n cosets are defined using the coset table-based method, and so on. The values of m and n can either be set directly by the user, or are determined by other parameters. Whether or not PROCESSDEDUCTIONS is used is optional and, if it is, then the size of the deduction stack can be set. By default $Y \cup R$ is used in place of Y, as suggested above. Another option, which tends to make enumerations much slower, but can enable a very difficult enumeration to complete, is to replace R by the set R^c of all cyclic conjugates of all relators. The user should consult the ACE3 manual for more details or, alternatively, consult the section on advanced features of finitely presented groups in the MAGMA manual ([BC93] or [Mag04]).

Finally, we mention the interactive coset enumeration package ITC written by V. Felsch, L. Hippe, and J. Neubüser, which is available as a GAP package [GAP04]. This allows the user to intervene with specific definitions during the course of an enumeration, and also to rerun an enumeration retroactively in an attempt to find shorter routes to completion.

5.2.5   Implementation issues

Most implementations of coset enumeration to date have started with a fixed amount of memory, which the user may specify at runtime, but they have not attempted to allocate extra memory in mid-enumeration. This means that there will a maximum size M of Ω.

The coset table is generally stored as a two-dimensional array with rows in the range $[1 \mathinner{\ldotp\ldotp} n]$ representing elements of Ω, and columns representing the elements of A in some order that the user might be allowed to specify. Then α^x is equal to the element in row α and column x of the array, where a 0 entry means that α^x is undefined.

As we have already explained, if DEFINE aborts and some cosets have been eliminated by calls of COINCIDENCE, then the procedure COMPRESS defined in Subsection 5.2.3 can be used to reclaim lost space. If we are using a strategy that involves calling LOOKAHEAD, then we may wish to call LOOKAHEAD before calling COMPRESS.

If we still have $M = n$ after calling COMPRESS, then we have to give up. Otherwise, we can resume the enumeration and make a new definition. In an implementation, there are some other housekeeping details to be handled. For example, when we resume an enumeration after a call of COMPRESS, the value of the main loop variable α in COSETENUMERATIONR or COSETENU-MERATIONC needs to be replaced by its new value; so this new value needs to be returned by COMPRESS. One way of handling this is to introduce an extra component to the coset table itself, which would represent the current coset being processed.

It is also possible to use linked lists to keep track of the integers representing live cosets; integers representing cosets to be eliminated can easily be unlinked, and the corresponding row in the coset table adjoined to another linked list of available coset numbers. This approach is used in the implementation described in [CDHW73].

The versions of coset enumeration that we have described involve two arrays p and η that are used in COINCIDENCE, and both of these must have space for M integers. We may also need some space, say L, for the deduction stack. The total amount of space required for the enumeration then amounts to $(2 + |A|)M + L$ integers, plus a small amount of space for other variables. The implementation described in [CDHW73] uses the same space for the arrays p and η and for the pointers needed for the linked lists described above.

Since it is important to use space as economically as possible, the more recent implementations, such as [HR99] have made use of an idea originally outlined by Beetham in [Bee84], which enables the information stored in p and η to be stored in the first two columns of the coset table. Since rows of the table are being continually made free during calls of COINCIDENCE, it is reasonable to hope that something of this kind might be possible. It is, however, not straightforward to achieve this in practice, because the coincidence queue can grow much faster than the rate at which rows of the coset table are made free.

To overcome this difficulty, it is necessary to handle the first two columns separately and in a different manner from the other columns when processing coincidences. We shall not give any further details here, but refer the reader to the Appendix of [Sim94] where the detailed pseudocode for this process is included. This means that the total space required for the enumeration is reduced to about $2|A|M + L$ signed integers. (The integers need to be signed because, with this approach, eliminated cosets are recognized by the fact that their entry in the first column of the coset table is negative.)

5.2.6   The use of coset enumeration in practice

After a successful call of COSETENUMERATIONR or COSETENUMERATIONC on a subgroup H of a group G, we can immediately define the associated permutation representation of G on Ω, which Theorem 5.2 tells us is equivalent to the action of G by multiplication on the right cosets of H, as defined in Example 2.3. To do this, it is sensible first to call COMPRESS in order to make Ω equal to $[1..n]$, where $n = |G:H|$. The permutation action is then the homomorphism $\varphi : G \to \mathrm{Sym}(\Omega)$ where, for each $x \in X$, $\varphi(x)$ is the permutation defined already by our notation; that is, by $\alpha \mapsto \alpha^x$, for $\alpha \in \Omega$. We can therefore perform computations in the permutation group $\mathrm{im}(\varphi)$. This group is isomorphic to $G/\ker(\varphi) = G/H_G$, where H_G is the core of H in G; see Example 2.4.

If $H_G = \{1\}$, then we have a permutation group isomorphic to G, and so we can easily compute the order of G and its group-theoretical properties. We can achieve this by putting $H = \{1\}$, but is more efficient if we can find a larger subgroup with $H_G = \{1\}$. One painless way of attempting this is to try $H = \langle v \rangle$ for some $v \in A^*$. This is particularly promising if one of the defining relators of G is v^m for some $m > 0$. In that case, if $\varphi(v)$ has order exactly m, then $\varphi|_H$ is clearly an isomorphism and, since $\ker(\varphi) \le H$, this implies that $\ker(\varphi) = \{1\}$. For example, in Example 5.2, we had $G = \langle\, x, y \mid x^2, y^3, (xy)^3\,\rangle$ with $H = \langle xy \rangle$, and we computed the coset table:

	x	y	y^{-1}
1	2	3	2
2	1	1	3
3	4	2	1
4	3	4	4

In the associated permutation representation $\varphi : G \to \mathrm{Sym}(\Omega)$, we have $\varphi(x) = (1,2)(3,4)$ and $\varphi(y) = (1,3,2)$. Hence $\varphi(xy) = (2,3,4)$ has order 3. Since G has a relator $(xy)^3$, we can conclude that xy has order exactly 3 in G and hence that $|H| = 3$, and so $|G| = 12$ and φ is faithful representation.

This will not always work, even when there are promising-looking group relators. For example, the group $\langle\, x, y \mid x^4, y^4, (xy)^4, yxyx^{-1}\,\rangle$ has order 16, but trying $H = \langle x \rangle$, $\langle y \rangle$ and $\langle xy \rangle$ yields permutation group images of orders 8, 4, and 8, respectively. And if there are no defining relators that are proper powers, then the approach cannot work. There is, however, a slightly more sophisticated possibility, which is to choose $H = \langle x \rangle$ for some $x \in A$, and also compute a presentation of the subgroup H. Computation of presentations of subgroups will be discussed in the next section. Experience has shown that, if our aim is to find the order of G, then this technique is usually more efficient than choosing $H = \{1\}$.

Exercises

1. Let $G := \langle\, x, y \mid x^2, y^3, (xy)^4\,\rangle$. Use coset enumeration to show that $|G:H| = 6$, where $H := \langle xy \rangle$ and use this to prove, using the techniques of Subsection 5.2.6, that $G \cong \mathrm{Sym}(4)$.

2. Let $G := \langle\, x, y \mid x^3, y^5, (xy)^2\,\rangle$ and let $H := \langle\, x, yx^{-1}y^2\,\rangle$.

   (i) Show by coset enumeration that H has index 5 in G, and that the image of the corresponding coset action is $\mathrm{Alt}(5)$.

   (ii) Show, by using the group relators, that the elements $a := x$ and $b := yx^{-1}y^2$ of G satisfy the relations $a^3 = b^3 = (ab)^2 = 1$.

   (iii) Deduce that $|H| \le 12$ and hence that $|G| = 60$ and $G \cong \mathrm{Alt}(5)$.

3. This example is from [War74]. Let $G := \langle\, x,y \mid x^3yxyx^{-3}, y^3xyxy^{-3}\,\rangle$. Show without using coset enumeration that $|G| = 3$.

   Suppose that we carry out COSETENUMERATIONR on this group with subgroup $H = 1$, but with lines 9 and 10 removed, so that Property 5(i) is no longer guaranteed to hold. Show that every application of SCANANDFILL on each coset α will define a new coset representing α^{x^3} or α^{y^3}, and hence the procedure will never terminate.

4. We could avoid the problem with the last example by replacing all group relators by cyclically reduced conjugates before we start. Find an example to show that, even with this modification, the procedure cannot be guaranteed to terminate on a finite group if we remove lines 9 and 10.

5.3 Presentations of subgroups

In this section, we shall present two algorithms for computing presentations of subgroups H of finite index in a group G defined by a finite presentation. The first of these, known generally as the Reidemeister-Schreier algorithm, computes a presentation on a set of Schreier generators of H, and the second one computes a presentation on the given set Y of generators. The reader might wish to review Section 2.5 before reading on.

Both algorithms are inclined to compute large and unwieldy presentations, particularly when $|G : H|$ is large. We shall conclude the section by briefly describing ways of using Tietze transformations to simplify large presentations by eliminating redundant generators and relators, and by shortening long relators.

5.3.1 Computing a presentation on Schreier generators

The input for this procedure is a completed coset table for G over H or, equivalently (cf Theorem 5.2), the permutation action of G by right multiplication on the right cosets of G. The coset table may have been computed by coset enumeration or by some alternative method. We shall assume, however, that $\Omega = [1 .. n]$, where $n = |G : H|$, and that the table has been standardized by a call of STANDARDIZE if necessary. The latter assumption is not strictly necessary, but it renders the details of the code simpler.

Example 5.5

As we describe the procedure, we shall work through the following example. Let
$$G := \langle\, x,y \mid x^2, y^3, (xy)^6\,\rangle.$$

Then it is easily checked, using Theorem 2.52, that the map $\varphi : X \to \mathrm{Sym}(\Omega)$ with $\Omega = \{1, 2, 3, 4, 5, 6\}$ defined by

$$\varphi(x) = (1, 2)(3, 5)(4, 6), \qquad \varphi(y) = (1, 3, 4)(2, 5, 6)$$

extends to a homomorphism $\varphi : G \to \mathrm{Sym}(\Omega)$, and φ is clearly a transitive permutation representation of G. Since $\varphi(x)\varphi(y) = \varphi(y)\varphi(x)$, we have $|\mathrm{im}(\varphi)| = 6$. Hence the action of G on Ω defined by φ is regular, and the stabilizer H of 1 in this action is equal to the kernel of the action. In fact it is not hard to show that $H = [G, G]$. Notice that we have not been given generators of H in this example. We have chosen $\varphi(x)$ and $\varphi(y)$ such that the associated coset table below is already standardized.

	x	x^{-1}	y	y^{-1}
1	2	2	3	4
2	1	1	5	6
3	5	5	4	1
4	6	6	1	3
5	3	3	6	2
6	4	4	2	5

Notice that, although there is a relator x^2, we have still included a column for x^{-1}. This is necessary for the computation of a presentation of H on the set of Schreier generators. In fact the relators of H corresponding to conjugates of x^2 will tell us that some of these Schreier generators are trivial in H, and, if we are happy to omit these from the generating set, then we could save a little time by not including the column for x^{-1}. ⬜

As in Section 2.5, let $G = \langle X \mid R \rangle$ where $G = F/N$ with $F = F_X$ and $N = \langle R^F \rangle$, and let $H = E/N$. We start by choosing our Schreier transversal T of E in F and defining the set Y of Schreier generators y_{tx} of H with respect to this transversal. We can do this by a scan of the coset table. As in Section 5.1, we shall denote the element of T representing $\alpha \in \Omega$ by $\tau(\alpha)$, but we do not need to store the words $\tau(\alpha)$.

The procedure DEFINESCHREIERGENERATORS defines Y, and computes a word $P[\alpha, x]$ in $(Y \cup Y^{-1})^*$ for each $\alpha \in \Omega$ and $x \in A := X \cup X^{-1}$. This word satisfies $\tau(\alpha)x =_F P[\alpha, x]\tau(\alpha^x)$. It is either the empty word (if it arises when $\tau(\alpha^x)$ is defined as $\tau(\alpha)x$), or a Schreier generator in Y (when $x \in X$), or an inverse of a Schreier generator (when $x \in X^{-1}$). Since each $\tau(\alpha^x)$ is defined as $\tau(\alpha)x$ for some existing element $\tau(\alpha)$ of T, it is clear that T is prefix closed, and hence a Schreier transversal.

We regard Y and P as being new components of the coset table $\mathcal{C}$, and define a *modified coset table* $\mathcal{C}_M$ to be an ordinary coset table endowed with these two extra components.

Let us apply DEFINESCHREIERGENERATORS to Example 5.5. To ease the notation, we shall introduce new names $a, b, c, \ldots$ for the Schreier generators.

DEFINESCHREIERGENERATORS($\sim \mathcal{C}$)

    Input: A complete standardized coset table on $\Omega = [1 \mathinner{.\,.} n]$

    $(* \ \tau(1) := \varepsilon \ *)$

1   $Y := \{\}$; $\gamma := 2$;

    $(* \ \gamma$ is the next new entry that we will find in the standardized table $*)$

2   **for** $\alpha \in [1 \mathinner{.\,.} n], \ x \in A$

3   **do** $\beta := \alpha^x$;

4   **if** $\beta = \gamma$

5   **then** $(* $ Choose coset representative corresponding to $\tau \ *)$

6                    $P[\alpha, x] := \varepsilon$; $P[\beta, x^{-1}] := \varepsilon$; $\gamma := \gamma{+}{+}1$;

7                    $(* \ \tau(\beta) := \tau(\alpha)x \ *)$

8   **elseif** $x \in X$ and $P[\alpha, x]$ is undefined

9   **then** $(* $ Adjoin new Schreier generator to $Y \ *)$

10                  $Y := Y \cup \{y_{\alpha x}\}$; $P[\alpha, x] := y_{\alpha x}$; $P[\beta, x^{-1}] := y_{\alpha x}^{-1}$;

When $\alpha = 1$, we define $\tau(2) := x$, $\tau(3) := y$ and $\tau(4) := y^{-1}$, and set $P[1, x]$, $P[2, x^{-1}]$, $P[1, y]$, $P[1, y^{-1}]$, $P[3, y^{-1}]$ and $P[4, y]$ equal to ε. When $\alpha = 2$, we introduce the Schreier generator $a := y_{2x}$ and set $P[2, x]$ to a and $P[1, x^{-1}]$ to a^{-1}. The next two entries give rise to the definitions $\tau(5) := \tau(2)y = xy$ and $\tau(6) := \tau(2)^{y^{-1}} = xy^{-1}$, together with the corresponding trivial entries in the table P. So we have now completed the definition of $T = \{\varepsilon, x, y, y^{-1}, xy, xy^{-1}\}$. The remaining entries α^x give rise to the Schreier generators $b := y_{3x}$, $c := y_{3y}$, $d := y_{4x}$, $e := y_{5x}$, $f := y_{5y}$, and $g := y_{6x}$.

The table below shows both $P[\alpha, x]$ and α^x for $\alpha \in \Omega$, $x \in A$. This is helpful when working through examples by hand, because it emphasizes the relationship $\tau(\alpha)x =_F P[\alpha, x]\tau(\alpha^x)$.

	x		x^{-1}		y		y^{-1}	
1		2	a^{-1}	2		3		4
2	a	1		1		5		6
3	b	5	e^{-1}	5	c	4		1
4	d	6	g^{-1}	6		1	c^{-1}	3
5	e	3	b^{-1}	3	f	6		2
6	g	4	d^{-1}	4		2	f^{-1}	5

We saw in the remark following Theorem 2.62 that

$$S := \{\, \rho(t, w) \mid t \in T, w \in R \,\}$$

is a set of defining relators for a group H', isomorphic to H, on the generating set Y. The function REWRITE computes $\rho(t, w)$ for $t \in T$ and $w \in W$. It takes the element $\alpha \in \Omega$ with $1^t = \alpha$ as an argument. So, to compute S, we just need to call REWRITE on all $\alpha \in \Omega$ and $w \in W$.

REWRITE($\mathcal{C}_M, \alpha, w$)

   Input: A complete modified coset table $\mathcal{C}_M$ on $\Omega = [1 \mathinner{\ldotp\ldotp} n]$,
      $\alpha \in \Omega$, $w = x_1 x_2 \cdots x_r \in A^*$.
   Output: $\rho(\tau(\alpha), w)$.

1   $v := \varepsilon$;
2   **for** $i \in [1 \mathinner{\ldotp\ldotp} r]$ **do** $v := vP[\alpha, x_i]$; $\alpha := \alpha^{x_i}$;
3   **return** v;

We shall now use this function to calculate the set S in Example 5.5. Calls of REWRITE with $\alpha = 1$ and $w = x^2$ or y^3 return a and c, respectively. Let us work through the call REWRITE($\mathcal{C}_M, 1, (xy)^6$) in detail. We start with $\alpha = 1$, $v = \varepsilon$. Working through the 'For' loop, produces the following sequence of assignments:

$$
\begin{aligned}
\alpha := \alpha^x = 2, \quad & v := vP[1, x] = \varepsilon; \\
\alpha := \alpha^y = 5, \quad & v := vP[2, y] = \varepsilon; \\
\alpha := \alpha^x = 3, \quad & v := vP[5, x] = e; \\
\alpha := \alpha^y = 4, \quad & v := vP[3, y] = ec; \\
\alpha := \alpha^x = 6, \quad & v := vP[4, x] = ecd; \\
\alpha := \alpha^y = 2, \quad & v := vP[6, y] = ecd; \\
\alpha := \alpha^x = 1, \quad & v := vP[2, x] = ecda; \\
\alpha := \alpha^y = 3, \quad & v := vP[1, y] = ecda; \\
\alpha := \alpha^x = 5, \quad & v := vP[3, x] = ecdab; \\
\alpha := \alpha^y = 6, \quad & v := vP[5, y] = ecdabf; \\
\alpha := \alpha^x = 4, \quad & v := vP[6, x] = ecdabfg; \\
\alpha := \alpha^y = 1, \quad & v := vP[4, y] = ecdabfg.
\end{aligned}
$$

So the relator returned is $ecdabfg$. The complete set of results returned by REWRITE($\mathcal{C}_M, \alpha, w$) is listed in the following table; we can hopefully leave the reader to check these!

	x^2	y^3	$(xy)^6$
1	a	c	$ecdabfg$
2	a	f	$abfgecd$
3	be	c	$bfgecda$
4	dg	c	$dabfgec$
5	eb	f	$ecdabfg$
6	gd	f	$gecdabf$

Looking at the complete set S of relators of H', we notice that there is some repetition, and also that the relators arising from $(xy)^6$ are all cyclic conjugates of each other, and hence only one of them is needed. So the presentation on the Schreier generators simplifies to

$$\langle a, b, c, d, e, f, g \mid a, c, f, be, dg, ecdabfg \rangle.$$

We can further simplify this presentation without much effort. The generators a, c, f are all trivial and so can be omitted, and we can use the relators be and dg to eliminate the generators e and g by replacing them by b^{-1} and d^{-1}, respectively, as described in Subsection 2.4.4. The result is the presentation $\langle\, b,d \mid b^{-1}dbd^{-1}\,\rangle$, which defines a free abelian group on two generators. Incidentally, we have now proved that H and hence G is an infinite virtually abelian group.

A moment's thought should convince the reader that the fact that we are getting lots of repetition of relators via cyclic conjugates is no accident. In general, if we have a relator of the form w^r for some $r > 1$, and if $\alpha^{w^i} = \beta$ for some $\alpha, \beta \in \Omega$ and $1 \le i < r$, then the words returned by $\textsc{Rewrite}(\mathcal{C}_M, \alpha, w^r)$ and $\textsc{Rewrite}(\mathcal{C}_M, \beta, w^r)$ will be cyclic conjugates of one another. Being aware of this can save time, particularly when working examples by hand.

Let us work through another example.

Example 5.6

Let $G := \langle\, x, y \mid x^3, y^5, (xy)^2\,\rangle$, and $H := \langle\, xy, x^{-1}y^{-1}xyx\,\rangle$. We leave it to the reader to carry out a coset enumeration and to calculate that $|G\!:\!H| = 5$. As a hint, the sequence of definitions $1^x = 2$, $2^x = 3$, $3^{y^{-1}} = 4$, $3^y = 5$ will lead to the enumeration completing with no coincidences, and the resulting coset table, which fortunately is already standardized, is:

	x	x^{-1}	y	y^{-1}
1	2	3	4	2
2	3	1	1	5
3	1	2	5	4
4	4	4	3	1
5	5	5	2	3

As in the previous example, we call $\textsc{DefineSchreierGenerators}$ to compute the function P, and the modified coset table is:

	x		x^{-1}		y		y^{-1}	
1		2		3		4	b^{-1}	2
2	a	3		1	b	1		5
3		1	a^{-1}	2	c	5	e^{-1}	4
4	d	4	d^{-1}	4	e	3		1
5	f	5	f^{-1}	5		2	c^{-1}	3

Now, calling $\textsc{Rewrite}(\mathcal{C}_M, \alpha, x^3)$ for $\alpha \in [1..5]$ and omitting cyclic conjugates, yields relators a, d^3, f^3. Doing the same with y^5 results in just ecb, and with $(xy)^2$ we get b^2, acf, and de. So the presentation of $H' \cong H$ on the Schreier generators is

$$\langle\, a, b, c, d, e, f \mid a, d^3, f^3, ecb, b^2, acf, de \,\rangle.$$

We can immediately eliminate a, e (by replacing e by d^{-1}), and f (by replacing f by c^{-1}). We now have $\langle\, b, c, d \mid d^3, c^{-3}, d^{-1}cb, b^2, \,\rangle$. Eliminating d, by replacing it by cb, and replacing the relator c^{-3} by the equivalent c^3, we get $\langle\, b, c \mid b^2, c^3, (bc)^3 \,\rangle$. In fact we have already studied the group defined by this presentation, and we saw in Subsection 5.2.6 that it has order 24, so $|H| = 24$ and $|G| = 5|H| = 60$. The image of the associated permutation representation φ, which is generated by $\varphi(x) = (1,2,3)$ and $\varphi(y) = (1,4,3,5,2)$ is Alt(5), so φ is a monomorphism, and $G \cong \mathrm{Alt}(5)$. □

The algorithm that we have described in this subsection to compute a presentation of a subgroup on its Schreier generators is one of the easiest algorithms in CGT both to implement and to use to work through examples by hand. Such an implementation is described by Havas in [Hav74]. In an implementation, elements of Y would typically be represented by positive integers, and then P could be a two-dimensional array of integers, where 0 is used to represent the empty string, and the inverse of the generator number i is represented by $-i$. The disadvantage of the algorithm is that, for $|G:H| = n$, we have $|Y| = (n-1)|X| + 1$ and $|S| = n|R|$, so the resulting presentations can be highly unwieldy when n is large. We shall discuss how to simplify such presentations in Subsection 5.3.3 below.

5.3.2   Computing a presentation on the user generators

If we are given a finitely presented group $G = \langle X \mid R \rangle$ and a subgroup $H = \langle \hat{Y} \rangle$ of G, then we may wish to compute a presentation of H on its given generating set, rather than on the set of Schreier generators. This can be done, but it is more complicated than the algorithm described in Subsection 5.3.1, and involves a process known as *modified coset enumeration* or the *modified Todd-Coxeter procedure* in which we carry out the coset enumeration using a modified coset table.

We introduce a new set of symbols Y corresponding to $\hat{Y}$. So we have a bijection $\varphi : Y \to \hat{Y}$, which extends to a homomorphism $\varphi : F_Y \to F$, where $\mathrm{im}(\varphi) = E$, with $G = F/N$ and $H = E/N$ as in Subsection 5.3.1. Assuming that $|G:H|$ is finite, we can use coset enumeration to calculate the associated coset table and (implicitly) a transversal T and a bijection $\tau : \Omega \to T$.

In modified coset enumeration, we compute elements y_{tx} of F_Y with the property that $tx =_G \varphi(y_{tx})\overline{tx}$ for all $t \in T$, $x \in X$ at the same time as we compute the coset table itself. As usual, we will take $y_{tx^{-1}}$ to be the inverse of the element y_{ux}, where $\overline{ux} = t$. We first need to describe how to compute the words y_{tx}. Once we have done that, computing the presentation of H will be straightforward, and will be similar to the method used in Subsection 5.3.1.

As in Subsection 5.3.1, we will store y_{tx} as a word $P[\alpha, x] \in (Y \cup Y^{-1})^*$, where $t = \tau(\alpha)$, but now we will compute these words as we carry out the coset enumeration. In fact $P[\alpha, x]$ will be defined if and only if α^x is defined, and we will always define $P[\alpha, x]$ such that it satisfies the required equation $\tau(\alpha)x =_G \varphi(P[\alpha, x])\tau(\beta)$. To do this, we need to modify our coset enumeration procedures DEFINE, SCAN, SCANANDFILL and COINCIDENCE so that they assign $P[\alpha, x]$ whenever they assign α^x.

We shall not prove the correctness of what we are doing formally, like we did for normal coset enumeration in Section 5.1, although this would not be particularly difficult. Instead, we shall attempt to explain what we are doing, so that the reader can construct formal proofs if necessary.

The change needed to DEFINE is easy. Since we are defining $\tau(\beta)$ to be $\tau(\alpha)x$, we just need to set $P[\alpha, x]$ and $P[\beta, x^{-1}]$ to be ε. The modified coset tables $\mathcal{C}_M$ have an additional component, which is a function p_P from $[1 .. n]$ to A^*. It is used in connection with the modified coincidence procedure, which we shall discuss shortly.

MODIFIEDDEFINE$(\sim \mathcal{C}_M, \alpha, x)$

  Input: $\alpha \in \Omega$, $x \in A$, with α^x undefined.

1 $n := n+1$; $\beta := n$; $p[\beta] := \beta$;
2 $\alpha^x := \beta$; $\beta^{x^{-1}} := \alpha$;
3 $P[\alpha, x] := \varepsilon$; $P[\beta, x^{-1}] := \varepsilon$; $p_P[\beta] := \varepsilon$;
4 (* Set $\tau(\beta)$ equal to $\tau(\alpha)x$. *)

The MODIFIEDSCAN procedure takes arguments α, w, and y. When we call it, the extra argument y is put equal to ε when $w \in R$, and to the corresponding $y \in Y$, when $\alpha = 1$ and w is a subgroup generator. Thus, in all cases, we have $\tau(\alpha)w =_G \varphi(y)\tau(\alpha)$, and the correctness of the procedure depends on that fact. The MODIFIEDCOINCIDENCE routine also takes three arguments, $\kappa, \lambda \in \Omega$ and a word $w \in Y^*$, and we call it in situations where we know that $\tau(\kappa) =_G \varphi(w)\tau(\lambda)$.

Throughout MODIFIEDSCAN, we maintain words f_P and b_P along with the coset numbers f and b that we encounter as we scan forwards from the beginning of the word, and backwards from the end of the word, respectively. At a given point we have $w = uzv$, where u and v are the subwords that have been scanned in the forwards and backwards traces, and then $f = \alpha^u$ and $b^v = \alpha$. The words f_P and b_P will be defined such that they satisfy $\tau(\alpha)u =_G \varphi(f_P)\tau(f)$ and $\tau(b)^u =_G \varphi(b_P^{-1})\varphi(y)\tau(\alpha)$.

The point of this is that, if z is a single generator $x \in A$, then we can make the deduction $f^z = b$ and $P[f, z] = f_P^{-1}b_P$, and the required property $\tau(f)z =_G \varphi(P[f, z])\tau(b)$ follows from the assumption $\tau(\alpha)w =_G \varphi(y)\tau(\alpha)$. On the other hand, if z is the empty word, then $\tau(\alpha)w =_G \varphi(y)\tau(\alpha)$ implies that $\tau(f) =_G \varphi(f_P^{-1}b_P)\tau(b)$.

We shall not write down a formal proof of correctness, but to do that we would simply need to check that the correctness of all of the properties that we have asserted in this paragraph is maintained by all of the statements in the procedure.

MODIFIEDSCAN($\sim \mathcal{C}_M, \alpha, w, y$)

Input: $\alpha \in \Omega$, $w \in A^*$, $y \in (Y \cup Y^{-1})^*$ with $\tau(\alpha)w =_G \varphi(y)\tau(\alpha)$.

(* Scan forwards *)

1　Let $w = x_1 x_2 \cdots x_r$, with $x_i \in A$;

2　$f := \alpha$; $f_P := \varepsilon$; $i := 1$;

3　**while** $i \leq r$ **and** f^{x_i} is defined

4　　　　**do** $f_P := f_P P[f, x_i]$; $f := f^{x_i}$; $i := i+1$;

5　**if** $i > r$

6　　　**then if** $f \neq \alpha$ **then** MODIFIEDCOINCIDENCE($\sim \mathcal{C}_M, f, \alpha, f_P^{-1}y$);

7　　　　　　**return** ;

(* Forward scan was incomplete. Scan backwards *)

8　$b := \alpha$; $b_P := y$; $j := r$;

9　**while** $j \geq i$ **and** $b^{x_i^{-1}}$ is defined

10　　　　**do** $b_P := b_P P[b, x_i^{-1}]$; $b := b^{x_i^{-1}}$; $j := j-1$;

11　**if** $j < i$

12　　　**then** MODIFIEDCOINCIDENCE($\sim \mathcal{C}_M, f, b, f_P^{-1}b_P$);

13　**elseif** $j = i$

14　　　**then** (* Deduction *)

15　　　　　$f^{x_i} := b$; $b^{x_i^{-1}} := f$;

16　　　　　$P[f, x_i] := f_P^{-1}b_P$; $P[b, x_i^{-1}] := b_P^{-1}f_P$;

17　(* Otherwise $j > i$ – scan is incomplete and yields no information *).

In MODIFIEDCOINCIDENCE, as well as the array p used to store the right congruence generated by the coincidence, we maintain an array p_P of words over $Y \cup Y^{-1}$, such that $p[\alpha] = \beta$ implies that $\tau(\alpha) =_G \varphi(p_P[\alpha])\tau(\beta)$. Again, we shall not write down a formal correctness proof. To do that, we would just need to verify that every assignment to $p_P[\alpha]$ or to $P[\alpha, x]$ maintains the correctness of the required properties of p_P and of P: namely that $\alpha^x = \beta$ implies $\tau(\alpha) =_G \varphi(P[\alpha, x])\tau(\beta)$ and that $p[\alpha] = \beta$ implies that $\tau(\alpha) =_G \varphi(p_P[\alpha])\tau(\beta)$.

MODIFIEDCOINCIDENCE uses modified versions M-REP and M-MERGE of REP and MERGE.

M-REP($\kappa, \sim p, \sim p_P$)

Input: $\kappa \in \Omega$, arrays p, p_P

1　$\lambda := \kappa$; $\rho := p[\lambda]$;

(* We introduce a new array s to trace back the compression path *)

2　**while** $\rho \neq \lambda$ **do** $s[\rho] := \lambda$; $\lambda := \rho$; $\rho := p[\lambda]$;

3　$\rho := s[\lambda]$;

4　**while** $\rho \neq \kappa$

5　　　　**do** $\mu := \rho$; $\rho := s[\mu]$; $p[\rho] := \lambda$; $p_P[\rho] := p_p[\rho]p_P[\mu]$;

6　**return** λ;

M-MERGE($\kappa, \lambda, w, \sim p, \sim p_P, \sim q, \sim l$)

    Input: $\kappa, \lambda \in \Omega$, $w \in (Y \cup Y^{-1})^*$ with $\tau(\kappa) =_G \varphi(w)\tau(\lambda)$, p, p_P, q, l
    (* q is a queue of length l of elements to be deleted from Ω *)
1 $\varphi := \text{REP}(\kappa, \sim p, \sim p_P)$; $\psi := \text{REP}(\lambda, \sim p, \sim p_P)$;
2 **if** $\varphi > \psi$
3   **then** $p[\varphi] := \psi$; $p_P[\varphi] := p_P[\kappa]^{-1} w p_P[\lambda]$;
4          $l := l+1$; $q[l] := \varphi$;
5 **elseif** $\psi > \varphi$
6   **then** $p[\psi] := \varphi$; $p_P[\psi] := p_P[\lambda]^{-1} w^{-1} p_P[\kappa]$;
7          $l := l+1$; $q[l] := \psi$;

MODIFIEDCOINCIDENCE($\sim \mathcal{C}_M, \alpha, \beta, w$)

    Input: A coincident pair $\alpha, \beta \in \Omega$, $w \in (Y \cup Y^{-1})^*$
            with $\tau(\alpha) =_G \varphi(w)\tau(\beta)$.
1 $l := 0$; M-MERGE($\alpha, \beta, w, \sim p, \sim p_P, \sim q, \sim l$); $i := 1$;
2 **while** $i \le l$
3   **do** $\gamma := q[i]$; $i := i+1$;
4   **for** $x \in A$ **do if** γ^x is defined and equal to δ
5   **then** (* Remove the entry $\delta^{x^{-1}} = \gamma$ from the table *)
6   Undefine $\delta^{x^{-1}}$;
                  (* Queue any new coincidences *)
7                  $\mu := \text{REP}(\gamma, \sim p, \sim p_P)$; $\nu := \text{REP}(\delta, \sim p, \sim p_P)$;
8   **if** μ^x is defined
9   **then** $v := p_P[\delta]^{-1} P[\gamma, x]^{-1} p_P[\gamma] P[\mu, x]$,
10   M-MERGE($\nu, \mu^x, v, \sim p, \sim p_P, \sim q, \sim l$);
11   **elseif** $\nu^{x^{-1}}$ is defined
12   **then** $v := p_P[\gamma]^{-1} P[\gamma, x] p_P[\delta] P[\mu, x^{-1}]$;
13   M-MERGE($\mu, \nu^{x^{-1}}, v, \sim p, \sim p_P, \sim q, \sim l$);
14   **else** $\mu^x := \nu$; $\nu^{x^{-1}} := \mu$;
15                      $v := p_P[\gamma]^{-1} P[\gamma, x] p_P[\delta]$;
16                      $P[\mu, x] := v$; $P[\nu, x^{-1}] := v^{-1}$;

The modifications required in the remaining coset enumeration procedures are straightforward: SCANANDFILL requires the changes corresponding to those of SCAN; COSETENUMERATIONR, COSETENUMERATIONC, and LOOK-AHEAD need changing only to call the modified versions of the other procedures; COMPRESS and STANDARDIZE, which redefine elements of Ω, require corresponding changes to be made to $P[\alpha, x]$ whenever α^x is redefined.

We shall now illustrate the modified coset enumeration process by working through some examples. As usual, when doing examples by hand, we do not make definitions and carry out scans in the precise order specified by COSETENUMERATIONR or COSETENUMERATIONC; we prefer to select those definitions and scans that result in the enumeration completing as quickly as possible.

Example 5.7

Let $G := \langle\, x, y \mid x^3, y^5, (xy)^2 \,\rangle$ and $H := \langle\, xy, x^{-1}y^{-1}xyx \,\rangle$.

(This was also Example 5.6.) First we introduce the set of new symbols $Y := \{a, b\}$ and define $\varphi : F_Y \to F$ by $\varphi(a) = xy$, $\varphi(b) = x^{-1}y^{-1}xyx$.

When working by hand, it is convenient to use a similar notation as we did for displaying modified coset tables, and write $\alpha^x = w\beta$ to mean $\alpha^x = \beta$ and $P[\alpha, x] = w$. In the following description, and also in subsequent examples, we have written MS for MODIFIEDSCAN.

First define $1^x = 2$, $2^x = 3$, and deduce $3^x = 1$ from $\mathrm{MS}(\sim \mathcal{C}_M, 1, x^3, \varepsilon)$. Then deduce $2^y = a1$ from $\mathrm{MS}(\sim \mathcal{C}_M, 1, xy, a)$. Then define $3^{y^{-1}} = 4$, and deduce $4^x = b4$ from $\mathrm{MS}(\sim \mathcal{C}_M, 1, x^{-1}y^{-1}xyx, b)$, and then $1^y = b^{-1}4$ from $\mathrm{MS}(\sim \mathcal{C}_M, 3, (xy)^2, \varepsilon)$. Then define $3^y = 5$ and deduce $5^y = ba^{-1}2$ from $\mathrm{MS}(\sim \mathcal{C}_M, 1, y^5, \varepsilon)$. Finally, we deduce $5^x = ab^{-1}5$ from $\mathrm{MS}(\sim \mathcal{C}_M, 2, (xy)^2)$, and then all other scans complete consistently. The complete modified coset table is:

	x		x^{-1}		y		y^{-1}	
1		2		3	b^{-1}	4	a^{-1}	2
2		3		1	a	1	ab^{-1}	5
3		1		2		5		4
4	b	4	b^{-1}	4		3	b	1
5	ab^{-1}	5	ba^{-1}	5	ba^{-1}	2		3

The procedure for computing the relations from the modified coset table is derived from Theorem 2.63 and the remark following it. The set of relators, $\{\,\rho(t, w) \mid t \in T,\ w \in R\,\}$, is computed exactly as in Subsection 5.3.1 by applying REWRITE$(\mathcal{C}_M, \alpha, w)$ to all $\alpha \in \Omega$ and $w \in R$. Unlike in Subsection 5.3.1, however, we must include the additional relators coming from the set S_2 of Theorem 2.63. These are computed as REWRITE$(\mathcal{C}_M, 1, \varphi(y))y^{-1}$ for $y \in Y$.

Carrying this out in the example above, the two relators in S_2 both freely reduce to ε, so we do not need them. We find that REWRITE$(\mathcal{C}_M, 4, x^3)$, REWRITE$(\mathcal{C}_M, 5, x^3)$ and REWRITE$(\mathcal{C}_M, 1, (xy)^2)$ yield relators b^3, $(ab^{-1})^3$, and a^2, respectively. All other relators in S_1 freely reduce to ε, so the presentation computed for H is $\langle\, a, b \mid a^2, b^3, (ab^{-1})^3 \,\rangle$. By replacing b by b^{-1} in the presentation, we can replace the third relator by $(ab)^3$, and so we now have the same presentation for H as we computed in Subsection 5.3.1. (Strictly speaking, replacing b by b^{-1} also involves changing the relator b^3 to b^{-3}, but this is equivalent to b^3 anyway!)　　□

By applying modified coset enumeration in the case $H = G$, we can compute a presentation of G on an arbitrary generating set of G. This technique is illustrated in the following example, which also provides an easy application of MODIFIEDCOINCIDENCE.

Example 5.8

Let $G := \langle\, x, y \mid x^3, y^3, (xy)^3 \,\rangle$ and $H := \langle xy, xy^{-1} \rangle$.

(Easy exercise: show directly that $H = G$.) As in the preceding example, put $Y = \{a, b\}$, $\varphi(a) = xy$, $\varphi(b) = xy^{-1}$. Define $1^x = 2$, and deduce $2^y = a1$ from $\mathrm{MS}(\sim \mathcal{C}_M, 1, xy, a)$ and $2^{y^{-1}} = b1$ from $\mathrm{MS}(\sim \mathcal{C}_M, 1, xy^{-1}, b)$. Applying $\mathrm{MS}(\sim \mathcal{C}_M, 1, y^3, \varepsilon)$ now results in a call of $\mathrm{MODIFIEDCOINCIDENCE}(\sim \mathcal{C}_M, 2, 1, ba^{-1}b)$, and the completed modified coset table after this call is:

	x		x^{-1}		y		y^{-1}	
1	$ba^{-1}b$	1	$b^{-1}ab^{-1}$	1	$a^{-1}b$	1	$b^{-1}a$	1

We find $\mathrm{REWRITE}(\mathcal{C}_M, 1, xy)a^{-1} = (ba^{-1})^3$ and $\mathrm{REWRITE}(\mathcal{C}_M, 1, xy^{-1})b^{-1}$ freely reduces to ε. For the relators in S_1, we get $\mathrm{REWRITE}(\mathcal{C}_M, 1, x^3) = (ba^{-1}b)^3$, $\mathrm{REWRITE}(\mathcal{C}_M, 1, y^3) = (a^{-1}b)^3$, and $\mathrm{REWRITE}(\mathcal{C}_M, 1, (xy)^3) = (ba^{-1}ba^{-1}b)^3$. We see that $(a^{-1}b)^3$ is a cyclic conjugate of $(ba^{-1})^3$, and $(ba^{-1}ba^{-1}b)^3$ simplifies to a^3 using $(ba^{-1})^3$, so the computed presentation simplifies to

$$\langle\, a, b \mid a^3, (ba^{-1})^3, (ba^{-1}b)^3 \,\rangle.$$

One of the difficulties with modified coset enumeration in practice, which is not apparent from the small examples that we have worked through, is that the words $P[\alpha, x]$ can grow very long during the course of the enumeration. A number of methods have been tried for keeping the lengths of these words down in practical implementations. For example, a method involving storing the words implicitly using a tree structure is discussed in [AMW82]; see [Neu82] for references to other implementations of this method. Another possibility is to collect the relators in the subgroup presentation during the enumeration itself, rather than at the end of the enumeration, and to attempt to use known relators to shorten the words $P[\alpha, x]$.

One situation in which modified coset enumeration is frequently applied is when H has a single generator. We mentioned this possibility in Subsection 5.2.6 in connection with finding the order of a finitely presented group G. In this case, the problem of long words $P[\alpha, x]$ can be ameliorated by storing these words as the integral exponent of the single generator.

Example 5.9

Let $G := \langle\, x, y \mid x^3y^{-3}, (xy)^3, (xy^{-1})^2 \,\rangle$ and $H := \langle x \rangle$.

We put $Y = \{a\}$ with $\varphi(a) = x$. Then $\mathrm{MS}(\sim \mathcal{C}_M, 1, x, a)$ results in the deduction $1^x = a1$. Define $1^y = 2$ and $1^{y^{-1}} = 3$. Then $\mathrm{MS}(\sim\mathcal{C}_M, 1, x^3y^{-3}, \varepsilon)$ and $\mathrm{MS}(\sim \mathcal{C}_M, 1, (xy^{-1})^2, \varepsilon)$ yield the deductions $2^y = a33$ and $3^x = a^{-1}2$. Define $2^x = 4$. Then $\mathrm{MS}(\sim \mathcal{C}_M, 2, x^3y^{-3}, \varepsilon)$ and $\mathrm{MS}(\sim \mathcal{C}_M, 2, (xy^{-1})^2, \varepsilon)$ give the deductions $4^x = a^43$ and $4^y = a4$. The table is now complete, and the remaining scans produce no coincidences. The relator in S_2 is trivial,

REWRITE($\mathcal{C}_M, i, (xy)^3$) $= a^6$ for $i \in [1..4]$, and the other relators in S_1 reduce to ε. Hence H is cyclic of order 6 and $|G| = 24$.

This is probably the easiest way of calculating $|G|$. We could also run a normal coset enumeration with $H = \langle xy \rangle$, but then we would have $|G:H| = 8$.
□

5.3.3   Simplifying presentations

As a measure of the complexity of a group presentation, we formally define its size to be the triple $(|X|, |R|, l)$, where l is the total relator length; that is, $l := \sum_{w \in R} |w|$.

Let $n := |G:H|$. Then the presentation $\langle Y \mid S \rangle$ for the group $H' \cong H$ on the set of Schreier generators as described in Subsection 5.3.1 has size $((n-1)|X| + 1, n|R|, nl)$. In practice, we replace relators by cyclically reduced conjugates and omit trivial relators and repetitions of relators (including repetition by cyclic conjugates), so the numbers $n|R|$ and nl are upper bounds. Such simplifications do not change the group H'.

The presentation described in Subsection 5.3.2 on the set of user-defined generators $\hat{Y}$ has size $(\hat{Y}, n|R|, ?)$. Although $\hat{Y}$ is typically much smaller than the number $(n-1)|X| + 1$ of Schreier generators, when n is large, it often happens that the total relator length is much larger and less predictable than nl. For this reason, if, as is often the case, the generating set for H is not of particular importance to the user, then for subgroups of large index, it can turn out to be more effective or feasible to compute the presentation on the Schreier generators.

But in either case the user will usually want to attempt to simplify the initially computed presentation, by shortening the total relator length, removing redundant relators and, for presentation on Schreier generators, eliminating generators. Simplifying presentations in this manner is the subject of this subsection. These methods are most often applied to the output of subgroup presentation programs, but they can be applied in principal to any group presentation, so we shall denote the presentation to be simplified by $G = \langle X \mid R \rangle$. This technique is used frequently in certain applications of CGT, which we shall discuss later in Subsection 9.3.3.

Implementations of these methods are generally known as *Tietze transformation programs*. One such implementation is described by Havas in [Hav74]. Unfortunately, they are rather difficult to implement, because they involve moderately complicated string searching and manipulation, and the programmer needs to bear in mind that computer memory may be at a premium. One might be tempted, for example, to compute and store all the inverses and all cyclic conjugates of all of the defining relators, but this could prove fatally extravagant for a presentation with total relator length equal to a seven-figure number. We shall not go deeply into implementation issues here, but we shall mention a few ideas that have proved useful.

The processes of *eliminating generators* and performing *common substring replacement* were defined in Subsection 2.4.4 in terms of the basic Tietze transformations. There are three principal subroutines that are required in a Tietze transformation program, namely:

1. use relators of length 1 and 2 to eliminate redundant generators;

2. eliminate some generators;

3. perform common substring replacement to shorten the total relator length and remove redundant relators.

We shall refer to these subroutines, respectively, as *elimination by short relators*, *elimination*, and *simplification*. The first is of course a special case of the second, but it is used more frequently than general elimination. Elimination by short relators always reduces all three parameters in the presentation size. General elimination reduces $|X|$ and $|R|$, but will typically increase the total relator length. Simplification which, unlike elimination, does not change the group defined by the presentation, leaves $|X|$ unchanged, but reduces the total relator length and possibly also $|R|$.

A flexible implementation, such as the ones in GAP and MAGMA, will allow the user to call these three subroutines in any desired order, and will also provide some default calling sequences, but still with some optional parameters. We shall assume that all relators are always cyclically reduced; that is, that whenever a relator is changed, it is immediately replaced by its cyclic reduction.

As we shall see shortly, simplification is performed in a series of *simplification rounds*. A typical calling sequence TIETZE(e, m, p), say, is to repeat the following two steps as many times as possible. Here e and m are positive integers, and $0 \le p \le 1$. So a possible call of this would be TIETZE$(100, 100000, 0.01)$.

(i) Eliminate generators until e generators have been eliminated, or until the total relator length exceeds m.

(ii) Perform rounds of simplification until the percentage decrease in total relator length resulting from a round is less that $100p$. Perform elimination by short relators after each simplification round.

This process would stop when neither (i) nor (ii) could be performed. The user might also want to put an upper bound on the number of generators eliminated altogether, or to limit the number of times that (i) and (ii) are repeated.

Let us now describe the elimination and simplification subroutines in more detail. Eliminating a generator consists of locating a word $w \in R$, such that some cyclic conjugate of w or w^{-1} has the form $x_1^{-1} x_2 \cdots x_r$, where $x := x_1 \in X$, and $x_i \ne x^{\pm 1}$ for $2 \le i \le r$. We then delete this relator from R, replace all occurrences of $x^{\pm 1}$ by $(x_2 \cdots x_r)^{\pm 1}$ in all other relators, and delete x from X. When the relator has length 1 or 2, this process decreases all three parameters of the presentation, and there is every advantage in carrying it out immediately. This is why we use elimination by short relators more frequently than general elimination.

Other eliminations typically increase the total relator length. Usually, and particularly early on in the process of simplifying a large presentation, there will be a large number of candidates for x and w, and it is of critical importance to choose the right one. Examples have been encountered of large presentations that can be simplified down to a very small size by the correct choices of generators to eliminate, and which remain completely unmanageable with other choices. In fact this choice is probably much more crucial than the choice of the parameters e, m, p in TIETZE. Unfortunately, there seem to be very few satisfactory heuristics known for guiding the choice. The increase in the total relator length resulting from the elimination of x_1 using w as described above is at most $(r-2)(o(x_1)-1) - r$, where $o(x_1)$ is the total number of occurrences of the generator x_1 in all relators (including w itself). The standard default approach is to choose x_1 and w so as to minimize this increase. For this we need to know $o(x)$ for all $x \in X$, but it is not difficult to calculate these values initially and then to adjust them each time that the presentation is changed. In any event, the user should be allowed to specify a particular elimination to make at any stage.

A round of simplification consists of looking at each pair $\{v, w\}$ of distinct elements of R in turn and, assuming that $|v| \leq |w|$, checking whether we can write cyclic conjugates of v and w or of their inverses as uv_1 and uw_1, respectively, where $|u| > |v_1|$. If so, then we do this with u as long as possible, and replace w by $v_1^{-1}w_1$. If this results in w freely reducing to ε, then we remove it from R.

A casual implementation of this process is likely to run very slowly, because there is a lot of checking of matching substrings to be done. There a number of simple ways of speeding this up. Before each round, the relators should be sorted into length-increasing order. For each relator, two one-bit flags should be maintained, recording whether this relator was changed during the previous simplification round, and whether it has been changed in the current round. If a particular pair $\{v, w\}$ was checked for matching substrings in the last round and neither v nor w has changed since that check, then it is not necessary to check that pair again. Suppose that we are about to check a particular pair $\{v, w\}$ with $|v| \leq |w|$ for matching substrings, let $v = x_1 x_2 \cdots x_r$, and let $s := \lfloor (r+2)/2 \rfloor$. If a cyclic conjugate of v can be written as uv_1 with $|u| \geq |v_1|$, then u must contain at least one of the generators x_1 and x_s in v. Similarly, for a cyclic conjugate of v^{-1}, u must contain at least one of x_1^{-1} and x_s^{-1}. Thus, when searching for matching substrings, we can begin our searches of v and v^{-1} at x_1, x_s, x_1^{-1} and x_s^{-1}, and work outwards in both directions from these starting points. In particular, if w contains none of x_1, x_s, x_1^{-1}, x_s^{-1}, then there is no possibility of a simplification.

In the simplification process, there are situations where it is advantageous to replace $w = uw_1$ by $v_1^{-1}w_1$ even when $|u| = |v_1|$, and the user should have some control as to whether or not this is done. Of course it should not be done on every possible occasion, or the process will go into an infinite loop! There is one specific situation where this should happen by default, and this is when $|v| = 4$ and v is a commutator of two generators. This means that, for some $x, y \in X$, we have $xy =_G yx$. Then, whenever $x^{\pm 1}$ and $y^{\pm 1}$ are adjacent in a relator, they should always be put into the same order, which would usually be to put the lower-numbered generator, x, say, before the higher-numbered generator y. Then any subword of any relator involving only x and y and their inverses would be put in the normal form $x^i y^j$, for $i, j \in \mathbb{Z}$.

There are even some situations where it would turn out to be a good idea to substitute longer strings for shorter strings, but it is probably not feasible to hope to be able to recognize such situations in a computer implementation. For example, if three generators x, y, z satisfy relations $[y, x] =_G z$, $xz =_G zx$, $yz =_G zy$ (which implies that they generate a nilpotent subgroup), then it might be a good idea to use the standard normal form $x^i y^j z^k$ for nilpotent groups for subwords involving just these generators, but that would involve replacing yx by xyz, resulting in a lengthening of the relators. In general, it is probably fair to say that Tietze transformation programs do not work very successfully on presentations that involve polycyclic-type relators.

For presentations that are not too large, there is an alternative approach to the manipulation of group presentations involving rewriting systems and the Knuth-Bendix completion process. That will be the main topic of Chapter 12. It involves systematically generating all consequences of the relations in a group presentation, and so in principal it can eventually find any short relations that hold in the presentation, and determine whether any of the defining relators are redundant. But it is incapable of handling the very large presentations, with perhaps hundreds of thousands of generators and relators, to which we typically apply Tietze transformation programs.

Exercises

1. Let $G := \langle x, y \mid x^2 y^2, y^{-1} xyx^{-3} \rangle$, as in Exercise 2 at the end of Section 5.1. Show by using the modified coset enumeration procedure with $H := \langle x \rangle$, that $|G:H| = 2$ and that H is cyclic of order 4.

2. Let $G := \langle x, y \mid x^2, y^4, (xy)^4 \rangle$.

   (i) Show that $G/[G, G] = C_2 \times C_4$.

   (ii) Find a presentation of the subgroup $[G, G]$ of G and use it to show that $[G, G]$ is a free abelian group of rank 2.

   (iii) Write down the generators used in your presentation of $[G, G]$ (after eliminating any redundant generators) as words in the generators of G and their inverses.

3. Let $G := \langle x, y \mid x^3, y^7, (xy)^2, (x^{-1}y^{-1}xy)^4 \rangle$ and $H := \langle x, y^{-1}xy^2 \rangle$.

(i) Show by coset enumeration that $|G:H| = 7$.

(ii) Calculate a presentation of H.

(iii) Show by eliminating generators from this presentation, and then manipulating it, that H is isomorphic to a quotient group of the group $\langle a, b \mid a^2, b^3, (ab)^4 \rangle$, and hence $|H| \leq 24$ and $|G| \leq 168$.

(iv) Show by using the permutation representation arising from (i) that $|G| \geq 168$, and so $|G| = 168$. (*Hint*: In the permutation representation, show that $H = G_1$ is transitive on $\Omega \backslash \{1\}$, hence $|H| = 6|G_{12}|$, and consider $(xz)^2$ and $xz(xzx)^2$ with $z := y^{-1}xy^2$ to get $|G_{12}| \geq 4$.)

5.4 Finding all subgroups up to a given index

In the final section of this chapter, we address a problem that is, in some respects, the converse of the basic problem of coset enumeration. Rather than finding the index of a given subgroup H of a finitely presented group G, we want to find all subgroups H such that the index $|G:H|$ is at most a specified number $N > 0$. This has traditionally been known as the *low-index subgroups* problem. For many presentations, particularly for groups with more than two defining generators, solving this problem only seems to be feasible for rather small values of N, which explains the name.

To give the reader an immediate rough idea of the current scope of algorithms to solve this problem, we observe that the $(2, 3, 7)$-triangle group $\langle x, y \mid x^2, y^3, (xy)^7 \rangle$ is an easy example, in which the subgroups up to index about 50 or 60 can be found within 10 seconds, whereas the Heineken group

$$\langle x, y, z \mid [x, [x, y]] = z, [y, [y, z]] = x, [z, [z, x]] = y \rangle$$

is a difficult example. Even after simplifying the presentation and removing a redundant generator, it takes about 60 seconds to find the subgroups up to index 10, and over 1000 seconds to do this while working with the initial 3-generator presentation. Of course, these times depend on which implementation one is using on which computer, and they are only intended to give an idea of the scope of the methods. In general, the observed behaviour of all current implementations is that, for a given example, the growth of the time required is more than exponential in N, so we expect there to be a critical value of N beyond which it is not practical to proceed. There is, however, considerable scope for further research into improvements of low-index subgroup algorithms.

The first practical methods for solving the low-index problem were developed independently by Dietze, Schaps [DS74], and Sims [Sim74]. For many years, the principal implementations, such as those used in CAYLEY and GAP were based on the ideas described in these papers. The method consists in running coset enumeration, initially over the identity subgroup, but allowing only $f(N)$ cosets to be defined, where $f(N)$ must satisfy $f(N) \geq N$, and usually $f(N) \leq 2N$. Then coincidences are systematically imposed between the defined cosets; each such coincidence is equivalent to adjoining a new generator to the subgroup H over which the coset enumeration is taking place. After a forced coincidence, the coset enumeration is resumed, but still allowing only $f(N)$ cosets to be defined. Whenever an enumeration completes with at most N cosets, a subgroup of index at most N has been found. Each forced coincidence corresponds to a node in a tree, and the complete procedure consists of a backtrack search through the nodes.

More recently, a rather different method has been proposed by Sims; it still involves a backtrack search, but over incomplete coset tables rather than over forced coincidences. This newer method is described in Section 5.6 (plus some earlier sections) of [Sim94], and has been implemented in MAGMA. Since the current evidence suggests that (with one caveat, which we shall mention later) it performs better on most typical examples than the earlier method, we shall describe the newer method of Sims here, and refer the reader to [Neu82] for an accessible description of the forced coincidence approach.

5.4.1 Coset tables for a group presentation

As usual, let $G = \langle X \mid R \rangle$ and let $A := X \cup X^{-1}$. Let us call a complete coset table with columns indexed by A in which Property 2 ($\alpha^x = \beta \iff \beta^{x^{-1}} = \alpha$) holds, and in which all $\alpha \in \Omega$ scan correctly under all $w \in R$, a *complete coset table for G*. As in the proof of Theorem 5.2, it follows from Theorem 2.52 that a complete coset table for G is equivalent to a transitive action of G on the set Ω of cosets in the coset table.

By Proposition 2.23, this action is equivalent to the action of G by right multiplication on the cosets of $H := G_1$. Conversely, for any subgroup H of G of finite index, the action of G by right multiplication on the cosets of H defines a complete coset table for G in which $H = G_1$. By Proposition 2.23, if we have two group actions in which $H = G_1$, then the actions are equivalent under an equivalence that fixes the point 1. Now a standardized coset table, as defined in Subsection 5.2.3, provides us with a canonical representative of such an equivalence class. Summing up, we have:

PROPOSITION 5.6 *For any $n > 0$, there is a one-one correspondence between standardized complete coset tables for G with $|\Omega| = n$ and subgroups of G of index n, in which the subgroup H corresponding to a coset table $\mathcal{C}$ is the stabilizer G_1 in the group action corresponding to $\mathcal{C}$.*

Our algorithm for producing subgroups of index up to N will proceed by systematically constructing a list of all complete standardized coset tables for G with $|\Omega| \leq N$. In practice, we save space and time with no genuine loss of information if we return coset tables corresponding to representatives of the conjugacy classes of subgroups of G rather than all subgroups. To do this, we shall make use of the fact that the conjugates of H are just the stabilizers G_α of the points $\alpha \in \Omega$ in the associated group action. (See proof of Proposition 2.23.)

If generators are required for the subgroups $H = G_1$ corresponding to the coset tables, then the Schreier generators, as defined in Subsection 5.3.1, can easily be computed. A presentation of H on these generators could also be computed, which could then be used, as described in Subsection 5.3.3, to eliminate redundant Schreier generators if a smaller generating set were deemed to be desirable.

5.4.2   Details of the procedure

We have been using 'bottom-up' descriptions of our algorithms so far in this chapter, starting with the lower-level routines. So, to make a change, we shall describe the procedure from the the top down, and use recursion to avoid the details of the backtrack search.

Sims suggests splitting the group relators R into two parts R_1 and R_2, where R_1 consists of shorter relators and R_2 of longer relators. The relators of R_1 will be used in the inner parts of the process to prune branches of the search tree, whereas those of R_2 will be only be checked at the last step for complete coset tables. During most parts of the search, the coset tables being manipulated are incomplete, and experience has shown that time is wasted by scanning the longer relators frequently on incomplete coset tables, because the scans rarely complete, and so no information is gained. Unfortunately there appear to be currently no reliable heuristics for deciding exactly how short is short!

Our top-level procedure performs this subdivision and computes the sets R_{1x}^c of the cyclic conjugates w of the relators in R_1 for which w has x as its first letter. These will be required for input to the procedure PROCESSEDE-DUCTIONS, which was described in Subsection 5.2.2. The main task is then passed over to the procedure DESCENDANTSUBGROUPS, which will be called recursively. The complete standardized coset tables to be returned eventually will be kept throughout in the list S.

Our base point for the search is the empty coset table with a single coset 1, and no transitions defined. Unlike in previous sections in this chapter, we shall be manipulating more than one coset table C, and so we shall denote the set Ω and the array p associated with C by Ω_C and p_C, and transitions $\alpha^x = \beta$ in C by $\alpha_C^x = \beta$. Since we shall never allow cosets to be eliminated as a result of coincidences, we shall always have $\Omega_C = [1 \mathrel{..} n_C]$ for some $n_C \leq N$. All of the coset tables will be with respect to the same set A of generators.

LowIndexSubgroups($\langle X|R \rangle, N$)

Input: A finite group presentation $\langle X|R \rangle$, $N > 0$
Output: Coset tables for representatives of classes of subgroups
of index at most N in G

1   Initialize a coset table $\mathcal{C}$ for $\langle X|R \rangle$;
2   $p_{\mathcal{C}}[1] := 1$; $n_{\mathcal{C}} := 1$;
3   Let $R := R_1 \cup R_2$ with 'short' relators in R_1 and 'long' relators in R_2;
4   Replace all elements of R_1 by their cyclic reductions;
5   Compute the set R_1^c of all cyclic conjugates of all elements of R_1;
6   **for** $x \in A$
7   **do** compute set R_{1x}^c of elements of R_1^c having first letter x;
8   $\mathcal{S} := []$;
9   DescendantSubgroups($\sim \mathcal{S}, \mathcal{C}, \{R_{1x}^c\}, R_2, N$);
10 **return** $\mathcal{S}$;

The next procedure, DescendantSubgroups, takes a complete or incomplete coset table $\mathcal{C}$ with $n_{\mathcal{C}} \leq N$ as input. It first checks whether $\mathcal{C}$ is complete; that is, whether α^x is defined for all $\alpha \in \Omega_{\mathcal{C}}$ and $x \in A$.

If $\mathcal{C}$ is complete then, as will be explained in detail later, we shall know already that α scans correctly under all $w \in R_1$ and also that, if $\mathcal{C}$ is a coset table for G, then the corresponding subgroup G_1 is a canonical representative of its conjugacy class of subgroups. So it only remains to test whether all $\alpha \in \Omega_{\mathcal{C}}$ scan correctly under all $w \in R_2$. If so, then we adjoin $\mathcal{C}$ to the list $\mathcal{S}$. If not, then we discard $\mathcal{C}$ and return.

To check whether α scans correctly under w, we have used a procedure ScanCheck, which we have not displayed, since it is a straightforward modification of Scan. ScanCheck returns `false` (rather than calling Coincidence) if the scan completes incorrectly; otherwise it returns `true`.

If $\mathcal{C}$ is incomplete, then we locate the first undefined entry $\alpha_{\mathcal{C}}^x$. We then need to try all possible values for $\alpha_{\mathcal{C}}^x$. These will consist of all $\beta \in \Omega_{\mathcal{C}}$ such that $\beta_{\mathcal{C}}^{x^{-1}}$ is undefined, and also a new coset $n_{\mathcal{C}}+1$ provided that $n_{\mathcal{C}}$ is not equal to N. We pass the problem of trying out each individual possibility for $\alpha_{\mathcal{C}}^x$ over to the next procedure TryDescendant.

The current coset table $\mathcal{C}$, represents a node in the search tree, and its descendants are the coset tables resulting from assigning the various possible values to the first undefined entry $\alpha_{\mathcal{C}}^x$. We do not want to alter $\mathcal{C}$ itself, so the procedure TryDescendant starts by making a copy $\mathcal{D}$ of $\mathcal{C}$, and then assigning the new entries $\alpha_{\mathcal{D}}^x := \beta$ and $\beta_{\mathcal{D}}^{x^{-1}} := \alpha$ in $\mathcal{D}$. Furthermore, in the case $\beta = n_{\mathcal{C}}+1$, we need to increment $n_{\mathcal{D}}$.

We then want to check that these new entries are consistent with the requirement that all $\alpha \in \Omega_{\mathcal{D}}$ scan correctly under all $w \in R_1$, and to insert any new deductions into $\mathcal{D}$ that follow from that requirement. The procedure ProcessDeductions defined in Subsection 5.2.2 does exactly that, once we have put (α, x) onto the deduction stack. (We use the deduction stack only

DESCENDANTSUBGROUPS($\sim\mathcal{S}, \mathcal{C}, \{R_{1x}^c\}, R_2, N$)

1   **if** $\mathcal{C}$ is complete
2   **then** ($*$ Check whether relators in R_2 are satisfied $*$)
3   **for** $w \in R_2$, $\alpha \in \Omega_\mathcal{C}$ **do if not** SCANCHECK($\mathcal{C}, \alpha, w$)
4   **then return**;
           ($*$ Relators in R_2 are satisfied $*$)
5   APPEND($\sim\mathcal{S}, \mathcal{C}$);
6   **else** Let $\alpha_\mathcal{C}^x$ be the first undefined entry in $\mathcal{C}$;
7   **for** $\beta \in \Omega_\mathcal{C} \cup \{n_\mathcal{C}+1\}$ **do if** $\beta \leq N$ and $\beta_\mathcal{C}^{x^{-1}}$ is undefined
8   **then** TRYDESCENDANT ($\sim\mathcal{S}, \mathcal{C}, \{R_{1x}^c\}, R_2, N, \alpha, x, \beta$);

at this stage, so all of the coset tables being manipulated can share the same deduction stack provided that it is emptied after use!)

However, since we do not want to call COINCIDENCE, we use an (undisplayed) variation of PROCESSDEDUCTIONS, which we shall call PROCESSDEDUCTIONSCHECK. This calls SCANCHECK wherever PROCESSDEDUCTIONS calls SCAN. If SCANCHECK returns false, then it records that we have encountered an inconsistency, which means that this particular coset table $\mathcal{D}$ can be abandoned. Let us assume, just for the convenience of writing down the code for TRYDESCENDANT, that PROCESSDEDUCTIONSCHECK records the fact that SCANCHECK has returned false by setting $n_\mathcal{D}$ to 0 and aborting.

If no inconsistency is encountered, then we run a function FIRSTINCLASS, to be described later, which checks whether the subgroup $H = G_1$ corresponding to $\mathcal{D}$ could possibly be the canonical representative of its conjugacy class. This test will be conclusive if $\mathcal{D}$ is complete. If FIRSTINCLASS returns false, then no descendant of $\mathcal{D}$ can have that property, and so we can abandon $\mathcal{D}$. If it returns true, then we need to process further the node of the search tree corresponding to $\mathcal{D}$, and so we call DESCENDANTSUBGROUPS recursively on $\mathcal{D}$.

TRYDESCENDANT($\sim\mathcal{S}, \mathcal{C}, \{R_{1x}^c\}, R_2, N, \alpha, x, \beta$)

1   $\mathcal{D} := \mathcal{C}$;
2   **if** $\beta = n_\mathcal{D}+1$ **then** $n_\mathcal{D} := n_\mathcal{D}+1$;
3   $\alpha_\mathcal{D}^x := \beta$; $\beta_\mathcal{D}^{x^{-1}} := \alpha$;
4   Push (α, x) onto the deduction stack;
5   PROCESSDEDUCTIONSCHECK($\mathcal{D}, \{R_{1x}^c\}$);
6   **if** $n_\mathcal{D} = 0$ **then return**;
7   **if** FIRSTINCLASS($\mathcal{D}$)
8   **then** DESCENDANTSUBGROUPS($\sim\mathcal{S}, \mathcal{D}, \{R_{1x}^c\}, R_2, N$);

It remains to describe FIRSTINCLASS, which is the most complicated of our procedures. For a given $n > 0$, let us order the set of all possibly incomplete coset tables C with $\Omega_C = [1 .. n]$ and a fixed ordered generating set A lexicographically by their transition matrix. For this purpose, we will think of undefined entries α_C^x as being equal to 0.

To be precise, if C_1 and C_2 are two such coset tables, then we define $C_1 < C_2$ if, for some $\alpha \in [1 .. n]$ and some generator $x \in A$, we have $\beta_{C_1}^y = \beta_{C_2}^y$ for all $y \in A$ and $1 \le \beta < \alpha$ and $\alpha_{C_1}^y = \alpha_{C_2}^y$ for all $y \in A$ with $y < x$, but $\alpha_{C_1}^x < \alpha_{C_2}^x$.

Then, from the discussion preceding Proposition 5.6, we see that a standardized complete coset table C for G is the least complete table under this ordering that corresponds to a particular subgroup $H = G_1$ of G of index n_C. It is natural to choose as canonical representative of a conjugacy class of subgroups the least complete coset table for G under this ordering that corresponds to a subgroup in that class. This representative will necessarily be standardized.

For a possibly incomplete coset table C, we want to decide whether it is possible that a complete extension of C might be the required canonical representative, and we would like this test to be conclusive whenever C is complete. Now, if the complete standardized coset table C for G corresponds to the subgroup $H = G_1$, then the conjugates of H are just the stabilizers H_α of the points $\alpha \in [1 .. n_C]$.

So we proceed as follows. We call two coset tables with $\Omega = [1 .. n]$ *equivalent* if one can be obtained from the other by renumbering the points of Ω. For each $1 \ne \alpha \in [1 .. n_C]$, we start by constructing the equivalent standardized coset table C_α corresponding to H_α. We stop if two corresponding entries β_C^x and $\beta_{C_\alpha}^x$ of C and C_α are either unequal or are not both defined.

If either of these two entries are undefined, then we cannot draw any conclusions, so we move onto the next $\alpha \in [1 .. n_C]$. We do the same if $\beta_C^x < \beta_{C_\alpha}^x$, since that indicates that C is definitely less than C_α. If $\beta_{C_\alpha}^x < \beta_C^x$ however, then $C_\alpha < C$, so C is not the required canonical representative, and we can exit the function immediately and return false.

The code below is taken directly from the function on page 208 of [Sim94]. For $\alpha \in \Omega_C$, $\nu[\alpha]$ is the point in Ω_{C_α} corresponding to α and, for $\alpha \in \Omega_{C_\alpha}$, $\mu[\alpha]$ is the point in Ω_C corresponding to α. In other words, ν and μ are the mutually-inverse equivalence maps between Ω_C and Ω_{C_α}. At any time λ is the largest numbered point in Ω_{C_α} which is currently defined.

Sims points out in [Sim94] that performance can be improved by remembering some of the information computed by FIRSTINCLASS. If the "continue α" statement is executed at line 14, then the same thing will happen for that value of α in any descendant of the table C, and so the values the values of α for which this occurs could profitably be stored and passed through to the descendants of C. Of course this would make the code more complicated.

As is usual with backtrack searches, working through examples by hand is very tedious, but we shall nevertheless do this with a small example.

FIRSTINCLASS($\mathcal{C}$)

```
1   n := n_C;  λ := 0;
2   for α ∈ [1..n] do ν[α] := 0;
3   for α ∈ [2..n]
4       do (* Reset ν to 0 after previous value of α *)
5           for β ∈ [1..λ] do ν[μ[β]] := 0;
            (* Try α as the new point 1 in Ω_{C_α} *)
6           μ[1] := α;  ν[α] := 1;  λ := 1;
            (* Compare corresponding entries in C and C_α *)
7           for β ∈ [1..n],  x ∈ A
8               do If β_C^x or μ[β]_C^x is undefined then continue α;
9                   γ := β_C^x;  δ := μ[β]_C^x;
10                  if ν[δ] = 0
11                      then (* δ becomes the next point in Ω_{C_α} *)
12                          λ := λ+1;  ν[δ] := λ;  μ[λ] := δ;
13                  if ν[δ] < γ then return false;
14                  if ν[δ] > γ then continue α;
15  return true;
```

Example 5.10

Let $G := \langle x, y \mid x^2, y^3, (xy)^4 \rangle$ and $N := 4$. It is not hard to prove that $G \cong \mathrm{Sym}(4)$ – indeed, the reader will presumably have already proved this while working through the exercises at the end of Section 5.2! But we shall not be making any use of that fact.

Because of the relator x^2, we take $A := \{x, y, y^{-1}\}$, and we shall use the obvious ordering $x < y < y^{-1}$ of A. The root of the search tree is the empty coset table $\mathcal{C}_0$ with a single coset. We shall denote the descendants of $\mathcal{C}_0$ by $\mathcal{C}_1, \mathcal{C}_2, \ldots$, and then the descendants of $\mathcal{C}_1$ by $\mathcal{C}_{11}, \mathcal{C}_{12}, \ldots$, and so on. The algorithm, as presented above, carries out a depth-first search through the search tree, which is sensible, because it minimizes the number of coset tables that need to be stored at any particular time. We shall do the same here.

Let us put x^2 and y^3 into R_1 and $(xy)^4$ into R_2. This means, of course, that we are really finding all subgroups of $\langle x, y \mid x^2, y^3 \rangle$ up to index 4, and then just checking which of these map onto subgroups of G.

In the first descendant $\mathcal{C}_1$, we set $1^x = 1$, and then in $\mathcal{C}_{11}$ we set $1^y = 1^{y^{-1}} = 1$, which yields the complete standardized coset table corresponding to $G = H$. (The same thing will happen initially for any group G.) So we have found our first subgroup.

In $\mathcal{C}_{12}$, we set $1^y = 2$, $2^{y^{-1}} = 1$. If we set $1^{y^{-1}} = 2$ then 1 does not scan correctly under y^3, so in $\mathcal{C}_{121}$ we set $1^{y^{-1}} = 3$, $3^y = 1$, and then scanning 1 under y^3 yields the deduction $2^y = 3$, $3^{y^{-1}} = 2$.

Now, in $\mathcal{C}_{1211}$ we put $2^x = 2$ and in $\mathcal{C}_{12111}$ we put $3^x = 3$. We now have a complete standardized coset table for which FIRSTINCLASS returns true, but then we find that 1 scans incorrectly under the relator $(xy)^4 \in R_2$.

In C_{12112}, we put $3^x = 4$, $4^x = 3$, and, since we now have $n_C = 4$, the only possibility in C_{121121} is to put $4^y = 4^{y^{-1}} = 4$. This gives us the following complete standardized coset table, for which FIRSTINCLASS returns true, and for which 1 scans correctly under the relator $(xy)^4 \in R_2$, so we have found our second subgroup of G.

	x	y	y^{-1}
1	1	2	3
2	2	3	1
3	4	1	2
4	3	4	4

Now we backtrack to C_{121} and put $2^x = 3$, $3^x = 2$ in C_{1212}. This yields the following complete table, and our third subgroup of G.

	x	y	y^{-1}
1	1	2	3
2	3	3	1
3	2	1	2

In C_{1213}, we set $2^x = 4$, $4^x = 2$, and then in C_{12131} we are forced to put $3^x = 3$, but then FIRSTINCLASS returns false when we substitute 1 for 3. In fact C_{1213} completes to a standardized table equivalent to but greater than C_{121121}.

At this stage, we backtrack all the way to C_0 and set $1^x = 2$, $2^x = 1$ in C_2. Then we put $1^y = 1^{y^{-1}} = 1$ in C_{21}, and $2^y = 2^{y^{-1}} = 2$ in C_{211}, yielding the following complete table, and our fourth subgroup of G.

	x	y	y^{-1}
1	2	1	1
2	1	2	2

Next we put $2^y = 3$, $3^{y^{-1}} = 2$ in C_{212}. Putting $2^{y^{-1}} = 3$ leads to an incorrect scan, so we are forced to put $2^{y^{-1}} = 4$, $4^y = 2$ in C_{2121}, which leads to the deduction $3^y = 4$, $4^{y^{-1}} = 3$. If we put $3^x = 3$ then FIRSTINCLASS returns false on replacing 3 by 1, so we must put $3^x = 4$, $4^x = 3$, which gives the complete following complete standardized coset table C_{21211}, for which FIRSTINCLASS returns true. But 2 scans incorrectly under $(xy)^4$, so this does not correspond to a subgroup of G.

	x	y	y^{-1}
1	2	1	1
2	1	3	4
3	4	4	2
4	3	2	3

Now we backtrack to $\mathcal{C}_2$ and put $1^y = 2$, $2^{y^{-1}} = 1$ in $\mathcal{C}_{22}$. To get a correct scan of 1 under y^3, we must put $1^{y^{-1}} = 3$, $3^y = 1$ in $\mathcal{C}_{221}$, and deduce $2^y = 1$, $3^{y^{-1}} = 2$. As before, FIRSTINCLASS returns `false` if we put $3^x = 3$, so we put $3^x = 4$, $4^x = 3$ in $\mathcal{C}_{2211}$, and we are forced to put $4^y = 4^{y^{-1}} = 4$ in $\mathcal{C}_{22111}$, and now FIRSTINCLASS returns `false` when we replace 4 by 1. In fact, $\mathcal{C}_{22111}$ is equivalent to $\mathcal{C}_{21211}$, which did not correspond to a subgroup of G anyway!

Backtracking again to $\mathcal{C}_2$, we put $1^y = 3$, $3^{y^{-1}} = 1$ in $\mathcal{C}_{23}$. In $\mathcal{C}_{231}$, we set $1^{y^{-1}} = 2$, $2^y = 1$, but then FIRSTINCLASS returns `false` when we replace 2 by 1. Finally, in $\mathcal{C}_{232}$, we put $1^{y^{-1}} = 4$, $4^y = 1$, and deduce $3^y = 4$, $4^{y^{-1}} = 3$, and then we are forced to put $2^y = 2^{y^{-1}} = 2$ in $\mathcal{C}_{2321}$, but then FIRSTINCLASS returns `false` on replacing 2 by 1.

This completes the search, and we have four conjugacy classes of subgroups of G up to index 4, one each of index 1,2,3 and 4.   ⬚

5.4.3   Variations and improvements

One commonly encountered variant of the basic low-index subgroups problem is to ask for all subgroups of index at most N that contain a given finitely generated subgroup $\langle Y \rangle$ of G. It is straightforward to adapt the code in the preceding subsection to handle this variant. We have to pass the list Y of subgroup generators as an extra parameter to the procedures. In TRYDE-SCENDANT, immediately before the call of FIRSTINCLASS, we reject a coset table $\mathcal{D}$ for which no subgroup corresponding to a completion of $\mathcal{D}$ could possibly contain $\langle Y \rangle$, by inserting the line:

for $w \in Y$ do if not SCANCHECK$(\mathcal{D}, 1, w)$ **then return**;

In FIRSTINCLASS, we only want to reject our current table in favour of a preceding table in the ordering in which 1 is replaced by α, if the subgroup G_α corresponding to this preceding table definitely contains the given subgroup $\langle Y \rangle$. We therefore insert the line

for $w \in Y$ do if $\alpha_{\mathbf{C}}^w \neq \alpha$ then continue α;

immediately after line 4 of FIRSTINCLASS. No other modifications are required.

A less frequently encountered variant is a request for subgroups that do *not* contain a particular element or elements. There seems to be no easy way of incorporating such conditions into the body of the search, so the only way to handle this seems to be to carry out such tests on the complete coset tables in DESCENDANTSUBGROUPS.

The performance of LOWINDEXSUBGROUPS has been observed to be satisfactory for 'straightforward' group presentations where, by a presentation that is not straightforward, we mean one in which there are relatively short group relators that cannot be derived easily from the defining relators.

For example, a nonobvious presentation of the trivial group, such as the well-known presentation of B.H. Neumann:

$$\langle\, x,y,z \mid x^y = x^2, y^z = y^2, z^x = z^2\,\rangle$$

is not straightforward in this sense.

Our procedure does not perform optimally on such examples, because it is unable to make any use of these hidden group relators. In fact the older low-index subgroups algorithm, which uses standard coset enumeration, may perform better on such examples, particularly as N grows larger, because it might discover these hidden relators in the course of the coset enumeration. If the derivation of the hidden relators is not inordinately difficult, and can be accomplished by a short run of coset enumeration, then we can attempt to remedy this deficiency as follows.

In TRYDESCENDANT, immediately after the run of PROCESSDEDUCTIONS, we carry out a short run of normal coset enumeration of G over the trivial subgroup (or over the subgroup $\langle Y \rangle$ discussed at the beginning of this subsection). We start with the current table $\mathcal{D}$. The code for this enumeration needs to be modified such that it aborts immediately if any two cosets in $\Omega_\mathcal{D} = [1\,..\,n_\mathcal{D}]$ ever become coincident, because we want to reject $\mathcal{D}$ if that happens. After this enumeration, all cosets other than those in the original $\Omega_\mathcal{D}$ are removed. The point of this is that the coset enumeration might have discovered new coset table entries within the original $\Omega_\mathcal{D}$.

To the author's knowledge, there has been little if any experimentation with this approach to date, and so it is difficult to assess its usefulness. Presumably the effort devoted to the coset enumeration should be restricted by imposing a rather small limit $f(N)$, perhaps $f(N) = 2N$, on the maximum number of cosets allowed.

There is certainly a danger that on straightforward examples it would simply waste time without achieving much! On the other hand, a coset table-based coset enumeration needs a maximum of about 160 cosets to prove that the example above is trivial, whereas a relator-based enumeration requires about 1600, so if the restrictions imposed on the coset enumeration were too harsh, then it would achieve nothing, even in this example.

Exercise

1. Four subgroups were found in Example 5.10. Find a set of Schreier generators for each of these subgroups, and then show that the subgroups are $\langle x, y \rangle$ (index 1), $\langle x, yxy^{-1} \rangle$ (index 4), $\langle x, yxy \rangle$ (index 3), and $\langle y, xyx \rangle$ (index 2).

5.5   Applications

The low-index subgroups, Reidemeister-Schreier and Tietze transformations algorithms have found many applications to other areas of mathematics, particularly to topology. Some of these applications also involve computing abelian quotients, which we shall be discussing later, in Section 9.2, so we shall defer consideration of these particular applications until Subsection 9.3.3.

The low-index subgroups algorithm can be used in the enumeration of certain types of finite graphs, and we shall now give a brief description of this process. See the paper by Conder and Dobcsányi [CD05] for the general theory and an overview, or [CD02] for more details.

The application is to the enumeration of undirected graphs that are *arc-transitive*; that is, graphs for which the automorphism group is transitive on directed edges. Such graphs are necessary *regular*, which means that every vertex has the same *valency*. Regular graphs of valencies 1 or 2 are easily described (exercise), so the first interesting case is valency 3. It has been known for some time that the automorphism group of every finite arc-transitive trivalent graph is a factor group of one of a known list of seven finitely presented groups and that, conversely, any finite nondegenerate homomorphic image of one of these seven groups acts arc-transitively on a suitable arc-transitive trivalent graph.

So the classification of these graphs is essentially equivalent to finding the low-index subgroups of these groups. For example, one of the groups is

$$G_3 := \langle\, h, a, p \mid h^3,\, a^2,\, p^2,\, [h,p],\, [p, apa],\, (apah)^2 \,\rangle.$$

As is explained in the above papers, it turns out that the stabilizer of a point in the automorphism group of one of the associated graphs is the homomorphic image of the subgroup $\langle h, p, apa \rangle$ of G_3, which is of order 12 and isomorphic to $\mathrm{Sym}(3) \times C_2$. Conder and Dobcsányi found that the most effective approach to finding all arc-transitive trivalent graphs up to valency n was to search for low-index *normal* subgroups K of G_3 up to index $12n$. Then, for any such K with $|G : K| = 12|G : HK|$, G/K arises as a group of automorphisms of such a graph with degree $|G : HK|$. To do this, they adapted the low-index subgroups procedure to enumerate only normal subgroups, and also saved time by implementing a parallel version of the program. They were thereby enabled to enumerate all such graphs having degree up to 768.

Exercise

1. Classify finite regular graphs that are regular with valency one or two.

Chapter 6

Presentations of Given Groups

To progress further in our study of algorithms for finite groups and, in particular, for finite permutation groups, we need to be able to compute a presentation of a given finite group. We have already seen, in Subsection 3.4.1, that a presentation of a group G is necessary, in general, for checking whether a map φ from a generating set X of G to another group H extends to a homomorphism $G \to H$. An alternative method for doing this when G and H are both permutation groups was described in Subsection 4.5.3 but, even in that situation, it is more efficient to use a presentation of G on the generating set X if one is available.

For permutation groups G of large degree, there are also dramatic improvements possible in the Schreier-Sims algorithm for computing a base and strong generating set (BSGS) of G, which involve a presentation for G. The idea is first to use RANDOMSCHREIER to compute a candidate BSGS, and then to compute and use a presentation of G to verify that it really is a BSGS for G.

In the first section of this chapter, we shall describe algorithms for finding presentations of a finite group that is defined by means of a 'concrete' representation, such as a permutation or matrix group. The basic method works by adjoining new relators repeatedly to a candidate presentation, and applying coset enumeration to check whether we really do have a presentation of the group. In the following section, we explain how a closely related method, known as *Todd-Coxeter Schreier-Sims*, can be used to improve the basic Schreier-Sims algorithm. Then we discuss an alternative method for the same problem, the *Sims Verify algorithm* in which, under certain conditions, the relators in a suitable presentation can be calculated theoretically, without recourse to coset enumeration.

6.1 Finding a presentation of a given group

A method for finding a presentation of a finite group G given by its Cayley graph (or, equivalently, its regular permutation representation; see Definition 2.13) was described by John Cannon in [Can72] and [Can73]. Since the regular permutation representation of G needs to be stored, this method is limited to groups of order up to about 10^7.

For larger groups, Cannon described a two-step version of the algorithm. For the two-step version, we assume that a presentation $\langle Y \mid S \rangle$ of a suitable subgroup H of $G = \langle X \rangle$ has already been calculated, and we assume also that words w_y that express the elements of Y as words in $(X \cup X^{-1})^*$ are known. We aim to find a presentation $\langle X \cup Y \mid R \rangle$ of G on the generating set $X \cup Y$. We will choose $R = R_1 \cup R_2 \cup S$, where $R_2 = \{y^{-1}w_y \mid y \in Y\}$, and R_1 is a set of additional relators, which will suffice to prove, using coset enumeration, that the index of $\langle Y \rangle$ in $\langle X \cup Y \mid R \rangle$ is $|G:H|$.

Since the presentation already contains a presentation of H on Y, we know that the subgroup $\langle Y \rangle$ of $\langle X \cup Y \mid R \rangle$ has order at most $|H|$ and so, provided that the relators in R_1 map onto relators of G, this will prove that $\langle X \cup Y \mid R \rangle$ really is a presentation of G. If we want to, then we can use the relators in R_2 to eliminate the generators in Y from the presentation, to give a presentation on the original generators X of G, although that might result in an increased total relator length.

The method is most effective when $|G|$ is approximately equal to $|H|^2$, because then we are dividing the work involved in solving the overall problem equally between the two subtasks of finding the presentation of H, and then finding the presentation of G over the subgroup H. For groups that are even larger, we could derive the presentation of H using a multistep version.

Here, we shall describe an algorithm that can be thought of as a mild generalization of Cannon's two-step process. To simplify the description, we shall assume that the generating set Y of the subgroup H is already a subset of X. Then we do not need the relators in R_2, because they would all be trivial. As we shall see later, this method is tailor-made for application to presentations of permutation groups on strong generating sets.

So, let $G = \langle x_1, \ldots, x_r \rangle$ be a finite group, let $H = \langle x_1, \ldots, x_s \rangle$ with $0 \le s \le r$, let $n := |G:H|$, and suppose that we are able to compute a complete coset table $\mathcal{C}$ on $\Omega := [1 \mathbin{..} n]$ corresponding to the action of G by right multiplication on the right cosets of H in G. For example, if G is a finite permutation group, then we could do this by the methods described in Subsection 4.6.7. We assume that we can solve the rewriting problem for H; that is, that we are able to write elements of H as words in the generators x_i ($1 \le i \le s$) of H and their inverses.

Now we let $\hat{G} = \langle \hat{x}_1, \ldots, \hat{x}_r \mid R \rangle$ be a finitely presented group. The set R of defining relators will be increased during the algorithm, but the map φ with $\varphi(\hat{x}_1) = x_i$ for $1 \le i \le r$ will always satisfy the hypothesis of Theorem 2.52 (that is, it will map all elements of R to the identity in G), and so it will always define an epimorphism, which we shall also denote by φ, from $\hat{G}$ to G. Initially, R consists of a set of words in the generators $\hat{x}_1, \ldots, \hat{x}_s$ and their inverses only, with the property that the restriction of φ to the subgroup $\hat{H} := \langle \hat{x}_1, \ldots, \hat{x}_s \rangle$ of $\hat{G}$ is an isomorphism.

The setup that we have described is the input to the algorithm and it is assumed that $\varphi_{\hat{H}} : \hat{H} \to H$ is an isomorphism. Since, we never remove relators from R during the algorithm, we will always have $|\hat{H}| \leq |H|$ and so, if we can show by coset enumeration that $|\hat{G}:\hat{H}| = n$, then we will have shown that $|\hat{G}| \leq |G|$. But since φ is an epimorphism, that will prove that φ is an isomorphism, and we will be done. This proves the correctness of the answer returned.

We shall now describe the algorithm itself. It a good idea to start by applying the procedure STANDARDIZE from Subsection 5.2.3 to $\mathcal{C}$. This means that entries corresponding to short words will be encountered first when we scan $\mathcal{C}$ in its natural order, which will tend to make the relators in the resulting presentation shorter.

We then proceed to initialize a coset table $\hat{\mathcal{C}}$ for a coset enumeration of $\hat{G}$ over the subgroup $\hat{H}$, and we can immediately set $1^{\hat{x}_i} = 1^{\hat{x}_i^{-1}} = 1$ for the generators $\hat{x}_i$ $(1 \leq i \leq s)$ of $\hat{H}$.

Let $X := \{x_1, \ldots, x_r\}$ and $A := X \cup X^{-1}$. Next we define a right transversal of H in G consisting of words $\tau(\beta) \in A^*$, where $\tau(1) = \varepsilon$ and $1^{\tau(\beta)} = \beta$ in $\mathcal{C}$ for $1 \leq \beta \leq n$. This is done in the same way as in the procedure DEFINESCHREIERGENERATORS in Section 5.3. The words $\tau(\beta)$ can either be stored explicitly as words, or else a Schreier vector for 1^G can be computed, enabling the $\tau(\beta)$ to be recomputed when required. Each time that we define a new transversal element $\tau(\beta)$ as $\tau(\alpha)x$ for some $1 \leq \alpha < \beta$ and $x \in A$, we make the corresponding definition $\alpha^{\hat{x}} = \beta$ in $\hat{\mathcal{C}}$. This ensures that the transversal elements $\hat{\tau}(\beta)$ associated with $\hat{\mathcal{C}}$ are mapped by φ to $\tau(\beta)$ for $1 \leq \beta \leq n$.

Now we proceed to the main part of the algorithm. This consists of a loop, which is repeated until we exit with the required presentation. Assuming that $r > s$, we certainly do not have a presentation of G initially, because none of the relators in R involve any $\hat{x}_i$ for $i > s$. (Initially $\hat{G}$ is the free product of $\hat{H} \cong H$ with the free group on $\hat{x}_{s+1}, \ldots, \hat{x}_r$, so it is infinite.) So we start the main loop by adjoining a new relator.

Whenever we need to adjoin a new relator to R, we proceed as follows. We scan the coset tables $\mathcal{C}$ and $\hat{\mathcal{C}}$ together, in their standard order, first by elements of Ω, and then by elements of A. We must, at some stage, encounter a difference where β^x in $\mathcal{C}$ is not equal to the corresponding $\beta^{\hat{x}}$ in $\hat{\mathcal{C}}$ for some β with $1 \leq \beta \leq n$ and $x \in A$, because otherwise $\hat{\mathcal{C}}$ would be complete with n cosets, and we are assuming that it is not.

The difference could be either because $\beta^{\hat{x}}$ is undefined, or because it is defined but equal to some $\gamma > n$. In either case, we now write the Schreier generator (or inverse) $\tau(\beta)x\tau(\beta^x)^{-1}$ of H as a word u over the generators x_i $(1 \leq i \leq s)$ of H, and let $w := \hat{\tau}(\beta)\hat{x}\hat{\tau}(\beta^x)^{-1}\hat{u}^{-1}$ be the word over the $\hat{x}_i$ that is mapped to $\tau(\beta)x\tau(\beta^x)^{-1}u^{-1}$ by φ. This word w is adjoined to R as a new relator of $\hat{G}$. Since $\tau(\beta)x\tau(\beta^x)^{-1}u^{-1} =_G 1$, we maintain our hypothesis that φ maps relators in R to the identity in G.

Next, we start or resume a coset enumeration on the coset table $\hat{\mathcal{C}}$, with a

PRESENTATION$(G, H, \sim \hat{G}, M)$

Input: Group $H = \langle x_1, \ldots, x_s \rangle \leq G = \langle x_1, \ldots, x_r \rangle$.

Group $\hat{G} = \langle \hat{x}_1, \ldots, \hat{x}_r \mid R \rangle$, such that $\langle \hat{x}_1, \ldots, \hat{x}_s \mid R \rangle \cong H$;

Standardized coset table $\mathcal{C}$ for G acting on cosets of H in G;

$M \in \mathbb{Z}$ with $M \geq |G:H|$;

Output: $\langle \hat{x}_1, \ldots, \hat{x}_r \mid R \rangle$ with R extended to make

$\langle \hat{x}_1, \ldots, \hat{x}_r \mid R \rangle \cong G$;

Notation: Let $n := |G:H|$; $X := [x_1, \ldots, x_r]$; $\hat{X} := [\hat{x}_1, \ldots, \hat{x}_r]$;

$A := X \cup X^{-1}$; $\hat{A} := \hat{X} \cup \hat{X}^{-1}$;

$\varphi : \hat{G} \to G$ with $\varphi(\hat{x}_i) = x_i$ $(1 \leq i \leq r)$;

1 **if** $r = s$ **then return** $\hat{G}$;

2 Initialize coset table $\hat{\mathcal{C}}$ of $\hat{H} := \langle \hat{x}_1, \ldots, \hat{x}_s \rangle$ in $\hat{G}$;

3 In $\hat{\mathcal{C}}$, set $1^{\hat{x}_i} = 1^{\hat{x}_i^{-1}} = 1$ for $1 \leq i \leq s$;

4 Define coset representatives $\tau(\beta) \in A^*$ $(1 \leq \beta \leq n)$ for H in G, and make corresponding definitions in $\hat{\mathcal{C}}$;

5 **while true**

6   **do** Find smallest $\beta \in [1 .. n]$ and first $x \in A$ such that the entries β^x in $\mathcal{C}$ and $\beta^{\hat{x}}$ in $\hat{\mathcal{C}}$ are different;

7   Write $\tau(\beta) x \tau(\beta^x)^{-1} \in H$ as word u over $x_i^{\pm 1}$ $(1 \leq i \leq s)$;

8   Adjoin $\varphi^{-1}(\tau(\beta) x \tau(\beta^x)^{-1} u^{-1})$ to R;

9   Start or resume coset enumeration on $\hat{\mathcal{C}}$ with maximum number of cosets set to M;

10   **if** $\hat{\mathcal{C}}$ is complete with $|\hat{\Omega}| = n$

11   **then return** $\hat{G} = \langle \hat{x}_1, \ldots, \hat{x}_r \mid R \rangle$;

prescribed upper limit of $M > n$ on the number of cosets that can be defined. This enumeration should be designed such that the first thing that it does is to scan the new relator w under the coset 1. Under the forward scan, 1 will be mapped to β under $\hat{\tau}(\beta)$ and under the backwards scan, 1 will be mapped to 1 under the word $\hat{u}$ over the $\hat{x}_i^{\pm 1}$ for $1 \leq i \leq s$ that maps onto u, and then 1 will be mapped to β^x under $\hat{\tau}(\beta^x)$. So, in the complete scan, either $\beta^{\hat{x}}$ was undefined initially, in which case we get the deduction $\beta^{\hat{x}} = \beta^x$, or $\beta^{\hat{x}}$ was defined and equal to some $\gamma > n$, in which case we will get a coincidence somewhere in the scan. After processing this and any resulting coincidences, 1 will scan consistently under w, and so we will again have $\beta^{\hat{x}} = \beta^x$.

If the coset enumeration terminates with a complete coset table with exactly n cosets, then we will have shown that $|\hat{G} : \hat{H}| = n$, and so, by the remarks above, we will have found the required presentation of G, and we can halt. Otherwise, if the enumeration either stops because it can go no further with the prescribed limit of M cosets, or if it completes but with more than n cosets, then we go back to the beginning of the main loop at line 5 of the algorithm, and adjoin another new relator.

In either case, when the coset enumeration process stops, the entries in $\mathcal{C}$ and $\hat{\mathcal{C}}$ that were different beforehand, and used to define the new relator, will be equal. So, eventually, all entries in $\hat{\mathcal{C}}$ for $1 \leq \beta \leq n$ will be the same as those in $\mathcal{C}$, and at that stage we will have a complete coset table with n cosets. So the algorithm must terminate, and we have an upper bound of $n(r-1)+1-s$, which is half of the number of entries of $\hat{\mathcal{C}}$ that were undefined immediately before adjoining the first new relator, on the number of new relators that are adjoined to R.

The displayed procedure PRESENTATION is a summary of the complete algorithm.

Cannon's original version took $M = n$, so no extra definitions were allowed in the coset enumeration. In that case, it is important to use a coset table style enumeration, where all consequences of all deductions are made before giving up. An advantage of doing that is that the same storage space can be used for both coset tables $\mathcal{C}$ and $\hat{\mathcal{C}}$, because an entry in $\hat{\mathcal{C}}$ is either the same as the corresponding entry in $\mathcal{C}$ or is undefined, so the defined entries in $\hat{\mathcal{C}}$ can be specified by using a negative integer rather than a positive, for example.

Example 6.1

Let G be the quaternion group Q_8 of order 8, generated by the permutations $a := (1,2,6,3)(4,8,5,7)$ and $b := (1,4,6,5)(2,7,3,8)$ of degree 8, and $H = 1$. We have carefully numbered the points such that the associated coset table $\mathcal{C}$ below is already in standard form.

	a	a^{-1}	b	b^{-1}
1	2	3	4	5
2	6	1	7	8
3	1	6	8	7
4	8	7	6	1
5	7	8	1	6
6	3	2	5	4
7	4	5	3	2
8	5	4	2	3

To limit the proliferation of hats, we shall use x and y rather than $\hat{a}$ and $\hat{b}$ as our generators of the finitely presented group $\hat{G}$ in this example. We define the coset representatives $\tau(\beta)$ for $\beta \in [1..8]$ to be ε, a, a^{-1}, b, b^{-1}, a^2, ab, ab^{-1}, respectively, while making the corresponding definitions $1^x = 2$, $1^{x^{-1}} = 3$, $1^y = 4$, $1^{y^{-1}} = 5$, $2^a = 6$, $2^b = 7$, $2^{b^{-1}} = 8$ in $\hat{\mathcal{C}}$.

The first undefined entry in $\hat{\mathcal{C}}$ is $3^{x^{-1}}$, and since since $3^{a^{-1}} = 6$ in $\mathcal{C}$, we adjoin the relator corresponding to $\tau(3)a^{-1}\tau(6)^{-1} = a^{-4}$, which is x^{-4} (or, equivalently, x^4) to R.

Coset enumeration yields only the forced deduction $3^{x^{-1}} = 6$, so we move on to the next undefined entry 3^y, and find $\tau(3)b\tau(8)^{-1} = a^{-1}b^2a^{-1}$, so we adjoin $x^{-1}y^2x^{-1}$ to R.

This time, coset enumeration results in deductions $3^y = 8$, $4^y = 6$ (from scanning 2 under $x^{-1}y^2x^{-1}$), $6^y = 5$ (from scanning 3 under $x^{-1}y^2x^{-1}$), and $7^y = 3$ (from scanning 6 under $x^{-1}y^2x^{-1}$). The next undefined entry is 4^x, with $\tau(4)a\tau(8)^{-1} = baba^{-1}$, so we adjoin $yxyx^{-1}$ to R.

Now coset enumeration finds all remaining deductions (check!), and so the process halts, and we have proved that $G \cong \langle x, y \mid x^4, x^{-1}y^2x^{-1}, yxyx^{-1} \rangle$. In fact, the first relator is redundant, and $G \cong \langle x, y \mid x^{-1}y^2x^{-1}, yxyx^{-1} \rangle$.
□

It is possible to derive a two-generator presentation of this group by using a different ordering of the generators from the one that we have used here; see the exercise below. It is a general feature of this algorithm that the earlier relators produced often turn out to be redundant. This happens more frequently with larger groups, and one is sometimes forced to wait a frustratingly long time before a single crucial relator is found by the program.

Although current implementations do not do this by default, if it is important to produce presentations that are as short as possible, then it would be a good idea to try repeating the coset enumerations with some of the earlier relators omitted, to see whether they are redundant. Of course, the user needs to balance the time taken in doing this against the advantage to be gained from a shorter presentation.

If a relator-based method is to be used for the coset enumeration, then it is necessary to have M rather larger than n. This approach was studied extensively by J.S. Leon, who used it in his implementation of the Todd-Coxeter-Schrier-Sims algorithm, which we shall be studying in the next section. He reports in [Leo80b, Leo80a] that $M = 1.2n + 50$ seems to work well for examples with n of order 10^4. He also found that it is important to choose the correct policy when using the LOOKAHEAD algorithm, which differs from the policy that one would generally adopt in a standard coset enumeration.

Given the not totally satisfactory behaviour of current versions of the algorithm, which tend to produce many redundant relators with larger groups, there would seem to be room for further research in this area.

Exercises

1. We have $\mathrm{Sym}(5) = \langle a_1, a_2, a_3, a_4 \rangle$, with $a_1 = (1,2)$, $a_2 = (2,3)$, $a_3 = (3,4)$, $a_4 = (4,5)$. Use the algorithm above to compute presentations, in turn, of the subgroups $\langle a_1 \rangle$, $\langle a_1, a_2 \rangle$, $\langle a_1, a_2, a_3 \rangle$, $\langle a_1, a_2, a_3, a_4 \rangle$ of $\mathrm{Sym}(5)$, using the preceding subgroup as the group H in the algorithm each time.

2. Repeat the calculation of a presentation of Q_8, but this time use the order of generators $a < b < a^{-1} < b^{-1}$, for which the standardized coset table comes from $a := (1,2,6,4)(3,8,5,7)$, $b := (1,3,6,5)(2,7,4,8)$. You should get a presentation with just two relators.

6.2 Finding a presentation on a set of strong generators

In this section, we let G be a permutation group on $[1 .. n]$, and let (B, S) be either a known BSGS or a candidate for a BSGS of G. We shall show that the algorithm described in the previous section can be applied to compute what we shall call a *strong presentation* of G.

We shall use the notation that we introduced in Section 4.4. In particular, $B = [\beta_1, \ldots, \beta_k]$, $G^{(i)} := G_{\beta_1, \ldots, \beta_{i-1}}$, $S^{(i)} := S \cap G^{(i)}$, and $H^{(i)} := \langle S^{(i)} \rangle$ for $1 \leq i \leq k+1$, where (B, S) is a BSGS for G if and only if $G^{(k+1)} = 1$ and $G^{(i)} = H^{(i)}$ for $1 \leq i \leq k$.

DEFINITION 6.1 If (B, S) is a BSGS for G, then a presentation $\langle S \mid R \rangle$ of G on S is said to be a *strong presentation* if, for each i with $1 \leq i \leq k$, $\langle S^{(i)} \mid R^{(i)} \rangle$ is a presentation of $G^{(i)}$, where $R^{(i)}$ is defined to be the subset of R consisting of those relators in R that involve only generators in $S^{(i)}$ and their inverses.

6.2.1 The known BSGS case

Suppose first that (B, S) is known to be a BSGS of G. So $G^{(k+1)} = 1$ and $|G^{(i)} : G^{(i+1)}| = |\Delta^{(i)}|$, where $\Delta^{(i)} := \beta_i^{H^{(i)}}$ for $1 \leq i \leq k$. By Proposition 2.23, $(G^{(i)})^{\Delta^{(i)}}$ is equivalent to the action of $G^{(i)}$ by right multiplication on the cosets of $G^{(i+1)}$ in $G^{(i)}$. So we can use $(G^{(i)})^{\Delta^{(i)}}$ to write down the coset table of this action, and then we just need to apply the algorithm PRESENTATION described in the previous section k times, working up the series of basic stabilizers.

First we run PRESENTATION on $G^{(k)}$ with generators $S^{(k)}$ and trivial subgroup. On subsequent runs, we compute a presentation $\langle S^{(i)} \mid R^{(i)} \rangle$ of $G^{(i)}$ on generators $S^{(i)}$, using the subgroup $G^{(i+1)}$ on generators $S^{(i+1)}$, with $R^{(i)}$ initialized to the relators $R^{(i+1)}$ of $G^{(i+1)}$ computed in the previous run. The function PRESENTATION assumes that we can write elements of the subgroup as words over the generators of the subgroup, but the STRIP algorithm, which is part of the basic BSGS machinery, enables us to do precisely that. Notice that our coset representatives $\tau(\beta)$ of $G^{(i+1)}$ in $G^{(i)}$ are the same as the elements $u_\beta^{(i)}$ defined in Section 4.4.

Example 6.2

Let $G := \text{Alt}(5)$ with $B := [1, 2, 3]$ and $S := [a, b, c]$, where $a := (1, 2, 3)$, $b := (2, 3, 4)$, $c := (3, 4, 5)$. In the corresponding finitely presented groups, we shall let x, y, z be generators that map onto a, b, c, respectively.

First, we compute a presentation of $G^{(3)}$ on $S^{(3)} = [c]$. The coset table comes from the action of $G^{(3)}$ on $\Delta^{(3)} = [3, 4, 5]$. To apply PRESENTATION in an implementation, we would compute this coset table with the points of $\Delta^{(3)}$ renumbered as $[1, 2, 3]$, but it makes it clearer while working through the example, if we break our normal convention and allow the domain of the coset tables $\mathcal{C}$ and $\hat{\mathcal{C}}$ to be $[3, 4, 5]$. Our definitions are $3^z = 4$ and $3^{z^{-1}} = 5$, our first relator is z^3, coming from the blank entry 4^z in $\hat{\mathcal{C}}$ and, with that single relator, the coset enumeration completes, so (not surprisingly!) we get the presentation $\langle\, z \mid z^3 \,\rangle = \langle\, S^{(3)} \mid R^{(3)} \,\rangle$ of $G^{(3)}$ with $R^{(3)} = \{z^3\}$.

Next we move on to $G^{(2)}$ on $S^{(2)} = [b, c]$, with subgroup $G^{(3)}$ and coset table coming from the action of $G^{(2)}$ on $\Delta^{(2)} = [2, 3, 4, 5]$. Our relator set $R^{(2)}$ is initialized to $R^{(3)}$. The coset representatives are $\tau(2) = \varepsilon$, $\tau(3) = b$, $\tau(4) = b^{-1}$, $\tau(5) = bc^{-1}$ with corresponding definitions $2^y = 3$, $2^{y^{-1}} = 4$, $3^{z^{-1}} = 5$, and we have $2^z = 2^{z^{-1}} = 1$ initially. The first blank entry in $\hat{\mathcal{C}}$ is 3^y, which yields the new relator y^3 and deduction $3^y = 4$. The first blank entry is now 3^z. The corresponding Schreier generator $\tau(3)c\tau(3^c)^{-1} = bcb = (3, 5, 4)$. We can write this as a word in the generators $S^{(3)}$ of $G^{(3)}$ using STRIP, and $bcb = c^{-1}$, so the new relator is the word mapping onto $bcbc$, which is $yzyz$. The coset enumeration now completes, so we have the presentation $\langle\, S^{(2)} \mid R^{(2)} \,\rangle$ of $G^{(2)}$ with $R^{(2)} = \{\, z^3,\, y^3,\, (yz)^2 \,\}$.

Finally we want a presentation of $G^{(1)}$ on $S^{(1)} = [a, b, c]$, with subgroup $G^{(2)}$ and coset table coming from the action of $G^{(1)}$ on $\Delta^{(1)} = [1, 2, 3, 4, 5]$. Our relator set $R^{(1)}$ is initialized to $R^{(2)}$. The coset representatives are $\tau(1) = \varepsilon$, $\tau(2) = a$, $\tau(3) = a^{-1}$, $\tau(4) = ab^{-1}$, $\tau(5) = a^{-1}c^{-1}$, with corresponding definitions $1^x = 2$, $1^{x^{-1}} = 3$, $2^{y^{-1}} = 4$, $3^{z^{-1}} = 5$ and we have $1^y = 1^z = 1$ initially. The first blank entry in $\hat{\mathcal{C}}$ is 2^x, which yields the new relator x^3 and deduction $2^x = 3$. The first blank entry is now 2^y. The corresponding Schreier generator $\tau(2)b\tau(2^b)^{-1} = aba = (2, 4, 3)$, which rewrites, using STRIP to b^{-1}, yielding the new relator $xyxy$. This results in deductions $2^y = 3$, $3^y = 4$ and $4^x = 4$. Now the first blank entry is 2^c, with corresponding Schreier generator $\tau(2)c\tau(2^c)^{-1} = aca^{-1} = (2, 4, 5)$, which rewrites with STRIP to cb^{-1}, so we adjoin the new relator $xzx^{-1}yz^{-1}$. The coset enumeration now completes, so we have the final presentation $\langle\, S^{(1)} \mid R^{(1)} \,\rangle$ of $G = G^{(1)}$ with $R^{(1)} = \{\, z^3,\, y^3,\, (yz)^2,\, x^3,\, (xy)^2,\, xzx^{-1}yz^{-1} \,\}$. $\Box$

This method is generally effective and, when combined with the BSGS facilities already described, it enables us to check whether a map φ from a permutation group G to a group H defined on the initial generators of G extends to a homomorphism. First we compute a BSGS for G, and use the resulting SLPs as described in Section 4.5 to compute the images of the putative extension of φ on the strong generators S of G. We can then check whether φ satisfies the relators of a strong presentation computed as above.

It suffers from the general problem of PRESENTATION mentioned earlier, that the relators that are found early on often turn out to be redundant, although this does not happen in the example above. The relator y^3 is actually redundant in the presentation $\langle S^{(1)} \mid R^{(1)} \rangle$ of G but, if we removed y^3 from R, then we would no longer have a strong presentation, because $\langle S^{(2)} \mid R^{(2)} \rangle$ would not be a presentation of $G^{(2)}$. If we leave out either (but not both!) of $(xy)^2$ or $(yz)^2$, then we are left with a presentation of the group Alt(5) $\times C_3$ of order 180.

6.2.2 The Todd-Coxeter-Schreier-Sims algorithm

Suppose now that (B, S) is a suspected rather than a known BSGS of the permutation group G on Ω. So we are not certain that $G^{(k+1)} = 1$ or that $H^{(i)} = G^{(i)}$ for all i with $1 \leq i \leq k+1$. In this case, by using essentially the same approach as the one described in the previous subsection, we can simultaneously calculate a strong presentation of G and check that (B, S) really is a BSGS of G. This procedure is known as the *Todd-Coxeter-Schreier-Sims (TCSS) algorithm*, because it combines coset enumeration with the basic Schreier-Sims method. Implementations by Leon using relator-based coset enumeration are described in detail in [Leo80b] and [Leo80a]. The current MAGMA implementation uses coset table-based coset enumeration, however.

If it turns out that, if (B, S) is not a BSGS then, as in SCHREIERSIMS, a new strong generator and possibly a new base point, will be found, and so we can update (B, S) and restart. However, this method runs much more effectively when (B, S) really is a BSGS, so the best policy is to run RANDOMSCHREIER first to calculate a probable BSGS rather than trying to run TCSS on the initial generating set.

We adopt exactly the same method as in the previous subsection, and apply PRESENTATION to $H^{(i)}$ with the subgroup $H^{(i+1)}$ for $i = k, k-1, \ldots$. However, when we need to add a new relator w to $R^{(i)}$, we check that the image $\varphi(w)$ of w in $H^{(i)}$ is equal to the identity in $H^{(i)}$. This was true by hypothesis in Subsection 6.2.1, but that is no longer the case. In fact, if the new relator arises from the entry $\beta^{\hat{x}}$ in $\hat{\mathcal{C}}$, then we apply STRIP to the element $\tau(\beta)x\tau(\beta^x)^{-1}$ of $H^{(i)}$ in an attempt to write this element as a word in the generators of $H^{(i+1)}$.

We are doing exactly the same thing here as we did in line 13 of SCHREIERSIMS; the coset representative $\tau(\beta)$ is the same as the u_β there. Just as in SCHREIERSIMS, if STRIP returns h, j with $j \leq k$, then h needs to be added as a new strong generator to $S^{(j)}$, whereas if $j = k+1$ and $h \neq 1$, then we need to add a new base point to B as well as adding h to S. If B is already known to be a base of G, perhaps because G is a subgroup of a larger subgroup of Sym(Ω) with base B, then we can omit the check that $h = 1$, thereby producing a faster version in the known-base case.

Eventually, it will happen that $\varphi(w)$ evaluates to the identity for all new relators w that we adjoin to $R^{(i)}$, and the coset enumeration will complete with $|H^{(i)} : H^{(i+1)}| = |\Delta^{(i)}|$. Since $|H^{(i)} : H^{(i)}_{\beta_i}| = |\Delta^{(i)}|$, this proves that $H^{(i+1)} = H^{(i)}_{\beta_i}$ and, if this is true for $1 \leq i \leq k$, then we have proved, by

Lemma 4.5, that (B, S) is a BSGS for G.

The TCSS algorithm carries out the same tests as in line 13 of SCHREIER-SIMS, but for far fewer pairs (u_β, x). In the example in the previous subsection, we would carry out this test using STRIP once for each new relator used; that is, 6 times altogether. If we apply SCHREIERSIMS to this example, then 18 such checks will be made, corresponding to the total numbers of Schreier generators that are not trivial by definition; that is, 2 checks for $i = 3$, 5 checks for $i = 2$, and 11 checks for $i = 1$. Of course, extra time is taken by the coset enumeration. For permutation degrees of degree less than about 100, the overhead resulting from the coset enumeration often makes it faster to use SCHREIERSIMS than TCSS but, for examples with larger degrees, TCSS is typically significantly faster.

6.3   The Sims 'Verify' algorithm

The algorithm VERIFY to be described in this section is an alternative to the use of PRESENTATION for calculating a strong presentation of a finite permutation group. As was the case in Section 6.2, we can use this method both in the situation when (B, S) is already known to be a BSGS of G, and when it is only suspected of being a BSGS. In the latter case, just as with the Todd-Coxeter-Schreier-Sims (TCSS) algorithm, it can be used to verify that (B, S) really is a BSGS. (This is the reason for the name VERIFY.)

Rather than use coset enumeration to find the presentation, we use the orbit structure of the subgroups $H^{(i)}$ to write down a presentation of $H^{(i)}$ on the generators $S^{(i)}$, on the assumption that such a presentation is already known for $H^{(i+1)}$ on $S^{(i+1)}$. As in TCSS, if we need to verify that (B, S) is a BSGS, then we need to check that the images in $H^{(i)}$ of the relators of the presentation are all equal to the identity. If this check fails, then (B, S) is not a BSGS, and we have a new strong generator in $H^{(i+1)}$. We then need to interrupt the algorithm and resume the computation of a BSGS, which (as in TSCC) is inefficient, so it is preferable only to apply VERIFY when we are reasonably sure that (B, S) is a BSGS.

VERIFY formed the basis of the methods used by Sims to prove the existence of the finite simple groups Ly and BM (see [Sim73] and [Sim80].) It is interesting to note that the calculation of the order of the Lyons group now can be done in a few hours in MAGMA, by a call of a standard function.

As with TCSS, we apply VERIFY to each $H^{(i)}$ separately, in reverse order, starting with $i = k$. In fact, provided that we process the i in this order, we can use TCSS for some values of i and VERIFY for others (see Example 6.4 below).

Since we shall be concerned with a fixed i in our description of VERIFY, we set $H := H^{(i)}$, $K := H^{(i+1)}$, $\alpha := \beta_i$ and $\Delta := \Delta^{(i)} = \alpha^H$. In addition, let $X := S^{(i)}$, $Y := S^{(i+1)}$, $A := (X \cup X^{-1})^*$, and $B := (Y \cup Y^{-1})^*$. As in PRESENTATION and TCSS, we let $\hat{H}$ be a finitely presented group with generators $\hat{X}$ and a bijection $\varphi : \hat{X} \to X$ inducing an epimorphism, $\varphi : \hat{H} \to H$, and we suppose that we already have a presentation of K; that is, we have a set R of words in $\hat{B}^*$ such that φ induces an isomorphism $\langle \hat{Y} \mid R \rangle \to K$.

Our objective is to decide whether $H_\alpha = K$. If so, then we want to extend the set R to make $\langle \hat{X} \mid R \rangle$ a presentation of G. If not, then we want to output an element of $H_\alpha \setminus K$, so that we can extend our strong generating sequence of G and restart the BSGS verification.

6.3.1   The single-generator case

We shall first discuss the case in which $X \setminus Y$ consists of a single generator z, so $H = \langle K, z \rangle$, and deal with the general case later. Let $\beta := \alpha^{z^{-1}}$. (It will become clear in the proof of Theorem 6.2 why we use z^{-1} rather than z.)

We start by calculating β^K. We then know the order $|K|/|\beta^K|$ of K_β, so we can find generators for K_β quickly by using RANDOMSTAB, and we choose orbit representatives $[\gamma_1, \ldots, \gamma_m]$ of K_β on $\Delta = \alpha^H$. These representatives include α and β, which we assume to be γ_1 and γ_2, respectively. The number m of orbits of K_β on Δ is critical for the performance of the algorithm, with larger values of m corresponding to slower running times.

We also need orbit representatives of K on Δ. We choose $\beta_0 := \alpha$ and $\beta_1 := \beta$ to be the first two of these. Then, in general, if, for some $i \geq 1$, orbit representatives β_j have been chosen for $0 \leq j \leq i$ and $\cup_{j=0}^i \beta_j^K \neq \Delta$, we choose β_{i+1} to be γ^z for some $\gamma \in \beta_j^K$ with $j \leq i$. Let the complete set of orbit representatives of K on Δ be $[\beta_0, \beta_1, \ldots, \beta_l]$.

As was the case with the function PRESENTATION, we choose a transversal $\{ \tau(\gamma) \mid \gamma \in \Delta \}$ of H_α in H, where each $\tau(\gamma)$ is stored, either explicitly or via a Schreier vector, as a word in A^*, and $\alpha^{\tau(\gamma)} = \gamma$. This transversal is chosen slightly differently from the method used in ORBITSTABILIZER.

As usual, we start with $\tau(\alpha) := \varepsilon$, and we put $\tau(\beta) := z^{-1}$. After defining $\tau(\beta_i)$ for any i with $1 \leq i \leq l$, we apply the generators in Y to β_i, so that all $\tau(\gamma)$ with $\gamma \in \beta_i^K$ have the form $\tau(\beta_i)w$, with $w \in B^*$. For $1 \leq i < l$, we define $\tau(\beta_{i+1}) := \tau(\gamma)z$, where γ was used, as described above, to define $\beta_{i+1} = \gamma^z$. In practice, we would define the coset representatives $\tau(\gamma)$ and the orbit representatives β_i together, rather than in separate stages as we have described it.

Now we are ready to write down the relations in our presentation of H. Our final relator set R will be a union of three sets R_1, R_2, R_3, where R_1 is the initial set of relators of K, and R_2 and R_3 are two sets of relators that we append to R during the algorithm.

First we describe R_2. For each β_i with $1 \leq i \leq l$, we compute generators $y_1, \ldots, y_t$ of K_{β_i}. In fact we have already done this for $\beta = \beta_1$ and, as was the

VERIFY($H, K, \sim \hat{H}, z, \alpha$)

    Input: Groups $K = \langle x_1, \ldots, x_s \rangle \leq H = \langle K, z \rangle$ with H^Δ transitive,

                Group $\hat{H} = \langle \hat{x}_1, \ldots, \hat{x}_s, \hat{z} \mid R \rangle$, with $\langle \hat{x}_1, \ldots, \hat{x}_s \mid R \rangle \cong K$,

                $\alpha \in \Delta$ with $K \leq H_\alpha$ and $z \notin H_\alpha$.

    Output: true if $K = H_\alpha$, otherwise false

                If **true**, then $\langle \hat{x}_1, \ldots, \hat{x}_s, \hat{z} \mid R \rangle$

                with R extended to make $\langle \hat{x}_1, \ldots, \hat{x}_s, \hat{z} \mid R \rangle \cong H$.

                If **false**, then an element of $H_\alpha \setminus K$.

    Notation: Let $X := [x_1, \ldots, x_s, z]$; $A := X \cup X^{-1}$;

                $Y := [x_1, \ldots, x_s]$; $B := Y \cup Y^{-1}$;

                $\varphi : \hat{H} \to H$ with $\varphi(\hat{x}_i) = x_i$ $(1 \leq i \leq s)$, $\varphi(\hat{z}) = z$.

1    $\beta := \alpha^{z^{-1}}$;

2   Choose orbit representatives $\alpha = \beta_0, \beta = \beta_1, \beta_2, \ldots, \beta_l$ of K on Δ, and

      orbit representatives $\gamma_1, \ldots, \gamma_m$ of K_β on Δ such that, for $1 \leq i < l$,

      we have $\beta_{i+1} = \gamma_k^z$ for some $\gamma_k \in \beta_j^K$ with $1 \leq j \leq i$;

      ($*$ Now choose coset reps. $\tau(\gamma)$ $(\gamma \in \Delta)$ of H_α in H as words in A^* $*$)

3    $\tau(\alpha) := \varepsilon$; $\tau(\beta) = z^{-1}$;

4   **for** $i \in [1 .. l]$

5   **do for** $\gamma \in \beta_i^K \setminus \{\beta_i\}$, choose $\tau(\gamma) := \tau(\beta_i)w$, for some $w \in B^*$;

6   **if** $i < l$ **then** choose $\tau(\beta_{i+1}) := \tau(\gamma_k)z$ for the appropriate

                $\gamma_k \in \beta_j^K$ with $0 \leq j \leq i$;

7   **for** $i \in [1 .. l]$

8   **do** Find generators $y_1, \ldots, y_t$ of K_{β_i};

9   **for** $j \in [1 .. t]$

10   **do if** $\tau(\beta_i)y_j\tau(\beta_i)^{-1} \notin K$

11   **then return** false, $\tau(\beta_i)y_j\tau(\beta_i)^{-1}$;

12   Write $\tau(\beta_i)y_j\tau(\beta_i)^{-1}$ as word u in generators of K;

13   APPEND($\sim R, \varphi^{-1}(\tau(\beta_i)y_j\tau(\beta_i)^{-1}u^{-1})$);

14   **for** $i \in [1 .. m]$

15   **do if** $\tau(\gamma_i)z\tau(\gamma_i^z)^{-1} \notin K$

16   **then return** false, $\tau(\gamma_i)z\tau(\gamma_i^z)^{-1}$;

17   Write $\tau(\gamma_i)z\tau(\gamma_i^z)^{-1}$ as word u in generators of K;

18   APPEND($\sim R, \varphi^{-1}(\tau(\gamma_i)z\tau(\gamma_i^z)^{-1}u^{-1})$);

19   **return** true, $\hat{H} = \langle \hat{x}_1, \ldots, \hat{x}_s, \hat{z} \mid R \rangle$;

case there, the fact that $|K|$ and $|\beta_i^K|$ are known allows us to do this easily by using RANDOMSTAB. In order to compute inverse images of the y_i in $\hat{H}$ under φ, we need the y_i as words in B^*, but that presents no problem, because we are assuming that Y is a strong generating set for K.

Now, for each such y_j, $\tau(\beta_i)y_j\tau(\beta_i)^{-1}$ fixes α. We test this element for membership of K and, if it is not in K, then we return **false** together with the offending element. If it is in K, then we can write it as a word $u \in B^*$, so $\tau(\beta_i)y_j\tau(\beta_i)^{-1}u^{-1}$ evaluates to 1_H and we adjoin its inverse image

$\varphi^{-1}(\tau(\beta_i)y_j\tau(\beta_i)^{-1}u^{-1})$ to R_2.

Now we describe R_3. For each γ_k with $1 \leq k \leq m$, the Schreier generator $\tau(\gamma_k)z\tau(\gamma_k^z)^{-1}$ fixes α. Once again, we test this element for membership of K and, if it is not in K, then we return false together with the element. If it is in K, then we can write it as a word $u \in B^*$, so $\tau(\gamma_k)z\tau(\gamma_k^z)^{-1}u^{-1}$ evaluates to 1_H and we put its inverse image $\varphi^{-1}(\tau(\gamma_k)z\tau(\gamma_k^z)^{-1}u^{-1})$ into R_3.

(*Technical note:* We gain a small increase in the efficiency of the algorithm if we choose our orbit representatives β_i of K on Γ for $i > 1$ as γ_k^z for some γ_k. We leave it as an exercise for the reader to prove that this is possible. If we do that, then we have $\tau(\beta_i) = \tau(\gamma_k^z)$, and the associated Schreier generator $\tau(\gamma_k)z\tau(\gamma_k^z)^{-1}$ is trivial by definition, and does not need to be tested for membership of K.)

The displayed function VERIFY is a summary of the algorithm. Let us prove its correctness.

THEOREM 6.2 *With the notation and assumptions of this subsection, if* VERIFY *returns* true, *then* $H_\alpha = K$ *and* $\varphi : \hat{H} = \langle \hat{X} \mid R_1 \cup R_2 \cup R_3 \rangle \to H$ *is an isomorphism.*

PROOF From the definitions of R_2 and R_3 we have $\varphi(w) = 1_H$ for all $w \in R_2 \cup R_3$, and this is true by assumption for $w \in R_1$, so $\varphi : \hat{H} \to H$ is certainly an epimorphism.

Let $n := |\Delta|$ and let $\hat{K}$ be the subgroup of $\hat{H}$ generated by $\hat{Y}$. If we can prove that $|\hat{H} : \hat{K}| \leq n$ then, since $\varphi(\hat{K}) = K$, we will have $|H : K| \leq n$. But $K \leq H_\alpha$ by assumption, and $|H : H_\alpha| = n$ by the orbit-stabilizer theorem, so we will then have $H_\alpha = K$ with $|H : K| = n$. Since $|\hat{K}| \leq |K|$ by assumption, $|\hat{H} : \hat{K}| \leq n$ will also imply $|\hat{H}| \leq |H|$ and so φ will then be an isomorphism with $|\hat{H}| = |H|$. So it suffices to prove that $|\hat{H} : \hat{K}| \leq n$.

For $\gamma \in \Delta$, let $\hat{\tau}(\gamma)$ be the inverse image in $\hat{H}$ under φ of $\tau(\gamma)$; that is, $\hat{\tau}(\gamma)$ is the word in $\hat{A}^*$ corresponding to the word in A^* for $\tau(\gamma)$. Let $\hat{H}_0$ be the union of the n cosets $\hat{K}\hat{\tau}(\gamma)$ ($\gamma \in \Delta$) of $\hat{K}$ in $\hat{H}$. It is enough to prove that $\hat{H}_0 = \hat{H}$ and, to do that, it is obviously enough to prove that $uw \in \hat{H}_0$ for all $u \in \hat{H}_0$ and $w \in \hat{H}$. By induction on $|w|$, it suffices to do this for $w \in \hat{A}$. But if this is true for some $w \in \hat{X}$ then, since there are only finitely many cosets in $\hat{H}_0$, the action on $\hat{H}_0$ by right multiplication by w must permute these cosets, and so the same will be true for w^{-1}. Hence it suffices to show that $uw \in \hat{H}_0$ for all $u \in \hat{H}_0$ and $w = \hat{x} \in \hat{X}$.

First suppose that $\hat{x}$ is in the set $\hat{Y}$ of generators of $\hat{K}$. From the way that we defined the elements $\tau(\gamma)$ as $\tau(\beta_i)v$ for a word $v \in B^*$ with $\beta_i^v = \gamma$, we have $u = u_0\hat{\tau}(\beta_i)\hat{v}$ with $u_0, \hat{v} \in \hat{B}^*$ and $\varphi(\hat{v}) = v$. Since $\gamma^x = \beta_i^{vx} \in \beta_i^K$, we have $\tau(\gamma^x) = \tau(\beta_i)v_1$ for some $v_1 \in B^*$ with $\beta_i^{v_1} = \gamma^x$. Hence $vx =_H v_0v_1$, with $v_0 \in K_{\beta_i}$.

We are assuming that our relators of $\hat{H}$ contain defining relators for K, and so we have the the corresponding equality $\hat{v}\hat{x} =_{\hat{H}} \hat{v}_0\hat{v}_1$ in $\hat{H}$, where $\varphi(\hat{v}_0) = v_0$ and $\varphi(\hat{v}_1) = v_1$.

Now $v_0 \in K_{\beta_i}$ can be written as a word in the generators $y_1, \ldots, y_t$ of K_{β_i} and their inverses that we computed in the algorithm. It follows once again from the assumption that $\hat{H}$ contains defining relators for K, that $\hat{v}_0$ is equal in $\hat{H}$ to the corresponding word in the elements $\hat{y}_j$ and their inverses with $\varphi(\hat{y}_j) = y_j$.

But for each such $\hat{y}_j$, there is a relator in R_2 of the form $\hat{\tau}(\beta_i)\hat{y}_j\hat{\tau}(\beta_i)^{-1}u_1$ with $u_1 \in \hat{B}^*$, and so $\hat{\tau}(\beta_i)\hat{y}_j =_{\hat{H}} u_1^{-1}\hat{\tau}(\beta_i)$ and $\hat{\tau}(\beta_i)\hat{y}_j^{-1} =_{\hat{H}} u_1\hat{\tau}(\beta_i)$. We can use these relations to write $\hat{\tau}(\beta_i)\hat{v}_0$ in $\hat{H}$ as $u_2\hat{\tau}(\beta_i)$ with $u_2 \in \hat{B}^*$, and we have now expressed $u\hat{x} =_{\hat{H}} u_0\hat{\tau}(\beta_i)\hat{v}_0\hat{v}_1$ as $u_0u_2\hat{\tau}(\beta_i)\hat{v}_1 =_{\hat{H}} u_0u_2\hat{\tau}(\beta_i^{vx}) \in \hat{H}_0$, as claimed.

The other case is when $\hat{x} = \hat{z}$. As in the previous case, let $\tau(\gamma) = \tau(\beta_i)v$ with $v \in B^*$. Let γ_k be the orbit representative of γ^{K_β}, and let $y_1, \ldots, y_t$ be the generators of $K_\beta = K_{\beta_1}$ that we used in the definition of R_2. Then $\gamma = \gamma_k^{v_0}$ for some word v_0 in the generators $y_1, \ldots, y_t$ and their inverses. Since $\gamma_k \in \beta_i^K$, we have $\tau(\gamma_k) = \tau(\beta_i)v_1$ with $v_1 \in B^*$ and $v_1v_0 =_K v$. We can once again use our assumption that $\hat{H}$ contains defining relators for K to infer that $\hat{\tau}(\gamma) =_{\hat{H}} \hat{\tau}(\gamma_k)\hat{v}_0$, where $\hat{v}_0$ is a word over $\hat{y}_1, \ldots, \hat{y}_t$ with $\phi(\hat{v}_0) = v_0$.

Since $\tau(\beta)$ was chosen to be z^{-1}, there are relations in R_2 of the form $\hat{z}^{-1}\hat{y}_j\hat{z}u_1$ for each such $\hat{y}_j$, and so we can use these to write $\hat{v}_0\hat{z} =_{\hat{H}} \hat{z}u_1$ with $u_1 \in \hat{B}^*$. Then we can use a relator in R_3 to write $\hat{\tau}(\gamma_k)\hat{z} =_{\hat{H}} u_2\hat{\tau}(\delta)$ for some $u_2 \in \hat{B}^*$ and $\delta \in \Delta$ and so, putting all of this together, we have $u_0\hat{\tau}(\gamma)\hat{z} =_{\hat{H}} u_0u_2\hat{\tau}(\delta)u_1$ with $u_0, u_1, u_2 \in \hat{B}^*$. It follows now from the previous case, $\hat{x} \in \hat{Y}$, that this element lies in $\hat{H}_0$. This completes the proof. ∎

Example 6.3

Let $H \leq \mathrm{Sym}(10)$ be generated by the permutations

$$a := (3,4)(5,8)(6,7)(9,10),$$
$$b := (2,5,8)(3,7,9)(4,10,6),$$
$$c := (1,2)(3,4)(5,6)(7,8),$$

let $K := \langle a, b \rangle$, and let $\alpha := 1$. Then $\Delta := [1 \,.\, 10]$. The generator c corresponds to z in the algorithm but, at the risk of causing confusion, we shall use x, y, and z as generators of our finitely presented group mapping on to a, b, and c, respectively. In fact $H \cong \mathrm{Alt}(5)$.

We see easily that K is dihedral of order 6, so we take our initial set of relators, which is used as input for VERIFY and is assumed to be a set of defining relators for K, as $R_3 := \{ x^2, y^3, (xy)^2 \}$. Now we let $\beta := 1^{c^{-1}} = 2$. Then K_β has the single generator a, and we can choose $1, 2, 3, 5, 6, 9$ as our orbit representatives γ_k of K_β. We have $\beta_1^K = \{2, 5, 8\}$, we choose $\beta_2 := 5^c = 6$, and find that $\beta_2^K = \{6, 7, 4, 10, 9, 3\}$, so α, β_1, β_2 are orbit representatives of K on Δ. Following the method described above, we choose

$$\tau(2) := c^{-1},\ \tau(5) := c^{-1}b,\ \tau(8) := c^{-1}b^{-1},\ \tau(6) := c^{-1}bc,\ \tau(7) := c^{-1}bca,$$

$$\tau(4) := c^{-1}bcb,\ \tau(10) := c^{-1}bcb^{-1},\ \tau(9) := c^{-1}bcab,\ \tau(3) := c^{-1}bcab^{-1}.$$

Since K_{β_1} has a single generator a and $K_{\beta_2} = 1$, we get only one relator in R_2. To calculate this, first we check whether $c^{-1}ac \in K$, and find that this is true, with $c^{-1}ac = a$, so we put $\varphi^{-1}(c^{-1}aca^{-1}) = z^{-1}xzx^{-1}$ into R_2.

The relators in R_3 correspond to the γ_k. From $\gamma_k = 1$, we get $z^2 \in R_3$, whereas the Schreier generators coming from $\gamma_k = 2$ and 5 are trivial by definition. When $\gamma_2 = 6$, we get the relator $z^{-1}yz^2y^{-1}z$. If we use the fact that z^2 is already a relator, then this becomes trivial; in fact, in general, when the generator z has order 2, the relators coming from $\gamma_k = \beta_i$ for $i > 1$ will collapse in this way, and this feature could be used in an implementation. For $\gamma_k = 9$ and 3 we get nontrivial new relators. In the first case, $9^c = 9$, so the Schreier generator is

$$\tau(9)c\tau(9)^{-1} = c^{-1}bcabcb^{-1}a^{-1}c^{-1}b^{-1}c = (2,8)(3,6)(4,9)(7,10) = a^{-1}b^{-1}$$

so we append $z^{-1}yzxyzy^{-1}x^{-1}z^{-1}y^{-1}zyx$ to R_3. Similarly, for $\gamma_k = 3$ we append $z^{-1}yzxy^{-1}zy^{-1}z^{-1}y^{-1}zxy$ to R_3, yielding the complete presentation

$$\langle x, y, z \mid x^2,\ y^3,\ (xy)^2,\ z^{-1}xzx^{-1},\ z^2,$$
$$z^{-1}yzxyzy^{-1}x^{-1}z^{-1}y^{-1}zyx,\ z^{-1}yzxy^{-1}zy^{-1}z^{-1}y^{-1}zxy \rangle.$$

Surprisingly, neither of the two long relators is redundant (omitting either of them results in a group of order 120), but z^2 and $zxz^{-1}x^{-1}$ are both redundant, and both can be removed from the presentation. TCSS produces a presentation with the same number of relators, but with the final two relators slightly shorter. $\square$

6.3.2 The general case

In the general case, we have $Y = [x_1, \ldots, x_t]$, $X = [x_1, \ldots, x_r]$ with $r - t > 1$. The algorithm in this case reduces to $t - r$ applications of the single-generator case and, for the most part, we shall just describe how this is done, without going into technical details. There is one step in the algorithm, however, in which we need to test whether a subset of Δ is a block, and this will require more detailed explanation.

For $t \le s \le r$, define $K(s) := \langle x_1, \ldots, x_s \rangle$; in particular, $K(t) = K$ and $K(r) = H$. For $s := t, t+1, \ldots, r-1$ in turn, we test whether $K(s+1)_\alpha = K$. If so, then we also extend our set of defining relators $R(s)$ of $K(s)$ on $[x_1, \ldots, x_s]$ to a set of defining relators $R(s+1)$ of $K(s+1)$ on $[x_1, \ldots, x_{s+1}]$. We are assuming here that the set $R(t)$ is given.

If the test fails for some s, then we output an element of $K(s+1)_\alpha \setminus K$, which must be appended to the strong generating set of G. Whenever this happens, we abort and restart the complete verification procedure on G, so it is important only to apply this procedure only when we are confident that our candidate BSGS really is a BSGS for G.

Let $\Delta(s) := \alpha^{K(s)}$ for $t \leq s \leq r$; so $\Delta(t) = \{\alpha\}$ and $\Delta(r) = \Delta$. Suppose, for some $s \geq t$, that we have successfully checked that $K(j)_\alpha = K$ and calculated the relators $R(j)$ for all j with $t \leq j \leq s$. This assumption is valid for $s = t$, because $K(t) = K \leq H_\alpha$. Let $L := K(s)$. We shall now describe how to carry out the test that $K(s+1)_\alpha = K$.

Suppose first that $\Delta(s+1) = \Delta(s)$. Let $\beta := \alpha^{x_{s+1}}$. Then there exists $\tau(\beta) \in L$ with $\alpha^{\tau(\beta)} = \beta$, and hence $x_{s+1}\tau(\beta)^{-1} \in K(s+1)_\alpha$. We test $x_{s+1}\tau(\beta)^{-1}$ for membership of K by using STRIP in the usual way. If this test fails, then the word output by STRIP is an element of $K(s+1)_\alpha \setminus K$ to be appended to the SGS of G. If it succeeds, then we have proved that $K(s+1) = L$, and hence that $K(s+1)_\alpha = K$. We can then write $x_{s+1}\tau(\beta)^{-1}$ as a word u in the generators of K, and we adjoin the corresponding relator $\varphi^{-1}(x_{s+1}\tau(\beta)^{-1}u^{-1})$ to $R(s+1)$. Since this relator expresses $\hat{x}_{s+1}$ in terms of the lower-numbered generators, it provides us with the required presentation of $K(s+1) = K(s)$.

So we can suppose that $\Delta(s)$ is properly contained in $\Delta(s+1)$. If $\Delta(s) = [\alpha]$, which is certainly the case when $s = t$, then we just apply the single-generator case of VERIFY to the action of $K(s+1)$ on $\Delta(s+1)$, and this yields the required result for $K(s+1)$.

Otherwise, $\{\alpha\} \subset \Delta(s)$ and $L_\alpha < L$. So, if the condition $K(s+1)_\alpha = K$ that we are trying to verify holds, then $K(s+1)_\alpha < L$ and, by Proposition 2.30, $\alpha^L = \Delta(s)$ is a block for the action of $K(s+1)$ on $\Delta(s+1)$.

The verification proceeds in two stages. In the first stage, we check whether $\Delta(s)$ really is a block for $K(s+1)$ on $\Delta(s+1)$. We shall describe the details of that check shortly. First, we shall describe the second and easier stage of the verification.

So suppose that we have checked that $\Delta(s)$ is a block for $K(s+1)$ on $\Delta(s+1)$, and that we have calculated the induced action $K(s+1)^\Sigma$ of $K(s+1)$ on the block system Σ containing $\Delta(s)$. Then, in this induced action, $L^\Sigma \leq K(s+1)^\Sigma_{\Delta(s)}$ and we apply the single-generator case of VERIFY to $K(s+1)^\Sigma$ to test whether $K(s+1)_{\Delta(s)} = L$. If so then, since $K(s+1)_\alpha$ certainly fixes the block $\Delta(s)$ and we already know that $L_\alpha = K$, we have $K(s+1)_\alpha = K$, as required, and the relators $R(s+1)$ computed by VERIFY will define a presentation of $K(s+1)$. If not, then we will find a new element of the strong generating set of G.

Note that, although we are applying VERIFY to an induced action on a block system Σ, at the places in the algorithm where we need to test an element for membership of L and, if possible, write it as a word in the generators of L, we carry out our computations in $K(s+1)$ rather than in its induced action $K(s+1)^\Sigma$. This can be done using standard homomorphism machinery, and presents no problem. These computations can be facilitated by replacing Ω

by $\Omega \cup \Sigma$, which enables the action on the block system to be handled in the same way as an action on an orbit, as is done in the algorithm displayed above for the one generator case.

Now we turn to the problem of checking whether $\Delta(s)$ is a block for the action of $K(s+1)$ on $\Delta(s+1)$. Of course, if we are in the situation where we know already that (B, S) is a BSGS for G, and we just want to compute a presentation, then we know already that $\Delta(s)$ is a block. So MINIMALBLOCK can be used to compute the complete block system, and then the method described in Subsection 4.5.2 can be used to find the induced action.

In fact MINIMALBLOCK can also be used to test whether $\Delta(s)$ is a block; but if it should turn out not to be a block, then we need to calculate an element of $K(s+1)_\alpha \setminus K$, which MINIMALBLOCK does not do as it stands. So we shall write a new function to accomplish this.

Since $\Delta(s) = \alpha^L$, the blocks of Σ (if it exists) are all of the form α^{Lw} for elements $w \in K(s+1)$. We proceed by constructing these blocks and calculating the induced action of the generators of $K(s+1)$ on the block system at the same time.

For each point γ in each block, we define a word u_γ over the generators of $K(s+1)$ with $\alpha^{u_\gamma} = \gamma$ and, as usual, these words u_γ can either be stored explicitly or by means of a Schreier vector. We choose these words u_γ so that they satisfy the following property: the block containing γ is α^{Lu_γ} and, if γ and δ are points in the same block, then $Lu_\gamma = Lu_\delta$.

We start by defining u_γ as words over the generators of L for all γ in the first block $\Delta(s)$. At this stage, the property above is clearly satisfied by all u_γ that have been defined so far.

In the main body of the algorithm, we consider each block Γ that has been defined so far in turn; so we start with $\Gamma = \Delta(s)$. Let β be the first point in Γ. Then, for each generator x_i of $K(s+1)$, we check whether $\delta := \beta^{x_i}$ is in a block that has been defined already. If not, then δ becomes the first point in a new block, and we put $u_\delta := u_\beta x_i$. In any case, we then apply x_i to all other points $\gamma \in \Gamma$.

If we defined a new block, and none of these γ^{x_i} are in an existing block, then they become the other points in the new block, and we define $u_\delta := u_\gamma x_i$ for each such $\delta := \gamma^{x_i}$. We may assume inductively that the property defined above for the words u_γ is satisfied for points γ in the existing blocks, and so we know that the block Γ containing β is α^{Lu_β}, and that $Lu_\gamma = Lu_\beta$ for all $\gamma \in \Gamma$. So the property is also satisfied by the words u_δ in the new block.

If we did not define a new block containing β^{x_i}, then the points γ^{x_i} should all be in the same existing block as β^{x_i}.

There are two ways in which it could emerge that $\Delta(s)$ is not a block. Let β be the first point and γ some other point in the block Γ to which we are applying the generator x_i. We might define a new block containing β^{x_i} but then find that γ^{x_i} is in an existing block. Alternatively, we might find that β^{x_i} is in an existing block, but that γ^{x_i} is either in no existing block or is in a different existing block. If neither of these eventualities occurs at any stage,

BLOCKVERIFY(H, L, α)

    Input: $L = \langle x_1, \ldots, x_s \rangle < H = \langle x_1, \ldots, x_r \rangle \leq \mathrm{Sym}(\Omega)$, $\alpha \in \Omega$

    Output: true if α^L is a block of H^{α^H}, false if not.

                If true, then induced actions $\bar{x}_i$ of x_i on blocks.

                If false, then an element of $H_\alpha \setminus L$.

```
 1    Δ := α^H;
 2    for β ∈ Δ do p[β] := 0;
 3    B[1] := ORBIT(α, [x₁, ..., xₛ]);
 4    for β ∈ B[1] do p[β] := 1;
 5    for β ∈ B[1] do define u_β ∈ L with α^{u_β} = β;
 6    ρ := 1;  m := 1;
 7    while ρ ≤ m
 8        do β := B[ρ][1];
 9            for i ∈ [1..r]
10                do δ := β^{xᵢ};  σ := p[δ];
11                    if σ = 0
12                        then (* New block *)
13                            m := m+1;  σ := m;  u_δ := u_β xᵢ;
14                            p[δ] := σ;  B[σ] := [δ];
15                            for j ∈ [2..|B[ρ]|]
16                                do γ := B[ρ][j];  δ := γ^{xᵢ};
17                                    if p[δ] = 0
18                                        then u_δ := u_γ xᵢ;  p[δ] := σ;
19                                            APPEND(∼B[σ], δ);
20                                        else return false, u_γ xᵢ u_δ^{-1};
21                        else (* Image in existing block *)
22                            for j ∈ [2..|B[ρ]|]
23                                do if p[B[ρ][j]^{xᵢ}] ≠ σ
24                                    then return false, u_β xᵢ u_δ^{-1};
25                    ρ^{x̄ᵢ} := σ;
26    return true, [x̄₁, ..., x̄ₖ];
```

then we will successfully compute the complete block system Σ containing $\Delta(s)$ together with the induced actions of the x_i on Σ, which will verify that $\Delta(s)$ is a block.

In the two negative situations above, we need to compute an element of $K(s+1)_\alpha \setminus K$. Suppose first that β^{x_i} is in a new block, but $\delta := \gamma^{x_i}$ is in an existing block, where β and γ are in the same existing block Γ. Then $u_\gamma x_i u_\delta^{-1} \in K(s+1)_\alpha$. By our assumed property of the words u_γ, $\alpha^{L u_\delta}$ is the existing block that contains δ, so $\beta^{x_i} = \alpha^{u_\beta x_i} \notin \alpha^{L u_\delta}$. Hence $u_\beta x_i \notin L u_\delta$. By the assumed property again, we have $L u_\beta = L u_\gamma$, so $u_\gamma x_i \notin L u_\delta$ and $u_\gamma x_i u_\delta^{-1} \notin L$. Hence we can return $u_\gamma x_i u_\delta^{-1}$ as our element in $K(s+1)_\alpha \setminus K$.

By a similar, argument, in the second situation, where $\delta := \beta^{x_i}$ is in an existing block but some γ^{x_i} is not in that block, we can return $u_\beta x_i u_\delta^{-1}$.

The algorithm is displayed as BLOCKVERIFY, which we apply with $L :=$ $K(s)$, $H := K(s{+}1)$ and $r = s{+}1$. In the code, the blocks being constructed are $B[1], \ldots, B[m]$, where m is the current number of blocks, $p[\beta]$ is the number of the block containing $\beta \in \Delta$, and ρ is the number of the block to which we are currently applying the generators x_i.

6.3.3 Examples

We shall now describe some statistical details of applications of VERIFY to two large permutation groups. In the first example, we also compare the performances of VERIFY and TCSS.

Example 6.4

Let G be the O'Nan sporadic simple group of order $460\,815\,505\,920$ as a permutation group of degree $122\,760$ on $\Omega := [1 \mathbin{..} 122\,760]$, let $K = G_\alpha$ with $\alpha = 1$ be the point stabilizer, and suppose that we are applying VERIFY to check that G_α really is equal to our current candidate subgroup $K \leq G_\alpha$.

Then, assuming that $K = G_\alpha$ as suspected, K has five orbits on $\Omega \setminus \{\alpha\}$ of lengths 5586, 6384, $52\,136$, $58\,653$. The corresponding two-point stabilizers have orders 672, 588, 72, 64, and the number of their orbits on Ω is 252, 261, 1779, 2076, respectively. The first three of these two-point stabilizers are two-generator groups, and the fourth one is a three-generator group, but RANDOMSTAB will frequently output three rather than two generators in the first three cases. So, when applying the single extra generator version of VERIFY listed above in Subsection 6.3.1, there will be about 12 elements in R_2, and either 252, 261, 1779, or 2076 elements in R_3. For each word in $R_2 \cup R_3$, an element of H_α is checked for membership in K. So, if $\alpha^{z^{-1}}$ is in the orbit of G_α of length 5586, then the algorithm will complete about 8 times more quickly than if it is in the orbit of length $58\,653$.

We found that TCSS took between 40 and 100 times longer than VERIFY to verify that $G_\alpha = H$. The set R_1 of relators computed by TCSS consisted of 272 relators of total length 8160 (this varied slightly from run to run), which is similar to that produced by VERIFY when $\alpha^{z^{-1}}$ is in a short orbit of G_α.

However, if we choose our second base point to be in the shortest orbit of G_α, of length 5586, then the verification at the second level of the stabilizer-chain was about three times faster with TCSS than with VERIFY. With the second base point in one of the two longer orbits of G_α, TCSS was faster by a much larger factor. This is generally because the stabilizers of three points in G are small and have very large numbers of orbits.

So, the optimal policy for the verification process on G is to use VERIFY for the first level and TCSS for the other levels of the stabilizer-chain. □

Example 6.5

Let G be the Lyons simple group of order $51\,765\,179\,004\,000\,000$ and degree $8\,835\,156$. The orbit lengths of $K := G_\alpha$ are $19\,530$, $968\,750$, $2\,034\,375$, and $5\,812\,500$. The associated two-point stabilizers have orders $300\,000$, 6048, 2880, 1008, and have 66, 1613, 3366, and 9127 orbits. All of the two-point stabilizers are 2-generator groups. $\quad\Box$

Unfortunately, if z is a random element of $G \setminus G_\alpha$ in Example 6.5, then $\alpha^{z^{-1}}$ is about 10 times more likely to be in one of the two longer orbits of G_α, so there appears to be a significant weakness in the default algorithm.

Fortunately, there is a remedy. We simply choose a new element z' with $\alpha^{z'}$ in the shortest orbit of G_α on $\Delta \setminus \{\alpha\}$, where $\Delta = \alpha^G$, and use that in place of z. In Example 6.5, G is primitive, so we would have $G = \langle K, z' \rangle$ but, in general, $\langle K, z' \rangle$ might not be transitive on Δ.

In that case, we proceed as in Subsection 6.3.2, and apply the single extra generator case of VERIFY u times, until we have $G = \langle K, z_1', \ldots, z_u' \rangle$, where each of the generators z_i' is selected to make the image of α lie in a shortest orbit. If each such application is successful, then we finish by applying STRIP to the original generators z_i in $G \setminus H$ to check that they can be rewritten as words in the new generators.

One minor disadvantage of this approach is that the choice of the new z_i' tends to result in a larger number of strong generators of G being required than there were initially. But experience shows that this disadvantage is by far outweighed on average by the gain in speed that comes from having fewer orbits of the two-point stabilizers. In any case, many of the longest and most difficult applications are to primitive groups, where only a single new generator is required.

There are some other housekeeping details that need to be taken care of with this approach, but these are routine. If, as is usually the case, we are computing a strong presentation of G for use with homomorphism algorithms, then we will need to store words for the new generators z_i' in the existing generators, so that they can be stored as SLP elements; see Subsection 3.4.1. If the original extra generators z_i were defining generators of G, then we will not want to discard them from the presentation, in which case we will also need to include relations that express these old generators as words in the new generators as part of the strong presentation of G.

Exercise

1. In the single-generator case of the VERIFY algorithm, when we need to choose a new orbit representative β_{i+1} $(i \geq 1)$ of K on Δ, prove that there must exist an orbit representative γ_k of K_β such that $\gamma_k \in \beta_j^K$ for some j with $1 \leq j \leq i$, and $\gamma_k^z \notin \cup_{0 \leq j \leq i} \beta_j^K$.

Chapter 7

Representation Theory, Cohomology, and Characters

This substantial chapter is devoted to computations involving representations and characters of finite groups, and computation within finite matrix groups. Apart from the computation of character tables, we shall be concerned almost entirely with representations over finite fields. This is mainly because the methods are much more highly developed for this case. Computing with representations over fields of characteristic zero, and over algebraic number fields in particular, is currently an active area of research, but there a number of formidable problems that do not arise over finite fields, and so this topic must unfortunately remain beyond the scope of this book.

After a short section devoted to computing in finite fields, we shall review some fundamental algorithms for basic linear algebra. It turns out that many of the most important techniques in computational representation theory over finite fields, including the fundamental 'Meataxe' algorithm, rely almost entirely on linear algebra, and involve very little knowledge of group representation theory. (A complexity analysis of these methods, which we shall not attempt to carry out here, does, however, require more theory.)

Later in the chapter, we shall discuss how to carry out some basic cohomological computations for group modules over finite fields. The Meataxe-based methods and the cohomological computations are important not just in their own right, but also as components of more advanced algorithms for structural computations in finite groups. For example, as we shall see in Sections 8.5 and 10.2, the Meataxe is used in the computation of the elementary abelian layers in chief series of finite groups, and in Section 10.4 we shall study an application of cohomological methods to the computation of the subgroups of a finite group.

The final section of the chapter, on computation in matrix groups over finite fields, will explain how the base and strong generator set methods introduced for permutation groups in Section 4.4 can be extended to matrix groups. This section will also discuss and provide references for further reading on recent work in this area, which is one of the most lively areas of current research in CGT.

7.1 Computation in finite fields

In general, if F is some field in which we can compute effectively, and $K :=$ $F(\alpha)$ is a simple algebraic extension of degree n over F for which the minimal polynomial f of α over F is known, then we can also compute effectively within K. Elements can be represented as polynomials of degree at most $n-1$ over F, and then addition is straightforward, multiplication is done modulo f, and multiplicative inverses can be computed by using the Euclidean algorithm.

This approach can be used for computing in finite extensions of $\mathbb{Q}$ (that is, in algebraic number fields) and in finite fields, which can all be defined as simple extensions of a suitable prime field $\mathbb{F}_p$.

It has a disadvantage when used for finite fields however. As we saw in Subsection 2.8.2, the multiplicative group of the finite field $\mathbb{F}_q$ of order q is cyclic, and is generated by a primitive element α. Let us assume that we can factorize $q-1$ easily. (It is well-known that factorizing very large integers is not easy; indeed modern cryptography systems are based on that fact. However, many factorizations of $q-1$ for large prime powers p^n have been precomputed and the results stored by computer algebra systems, so this assumption is not unreasonable.) Since the multiplicative order of any $\alpha \in \mathbb{F}_q^\#$ divides $q-1$, we can use the function ORDERBOUNDED displayed in Subsection 3.3.1 to compute the multiplicative order of g. Because a reasonably high proportion $(\Phi(q-1)/(q-1))$ of elements of $\mathbb{F}_q^\#$ are primitive elements, we can expect to find a primitive element α quickly by trying random elements of $\mathbb{F}_q^\#$.

However, the problem of expressing an arbitrary $\beta \in \mathbb{F}_q^\#$ as a power α^n of our primitive element α, which is known as the *discrete log problem*, is a notoriously difficult one, and is the main stumbling block for many computations involving large finite fields. See [SWD96] for details of recent methods for solving this problem.

The systems **GAP** and MAGMA avoid the discrete log problem for smaller fields by representing and storing field elements as powers of a primitive element α, rather than as polynomials. The element α is chosen to be a root of the *Conway polynomial* (see Subsection 2.8.3) for the particular $\mathbb{F}_q$. With this method of representing field elements, we have the problem that addition of field elements is not straightforward. But, since $\alpha^i + \alpha^j = \alpha^i(1 + \alpha^{j-i})$ when $j \geq i$, this problem reduces immediately to evaluating $1 + \alpha^i$ for $0 \leq i < q$ as a power of α. For reasonably sized q (up to about 10^6), it is feasible to precompute and store all of these values in lookup tables. These are known as *Zech logarithms*.

For larger finite fields, for which no such tables are stored, it is necessary to use the polynomial representation for field elements, and to put up with the difficulty of solving the discrete log problem.

For more details on computation in finite fields, the reader may consult the book by Shparlinski92 [Shp92].

Exercise

1. The Conway polynomials for the finite fields of orders 4, 8, and 9 are, respectively, $x^2 + x + 1$, $x^3 + x + 1$ and $x^2 + 2x + 2$. Calculate the tables of Zech logarithms for these fields with respect to the primitive elements defined by these polynomials.

7.2 Elementary computational linear algebra

In this section, we present straightforward procedures for performing the basic operations of linear algebra over a field K. This is intended to provide ourselves with the required functionality for use in later sections in this chapter on modules over algebras.

We are mainly interested in finite fields K in this chapter, but the basic algorithms are for the most part independent of K. There are, however, a number of problems involved in calculations over infinite fields, which do not arise for finite fields. For exact computations over the rationals or over an algebraic number field, there is the problem that the integers involved can grow very large and unwieldy in the course of straightforward manipulations. For calculations in the real or complex field on the other hand, there is the problem of floating point error.

The algorithms presented here will be straightforward, and will not be concerned with finer points of computational efficiency, but a few remarks on such issues are in order.

The systems MAGMA and GAP both include efficient implementations of basic linear algebra over finite fields. We mention also the stand-alone package [Rin92], which was implemented mainly by M. Ringe and is based on an earlier FORTRAN implementation by Parker (see [Par84]). A fundamental objective of this package is to manipulate very large matrices using as little core memory as possible, so matrices are stored in files and read in only when they are needed.

Another feature of the package, which is also present in MAGMA and GAP, is that matrix entries are packed as efficiently as possible into machine words; so, for example, a vector or matrix over the field of order 2 is stored using a single bit for each each entry. Furthermore, multiplication of matrices is speeded up by using moderately large stored lookup tables for the scalar products of all pairs of vectors up to some length that depends on $|K|$.

Divide-and-conquer methods for matrix multiplication, as introduced originally by Strassen, which reduce the complexity of this and other fundamental matrix manipulations from the standard $O(n^3)$ field operations, are used in the MAGMA implementation. See, for example, the paper by Coppersmith and

Winograd [CW90]. Experiments show that these methods lead to significant gains in running times for about $n \geq 100$.

To simplify our pseudocode, we shall adopt a number of conventions that would require more careful handling in a real implementation. Any sequence $[a_1, a_2, \ldots, a_d]$ of length d containing elements of the field K can be regarded as a vector $(a_1, a_2, \ldots, a_d) \in K^d$, and any sequence of c such sequences, all of the same length d, can be regarded as a $c \times d$ matrix over K. We allow the empty sequence to represent a vector or matrix over any field. We use numbers like 0, 1, 2, to represent the corresponding elements of any field K; so, we rarely need to refer explicitly to K in the pseudocode. We use the symbol $\cdot$ to represent vector or matrix multiplication. As explained earlier, in Section 2.2, we regard matrices as acting on vectors from the right. So, if $v \in K^d$, and A is a $d \times c$ matrix, then v^A is the same as $v \cdot A$. We denote the length of a vector v by $|v|$; so $|v| = d$ for $v \in K^d$.

We shall not write out code for straightforward manipulations such as matrix multiplication, but the reader should bear in mind the above remarks about how these operations might be implemented in practice.

For $v := [a_1, \ldots, a_d] \in K^d$, we denote the subsequence $[a_c, a_{c+1}, \ldots, a_{c+e}]$ in K^{e+1} by $v[c \mathinner{.\,.} c+e]$. Similarly, for a matrix α, $\alpha[c \mathinner{.\,.} c+e][d \mathinner{.\,.} d+f]$ represents the $(e+1) \times (f+1)$ submatrix with top-left entry $\alpha[c][d]$. For $c \times d$ and $c \times e$ matrices α, β, HORIZONTALJOIN(α, β) represents the $c \times (d+e)$ matrix formed by writing β to the right of α, and VERTICALJOIN(α, β) is defined similarly for $c \times e$ and $d \times e$ matrices α, β.

For $c \times d$ and $e \times f$ matrices α, β, DIRECT-SUM(α, β), or equivalently $\alpha \oplus \beta$, is defined to be the $(c + e) \times (d + f)$ matrix γ with $\gamma[1 \mathinner{.\,.} c][1 \mathinner{.\,.} d] = \alpha$, $\gamma[c+1 \mathinner{.\,.} c+e][d+1 \mathinner{.\,.} d+f] = \beta$ and all other entries zero. These three functions are all associative, and can also be applied to lists of matrices of compatible dimensions. The transpose of a matrix α is written as α^{T}.

We shall store a subspace W of $V = K^d$ as a record $\mathcal{W}$ with two components, β and γ, with the following properties:

(i) β is a basis of W;

(ii) the leading nonzero coefficient of each vector in β is 1;

(iii) the leading nonzero coefficients of the vectors in β all occur in different positions;

(iv) γ is a list of these positions.

For example, $\beta := [[0, 0, 1, 2, 0], [1, 1, 2, 0, -1], [0, 0, 0, 0, 1]]$, $\gamma := [3, 1, 5]$ satisfies these conditions for a subspace of dimension 3 in K^5. Notice that by suitably permuting the vectors in β we could get an echelonized matrix, but this is not usually necessary in our applications.

Our first function locates the leading nonzero coefficient a of a vector and returns its position and a itself, or $0, 0$ if the vector is zero.

DEPTHVECTOR(v)
1   **for** $i \in [1 .. |v|]$ **do if** $v[i] \neq 0$ **then return** $i, v[i]$;
2   **return** 0,0;

The next function, VECTORINSUBSPACE tests whether a vector $v \in K^d$ lies in a subspace W with associated record $\mathcal{W}$; it returns **true** or **false**, accordingly. If the third argument 'add' is **true** and $v \notin W$, then W is enlarged to $\langle W, v \rangle$ by adjoining an extra vector to β. If $v \in W$ or 'add' is **true**, then the function also returns a list c of coefficients, which satisfies

$$v = \sum_{i=1}^{|\beta|} c[i]\beta[i],$$

where β is the (possibly extended) basis sequence of $\mathcal{W}$.

VECTORINSUBSPACE($\sim \mathcal{W}, v$,add)
1     $\beta := \mathcal{V}.\beta$; $\gamma := \mathcal{V}.\gamma$;
2     $c := []$;
3   **for** $i \in [1 .. |\gamma|]$
4   **do** $a := v[\gamma[i]]$;   APPEND($\sim c, a$);
5   **if** $a \neq 0$ **then** $v := v - a\beta[i]$;
6     $d, a :=$ DEPTHVECTOR(v);
7   **if** $d = 0$ **then return true**, c;
8   **if** add
9   **then** APPEND($\sim \mathcal{V}.\beta, a^{-1}v$);   APPEND($\sim \mathcal{V}.\gamma, d$);   APPEND($\sim c, a$);
10   **return false**, c;
11   **return false**;

For completeness, we have written out code for inverting a matrix. This follows the well-known method of reducing the matrix to the identity using elementary row operations, while simultaneously performing the identical row operations, but starting with the identity matrix.

Next, we turn to the calculation of the nullspace of a $d \times c$ matrix A; that is, the subspace of K^d consisting of those $v \in K^d$ for which $v \cdot A = 0$. The method involves performing elementary column operations on A, which do not change the nullspace of A. But, since we prefer to use row operations in our code, we replace A by the $c \times d$ matrix $B := A^{\mathrm{T}}$, and then calculate the space of $v \in K^d$ such that $B \cdot v^{\mathrm{T}} = 0$.

We do this by transforming B to reduced echelon form, while maintaining a list l of the positions in which the leading nonzero entries of the rows of B occur. For example, if the echelonized form C of B is

$$\begin{pmatrix} 1 & 2 & 0 & 1 & 0 & 1 \\ 0 & 0 & 1 & 2 & 0 & 2 \\ 0 & 0 & 0 & 0 & 1 & 2 \end{pmatrix},$$

then $l = [1, 3, 5]$.

INVERTMATRIX(A)

    Input: An invertible square matrix A
    Output: A^{-1}
1 $d := |A[1]|$;
2 $I :=$ IDENTITY-MATRIX(FIELD(A), d);
3 **for** $c \in [1..d]$
4   **do for** $s \in [c..d]$
5   **do if** $A[s][c] \neq 0$
6   **then** $I[s] := A[s][c]^{-1} \cdot I[s]$; $A[s] := A[s][c]^{-1} \cdot A[s]$;
7   **for** $t \in [1..c-1]$ **cat** $[s+1..d]$
8   **do** $I[t] := I[t] - A[t][c] \cdot I[s]$;
9                       $A[t] := A[t] - A[t][c] \cdot A[s]$;
10   **if** $c \neq s$
11   **then** $z := I[s]$; $I[s] := I[c]$; $I[c] := z$;
12                    $z := A[s]$; $A[s] := A[c]$; $A[c] := z$;
13   **continue** c;
14   **Error** "Matrix is not Invertible!";
15 **return** I;

The rank of A is equal to the number $|l|$ of nonzero rows of C, and so the nullity of A is $d - |l|$. For each $s \in [1..|d|]$ with $s \notin l$, it is straightforward to write down a vector v_s of the nullspace of A with $v_s[s] = 1$, $v_s[t] = 0$ for all $t > s$, and also $v_s[t] = 0$ for all t with $t \neq s$ and $t \notin l$. These $d - |l|$ vectors v_s are linearly independent, and so they form a basis of the nullspace of A.

In the example above, the column numbers not in l are 2, 4, and 6, and we can choose the corresponding vectors v_s to be $v_2 := [-2, 1, 0, 0, 0, 0]$, $v_4 := [-1, 0, -2, 1, 0, 0]$, and $v_6 := [-1, 0, -2, 0, -2, 1]$.

In the code below, we calculate the vectors in the nullspace during the echelonization process rather than at the end of it, and we keep the nullspace itself in the subspace record $\mathcal{W}$.

Now we consider the calculation of the minimal polynomial of a $d \times d$ matrix A over a field K. Let v be any nonzero vector in K^d. Then, if the vectors $v \cdot A^j$ are linearly independent for $0 \leq j < r$, but $v \cdot A^r = \sum_{j=0}^{r-1} a_j v \cdot A^j$ for some $a_i \in K$, then clearly $\rho_v(x) := x^r - a_{r-1}x^{r-1} - \cdots - a_1 x - a_0$ is the unique monic polynomial of minimal degree over K such that $v \cdot \rho_v(A) = 0$. The matrix of A restricted to the subspace of K^d spanned by $v \cdot A^j$ ($0 \leq j < r$) and with respect to this basis of the subspace is

$$\begin{pmatrix} 0 & 1 & 0 & \cdots & 0 \\ 0 & 0 & 1 & \cdots & 0 \\ & & \cdots & & \\ 0 & 0 & 0 & \cdots & 1 \\ a_0 & a_1 & a_2 & \cdots & a_{r-1} \end{pmatrix}.$$

NULLSPACE(α, X)

  Input: Matrix A
  Output: A record $\mathcal{W}$ defining the nullspace of A
1 $d := |A|$; $c := |A[1]|$;
2 $\mathcal{W} := \mathsf{rec}\langle \beta := [\,], \gamma := [\,]\rangle$;
3 $B := A^{\mathrm{T}}$; (* B is a $c \times d$ matrix *)
4 $r := 1$; $l := [\,]$;
5 **for** $t \in [1 .. d]$
6 **do for** $s \in [r .. c]$
7   **do if** $B[s][t] \neq 0$
8   **then** $B[s] := B[s][t]^{-1} \cdot B[s]$;
9   **for** $k \in [1 .. r-1]$ cat $[s+1 .. c]$
10   **do** $B[k] := B[k] - B[k][t] \cdot B[s]$;
11   **if** $r \neq s$
12   **then** $z := B[s]$; $B[s] := B[r]$; $B[r] := z$;
13     $l[r] := t$; $r := r+1$;
14   **continue** t;
   (* This t will not occur in l. Calculate vector in nullspace *)
15 $v := [0 : i \in [1 .. d]]$; $v[t] := 1$
16 **for** $s \in [1 .. r-1]$ **do** $v[l[s]] := -B[s][t]$;
17 VECTORINSUBSPACE($\sim\mathcal{W}, v, \mathsf{true}$);
18 **return** $\mathcal{W}$;

This is known as the *companion matrix* of the polynomial ρ_v.

To compute ρ_v, we need to express $v \cdot A^r$ as a linear combination of $v \cdot A^j$ for $0 \leq j < r$. The function VECTORINSUBSPACE does not do exactly that; rather, it computes $v \cdot A^r$ as a linear combination of the echelonized basis of the subspace. But we can perform this computation by storing the lists returned by VECTORINSUBSPACE, which express each $v \cdot A^j$ in terms of the echelonized basis. The lists are stored in μ in lines 7, 12 of MINIMALPOLYNOMIAL, which is used in lines 16, 17 to compute the required list w of coefficients of $v \cdot A^j$.

For a polynomial $\rho \in K[x]$, we have $v \cdot \rho(A) = 0$ for all $v \in K^d$ if and only if $v_i \cdot \rho(A) = 0$ for all v_i in a basis of K^d, so we can compute the minimal polynomial ρ of A as the least common multiple of the polynomials ρ_{v_i} over a basis. But we do not usually need to compute ρ_{v_i} for all i. We start by computing ρ_{v_1}. Since $v \cdot \rho(A) = 0$ implies $v \cdot \sigma(A) \cdot \rho(A) = 0$ for any $\sigma \in K[x]$, if any v_i is in the subspace spanned by $v_1 \cdot A^j$ ($j \geq 0$), then $v_i \rho_{v_1}(A) = 0$, and we do not need to compute ρ_{v_i}. In general, if v_i is in the subspace spanned by $v_k \cdot A^j$ ($j \geq 0$) for all $k < i$, then we do not need to compute ρ_{v_i}. So, in the code below, we keep track of this subspace.

The function MINIMALPOLYNOMIAL uses three other straightforward undisplayed functions with the following specifications: BASIS(K^d) returns the natural basis of K^d; FILL-VECTOR($\sim l, r$) appends $r - |l|$ zeros to the vector l, to

MINIMALPOLYNOMIAL(A)

    Input: Square matrix A

    Output: The minimal polynomial of A

1   $d := |A|$; $K := \text{FIELD}(A)$;

2   $\rho := \text{POLYNOMIAL}([1_K])$;

    (* We accumulate the whole subspace spanned so far in $\mathcal{W}$ *)

3   $\mathcal{W} := \text{rec}\langle \beta := [], \gamma := [] \rangle$;

4   **for** $b \in \text{BASIS}(K^d)$

5   **do if not** VECTORINSUBSPACE($\sim\mathcal{W}, b,$ false)

6   **then** (* Calculate ρ_b and take LCM with ρ *)

7                    $v := b$; $\mu := []$;

8                    $\mathcal{X} := \text{rec}\langle \beta := [], \gamma := [] \rangle$;

9   **while true**

10   **do** $a, c := $ VECTORINSUBSPACE($\sim\mathcal{X}, v,$ true);

11   **if** a **then break**;

12   APPEND($\sim\mu, c$);

13   VECTORINSUBSPACE($\sim\mathcal{W}, v,$ true);

14                          $v := v \cdot A$;

15                    $r := |\mu|$;

                    (* Get the vector c in terms of $b \cdot A^j$ ($0 \le j < r$) *)

16                    $B := [\text{FILL-VECTOR}(\mu[i], r) : i \in [1 .. r]]$;

17                    $w := c \cdot B^{-1}$;

18                    $\rho := \text{LCM}(\rho, \text{POLYNOMIAL}(-w \text{ cat } [1_K]))$;

19 **return** ρ;

produce a vector in K^r; POLYNOMIAL(v), with $v = [a_0, a_1, \ldots, a_d] \in K^{d+1}$, returns the polynomial $a_0 + a_1 x + \cdots + a_d x^d$.

The algorithm for computing the characteristic polynomial of a matrix A is quite similar. But when considering a basis vector v_i which is not in the subspace W of K^d spanned by $v_k \cdot A^j$ for $k < i$, $j \ge 0$, we compute ρ_{v_i} for the action of A on K^d modulo W rather than on K^d, and multiply the resulting polynomials together, rather than taking their least common multiple. We leave the justification of this process as an exercise for the reader.

7.3   Factorizing polynomials over finite fields

The Meataxe algorithm for testing modules for irreducibility, which we shall be describing in detail in the next section, involves factorizing the characteristic polynomial of a matrix over a finite field. Polynomial factorization over finite fields, and also over other rings allowing exact computation, such as $\mathbb{Z}$, $\mathbb{Q}$, and

CHARACTERISTICPOLYNOMIAL(A)

    Input: Square matrix A
    Output: The characteristic polynomial of A

```
1   d := |A|;  K := FIELD(A);
2   ρ := POLYNOMIAL([1_K]);
        (* We accumulate the whole subspace spanned so far in W *)
3   W := rec⟨β := [], γ := []⟩;
4   d := 0;  (* d is just the dimension of W *)
5   for b ∈ BASIS(K^d)
6       do if not VECTORINSUBSPACE(~W, b, false)
7           then (* Calculate ρ_b on K^d mod W and multiply by ρ *)
8               v := b;  μ := [];
9               while true
10                  do a, c := VECTORINSUBSPACE(~W, v, true);
11                      if a then break;
12                      APPEND(~μ, c[d+1 .. |c|]);
13                      v := v·A;
14              r := |μ|;
15              B := [FILL-VECTOR(μ[i], r) : i ∈ [1 .. r]];
16              w := c·B^{-1};
17              ρ := ρ × POLYNOMIAL(-w cat [1_K]);
18              d := d + r;
19  return ρ;
```

algebraic number fields, is one of the central problems in computer algebra; for a detailed treatment of this topic see, for example, [GCL92, Chapter 8] or [Coh73, Chapter 3].

The algorithms for factorizing polynomials over infinite domains such as $\mathbb{Z}$, all involve reducing modulo one or more primes and then factorizing over the resulting finite field, so the finite field case is particularly important. There are two methods in common use, the Berlekamp algorithm [Ber67, Ber70] and the Cantor-Zassenhaus algorithm [CZ81]. Both of these are discussed in the two books cited above. The Berlekamp method is generally faster for very small fields and, by borrowing some of the ideas of the Cantor-Zassenhaus algorithm, which involves probabilistic methods, it can be adapted to perform well also for large finite fields.

A detailed treatment of this topic here would be straying too far from the main theme of this book but, since the Cantor-Zassenhaus algorithm is relatively easy to describe, we shall summarize it, but without giving any proofs. Programming the Euclidean and greatest common divisor algorithms for polynomials over finite fields is straightforward, so we shall assume that these facilities are available. The Cantor-Zassenhaus algorithm breaks up into three steps, which we shall describe in the following three subsections.

7.3.1   Reduction to the squarefree case

Let $f = f(x) \in \mathbb{F}_q[x]$ with q a power of the prime p, and suppose that $f = f_1^{\alpha_1} \cdots f_r^{\alpha_r}$, where the $f_i(x)$ are distinct irreducibles in $\mathbb{F}_q[x]$. We say that f is *squarefree* if $\alpha_i = 1$ for $1 \le i \le r$. The first step in the factorization algorithm is to reduce to the case when f is squarefree. This step is common to the Berlekamp and Cantor-Zassenhaus methods.

The basic idea is to divide f by the greatest common divisor $\gcd(f, f')$ of f with its formal derivative, f'. In characteristic zero, this would yield the squarefree polynomial $f_1 \cdots f_r$. There is a complication with finite fields, however, because the derivative of the p-th power g^p of any $g \in \mathbb{F}_q[x]$ is zero, and the reader can verify that $f/\gcd(f, f')$ is the product of those f_i for which p does not divide α_i. In particular, if $p|\alpha_i$ for all i, then $f = g^p$ is a p-th power, $f' = 0$ and this gets us nowhere! (Note that the gcd of any polynomial f with 0 is f.)

But, if $g = a_0 + a_1 x + \cdots + a_m x^m$, then $g^p = a_0^p + a_1^p x^p + \cdots + a_m^p x^{mp}$. Since $a^q = a$ for all $a \in \mathbb{F}_q$, we have $(a_m^p)^{q/p} = a$, and so a_m can easily be calculated from a_m^p, and hence g can be calculated from g^p.

So, we can proceed as follows. Calculate f'. If $f' = 0$, then compute $g := f^{1/p}$, which reduces the problem to factorizing g. Otherwise, f is the product of the squarefree polynomial $f/\gcd(f, f')$ and $\gcd(f, f')$. Since $\gcd(f, f')$ has lower degree than f, this effectively reduces the factorization problem to the squarefree case.

The following procedure does this a little more cleverly. It outputs a list of pairs (f_i, α_i), such that $f = f_1^{\alpha_1} \cdots f_r^{\alpha_r}$ and each f_i is squarefree, although not necessarily irreducible.

REDUCETOSQUAREFREE(f)

  Input: $f \in \mathbb{F}_q[x]$ with q a power of prime p
  Output: List L of pairs (f_i, α_i) as described above
 1 $L := [\,]; \;\; e := 1;$
 2 $g := \gcd(f, f');$
 3 $w := f/g;$
 4 **while** $w \ne 1$
 5   **do** $y := \gcd(w, g); \;\; z := w/y;$
 6   APPEND$(\sim L, (z, e));$
 7    $e := e + 1;$
 8    $w := y; \;\; g := g/y;$
 9 **if** $g \ne 1$
 10 **then** ($*$ g is a p-th power $*$)
 11   **for** $t \in$ REDUCETOSQUAREFREE$(g^{1/p})$
 12   **do** APPEND$(\sim L, (t[1], p\, t[2]));$
 13 **return** $L;$

7.3.2   Reduction to constant-degree irreducibles

Having reduced to the case when the polynomial $f \in \mathbb{F}_q[x]$ to be factorized is squarefree, the next step is to reduce to the situation where $f = f_1 \cdots f_r$, and the f_i are all irreducible of the same known degree.

Let $e \geq 1$, and let g be any irreducible polynomial of degree e over $\mathbb{F}_q$. Then $\mathbb{F}_{q^e} = \mathbb{F}_q(\alpha)$, where α is a root of g. Since $\alpha \in \mathbb{F}_{q^e}$ implies $\alpha^{q^e} = \alpha$, and g is the minimal polynomial of α over $\mathbb{F}_q$, we have $g \mid x^{q^e} - q$. On the other hand, since $x^{q^e} - x$ factorizes into q^e linear factors over $\mathbb{F}_{q^e}$, its roots all have degree at most e over $\mathbb{F}_q$, and so all of its irreducible factors have degree at most e over $\mathbb{F}_q$.

So, if $f \in \mathbb{F}_q[x]$ is squarefree, then $f_1 := \gcd(f, x^q - x)$ is the product of all of the linear irreducible factors of f, $f_2 := \gcd(f/f_1, x^{q^2} - x)$ is the product of all irreducible factors of degree 2, and so on. This gives rise to the following algorithm for reducing to the constant-degree case. Note that we do not need to work explicitly with the large-degree polynomials $x^{q^e} - x$. It suffices to compute them modulo f.

REDUCETOCONSTANTDEGREE(f)

    Input: $f \in \mathbb{F}_q[x]$, f squarefree

    Output: List $L = [f_1, f_2, \ldots, f_s]$, where $f = f_1 \cdots f_s$ and, for $1 \leq d \leq s$,
            all irreducible factors of f_d have the same degree d

1    $L := []$; $e := 1$;

2    $w := x^q \bmod f$;

3   **while** $e \leq \deg(f)$

4   **do** $g := \gcd(f, w - x)$;

5   APPEND($\sim L, g$);

6             $f := f/g$;

7             $w := w^q \bmod f$;

8             $e := e + 1$;

9   **return** L;

7.3.3   The constant-degree case

So now we may assume that $f = f_1 \cdots f_r$, where each f_i is irreducible of the same known degree d. If $\deg(f) = d$, then f is irreducible and we are done. Otherwise, we use a probabilistic method. We choose random polynomials $g \in \mathbb{F}_q[x]$ with $\deg(g) \leq 2d - 1$. If the characteristic p of the field is odd, then it turns out that, with probability almost $1/2$, $\gcd(g^{(q-1)/2} - 1, f)$ is a nontrivial proper factor of f; for a proof, see [GCL92, Theorem 8.11]. For $p = 2$, $q = p^k$, the same is true, but with $g + g^{2^1} + g^{2^2} + \cdots + g^{2^{kd-1}}$ in place of $g^{(q-1)/2} - 1$. So we have the following procedure for the final splitting.

Note that, in line 8, CDF is an abbreviation for a recursive call of the procedure CONSTANTDEGREEFACTORIZATION.

CONSTANTDEGREEFACTORIZATION(f, d)

    Input: $f \in \mathbb{F}_q[x]$, f squarefree, $\deg(f_i) = d$ for all irreducible factors of f

    Output: List of irreducible factors of f

1 **if** $\deg(f) = d$ **then return** $[f]$;

2 **while** true

3   **do** Choose random $g \in \mathbb{F}_q[x]$ with $\deg(g) \leq 2d - 1$;

4   **if** q is odd

5   **then** $g := g^{(q-1)/2} - 1$;

6   **else** $g := g + g^{2^1} + g^{2^2} + \cdots + g^{2^{kd-1}}$;

7        $g := \gcd(f, g)$;

8   **if** $0 < \deg(g) < \deg(f)$ **then return** CDF(g) cat CDF(f/g);

Exercises

    1. Improve REDUCETOCONSTANTDEGREE by showing that no further calculations are necessary once $2e < \deg(f)$.

    2. Apply the Cantor-Zassenhaus algorithm to factorize the polynomial $x^6 + x^5 + 2x^4 + 4x^3 + x^2 + 4x + 2$ over $\mathbb{F}_5$.

7.4   Testing *KG*-modules for irreducibility — the Meataxe

Let G be a finite group and K a field. We turn now to the fundamental problem of deciding whether a finite-dimensional module over the group algebra KG is irreducible, and of finding an explicit submodule when it is reducible. We shall restrict our attention to the case when $K = \mathbb{F}_q$ is a finite field. It is proved by Rònyai in [Rón90] that this problem can be solved in time polynomial in $d \log(q)$, where d is the dimension of the module, but the method described there does not lend itself to efficient implementation for large d.

7.4.1   The Meataxe algorithm

An algorithm that has become universally known as the 'Meataxe' was described by Parker in [Par84]. It turned out to be usable for very large values of d (in the thousands) for small fields, and was particularly suitable for computing explicit irreducible representations of the finite simple groups. In this section, we shall describe a generalization of Parker's original algorithm due to Holt and Rees [HR94]. The range of applicability of this version is less restricted in terms of the size of the field and the structure of the group involved.

In fact, it works for modules over any finite-dimensional matrix algebra over a finite field.

Let A be an associative algebra over a finite field K, and suppose that A has r generators $x_1, \ldots, x_r$. We are particularly interested in the case when A is a group algebra KG and the x_i are generators of the finite group G.

A representation φ of degree d of A is a homomorphism from A to the algebra of $d \times d$ matrices over K. We shall assume that such representations are defined by means of a list $\Lambda := [\alpha_1, \ldots, \alpha_r]$ of $d \times d$ matrices over K, where $\alpha_i = \varphi(x_i)$. These matrices specify the action of the generators of A on the (right) module M associated with the representation where, as vector space, $M = K^d$, and the action of the matrices on M is given by multiplication on the right. We shall call them the *action matrices* of the module, and we shall call the algebra that they generate the *matrix algebra* of the module. (It is also known as the *enveloping algebra*.)

Let $\Lambda^T := [\alpha_1^T, \ldots, \alpha_r^T]$ be the transposes of the matrices in Λ, and let M^D be the corresponding right module. This is called the *dual module* of M. (Since $(\alpha \cdot \beta)^T = \beta^T \cdot \alpha^T$, M^D is actually a module for the opposite algebra A^o of A rather than an A-algebra.) For a submodule L of M of dimension c, the orthogonal complement $L^\perp := \{ v \in K^d \mid w \cdot v^T = 0 \; \forall w \in L \}$ has dimension $d - c$, and it is straightforward to check that $L^\perp$ is a submodule of M^D.

As we gather information about a module, we wish to store it. We shall therefore introduce a module record-type. Such a record $\mathcal{M}$ for a module M will have as components a finite field $K = \mathbb{F}_q$, a degree d, and a list Λ of r $d \times d$ matrices over K that generate the algebra A over which M is defined. The list Λ^T of transposes of the generators may also be stored.

Before proceeding to the main MEATAXE function, we need a function for computing the submodule of M generated by a given nonzero vector. We compute this in the function SPINBASIS, by maintaining a basis of the submodule spanned so far, using a subspace record $\mathcal{W}$ as defined in Section 7.2, and applying each algebra generator to each basis vector in this subspace.

SPINBASIS(Λ, v)

> **Input**: List Λ of matrices defining module M, vector $v \in M$
> **Output**: Submodule of M generated by v

```
1   W := rec⟨β := [], γ := []⟩;
2   VECTORINSUBSPACE(∼W, v, true);
3   r := 1;
4   while r ≤ |W.β|
5       do w := W.β[r];
6           for α ∈ Λ do VECTORINSUBSPACE(∼W, w·α, true);
7           r := r + 1;
8   return W;
```

The MEATAXE function displayed below is probabilistic of the Las Vegas type; see Subsection 3.2.1. An answer may or may not be returned on any iteration of the main loop at line 2, and we shall prove later that a returned answer is always correct. An analysis of the algorithm should include an estimate of the probability that an answer is returned on a specific iteration, but for this we refer the reader to [HR94].

The expression RANDOMALGELT($\sim \mathcal{M}$) in line 3 of MEATAXE, denotes a random element of the matrix algebra of M. In practice, we choose this to be a random linear combination of a small number (usually just two or three) of products of matrices from $\mathcal{M}.\Lambda$ calculated using the product replacement algorithm (see Subsection 3.2.2). Although we have not made this explicit in the code, if $\mathcal{M}$ is irreducible, then the formula that defines θ as a function of the matrices in $\mathcal{M}.\Lambda$ is stored along with θ as a record component of $\mathcal{M}$ at line 28. We shall need this formula later, in Subsection 7.5.2, when computing module homomorphisms with domain M.

MEATAXE($\sim \mathcal{M}$)

    (* Decide whether the module M defined by $\sim \mathcal{M}$ is irreducible *)
1   $d := \dim(M)$;
2   **while** true
3   **do** $\theta :=$ RANDOMALGELT($\sim \mathcal{M}$);
4         $\chi :=$ CHARACTERISTICPOLYNOMIAL(θ);
5   **for** $\pi \in$ IRREDUCIBLEFACTORS(χ)
6   **do** $\xi := \pi(\theta)$;
7              $N :=$ NULLSPACE(ξ);
8   Let $v \in N \setminus \{0\}$;
9              $\mathcal{W} :=$ SPINBASIS($\mathcal{M}.\Lambda, v$);
10   **if** $|\mathcal{W}.\beta| < d$
11   **then** (* Proper submodule found *)
12                    $\mathcal{M}.\mathcal{W} := \mathcal{W}$;
13   **return** false;
14   **elseif** $\deg(\pi) =$ DIMENSION(N)
15   **then** (* Try transposing *)
16   Compute $\mathcal{M}.\Lambda^{\mathrm{T}} := \Lambda^{\mathrm{T}}$ if not already done;
17                    $N_T :=$ NULLSPACE(ξ^{T});
18   Let $w \in N_T \setminus \{0\}$;
19                    $\mathcal{W} :=$ SPINBASIS($\mathcal{M}.\Lambda^{\mathrm{T}}, w$);
20   **if** $|\mathcal{W}.\beta| < d$
21   **then** (* Proper submodule found *)
22   Let $w \in$
23   ORTHOGONALCOMPLEMENT($\mathcal{W}$);
24                      $\mathcal{W} :=$ SPINBASIS($\mathcal{M}.\Lambda, w$);
25                      $\mathcal{M}.\mathcal{W} := \mathcal{W}$;
26   **return** false;
27   (* The module is irreducible *)
28                $\mathcal{M}.\theta := \theta$; $\mathcal{M}.\pi := \pi$; $\mathcal{M}.v := v$;
29   **return** true;

On the other hand, if the module turns out to be reducible, then a subspace record $\mathcal{W}$ of a proper submodule is stored as a record component of $\mathcal{M}$.

For clarity, we have written DIMENSION(N) for the dimension of the N, but in fact, with our implementation of subspaces as records, this could be coded as $|\mathcal{N}.\beta|$, where $\mathcal{N}$ is the subspace record corresponding to N. Similarly, we could code "let v be a nonzero vector in N" by $v := \mathcal{N}.\beta[1]$.

In the 'For' loop at line 5 over the irreducible factors of χ, we choose factors in order of increasing degree. This makes the Cantor-Zassenhaus method of polynomial factorization described in Section 7.3 particularly appropriate, because it computes the irreducible factors in order of increasing degree. Among those factors with a given degree, we give preference to those that are unrepeated factors π of χ, because they are guaranteed to have degree equal to the dimension of the nullspace of $\pi(\theta)$.

Example 7.1

Before proving correctness of MEATAXE, let us work through a small example. We let M be a 5-dimensional module over $K = \mathbb{F}_3$ in which A has the two generating matrices:

$$
\alpha_1 := \begin{pmatrix} 0 & 0 & 0 & 0 & 1 \\ 2 & 1 & 1 & 2 & 0 \\ 0 & 1 & 0 & 0 & 0 \\ 2 & 1 & 0 & 0 & 0 \\ 0 & 1 & 0 & 2 & 0 \end{pmatrix}, \quad
\alpha_2 := \begin{pmatrix} 0 & 1 & 0 & 0 & 2 \\ 1 & 0 & 0 & 0 & 2 \\ 0 & 0 & 0 & 1 & 2 \\ 0 & 0 & 1 & 0 & 2 \\ 0 & 0 & 0 & 0 & 2 \end{pmatrix}.
$$

This algebra is a representation of the group algebra KG, where $G = \mathrm{Sym}(4)$, and the matrices are the images of the permutations $(1, 2, 3, 4)$ and $(1, 2)$.

As our random element, we choose

$$
\theta := \alpha_1 + 2\alpha_2 + 2\alpha_1\alpha_2 = \begin{pmatrix} 0 & 2 & 0 & 0 & 0 \\ 0 & 2 & 2 & 1 & 1 \\ 2 & 1 & 0 & 2 & 2 \\ 1 & 2 & 2 & 0 & 1 \\ 2 & 1 & 1 & 2 & 1 \end{pmatrix}.
$$

The characteristic polynomial is $\chi = x^5 + x^3 + x^2 + 1$, which factorizes over K to $(x + 1)^3(x^2 + 1)$. The first factor $(x + 1)$ yields no result in MEATAXE; we might in any case prefer to use the second factor $(x^2 + 1)$, because it is a nonrepeated factor and therefore certain to yield a result. So, putting $\pi := x^2 + 1$, we find that

$$
\xi := \pi(\theta) = \theta^2 + I_5 = \begin{pmatrix} 1 & 1 & 1 & 2 & 2 \\ 1 & 1 & 1 & 2 & 2 \\ 0 & 0 & 0 & 2 & 2 \\ 0 & 0 & 2 & 0 & 1 \\ 0 & 0 & 1 & 2 & 1 \end{pmatrix},
$$

of which the nullspace is generated by the vectors $(1, 2, 0, 0, 0)$, $(0, 0, 1, 2, 2)$. Applying SPINBASIS to the first of these vectors, yields the proper submodule M with basis $[(1, 2, 0, 0, 0), (0, 0, 1, 2, 2), (0, 0, 0, 1, 0)]$, and it is not hard to check directly that this is indeed a submodule.   □

7.4.2   Proof of correctness

We shall now prove that, if MEATAXE returns an answer, then it is correct; that is, the answer `true` means that the module is irreducible, and `false` means that it is reducible. MEATAXE returns `false` only when it has found a submodule, so correctness is clear in that case. So assume that `true` is returned.

Now `true` is only returned when the degree of the polynomial π is equal to the dimension of the nullspace N of $\xi := \pi(\theta)$. So, in order to prove correctness, we need only show that, if M is reducible, then a submodule of M will be found by MEATAXE in that situation. So assume that $\deg(\pi) = \dim(N)$ and that M has a proper nonzero submodule L.

Since N is the nullspace of $\xi = \pi(\theta)$ and π is irreducible, the minimal polynomial of the restriction $\theta|_N$ of θ to N must be π. But then $\dim(N) = \deg(\pi)$ implies that $\theta|_N$ acts irreducibly on N. It follows that the subspace $L \cap N$ of K^d, which is fixed by θ, must either be zero or equal to the whole of N. In the latter case, the chosen nonzero vector $v \in N$ lies in L, and so a proper submodule of M (contained in L) will be found and the answer `false` returned at line 13.

Suppose, on the other hand, that $L \cap N = \{0\}$, and that we proceed to line 16, where Λ^{T} and ξ^{T} are calculated. Let $e_1, e_2, \ldots, e_d$ be a basis of M such that $e_1, \ldots, e_c$ is a basis of L, and let P be the $d \times d$ matrix with rows $e_1, e_2, \ldots, e_d$. Then, writing ξ and ξ^{T} with respect to the basis $e_1, e_2, \ldots, e_d$, we get

$$P\xi P^{-1} = \begin{pmatrix} \zeta^{(1)} & 0 \\ \zeta^{(2)} & \zeta^{(3)} \end{pmatrix}, \quad Q\xi^{\mathrm{T}}Q^{-1} = \begin{pmatrix} \zeta^{(1)\,\mathrm{T}} & \zeta^{(2)\,\mathrm{T}} \\ 0 & \zeta^{(3)\,\mathrm{T}} \end{pmatrix},$$

where $Q := (P^{-1})^{\mathrm{T}}$, and $\zeta^{(1)}$, $\zeta^{(2)}$ and $\zeta^{(3)}$ are, respectively, $c \times c$, $(d-c) \times c$, and $(d-c) \times (d-c)$ matrices.

Now $L \cap N = \{0\}$ implies rank $\zeta^{(1)} = c$, and so nullity $\zeta^{(3)\,\mathrm{T}} = $ nullity ξ^{T}. Thus all vectors in the nullspace of $Q\xi^{\mathrm{T}}Q^{-1}$ lie in a proper submodule of the module for the algebra defined by the matrices $Q a_i^{\mathrm{T}} Q^{-1}$ ($1 \leq i \leq r$). Equivalently, all vectors in the nullspace of ξ^{T} lie in a proper submodule of M^{D}. So a proper submodule of M^{D} will be found at line 19, a vector in the orthogonal complement of this submodule will lie in a proper submodule of M, and `false` will be returned at line 26.

7.4.3 The Ivanyos-Lux extension

We have proved that any answer returned by MEATAXE is correct, and the proof is valid for any field K. However, it is only for finite fields that the probability of an answer being returned at all is reasonably high. It is proved in [HR94] that, for a finite field K, in almost all circumstances, an answer is returned for a particular choice of θ with probability at least 0.144.

There is unfortunately one exceptional configuration in which an answer is returned with very low probability, and that is when M is reducible, all compositions factors of M are isomorphic to the same irreducible module L, and the centralizing field of L is large compared with the base field K. This situation has occasionally been observed to arise in chief factors of solvable groups, so it needs to be dealt with. Two remedies have been proposed, an unpublished one by C.R. Leedham-Green, and one by Ivanyos and Lux in [IL00], which includes a proof that an answer will result from their modification in the exceptional situation with probability at least 0.08.

The Ivanyos-Lux extension is easy to describe, so we shall do so now, briefly, and without detailed explanations. If the first few (perhaps three or four) values of θ in MEATAXE produce no answer, then we proceed as follows, maintaining the current choice for θ, and with π equal to an irreducible factor of minimal degree of the characteristic polynomial χ of θ. Suppose that $\chi = \pi^l \rho$, where π does not divide ρ, find polynomials σ, τ such that $1 = \sigma\pi^l + \tau\rho$, and let $\iota := \tau\rho \pmod{\chi}$. (Then ι modulo χ is a primitive idempotent of the algebra $K[x]/\chi$.) Choose another random element $\eta \in A$ and a random vector $v \in M$, and calculate the submodule L of M spanned by $v \cdot [\theta, \iota(\theta)\eta\iota(\theta)]$ where $[\alpha, \beta]$ means $\alpha\beta - \beta\alpha$. If L is a proper submodule of M then return `false`.

7.4.4 Actions on submodules and quotient modules

If a module turns out to be reducible, then we might want to calculate matrices for the actions on the associated submodule and quotient module. Doing this is straightforward, and we shall briefly describe the method, without displaying detailed code. By iterating the process, we can then find a composition series, and all irreducible composition factors of the original module.

When MEATAXE returns `false` on a module with record $\mathcal{M}$, it sets a component $\mathcal{W}$ of $\mathcal{M}$, which contains a subspace record for a proper submodule. Let $\beta := \mathcal{W}.\beta$ be the basis returned for this subspace. We extend this to a basis, which we shall also denote by β, of K^d by appending those natural basis vectors of K^d for which the position of their nonzero entry does not occur in $\mathcal{W}.\gamma$.

Let $\Lambda := \mathcal{M}.\Lambda$. Then we compute the list $\Lambda_\beta := [\beta \cdot \alpha \cdot \beta^{-1} \mid \alpha \in \Lambda]$ of matrices for the module with respect to the basis β, and we obtain matrices for the actions on the submodule and quotient module as $[\alpha[1 .. c][1 .. c] \mid \alpha \in \Lambda_\beta]$ and $[\alpha[c+1 .. d][c+1 .. d] \mid \alpha \in \Lambda_\beta]$, respectively, where c is the dimension of the submodule.

In Example 7.1 above, we get

$$\beta = \begin{pmatrix} 1\,2\,0\,0\,0 \\ 0\,0\,1\,2\,2 \\ 0\,0\,0\,1\,0 \\ 0\,1\,0\,0\,0 \\ 0\,0\,0\,0\,1 \end{pmatrix}, \quad \beta\cdot\alpha_1\cdot\beta^{-1} = \begin{pmatrix} 1\,2\,0\,0\,0 \\ 1\,0\,1\,0\,0 \\ 2\,0\,0\,0\,0 \\ 2\,1\,0\,0\,1 \\ 0\,0\,2\,1\,0 \end{pmatrix}, \quad \beta\cdot\alpha_2\cdot\beta^{-1} = \begin{pmatrix} 2\,0\,0\,0\,0 \\ 0\,2\,0\,0\,0 \\ 0\,1\,1\,0\,0 \\ 1\,0\,0\,1\,2 \\ 0\,0\,0\,0\,2 \end{pmatrix},$$

so the lists of matrices for the action on the submodule and quotient module are, respectively,

$$\left[\begin{pmatrix} 1\,2\,0 \\ 1\,0\,1 \\ 2\,0\,0 \end{pmatrix}, \begin{pmatrix} 2\,0\,0 \\ 0\,2\,0 \\ 0\,1\,1 \end{pmatrix} \right] \quad \text{and} \quad \left[\begin{pmatrix} 0\,1 \\ 1\,0 \end{pmatrix}, \begin{pmatrix} 1\,2 \\ 0\,2 \end{pmatrix} \right].$$

7.4.5 Applications

Applications of the Meataxe algorithm are mostly within group theory and representation theory. It was used extensively in the construction of Brauer trees by Hiss and Lux in [HL89] and of Brauer character tables by Jansen, Lux, Parker, and Wilson in [JLPW95].

For the larger representations that arise in this type of computation, an additional technique, known as *condensation* is required; see, for example, the papers of Ryba, Lux, and Müller [Ryb90, LMR94, Ryb01]. The general idea is to replace the original module M by a condensed version M_C, say, which is of much smaller dimension and defined over a different algebra from C. In favourable circumstances, the irreducible constituents of M can be recovered from those of M_C. This approach has been developed mainly for the cases when M is a permutation module and when it is defined as a tensor product of pairs of smaller dimensional modules. For example, in [MNRW02], in order to compute the Brauer tree of the Lyons sporadic simple groups for the primes $p = 37$ and 67, the authors were able to condense a permutation module of dimension $1\,113\,229\,656$ to one of dimension 3207.

The Meataxe is also used as a component in a number of structural group-theoretical computations. As we shall see in Sections 8.5 and 10.2, it is used to to refine elementary abelian sections when computing chief series of finite groups. It plays a rôle in a number of other computations, such as the computation of automorphism groups of p-groups, which will be discussed briefly in Subsection 9.4.6.

Exercises

1. Prove that, if L is a submodule of M, then $L^\perp$ is a submodule of M^D.

2. Assume that MEATAXE returns `false` on a module record $\mathcal{M}$. Write some code to compute the matrices for the action on the submodule directly, without performing the basis change described in Subsection 7.4.4

7.5 Related computations

In the following subsections, we describe some algorithms for KG-modules with K and G finite, which follow on naturally from the Meataxe procedure.

7.5.1 Testing modules for absolute irreducibility

Our next problem is to take a module M over a finite field $K = \mathbb{F}_q$ that has been proved irreducible by a call of MEATAXE, and to compute its centralizing algebra C. As was explained in Subsection 2.7.1, C is isomorphic to an extension field $\mathbb{F}_{q^e}$ of $K = \mathbb{F}_q$, and M is absolutely irreducible if and only if $e = 1$. Let λ be a $d \times d$ matrix over K that generates C as an algebra or, equivalently, as a field. Then the minimal polynomial μ of λ has degree e. Let W be a nonzero λ-invariant subspace of K^d of minimal dimension; so $\lambda|_W$ is irreducible. Then, for any matrix α in the matrix algebra of M, W^α is also λ-invariant and, by minimality of W, we have either $W = W^\alpha$ or $W \cap W^\alpha = \{0\}$. Using this fact and the irreducibility of M, it is straightforward to show that K^d can be written as a direct sum of some subspaces W^α, where the minimal polynomial of $\lambda|_{W^\alpha}$ is the same for each W^α. So this minimal polynomial must be μ, and hence $\dim(W) = e$ and $e|d$.

Since we are assuming that M was proved irreducible using MEATAXE, we know that the following objects used in the proof have been stored as components of the module record $\mathcal{M}$:

(i) the element θ;

(ii) the irreducible factor π of the characteristic polynomial χ of θ;

(iii) a nonzero vector v in the nullspace N of $\pi(\theta)$, where $\dim(N) = \deg(\pi)$.

Let $\dim(N) = c$. Our centralizing matrix λ must leave N invariant, and so N can also be written as a direct sum of translates W^α of W, and hence $e|c$. Since $\theta|_N$ acts irreducibly on N, the centralizing algebra D, say, of $\theta|_N$ is isomorphic to $\mathbb{F}_{q^c}$, and includes $C|_N$, so $C|_N$ is the unique subfield of D isomorphic to $\mathbb{F}_{q^e}$. So, if we choose any element κ of D, then $\kappa^f \in C|_N$, where $f := (p^c - 1)/(p^e - 1)$, and for a suitable choice of κ (for example, a generator of the multiplicative group of D), κ^f will generate the field $C|_N$.

These considerations justify our algorithm for finding C, which proceeds as follows. We start by calculating the basis $\beta := [v \cdot \theta^i \mid i \in [0 .. c-1]]$ of N, with respect to which the matrix γ of $\theta|_N$ is the companion matrix of π. It easy to write down random elements κ of the centralizing algebra D of $\theta|_N$ with respect to this basis. The reader can verify that we get such a centralizing matrix by letting the first row of κ be an arbitrary vector in K^c (of which there are $|D| = q^c$), and then the remaining rows of κ are determined by

$\kappa[i] = \kappa[i-1]\cdot\gamma$ for $2 \leq i \leq c$. Although we have not done this in the code below, it can save time to choose $\theta|_N$ itself as the first choice for κ.

We consider the common divisors e of d and c in decreasing order and, for each e, we decide whether C includes the field $\mathbb{F}_{q^e}$. An affirmative answer will then imply that $C \cong \mathbb{F}_{q^e}$. If we reach $e = 1$ then we have proved that M is absolutely irreducible.

For a fixed value of e and an element $\kappa \in D$, we calculate $\lambda_N := \kappa^f$, where $f := (q^c - 1)/(q^e - 1)$ as above, and check whether λ_N generates the subfield $\mathbb{F}_{q^e}$ of D. This last condition is equivalent to the degree of the minimal polynomial μ of λ_N being e, which we can check. If not, then we keep trying new matrices κ until we find $\lambda_N := \kappa^f$ such that the condition holds.

From the theory above, we know that, if $|C| = q^e$, then this λ_N is the restriction to N of λ for a suitable $\lambda \in C$, so we just have to check whether this is the case.

If so, then K^d is the direct sum of translates of a minimal C-invariant subspace W, and we can take W to be a minimal λ_N-invariant subspace of N. In the code below, we first calculate a basis β_λ of W in terms of the basis β of N. This is because our matrix λ_N is written with respect to β. We get β_λ simply by applying powers λ_N^j of λ_N for $0 \leq j < e$ to the vector $w = [1, 0, \ldots, 0]$ of K^c. We can then write β_λ with respect to the original basis for the module by replacing β_λ by $\beta_\lambda\beta$. The matrix of $\lambda_N|_W$ with respect to the basis β_λ of W is then just the companion matrix ν of the minimal polynomial μ of λ.

The subroutine COMPLETEBASIS attempts to construct K^d as such a direct sum of translates of W. It may fail as a result of some translate W^α of W being neither contained in nor disjoint from the direct sum found so far. If that happens, then we know that $|C| \neq q^e$, so we can proceed to the next candidate for e. If COMPLETEBASIS succeeds then it extends the basis β_λ of W to a basis, also denoted by β_λ, of K^d, with respect to which the matrix for λ is the direct sum $\nu_{d/e}$ of d/e copies of ν. We then get the matrix of λ with respect to the original basis of the module as $\lambda := \beta_\lambda^{-1}\cdot\nu_{d/e}\cdot\beta_\lambda$. Now, it only remains to check whether the generating matrices of the module all commute with λ. If so, then we have verified that $|C| = q^e$ and, if not, we can proceed to the next e. Note that VECTORINSUBSPACE has been abbreviated to VIS in COMPLETEBASIS.

ABSOLUTELYIRREDUCIBLE must only be called on a module $\mathcal{M}$ for which MEATAXE($\mathcal{M}$) has already been called and returned **true**. The positive integer e, such that $|C| = q^e$ will be set as a record component of the module, which is absolutely irreducible if and only if $e = 1$. If $e > 1$, then the $d \times d$ matrix λ that generates C is also stored. (Note that λ will not necessarily have multiplicative order $q^n - 1$; that is, it will not necessarily generate the multiplicative group of C. But if such a matrix is required, then it can be found later by choosing random elements of C.)

COMPLETEBASIS($\sim \beta_\lambda, d, e, \Lambda$)

    Input: Partial basis β_λ for centralizing element

    (∗ Set up a subspace record $\mathcal{B}$ for the basis ∗)

1   $\mathcal{B} := \mathsf{rec}\langle \beta := [], \gamma := []\rangle$;

2   **for** $v \in \beta_\lambda$ **do** VIS($\sim \mathcal{B}, v$,true);

3   **for** $i \in [1 .. d/e]$, $\alpha \in \Lambda$

4   **do if not** VIS($\sim \mathcal{B}, \beta_\lambda[(i-1)e+1]\cdot\alpha$,false)

5   **then for** $j \in [(i-1)e+1 .. ei]$

6   **do** VIS($\sim \mathcal{B}, \beta_\lambda[j]\cdot\alpha$, true);

7   **if** $|\mathcal{B}.\beta| \bmod e \neq 0$ **then return** false;

8                $\beta_\lambda := \beta_\lambda$ cat $[\beta_\lambda[j]\cdot\alpha \mid j \in [e(i-1)+1 .. ei]]$;

9   **if** $|\beta_\lambda| = d$ **then return** true;

ABSOLUTELYIRREDUCIBLE($\sim \mathcal{M}$)

    (∗ Is the irreducible module M absolutely irreducible? ∗)

1   $\theta := \mathcal{M}.\theta$; $\pi := \mathcal{M}.\pi$; $v := \mathcal{M}.v$; $d := \dim(M)$;

    (∗ The components θ, π, v have been set by MEATAXE ∗)

2   $c := \deg(\pi)$; $\gamma := $ COMPANIONMATRIX(π);

3   $\beta := [v\cdot\theta^i \mid i \in [0 .. c-1]]$;

4   Let l be the list of common divisors of c and d in decreasing order;

5   **for** $e \in l$ (∗ Test whether $|\mathcal{C}(\mathcal{M})| = q^e$ ∗)

6   **do** $\mathcal{M}.e := e$;

7   **if** $e = 1$ **then return** true;

8        $f := (q^c - 1)/(q^e - 1)$;

9   **while** true

10   **do** (∗ Choose a random $c \times c$ matrix centralizing γ ∗)

11               $\kappa := [$RANDOMVECTOR(K, c)$]$;

12   **for** $i \in [2 .. c]$ **do** $\kappa[i] := \kappa[i-1]\cdot\gamma$;

13               $\lambda_N := \kappa^f$; $\mu := $ MINIMALPOLYNOMIAL(λ_N);

14   **if** $\deg(\mu) = e$ **then break**;

15           $w := [1]$ cat $[0 : i \in [1 .. c-1]]$;

16           $\beta_\lambda := [w\cdot\lambda_N^i \mid i \in [0 .. e-1]]$;

17           $\beta_\lambda := \beta_\lambda\cdot\beta$;

18   **if** COMPLETEBASIS($\sim \beta_\lambda, d, e, \mathcal{M}.\Lambda$)

19   **then** (∗ Test directly if we have a centralizing element ∗)

20                 $\nu := $ COMPANIONMATRIX(μ);

21                 $\nu_{d/e} := $ DIRECTSUM($[\nu \mid i \in [1 .. d/e]]$);

                 (∗ Put $\nu_{d/e}$ into the original basis ∗)

22                 $\lambda := \beta_\lambda^{-1}\cdot\nu_{d/e}\cdot\beta_\lambda$;

23   **for** $\alpha \in \Lambda$ **do if** $\alpha\cdot\lambda \neq \lambda\cdot\alpha$

24   **then continue** e;

                 (∗ Centralizing element found ∗)

25                 $\mathcal{M}.\lambda := \lambda$; **return** false;

26   **else continue** e;

Example 7.2

Let M be a 4-dimensional module over $K = \mathbb{F}_2$ in which A has the two generating matrices:

$$\alpha_1 := \begin{pmatrix} 1\,1\,1\,0 \\ 1\,0\,0\,1 \\ 0\,0\,0\,1 \\ 0\,0\,1\,1 \end{pmatrix}, \quad \alpha_2 := \begin{pmatrix} 0\,0\,1\,1 \\ 1\,1\,1\,0 \\ 0\,1\,1\,1 \\ 1\,1\,0\,0 \end{pmatrix}.$$

The algebra is an irreducible representation of the group algebra KG with $G = \mathrm{Alt}(5)$, where α_1 and α_2 are the images of $(1,2,3)$ and $(1,4,5)$.

Suppose that the random element of A chosen in the call of MEATAXE is

$$\theta := \alpha_1 + \alpha_2 \cdot \alpha_1 = \begin{pmatrix} 1\,1\,0\,0 \\ 1\,1\,1\,1 \\ 1\,0\,1\,0 \\ 0\,1\,0\,0 \end{pmatrix}.$$

Then θ has characteristic polynomial $x^4 + x^3 + x^2 + x + 1$, which is irreducible over K, and can be used by MEATAXE to prove that M is irreducible.

In the test for absolute irreducibility, we first calculate the basis β of the companion matrix of θ_N, which is the list of rows of the matrix:

$$\begin{pmatrix} 1\,0\,0\,0 \\ 1\,1\,0\,0 \\ 0\,0\,1\,1 \\ 1\,1\,1\,0 \end{pmatrix}.$$

The candidates for e are 4, 2, and 1. When $e = 4$, $f = 1$, and we can take $\lambda_N := \kappa := \gamma$ at line 13, which results in $\lambda = \theta$ at line 22. Since θ does not commute with α_1, this rules out $e = 4$.

So next we try $e = 2$. Here $f := (2^4 - 1)/(2^2 - 1) = 5$. Since γ^5 is the identity, we cannot use $\kappa = \gamma$. A suitable random matrix κ that centralizes the matrix γ of θ with respect to the base β is

$$\kappa := \begin{pmatrix} 1\,0\,0\,1 \\ 1\,0\,1\,1 \\ 1\,0\,1\,0 \\ 0\,1\,0\,1 \end{pmatrix}, \quad \lambda_N := \kappa^5 := \begin{pmatrix} 0\,0\,1\,1 \\ 1\,1\,1\,0 \\ 0\,1\,1\,1 \\ 1\,1\,0\,0 \end{pmatrix}.$$

(It is just a coincidence that $\lambda_N = \alpha_2$.) Now the minimal polynomial μ of λ_N is $x^2 + x + 1$, which has degree $e = 2$, so we can use this λ_N to test whether $C = \mathbb{F}_{2^2}$. The two vectors of β_λ with respect to the basis β are $[1,0,0,0]$ and $[1,0,0,0] \cdot \lambda_N = [0,0,1,1]$, and hence the two vectors with respect to the original basis are $[1,0,0,0]$, $[1,1,0,1]$. Now applying COMPLETEBASIS to this pair of vectors successfully completes the pair to the basis with matrix

$$\beta_\lambda := \begin{pmatrix} 1 & 0 & 0 & 0 \\ 1 & 1 & 0 & 1 \\ 1 & 1 & 1 & 0 \\ 0 & 1 & 0 & 0 \end{pmatrix}.$$

Now the companion matrix ν of μ is the 2×2 matrix with rows $[1, 0]$, $[1, 1]$, and if we let ν_2 be the direct sum of two copies of ν, then we find that

$$\lambda := \beta_\lambda^{-1} \cdot \nu_2 \cdot \beta_\lambda = \begin{pmatrix} 1 & 1 & 0 & 1 \\ 1 & 0 & 1 & 0 \\ 0 & 0 & 1 & 1 \\ 0 & 0 & 1 & 0 \end{pmatrix}.$$

We can check that this last matrix λ does indeed commute with α_1 and α_2, so we have proved that $e = 2$. So this module is not absolutely irreducible, and has $\mathbb{F}_4$ as centralizing field. $\quad\square$

7.5.2 Finding module homomorphisms

We shall now consider the problem of finding a basis over K of the vector space of algebra homomorphisms $\mathrm{Hom}_A(M, L)$ where M and L are modules over the same K-algebra A. We shall write out detailed code for the case when M is irreducible. By choosing $M = L$, we could use this approach for finding the endomorphism algebra of M although, when M is irreducible, the algorithm described in Subsection 7.5.1 is significantly more efficient.

The method for computing $\mathrm{Hom}_A(M, L)$ for general M and L that springs to mind is the following. Find a sequence of vectors $[v_1, \ldots, v_k]$ that generate M as an A-module, with k as small as possible. Let $[w_1, \ldots, w_t]$ be a K-basis of L. Then the images of v_i under a homomorphism $\varphi : M \to L$ can be written as linear combinations of the w_j, where the coefficients are unknowns in K. So we have a total of kt unknowns, and the condition that φ should be an algebra homomorphism results in a system of homogeneous linear equations in these unknowns. We solve this system by finding the nullspace of the matrix of coefficients, which gives us our basis of $\mathrm{Hom}_A(M, L)$.

Let $s := \dim(M)$ and $t := \dim(L)$. To construct the system of linear equations, we first apply the matrices in A to the vectors $v_1, \ldots, v_k$ to build up a basis $v_1, \ldots, v_s$ of M, and apply the same operations in L to find the images of the complete basis of M in terms of our kt unknowns. We then apply each algebra matrix α_i ($1 \le i \le r$) to each basis vector v_j of M. For each i and j, the algebra homomorphism condition $\varphi(v_j^{\alpha_i}) = \varphi(v_j)^{\alpha_i}$ is an equality between two vectors in L, and yields t equations in the kt unknowns. This results in a system of rst equations in kt unknowns, which is unpleasantly large, even over a finite field, if s and t are more than about 100.

There are better methods available when the dimensions are large, but we shall not discuss them any further here. One such method was proposed by

Schneider in [Sch90a], and further unpublished ideas have been proposed by C.R. Leedham-Green, and are used in the MAGMA implementation. In the same paper, Schneider shows how similar methods may be applied to test a module M for indecomposability.

In the special case when M is irreducible, significant improvements can be made to the basic idea described above, and we shall discuss these now. So, suppose that we have module records $\mathcal{M}$, $\mathcal{L}$ for the A-modules M and L, and that M has been proved irreducible by a call of MEATAXE($\sim \mathcal{M}$). The algorithm for computing a K-basis of $\mathrm{Hom}_A(M, L)$ is displayed in the function MODULEHOMOMORPHISMS below. Elements of this K-basis are output as $s \times t$-matrices over K, where $s := \dim(M)$ and $t := \dim(L)$.

Since M is irreducible, it is generated as an A-module by any of its nonzero vectors, and we choose the generator v_1 (in the notation above) to be the vector $\mathcal{M}.v$ in the nullspace of $\mathcal{M}.\xi$.

In a module homomorphism $\varphi : M \to L$, $\varphi(v_1)$ must lie in the nullspace of the element ξ_L in the matrix algebra of L corresponding to ξ. As we mentioned earlier, a formula for calculating $\mathcal{M}.\theta$ from the list of action matrices $\mathcal{M}.\Lambda$ of M is stored by MEATAXE as a component of $\mathcal{M}$, and so we can apply the same formula to the action matrices $\mathcal{L}.\Lambda$ of L to get the element θ_L corresponding to θ. We have called the function for doing this COPYTHETA in line 1 of MODULEHOMOMORPHISMS, and we set $\xi_L := \mathcal{M}.\pi(\theta_L)$.

Next we calculate the subspace record $\mathcal{N}$ of the nullspace N of ξ_L, and put $d := \dim(N)$. Let $w_\lambda := \mathcal{N}.\beta$ be the basis of N returned by NULLSPACE. Then a module homomorphism φ must map v_1 to a vector in N; that, is, there exist $\lambda_1, \ldots, \lambda_d \in K$ with $\varphi(v_1) = \sum_{j=1}^{d} \lambda_j w_\lambda[j]$.

In MODULEHOMOMORPHISMS, we construct a basis of M as the list $\beta := \mathcal{V}.\beta$ in a subspace record $\mathcal{V}$, and simultaneously define d lists of vectors $\tau_\lambda[1], \ldots, \tau_\lambda[d]$ in L with the property that $\varphi(\beta[i]) = \sum_{j=1}^{d} \lambda_j \tau_\lambda[j][i]$, for each $\beta[i]$ in the list β. The subroutine PVIS (which stands for 'parallelized vector in subspace') takes as arguments a vector $v \in M$ and a list ω of d vectors in L, which are known to satisfy the equation $\varphi(v) = \sum_{j=1}^{d} \lambda_j \omega[j]$.

PVIS starts by applying the code of VECTORINSUBSPACE to test membership of v in the subspace $\mathcal{V}$ that has been constructed so far. As it subtracts multiples of vectors in β from v, it simultaneously subtracts the same multiples of the corresponding vectors in $\tau_\lambda[j]$ from $\omega[j]$ for $1 \leq j \leq d$, thereby maintaining the relationship $\varphi(v) = \sum_{j=1}^{d} \lambda_j \omega[j]$.

If v turns out not to be in $\mathcal{V}$, then a new basis vector is appended to $\mathcal{V}.\beta$ and the corresponding vectors $\omega[j]$ are appended to $\tau_\lambda[j]$ for $1 \leq j \leq d$. On the other hand, if v is in $\mathcal{V}$ already, then v will be reduced to 0, and then we must also have $\sum_{j=1}^{d} \lambda_j \omega[j] = 0$, which yields $t = \dim(L)$ linear equations over K that are satisfied by $\lambda_1, \ldots, \lambda_d$. The coefficients of these linear equations are stored as the columns of the matrix χ, and any extra equations discovered in a call of PVIS are appended to χ by a call of HORIZONTALJOIN.

In the function MODULEHOMOMORPHISMS, we initialize $\mathcal{V}$ and τ_λ by calling

PVIS($v, \omega, \sim \mathcal{V}, \sim \tau_\lambda, \sim \chi$)

    Input: $v \in M$, list ω of vectors in L

1    $d := |w|$; $\beta := \mathcal{V}.\beta$; $\gamma := \mathcal{V}.\gamma$;
2   **for** $i \in [1 .. |\gamma|]$
3   **do** $a := v[\gamma[i]]$;
4   **if** $a \neq 0$
5   **then** $v := v - a \cdot \beta[i]$;
6   **for** $j \in [1 .. d]$ **do** $\omega[j] := \omega[j] - a \cdot \tau_\lambda[j][i]$;
7    $c, a := $ DEPTHVECTOR(v);
8   **if** $c \neq 0$
9   **then** APPEND($\sim \mathcal{V}.\beta, a^{-1} \cdot v$);   APPEND($\sim \mathcal{V}.\gamma, c$);
10   **for** $j \in [1 .. d]$ **do** APPEND($\sim \tau_\lambda[j], a^{-1} \cdot \omega[j]$);
11   **else** $\chi := $ HORIZONTALJOIN(χ, ω);

MODULEHOMOMORPHISMS($\mathcal{M}, \mathcal{L}$)

    Input: Module records $\mathcal{M}, \mathcal{L}$ with $\mathcal{M}$ proved irreducible by MEATAXE

    Output: A K-basis of $\mathrm{Hom}_A(M, L)$

1    $\xi_L := \mathcal{M}.\pi(\mathrm{COPYTHETA}(\mathcal{M}, \mathcal{L}))$;
2    $\mathcal{N} := \mathrm{NULLSPACE}(\xi)$; $d := |\mathcal{N}.\beta|$;
3   **if** $d = 0$ **then return** [];
4    $\Lambda_M := \mathcal{M}.\Lambda$; $\Lambda_L := \mathcal{L}.\Lambda$;
5    $w_\lambda := [\mathcal{N}.\beta[j] : j \in [1 .. d]]$;
6    $\mathcal{V} := \mathrm{rec}\langle \beta := [], \gamma := [] \rangle$; $\tau_\lambda := [[] : j \in [1 .. d]]$; $\chi := []$;
7   PVIS($\mathcal{M}.v, w_\lambda, \sim \mathcal{V}, \sim \tau_\lambda, \sim \chi$);
8    $i := 1$;
9   **while** $i \leq |\mathcal{V}.\beta|$
10   **do for** k **in** $[1 .. |\Lambda_M|]$ **do** PVIS(
11              $\mathcal{V}.\beta[i] \cdot \Lambda_M[k], [\tau_\lambda[j][i] \cdot \Lambda_L[k] \mid j \in [1 .. d]], \sim \mathcal{V}, \sim \tau_\lambda, \sim \chi$);
12            $i := i + 1$;
13    $\mathcal{N} := \mathrm{NULLSPACE}(\chi)$; $\mu := (\mathcal{V}.\beta)^{-1}$; $h := []$;
14   **for** $u \in \mathcal{N}.\beta$
15   **do** $\nu := \sum_{j=1}^{d} u[j]\tau_\lambda[j]$;
16   APPEND($\sim h, \mu \cdot \nu$);
17   **return** h;

PVIS($\mathcal{M}.v, w_\lambda, \sim \mathcal{V}, \sim \tau_\lambda, \sim \chi$) at line 7. Let $\mathcal{M}.\Lambda := [\alpha_1, \ldots, \alpha_r]$ and $\mathcal{L}.\Lambda := [\delta_1, \ldots, \delta_r]$ be the lists of action matrices of M and L, respectively. Then the condition for φ to be a module homomorphism is that

$$\varphi(\mathcal{V}.\beta[i]^{\alpha_k}) = \sum_{j=1}^{d} \lambda_j \tau_\lambda[j][i]^{\delta_k}$$

for all $\mathcal{V}.\beta[i] \in \mathcal{V}.\beta$, and $1 \leq k \leq r$. We enforce these conditions by making the corresponding calls of PVIS at line 11. Some of these calls will have the effect of appending a new basis element to $\mathcal{V}.\beta$ and the remainder will yield $\dim(L)$ linear equations satisfied by $\lambda_1, \ldots, \lambda_d$.

When this process is complete, we solve this system of linear equations in the λ_j by calculating the nullspace of χ at line 13. A basis of this nullspace can then be used to calculate and return a basis, called h in the code, of $\mathrm{Hom}_A(M, L)$.

One problem with the code as it stands is that we may generate a very large number of equations, and so the matrix χ can grow correspondingly large. We can avoid this by keeping χ in column-echelonized form throughout. Then it will never have more than d columns (or rather $d + \dim(N)$ columns before echelonization) and, in fact, if it should ever have d columns after echelonization, then we can conclude that the nullspace is trivial, which means that there are no nonzero module homomorphisms, and we can abort.

7.5.3 Testing irreducible modules for isomorphism

Testing two modules M and L over an algebra A for isomorphism is not straightforward in general because, even if we can compute $\mathrm{Hom}_A(M, L)$, it is not clear how to check whether this contains a surjective homomorphism. Some unpublished ideas for solving this problem have been proposed by C.R. Leedham-Green and implemented in MAGMA, but we shall not discuss them further here. In the situation where at least one of the modules to be tested, say M, has been proved to be irreducible by a call of MEATAXE, then there is an efficient test for isomorphism that is very similar to the function MODULEHOMOMORPHISMS in the preceding section.

Before applying this method, we call ABSOLUTELYIRREDUCIBLE($\mathcal{M}$) to calculate the degree e of the centralizing field of M. The method only works if the dimension d of the nullspace of $\mathcal{M}.\pi(\mathcal{M}.\theta)$ (which is the same as $\deg(\mathcal{M}.\pi)$) is equal to e (we saw in Subsection 7.5.1 that $d|e$). So, if $d \neq e$, then we choose new random elements θ from the matrix algebra of $\mathcal{M}$ and new irreducible factors π of the characteristic polynomial of θ until we find θ and π such that $d = e$. It is proved in [HR94] that the probability of a random θ giving rise to π with this property is at least 0.144, and so we can expect to find θ and π reasonably quickly. We shall assume for the remainder of this subsection that this has been done.

The point of having this condition satisfied is that it ensures that, for any two nonzero vectors in the nullspace N_M of $\mathcal{M}.\pi(\mathcal{M}.\theta)$, there is an element of the centralizing field of M mapping v to w and hence, if M and L are isomorphic, then the same property holds for the corresponding elements in L. Let N_L be the corresponding nullspace in L. Then, if $\varphi : M \to L$ is a module isomorphism, we must have $\varphi(N_M) = N_L$ and, for any $v \in N_M \setminus \{0\}$ and $w \in N_L \setminus \{0\}$, we can assume, by composing φ with a suitable module endomorphism of L, that $\varphi(v) = w$. This means that we do not need to maintain d separate lists $\tau_\lambda[1], \ldots, \tau_\lambda[d]$ of possible images of vectors in M under φ, as we did in MODULEHOMOMORPHISMS. We can also forget the

PVIS-I$(v, w, \sim\mathcal{V}, \sim\tau)$

    Input: Vectors v in M, w in L

1    $\beta := \mathcal{V}.\beta$; $\gamma := \mathcal{V}.\gamma$;

2   **for** $i \in [1 .. |\gamma|]$

3   **do** $a := v[\gamma[i]]$;

4   **if** $a \neq 0$

5   **then** $v := v - a{\cdot}\beta[i]$; $w := w - a{\cdot}\tau[i]$;

6    $c, a := \text{DEPTHVECTOR}(v)$;

7   **if** $c \neq 0$

8   **then** $\text{APPEND}(\sim\mathcal{V}.\beta, a^{-1}{\cdot}v)$; $\text{APPEND}(\sim\mathcal{V}.\gamma, c)$;

9           $\text{APPEND}(\sim\tau, a^{-1}{\cdot}w)$;

10 **elseif** $w \neq 0$ **then return** false;

11 **return** true;

MODULEISOMORPHISM$(\mathcal{M}, \mathcal{L})$

    Input: Module records $\mathcal{M}, \mathcal{L}$ with $\mathcal{M}$ proved irreducible by MEATAXE

          and $\mathcal{M}.e = \deg(\mathcal{M}.\pi)$

    Output: true or false; if true, a module isomorphism $M \to L$

1   **if** $\dim(\mathcal{M}) \neq \dim(\mathcal{N})$ **then return** false;

2    $\Lambda_M := \mathcal{M}.\Lambda$; $\Lambda_L := \mathcal{L}.\Lambda$;

3    $\xi_L := \mathcal{M}.\pi(\text{COPYTHETA}(\mathcal{M}, \mathcal{L}))$;

4    $\mathcal{N} := \text{NULLSPACE}(\xi_L)$; $d := |\mathcal{N}.\beta|$;

5   **if** $d \neq \deg(\mathcal{M}.\pi)$ **then return** false;

6    $w := \mathcal{N}.\beta[1]$; $\mathcal{V} := \text{rec}\langle \beta := [], \gamma := []\rangle$; $\tau := []$;

7   PVIS-I$(\mathcal{M}.v, w, \sim\mathcal{V}, \sim\tau)$;

8    $i := 1$;

9   **while** $i \leq |\mathcal{V}.\beta|$

10   **do for** $k \in [1 .. |\Lambda_M|]$

11   **do if not** PVIS-I$(\mathcal{V}.\beta[i]{\cdot}\Lambda_M[k], \tau[i]{\cdot}\Lambda_L[k], \sim\mathcal{V}, \sim\tau)$

12   **then return** false;

13      $i := i + 1$;

14 **return** true, $(\mathcal{V}.\beta^{-1}){\cdot}\tau$;

coefficients $\lambda_1, \ldots, \lambda_d$ and just work with a single basis of image vectors in L. If we encounter a contradiction of any kind, then we can conclude that M and L are not isomorphic. Otherwise, the function MODULEISOMORPHISM computes and outputs an explicit module isomorphism from M to L.

7.5.4 Application — invariant bilinear forms

An easy but important application of isomorphism testing is to the determination of invariant bilinear, sesquilinear, and quadratic forms for matrix groups over finite fields. As we shall see later in this chapter, in Subsection 7.8.2, computing with large classical groups in their natural representation over a finite field is one of the more challenging cases to be handled in the development of algorithms for finite matrix groups, mainly because the groups involved are very large.

The best approach seems to be first to identify the isomorphism type of the group $G \leq \mathrm{GL}(d,q)$ with high probability, perhaps by doing some kind of statistical analysis on the orders of its elements. We might then be reasonably certain that G is the symplectic group $\mathrm{Sp}(d,q)$, for example. But this only identifies G up to conjugacy in $\mathrm{GL}(d,q)$ and to progress further, we need to identify the symplectic bilinear form fixed by G. Then we can, if we wish, easily conjugate G to a group that fixes a standard symplectic form, such as the antidiagonal matrix with entries $\alpha_{i,d+1-i} = 1$ for $1 \leq i \leq d/2$ and $\alpha_{i,d+1-i} = -1$ for $d/2+1 \leq i \leq d$.

The matrix A of the bilinear form fixed by G satisfies $gAg^{\mathrm{T}} = A$ or, equivalently, $gA = A(g^{-1})^{\mathrm{T}}$, for all $g \in G$. But this just says that A is a module isomorphism from the d-dimensional module M over $\mathbb{F}_q$ defined by G to its dual module M^{D}. We can therefore find A by using the methods described in Subsection 7.5.3. Provided that G acts absolutely irreducibly on M, A will be uniquely determined up to a scalar multiple. We can use similar methods to find sesquilinear and quadratic forms that are fixed by absolutely irreducible matrix groups over finite fields.

7.5.5 Finding all irreducible representations over a finite field

There are situations in which we need to find explicitly certain irreducible representations of a finite group G over a finite field K. We might want to find representatives of all irreducibles up to isomorphism, or all irreducibles of degree less than a given bound. These may be required, for example, when compiling various databases of groups, such as the perfect groups databases and small groups databases to be discussed in Chapter 11.

Or perhaps, we know that a certain representation of some specific degree exists, and we want to find it explicitly. Indeed, the Meataxe algorithm was developed initially by Parker and others with this application in mind; the aim was typically to construct a specific representation of a sporadic simple group, usually over a very small finite field.

We shall briefly discuss some general methods for solving this problem here. It should be pointed out that, for finite solvable groups, alternative approaches are available (see Plesken's article [Ple87]), which probably perform better in practice if all representations are required.

The methods that we shall discuss here are based on an old result of Burnside (see Section 226 of [Bur02]), which has become known as the Burnside-Brauer theorem. This says that, if M is a faithful KG-module for a finite group G over a field K, then any irreducible KG-module occurs as a composition factor of the k-th tensor power $M^{\otimes k}$ of M for some $k \in \mathbb{N}$. Since, for KG-modules M, N, the composition factors of $M \otimes N$ all occur as composition factors of $M_i \otimes N_j$ as M_i, N_j range over the composition factors of M and N, this immediately suggests the following algorithm for finding all irreducible KG-modules.

Start with a small collection of KG-modules M_i, at least one of which is faithful. The easiest way to construct the M_i is as permutation modules coming from low-degree (faithful) permutation representations of G. More generally, they may be constructed as induced modules from small-degree representations of subgroups of G. Use the Meataxe and isomorphism testing algorithms to find all irreducible constituents of the M_i, and make a list of distinct irreducible KG-modules. Various related irreducibles, such as duals and conjugates under field automorphisms of the existing ones, can also be added to the list immediately. Then, for each pair N_i, N_j of irreducibles found, find the irreducible constituents of $N_i \otimes N_j$ and adjoin any new irreducibles to the list. Repeat these steps until no new irreducibles are found.

This approach can be adequate when a specific representation is being sought, or when the degrees of all of the irreducible modules are reasonably small. But if some of the irreducibles have large degree, then we wish to avoid tensoring these together whenever possible, because it becomes very expensive and possibly impractical to use the Meataxe on modules of very large degree (its complexity is at least $O(d^3)$).

As a first improvement, we can make use of the following standard result in modular representation theory; see, for example, Corollary 15.11 of [Isa76]. An element $g \in G$ is called *p-regular* if its order is coprime to p. Let $L \geq K$ be a splitting field for G (see Subsection 2.7.1). Then the total number of irreducible LG-modules is equal to the number of conjugacy classes of p-regular elements of G.

In many applications, it is relatively inexpensive to compute the conjugacy classes of G, and hence to determine in advance how many irreducible LG-modules there are. Let us assume that we can do that.

For each KG-irreducible module M that we find, we use the procedure ABSOLUTELYIRREDUCIBLE to calculate its centralizing field $C := \text{End}_{KG}(M)$. It is easy to see that M corresponds to $|C:K|$ distinct absolutely irreducible CG-modules, and hence also to $|C:K|$ distinct LG-irreducibles. We get one of these by making M into a CG-module just by using the action of C on M. The others are conjugates of this under the $|C:K|$ elements of $\text{Aut}_K(C)$, where the matrices for a conjugate are obtained by applying the field automorphism to the original matrices. These CG-modules can easily be computed explicitly from the output of ABSOLUTELYIRREDUCIBLE if required.

But whether or not we do this, we can keep count of the number of LG-irreducibles that we have found so far, and hence we will know when we have found them all. At that point, we will also have found all irreducible KG-modules, and we can stop. Note that there is no need to determine L explicitly, although it could be chosen to be the smallest finite field containing all of the centralizing fields C that have arisen.

Example 7.3

Let $G := \mathrm{Alt}(7)$ and $K := \mathbb{F}_3$. There are 6 classes of p-regular elements, with representatives $()$, $(1,2)(3,4)$, $(1,2)(3,4,5,6)$, $(1,2,3,4,5)$, $(1,2,3,4,5,6,7)$, $(1,7,6,5,4,3,2)$. We start with the natural 7-dimensional permutation module M over K. The irreducible constituents of M are M_1 and M_6, where the subscript indicates the dimension, and both are absolutely irreducible. We find that $M_6 \otimes M_6$ has irreducible constituents M_1, M_6, M_{13}, and M_{15}, the last two of which are new. Then $M_6 \otimes M_{13}$ has a new KG-irreducible constituent M_{20} but this has centralizing field $C := \mathbb{F}_9$, and so this corresponds to two conjugate absolutely irreducible CG-modules, both of dimension 10. So we have now found 6 absolutely irreducible LG-modules, with $L = \mathbb{F}_9$, and we are done. Of course, we could check this by applying the Meataxe to the other tensor products $M_i \otimes M_j$, but this would be a waste of effort. ☐

Unfortunately, there is no general result corresponding to the theorem that, in characteristic 0, the sums of the squares of the degrees of the absolutely irreducible representations is equal to $|G|$. So we have no information that might help us to predict the degrees of the LG-modules.

There are further ways of improving this algorithm, however, and this is still an ongoing area of research. One approach is to precompute the *Brauer character table* of G with respect to the characteristic p of K. The Brauer characters of the absolutely irreducible representations found so far can then be computed, and it becomes possible to predict in advance whether a given tensor product of irreducibles will yield new irreducibles. The condensation methods mentioned in Subsection 7.4.5 can sometimes be used on tensor products of large dimension. We omit the details, which have not yet been published. It is not always completely clear that the extra computations involving the Brauer characters will consume less resources than the unnecessary Meataxe applications that are avoided, but the indications are promising that this can be adapted into a worthwhile general-purpose approach.

7.6 Cohomology

The cohomology groups $H^n(G, M)$ for a group G acting on a KG-module M for some ring K are defined for all integers $n \geq 0$, but the groups $H^1(G, M)$ and $H^2(G, M)$ are the most relevant to group theory itself. As we saw in Section 2.7, $H^1(G, M)$ arises in connection with complements in semidirect products, whereas $H^2(G, M)$ is intimately connected with group extensions by abelian groups. For these reasons, it is important to have algorithms available to compute $H^1(G, M)$ and $H^2(G, M)$, at least for finite G and M. Such algorithms are the topic of this section.

7.6.1 Computing first cohomology groups

Let G be a group acting on a KG-module M, where K is a commutative ring with identity. We shall assume that M is free and finitely generated of rank d as a K-module, so the action of an element $g \in G$ on M can be represented as an element $\varphi(g) \in \mathrm{GL}(d, K)$, where matrices are written with respect to a fixed free basis $[b_1, \ldots, b_d]$ of M. We shall assume that the module is defined by a list Λ of action matrices for a generating sequence $X = [x_1, \ldots, x_r]$ of G; so $\Lambda[i] = \varphi(x_i)$ for $1 \leq i \leq r$. Elements of M will be represented as row vectors in K^d.

In this subsection, we shall describe a straightforward algorithm to compute the first cohomology group $H^1(G, M)$ as a K-module; see Subsection 2.7.2. Usually K is either a finite field or the integers $\mathbb{Z}$. The former case arises frequently in algorithms for investigating the structure of finite groups: in automorphism group computation (Section 8.9) and in the computation of all subgroups (Section 10.4), for example.

Occasionally, there is a need to carry out this calculation in cases when M is not free as a K-module; for example, when $K = \mathbb{Z}$ and M is a finite abelian group with invariant factors of different orders. The methods to be described can be generalized to handle that situation, but we shall not discuss it further here.

We shall compute $H^1(G, M)$ as the quotient $Z^1(G, M)/B^1(G, M)$. Let us start with the computation of the K-module $Z^1(G, M)$ of crossed homomorphisms (or derivations) $G \to M$.

For this calculation, we need G to be defined by a finite presentation. This is no problem if G is given as a PC-group (see Definition 8.7 below). For a permutation or matrix group G, we can use the methods described in Section 6.1 to compute a presentation, either on the original generating set X of G or, for larger groups, on a strong generating set. In the latter case, we can use the definition of the strong generators as SLPs in the original generators to compute the required matrices $\varphi(g_i)$ for the strong generators g_i. But, in any case, let us assume from now on that $G = \langle X \mid R \rangle$ is given as a finitely presented group, where $X = [x_1, \ldots, x_r]$ and $R = [w_1, \ldots, w_s]$.

Example 7.4

As an example, we shall use the presentation $G = \langle x, y \mid x^3, y^3, (x^{-1}y)^2 \rangle$ of the group $\mathrm{Alt}(4)$ (generators x, y correspond to $(1,3,2), (2,3,4) \in \mathrm{Alt}(4)$) acting on the permutation module over $\mathbb{F}_3$. So

$$\Lambda[1] = \begin{pmatrix} 0\,0\,1\,0 \\ 1\,0\,0\,0 \\ 0\,1\,0\,0 \\ 0\,0\,0\,1 \end{pmatrix}, \quad \Lambda[2] = \begin{pmatrix} 1\,0\,0\,0 \\ 0\,0\,1\,0 \\ 0\,0\,0\,1 \\ 0\,1\,0\,0 \end{pmatrix}.$$

⬜

By Proposition 2.73, a crossed homomorphism $\chi : G \to M$ is specified uniquely by the images $\chi(x_j)$ of the generators of G in M. We shall represent r-tuples of module elements $(m_1, \ldots, m_r)$ $(m_j \in M)$ by vectors $v \in K^{rd}$, where, for $1 \leq j \leq r$, the components of v in positions $d(j-1)+1 \ldots dj$ define m_j; that is, $v[d(j-1)+1 \ldots dj] = m_j$, written with respect to our fixed basis of M. In particular, we shall represent $\chi \in Z^1(G, M)$ by the corresponding vector in K^{rd} for $(\chi(x_1), \ldots, \chi(x_r))$.

For an arbitrary such r-tuple $(m_1, \ldots, m_r)$, we need to test whether there exists $\chi \in Z^1(G, M)$ with $\chi(x_j) = m_j$ for $1 \leq j \leq r$. We saw in Proposition 2.73 that such a χ must satisfy $\chi(x_j^{-1}) = -\chi(x_j)^{x_j^{-1}}$ and, for a general element $w = x_{k_1}^{\varepsilon_1} \cdots x_{k_l}^{\varepsilon_l} \in G$ with each $\varepsilon_i = \pm 1$,

$$\chi(w) = \sum_{i=1}^{l} \varepsilon_i \chi(x_{k_i})^{g_i}, \tag{$\dagger$}$$

where $g_i = x_{k_{i+1}}^{\varepsilon_{i+1}} \cdots x_{k_l}^{\varepsilon_l}$ or $x_{k_i}^{\varepsilon_i} \cdots x_{k_l}^{\varepsilon_l}$ when $\varepsilon_i = 1$ or -1, respectively. Furthermore, by Theorem 2.78, $\chi \in Z^1(G, M)$ if and only if the above expression evaluates to 0 for each $w \in R$.

Let $m_j = (z_{j1}, \ldots, z_{jd}) \in K^d$ for $1 \leq j \leq r$. Then, for a word w, the condition $\chi(w) = 0$ reduces to a system of d equations in the rd unknowns:

$$z_{11}, \ldots, z_{1d}, z_{21}, \ldots, z_{2d}, \ldots, z_{r1}, \ldots, z_{rd}.$$

To see this, let $\alpha(i) \in \mathrm{GL}(d, K)$ be the matrix $\varphi(g_i)$ in the above expression ($\dagger$) for $\chi(w)$. Then $\chi(x_{k_i})^{g_i}$ is just the vector $m_{k_i} \cdot \alpha(i)$, of which the i_2-th component (for $1 \leq i_2 \leq d$) is $\sum_{i_1=1}^{d} z_{k_i i_1} \alpha(i)_{i_1 i_2}$.

So, since there are s relators, each giving rise to d equations, the K-module $Z^1(G, M)$ is given by the solution set of a system of sd equations in rd unknowns; that is, by the nullspace of a certain $rd \times sd$ matrix Γ_1 over K. The function Z1-MATRIX calculates and returns this matrix Γ_1. For $1 \leq k \leq s$, the d equations arising from the relator $w_k \in R$ occupy the columns $d(k-1)+1 \ldots dk$ of Γ_1. For each individual w_k, we add the coefficients to Γ_1 arising from the terms $\chi(x_{k_i})^{g_i}$ in the expression ($\dagger$) in reverse order, $i := l, l-1, \ldots, 1$. This allows us to compute the action matrix $\alpha := \alpha(i)$ as we go along.

We compute $Z^1(G, M)$ as the nullspace of the matrix Z returned by Z1-MATRIX. If the ring K over which M is defined is a field, then we can use the function NULLSPACE presented in Section 7.2 to do this. Another case that arises frequently is $K = \mathbb{Z}$. The nullspace in that case can be computed from the matrix that transforms Z into Hermite normal form; see Proposition 9.6 below.

In Example 7.4, Z will be an 8×12 matrix over $\mathbb{F}_3$. Let us work through the computation on the third relator $x^{-1}yx^{-1}y$ of G, which will define columns $9 \ldots 12$ of this matrix. The loop beginning at line 12 of the algorithm will have the effect of adding a certain 4×4 matrix to the submatrix $Z[1 \ldots 4][9 \ldots 12]$

Z1-MATRIX(G, K, Λ)

      Input: $G = \langle X \mid R \rangle$, $X = [x_1, \ldots, x_r]$, $R = [w_1, \ldots, w_s]$;
              commutative ring K;
              list Λ of r matrices over K defining KG-module
      Output: Matrix with nullspace equal to $Z^1(G, M)$.

```
 1   r := |X|;  s := |R|;  d := |Λ[1]|;
 2   Z := ZERO-MATRIX(K, dr, ds);
 3   for k ∈ [1 .. s]
 4       do (* Fill in columns d(k−1)+1 .. dk of Z for relator wₖ *)
 5           c := d(k−1);
 6           α := IDENTITY-MATRIX(K, d);
             (* Scan relator k from right to left *)
 7           Suppose wₖ = x_{k₁}^{ε[1]} ⋯ x_{kₗ}^{ε[l]} with ε[i] = ±1;
 8           for i ∈ [l .. 1 by −1]
 9               do j := kᵢ;  e := ε[i];
10                   if e = −1 then α := Λ[j]⁻¹·α;
                     (* add α entries to Z in rows d(j−1)+1 .. dj *)
11                   r := d(j−1);
12                   for i₁ ∈ [1 .. d], i₂ ∈ [1 .. d]
13                       do Z[r+i₁][c+i₂] := Z[r+i₁][c+i₂] + eα[i₁][i₂];
14                   if e = 1 then α := Λ[j]·α;
15   return Z;
```

(when $j = 1$) or $Z[5 \mathinner{.\,.} 8][9 \mathinner{.\,.} 12]$ (when $j = 2$) of Z, so let us denote these submatrices as Z_1 and Z_2, respectively.

The i-loop at line 8 of the algorithm starts with $i = 4$, and the fourth generator of the relator is y, so we get $j = 2$, $e = 1$. The acting matrix α is the identity, so the loop at line 12 adds the identity matrix to the Z_2. We then multiply α on the left by $\Lambda[2]$ to give $\alpha = \lambda[2]$. Moving back to $i = 3$, we have $j = 1$, $e = -1$, so we multiply α on the left by $\Lambda[1]^{-1}$, to give

$$\alpha = \begin{pmatrix} 0 & 0 & 1 & 0 \\ 0 & 0 & 0 & 1 \\ 1 & 0 & 0 & 0 \\ 0 & 1 & 0 & 0 \end{pmatrix}$$

and add $-\alpha$ to Z_1. Proceeding to $i = 2$, we have $j = 2$, $e = 1$ again, and so we add the same α to Z_2, before multiplying α on the left by $\Lambda[2]$. Finally, with $i = 1$, we have $j = 1$, $e = -1$, so we multiply α on the left by $\Lambda[1]^{-1}$ to give $\alpha = I_4$, and we add $-\alpha$ to Z_1.

We have now defined columns $9 \mathinner{.\,.} 12$ of Z. Columns $1 \mathinner{.\,.} 4$ and $5 \mathinner{.\,.} 8$ are defined similarly using the first two relators x^3 and y^3, and the complete matrix Z over $\mathbb{F}_3$ returned is as shown below. The nullspace of Z is easily computed and is spanned by the rows of the matrix N below.

$$Z := \begin{pmatrix} 1\,1\,1\,0\,0\,0\,0\,0\,2\,0\,2\,0 \\ 1\,1\,1\,0\,0\,0\,0\,0\,0\,2\,0\,2 \\ 1\,1\,1\,0\,0\,0\,0\,0\,2\,0\,2\,0 \\ 0\,0\,0\,0\,0\,0\,0\,0\,0\,2\,0\,2 \\ 0\,0\,0\,0\,0\,0\,0\,0\,1\,0\,1\,0 \\ 0\,0\,0\,0\,0\,1\,1\,1\,0\,1\,0\,1 \\ 0\,0\,0\,0\,0\,1\,1\,1\,1\,0\,1\,0 \\ 0\,0\,0\,0\,0\,1\,1\,1\,0\,1\,0\,1 \end{pmatrix} \qquad N := \begin{pmatrix} 1\,0\,2\,0\,0\,0\,0\,0 \\ 0\,1\,2\,0\,0\,0\,2\,1 \\ 0\,0\,0\,1\,1\,0\,2\,1 \\ 0\,0\,0\,0\,0\,1\,0\,2 \end{pmatrix}$$

Returning to the general case, the K-module $B^1(G, M)$ is spanned by the derivations $\{\, g \mapsto b_i - b_i^g \mid 1 \le i \le d \,\}$, where $[b_1, \ldots, b_d]$ is our fixed K-basis of M. So it is equal to the row-space of the $d \times dr$ matrix B, where the entries in columns $d(j-1)+1 \ldots dj$ of the i-th row of B define the element $b_i - b_i^{x_j}$ of M. In other words, B is just the horizontal join of the matrices $I_d - \Lambda[j]$ for $1 \le j \le r$. The following function returns B.

B1-GM(G, K, Λ)

> **Input**: $G = \langle X \mid R \rangle$, $X = [x_1, \ldots, x_r]$, $R = [w_1, \ldots, w_s]$;
> commutative ring K;
> list Λ of r matrices over K defining KG-module M
> **Output**: Matrix with row-space $B^1(G, M)$

```
1   r := |X|;  s := |R|;  d := |Λ[1]|;
2   B := ZERO-MATRIX(K, d, dr);
3   for j ∈ [1 .. r]
4       do c := d(j−1);
5          for i₁ ∈ [1 .. d], i₂ ∈ [1 .. d]
6              do if i₁ = i₂ then B[i₁][c+i₂] := 1 − Λ[i₁][i₂];
7                 else B[i₁][c+i₂] := −Λ[i₁][i₂];
8   return B;
```

In Example 7.4, the matrix B returned is as below; we have also calculated a row echelonized matrix B_E with the same row-space as B:

$$B := \begin{pmatrix} 1\,0\,2\,0\,0\,0\,0\,0 \\ 2\,1\,0\,0\,0\,1\,2\,0 \\ 0\,2\,1\,0\,0\,0\,1\,2 \\ 0\,0\,0\,0\,0\,2\,0\,1 \end{pmatrix} \qquad B_E := \begin{pmatrix} 1\,0\,2\,0\,0\,0\,0\,0 \\ 0\,1\,2\,0\,0\,0\,2\,1 \\ 0\,0\,0\,0\,0\,1\,0\,2 \end{pmatrix}.$$

We see now that $H^1(G, M) = Z^1(G, M)/B^1(G, M)$ has dimension 1 over $\mathbb{F}_3$; that is, it has order 3. The derivation $\chi \in Z^1(G, M)$ defined by the third row of N, for which $\chi(x) = (0, 0, 0, 1)$ and $\chi(y) = (1, 0, 2, 1)$, maps onto a generator of $H^1(G, M)$.

7.6.2 Deciding whether an extension splits

We can use methods that are very similar to those described in the previous subsection to decide whether a given KG-module extension E of M by G splits. Indeed, we use the matrix Z returned by Z1-MATRIX once more.

As in Subsection 2.7.3, we assume that the extension E contains M as a subgroup, and that we have an epimorphism $\rho : E \to G$ with kernel M. We shall assume that G and M are defined in the same way as in Subsection 7.6.1; that is, we have a finite presentation $\langle X \mid R \rangle$ of G, and a list of matrices for the generators of G that define the module M. We assume also that we can carry out basic computations with elements of E, and that we can readily compute images and inverse images of elements under ρ.

In practice, E might be a finite permutation group with $M \trianglelefteq E$ and M an elementary abelian p-group, in which case we take $K = \mathbb{F}_p$. We could then work with a generating set $X \cup Y$ of E, where X generates E modulo M, and $Y = [b_1, \ldots, b_d]$ is a minimal generating sequence of M. A presentation of the quotient group $G = E/M$ could be computed by finding a presentation of E on $X \cup Y$ and then adding the generators of M as relators. The map ρ would then just map $x \in X \subseteq E$ to the corresponding generator x of G, and would map all elements of Y to 1_G. The action matrices $\Lambda[i]$ for $x_i \in X$ could be computed easily by expressing the elements y^x, for $x \in X$, $y \in Y$, as words in Y. This technique will be used as an essential part of our algorithm to find the conjugacy classes of subgroups of a finite permutation groups, which will be described in Section 10.4.

In any case, we start by choosing elements $\hat{x} \in E$ with $\rho(\hat{x}) = x$ for all $x \in X$. As in Subsection 2.7.3, let F be the free group on X and let $\theta : F \to E$ be the homomorphism defined by $\theta(x) = \hat{x}$ for $x \in X$. Then we can compute the elements $\theta(w)$ for $w \in R$ and, since $\rho(\theta(w)) = 1_G$, we have $\theta(w) \in M$ for all $w \in R$, so we can compute each $\theta(w)$ as a vector in M with respect to the free basis $[b_1, \ldots, b_d]$ of M. Let $v_E \in K^{sd}$ be defined by setting $v[d(k-1)+1 .. dk]$ to be equal to $\theta(w_k)$ for $1 \le k \le s$, where $R = [w_1, \ldots, w_s]$.

Let $\chi : X \to M$ be a map. Then we can use (i) and (ii) of Proposition 2.73 to extend χ to a derivation $\chi : F \to M$. By Proposition 2.77, the elements $\{\hat{x}\chi(x) \mid x \in X\}$ of E generate a complement of M in E if and only if $\chi(w) = -\theta(w)$ for all $w \in R$.

Let Z be the $dr \times ds$ matrix over K returned by Z1-MATRIX(G, K, Λ). In Subsection 7.6.1, we saw that the maps $\chi : X \to M$ for which $\chi(w) = 0$ for all $w \in R$ correspond to elements of the nullspace of Z. In the same way, the maps χ satisfying $\chi(w) = -\theta(w)$ for all $w \in R$ are given by the solutions u, if any, of the system of linear equations $-v_E = u \cdot Z$. As before, such solutions are elements $u \in R^{rd}$, where $u[d(j-1)+1 .. dj] = \chi(x_j)$ for $1 \le j \le r$. If u is one such solution, then the general solution is of the form $u + z$ for z in the nullspace of Z. If there is no solution, then the extension does not split.

So deciding whether an extension splits can be done in a very similar manner to finding all complements in a semidirect product. If K is a field, then this is just elementary linear algebra. Solving a system of linear equations $-v_E = uZ$ over $\mathbb{Z}$ is also possible, and can be done by first finding the Hermite normal form of Z. See Exercise 2 at the end of Section 9.2.

Example 7.5

Consider the following two subgroups of $\mathrm{Sym}(8)$. Let

$$x_1 = (1,5,3,2)(4,8,7,6),\ x_1' = (1,2,4,6)(3,5,7,8),$$
$$x_2 = (2,5)(6,8),\qquad x_3 = (1,3)(4,7),\qquad x_4 = (1,4)(2,6)(3,7)(5,8),$$

and let $E = \langle x_1, x_2, x_3, x_4 \rangle$, $E' = \langle x_1', x_2, x_3, x_4 \rangle$.

The reader can readily verify that both E and E' contain the group $M = \langle x_2, x_3, x_4 \rangle$ as a normal elementary abelian subgroup of order 8, where $x_2^{x_1} = x_2^{x_1'} = x_3$, $x_3^{x_1} = x_3^{x_1'} = x_2$, $x_4^{x_1} = x_4^{x_1'} = x_4$. Furthermore x_1^2 and $(x_1')^2$ both lie in M, so $|E| = |E'| = 16$. Hence, if $G = \langle x \mid x^2 \rangle$, then we can define epimorphisms $\rho : E \to G$ and $\rho' : E' \to G$ by setting $\rho(x_1) = \rho'(x_1') = x$ and $\rho(x_i) = \rho'(x_i) = 1_G$ for $2 \le i \le 4$.

In both cases, we have $r = s = 1$. The matrix $\Lambda[1]$, and the matrices Z and B returned by Z1-MATRIX and B1-GM are as follows:

$$\Lambda[1] = \begin{pmatrix} 0 & 1 & 0 \\ 1 & 0 & 0 \\ 0 & 0 & 1 \end{pmatrix}, \quad Z = \begin{pmatrix} 1 & 1 & 0 \\ 1 & 1 & 0 \\ 0 & 0 & 0 \end{pmatrix}, \quad B = \begin{pmatrix} 1 & 1 & 0 \\ 1 & 1 & 0 \\ 0 & 0 & 0 \end{pmatrix}.$$

The nullspace of Z is spanned by B and $(0,0,1)$, so $H^1(G, M)$ has dimension 1. In E, we have $x_1^2 = x_2 x_3$, so complements are given by solutions u of the equation $-v_E = u \cdot Z$ with $v_E = -v_E = (1,1,0)$. One such solution is $u = (1,0,0)$, so there is a complement generated by $x_1 x_2$. In E', however, we have $(x_1')^2 = x_4$ and the corresponding equation with $v_{E'} = (0,0,1)$ has no solution, so the extension is nonsplit. $\quad\square$

7.6.3 Computing second cohomology groups

To determine all of the KG-module extensions of a KG-module M by a group G, we need to compute $H^2(G, M)$. This problem is considerably more difficult than computing $H^1(G, M)$. The immediate difficulty is that a 2-cocycle $\tau : G \times G \to M$ is determined uniquely by its values $\tau(x, g)$ for $x \in X$ (a generating set of G) and $g \in G$, but not by $\tau(x, y)$ for $x, y \in X$. This means that the underlying vector space for a straightforward attempt to compute $Z^2(G, M)$ has dimension $r|G|d$, where $|X| = r$ and $d = \dim(M)$, which makes such a method impractical, even for groups of moderately large order.

Various applications, such as the algorithms to compile complete lists of perfect groups and all groups up to a specified order to be described in Sections 11.3 and 11.4, depend on the ability to compute $H^2(G, M)$, so it is important to devise practical methods for doing this.

Algorithms to solve this problem when and G is a finite permutation group and $K = \mathbb{F}_p$ are described in the three papers of Holt [Hol84, Hol85a, Hol85b]. These are beyond the scope of this book. According to a standard result, $H^2(G, M)$ is isomorphic to a certain subgroup of $H^2(P, M)$ for $P \in \mathrm{Syl}_p(G)$, and the algorithms first compute $H^2(P, M)$ and then proceed to identify the appropriate subgroup. They were used heavily in the compilation of the lists of perfect groups; see Section 11.3.

There is also an efficient algorithm to compute $H^2(G, M)$ when G is a finite PC-group. This will be described later in Subsection 8.7.2. It is essentially the same method as that used to compute $H^2(P, M)$ in the preceding paragraph. It was used in the construction of the library of small groups; see Section 11.4.

In both of the above applications, to help solve the isomorphism problem beween the groups constructed, the action of $\mathrm{Comp}(G, M)$ on $H^2(G, M)$ as defined in Subsection 2.7.4 was computed. This will also be discussed further in the polycyclic groups context in Section 8.9.

The author is currently actively working on the development of some more general methods to solve this problem, and some experimental implementations in MAGMA exist already. The group G must still be finite, but there is considerably more flexibility allowed for the module M.

7.7 Computing character tables

The first algorithm for computing character tables of finite groups was described by Burnside in [Bur11], although it could not really be thought of as a practical method at that time. He demonstrated it by using it to compute the character table of the dihedral group D_{10}, and even that was a tedious computation. An implementation by John McKay is described in [McK70]; he was able it to apply it in earnest to compute the character tables of the first two Janko sporadic simple groups J_1 and J_2. Improvements were introduced by Dixon in [Dix67] and later by Schneider in [Sch90b].

The resulting algorithm remains today the best general-purpose method available, although there are many improvements possible for skilled users who know how to make use of additional information that may be available. The Aachen-based CAS system [NPP84] was designed to give the user maximum flexibility with calculations involving group characters, although in recent years it has been superseded by GAP.

Here we shall content ourselves with providing a straightforward account of the Burnside-Dixon-Schneider algorithm. We shall assume that the reader has some familiarity with the rudiments of the character theory of finite groups, and we shall use [Isa76] as our main reference.

7.7.1   The basic method

Let $C_1, \ldots, C_r$ be the conjugacy classes of the finite group G with representatives $g_j \in C_j$, and let $h_j := |C_j| = |G : \mathbf{C}_G(g_j)|$. We choose $g_1 := 1$, so $h_1 = 1$.

We shall assume that the elements g_j and their centralizers $\mathbf{C}_G(g_j)$ and hence also the integers h_j have been computed. There are efficient methods for doing this in solvable groups defined by a PC-presentation, which will be discussed in Subsection 8.8.4. For permutation groups, the g_j were traditionally computed by choosing random elements of G and testing them for conjugacy with the elements already in the list until a complete set of class representatives had been found. Conjugacy testing of elements and computing centralizers was done using the methods described in Subsection 4.6.5. A newer method for finding class representatives, which uses the methodology to be discussed in Chapter 10, is described by Cannon and Souvignier in [CS97].

We must also assume that the *class map* of G is easily computable. That is, given $g \in G$, we must be able to quickly find the unique g_j that is conjugate to g in G. We do not, however, need to find a conjugating element, so in many cases this can be done quickly by considering the order $|g|$ of g or perhaps, if G is a permutation group, the cycle-type of g. In more difficult cases, it might be necessary to test g explicitly for conjugacy with some of the g_j that have the same order or cycle-type as g. Since large numbers of applications of the class map will generally be necessary for the computation of the class matrices M_j to be described below, if there is enough space available then it is worthwhile to precompute and store its values on all elements of G.

Let $\chi^1, \ldots, \chi^r$ be the irreducible characters of G, and let $\chi_j^i := \chi^i(g_j)$ for $1 \leq i, j \leq r$. Let $d_i := \chi_1^i$ be the degree of χ_i. For $1 \leq j, k, l \leq r$, fix $z \in C_l$, and let c_{jkl} be the number of pairs $(x, y) \in C_j \times C_k$ with $xy = z$. Then it is proved in Proposition 3.7 of [Isa76] that, for $1 \leq i, j, k \leq r$,

$$\frac{h_j \chi_j^i}{d_i} \frac{h_k \chi_k^i}{d_i} = \sum_{l=1}^{r} c_{jkl} \frac{h_l \chi_l^i}{d_i}. \tag{7.1}$$

For $1 \leq j \leq r$, let M_j be the $r \times r$ matrix whose (k, l)-entry is c_{jkl}. Then the Equations 7.1 say that, for each $1 \leq i \leq r$ and each $1 \leq j \leq r$, the column vector $[h_1 \chi_1^i / d_i, h_2 \chi_2^i / d_i, \ldots, h_r \chi_r^i / d_i]^{\mathrm{T}}$ is a right eigenvector of M_j.

In the Burnside and the Dixon versions of the algorithm, the M_j and their eigenvectors are computed, from which the values of $h_j \chi_j^i / d_i$ can be found. It is then not difficult to compute the d_i and, since the h_j are already known, we can find the required character values χ_j^i.

Schneider, however, found it more convenient to use a slightly different version of the Equations 7.1, so we shall do that here, too. Throughout this section, whenever $j \in [1 \mathbin{..} r]$, we shall use j' to denote the integer in $[1 \mathbin{..} r]$ such that $g_j^{-1} \in C_{j'}$; so $C_{j'} = C_j^{-1}$. Since $|C_l| = h_l$, the total number of triples $(x, y, z) \in C_j \times C_k \times C_l$ with $xy = z$ is equal to $h_l c_{jkl}$ But this is also equal to the total number of such triples with $y = x^{-1}z$, and so $h_l c_{jkl} = h_k c_{j'lk}$. Making this substitution in the right-hand side of the Equations 7.1, simplifying, and then interchanging j and j' yields

$$\frac{h_j \chi_{j'}^i}{d_i} \chi_k^i = \sum_{l=1}^{r} \chi_l^i c_{jlk} \tag{7.2}$$

for $1 \leq i, j, k \leq r$, which says that the row vectors $[\chi_1^i, \ldots, \chi_r^i]$ are left eigenvectors of M_j for $1 \leq i, j \leq r$.

We compute the matrices M_j column by column. To get column l of M_j, we fix $z = g_l$ and determine the conjugacy class of $y = x^{-1}z$, for each $x \in C_j$. Then the (k, l)-entry of M_j is just the number of these y that are in C_k. This calculation requires a large number of applications of the class map, which can be slow. As we shall see later, we can often avoid computing more than the first column of some of the M_j. To find all of the elements of C_j, we need a right transversal of $\mathbf{C}_G(g_j)$ in G, which we would normally compute and store at the same time as we find the g_j and $\mathbf{C}_G(g_j)$ themselves.

7.7.2 Working modulo a prime

A disadvantage of the original method as described by Burnside and as implemented by McKay [McK70] is that it involves floating point computations over the complex numbers, and the character values are returned as floating point approximations. If the character χ of G has degree d and $g \in G$, then $\chi(g)$ is a sum of $d |g|$-th roots of unity, and it is considerably more convenient to have it returned in that form. It would be possible in principal to perform the calculations exactly in an appropriate algebraic number field, but it turns out that the desired results can be found much more efficiently by performing the eigenvector computations modulo a suitable prime p, and then lifting the result back to the complex numbers. This was the principal improvement introduced by Dixon in [Dix67], and we shall describe it briefly in this subsection.

Let e be the least common multiple of the orders of the elements $g \in G$. Then, if ζ is a complex e-th root of unity, all character values χ_j^i lie in the ring $\mathbb{Z}[\zeta]$. By [Isa76, Theorem 3.7], the numbers $h_j \chi_j^i / d_i$ are algebraic integers. Since they lie in $\mathbb{Q}[\zeta]$, of which $\mathbb{Z}[\zeta]$ is the ring of integers [ST02, Theorem 3.5], we have $h_j \chi_j^i / d_i \in \mathbb{Z}[\zeta]$. Hence the Equations 7.2 are equations in $\mathbb{Z}[\zeta]$.

Now choose a prime p with $e|p-1$ and $p > 2d_i$ for $1 \leq i \leq r$. Of course, we do not know the d_i yet but, since $\sum_{i=1}^{r} d_i^2 = |G|$ [Isa76, Theorem 3.7], we can satisfy this condition by choosing $p > 2\lfloor \sqrt{|G|} \rfloor$. Since $e|p-1$, the field $\mathbb{F}_p$ has an element ω with multiplicative order e, and the mapping

$$\sum_{i=0}^{e-1} a_i \zeta^i \mapsto \sum_{i=0}^{e-1} a_i \omega^i$$

defines a ring homomorphism $\Theta : \mathbb{Z}[\zeta] \to \mathbb{F}_p$. Then, by applying Θ to the Equations 7.2, we obtain a corresponding system of equations over $\mathbb{F}_p$.

Let X be the $r \times r$ matrix with (i, j)-entry equal to χ_j^i – so X is the character table of G – and let Y be the $r \times r$ matrix with (j, i)-entry equal to $h_j \chi_{j'}^i$. Then, by the orthogonality relations of character theory [Isa76, (2.18)], we have $XY = |G|I_r$. In particular, the rows $X[i] = [\chi_1^i, \ldots, \chi_r^i]$ of X which, by the Equations 7.2 are eigenvectors of each matrix M_j for $1 \le i, j \le r$, are linearly independent. Since all primes q dividing $|G|$ divide e and $e | p - 1$, p does not divide $|G|$, and so the vectors $\Theta(X[i])$ are also linearly independent over $\mathbb{F}_p$.

For two rows $X[i]$ and $X[i']$ of X, the corresponding eigenvalues $h_j \chi_{j'}^i / d_i$ and $h_j \chi_{j'}^{i'} / d_{i'}$ cannot be equal for all j, for otherwise columns i and i' of Y would be linearly dependent. Again the same applies to their images under Θ. So the vectors $\Theta(X[i])$ are uniquely determined up to scalar multiples as r linearly independent common eigenvectors of the matrices $\Theta(M_j)$ $(1 \le j \le r)$.

Assuming that we have calculated the class matrices M_j for $1 \le j \le r$, we can apply Θ to obtain images $\Theta(M_j)$ over $\mathbb{F}_p$. We can use the methods described in Sections 7.2 and 7.3 to compute and factorize the characteristic polynomials of the $\Theta(M_j)$ to obtain their eigenvalues, and then find their eigenvectors. We can thus obtain r linearly independent common eigenvectors $v_1, \ldots, v_r$ over $\mathbb{F}_p$ of the matrices $\Theta(M_j)$. We shall return to this part of the computation in the next subsection, where we shall see that it is normally unnecessary to evaluate all of the M_j.

Let us normalize the v_i to make $v_i[1] = 1$ for all i. Each v_i is a scalar multiple of some row of $\Theta(X)$, which we can assume to be $\Theta(X[i])$. Since $\Theta(X[i][1]) = \Theta(\chi_1^i) = d_i$, we have $v_i = \Theta(X[i])/d_i$, and hence

$$\sum_{j=1}^{r} h_j v_i[j] v_i[j'] = |G|/d_i^2.$$

We can evaluate the left-hand side of this equation, and the fact that $p > 2d_i$ enables us to determine d_i uniquely from the value of d_i^2 modulo p. So we can find $\Theta(X[i]) = d_i v_i$ and hence the matrix $\Theta(X)$.

The final step of the algorithm is to recover X from $\Theta(X)$. Since each χ_j^i is a sum of d_i powers of ζ, we have $\chi_j^i = \sum_{k=0}^{e-1} m_{ijk} \zeta^k$ for integers m_{ijk} satisfying $0 \le m_{ijk} \le d_i$, and we need to find the m_{ijk}. The key to achieving this objective is the following lemma.

LEMMA 7.1   Let $\chi_j^i = \sum_{k=0}^{e-1} m_{ijk} \zeta^k$. Then

$$m_{ijk} = \frac{1}{e} \sum_{l=0}^{e-1} \chi^i(g_j^l) \zeta^{-kl}. \tag{7.3}$$

PROOF   For any $l \in \mathbb{Z}$, we have $\chi^i(g_j^l) = \sum_{k'=0}^{e-1} m_{ijk'} \zeta^{k'l}$. Substituting in the right-hand side of Equation 7.3 yields the expression

$$\frac{1}{e} \sum_{k'=0}^{e-1} m_{ijk'} \sum_{l=0}^{e-1} \zeta^{(k'-k)l}.$$

Since $\sum_{l=0}^{e-1} \zeta^l = 0$, the inner sum evaluates to e if $k' = k$ and to 0 otherwise, and the result follows.   ∎

Let us denote the number of the conjugacy class containing g_j^l by $j(l)$; so $\chi^i(g_j^l) = \chi^i_{j(l)}$. Then applying Θ to Equation 7.3 yields

$$\Theta(m_{ijk}) = \Theta(e)^{-1} \sum_{l=0}^{e-1} \Theta(\chi^i_{j(l)}) \omega^{-kl}.$$

Hence we can evaluate m_{ijk} modulo p but, since $0 \le m_{ijk} \le d_i < p$, this is sufficient to find the m_{ijk} themselves.

Here is a summary of the complete algorithm.

1. Let $e := \mathrm{lcm}(\{|g_j| \mid 1 \le j \le r\})$.

2. Choose a prime p with $e | p - 1$ and $p > 2\sqrt{|G|}$.

3. Calculate the matrices M_j for $1 \le j \le r$ and let $\Theta(M_j)$ be M_j reduced mod p.

4. Find r linearly independent common eigenvectors $v_1, \ldots, v_r$ of the matrices $\Theta(M_j)$, with $v_i[1] = 1$ for $1 \le i \le r$.

5. For $1 \le i \le r$, let d_i be the unique integer with $1 \le d_i < p/2$ that satisfies $\sum_{j=1}^{r} h_j v_i[j] v_i[j'] \equiv |G|/d_i^2 \pmod{p}$.

6. Compute the character values χ^i_j from $\Theta(\chi^i_j) = d_i v_i[j]$ as described above.

Example 7.6

Let $G := \mathrm{Alt}(4)$, and choose class representatives $g_1 = 1$, $g_2 = (1,2)(3,4)$, $g_3 = (1,2,3)$, $g_4 = (1,3,2)$, with $h_1 = 1$, $h_2 = 3$, $h_3 = h_4 = 4$. Then $e = 6$, and we can choose $p = 7$ and $w = 3$.

The first matrix M_1 is always the identity matrix, and so is not useful for calculation of eigenvectors. The remaining M_j are

$$M_2 = \begin{pmatrix} 0&1&0&0 \\ 3&2&0&0 \\ 0&0&3&0 \\ 0&0&0&3 \end{pmatrix}, \quad M_3 = \begin{pmatrix} 0&0&1&0 \\ 0&0&3&0 \\ 0&0&0&4 \\ 4&4&0&0 \end{pmatrix}, \quad M_4 = \begin{pmatrix} 0&0&0&1 \\ 0&0&0&3 \\ 4&4&0&0 \\ 0&0&4&0 \end{pmatrix}.$$

Now M_2 has an eigenvalue 3 with multiplicity 3, which is not very useful for finding linearly independent eigenvalues. The characteristic polynomial of M_3 (and also of M_4) is $x^4 + 6x$, which factorizes to $x(x+3)(x+5)(x+6)$ and so we can use M_3 alone to find the required eigenvectors. (We shall see in the next subsection that we cannot always use a single M_j for this purpose.) The normalized eigenvectors corresponding to eigenvalues $0, -3, -5, -6$ are

$$v_1 = (1\ 2\ 0\ 0),\ v_2 = (1\ 1\ 1\ 1),\ v_3 = (1\ 1\ 2\ 4),\ v_4 = (1\ 1\ 4\ 2),$$

from which we calculate $d_1 = 3$, $d_2 = d_3 = d_4 = 1$. We can now use the lemma to compute the character table X, which yields

$$\Theta(X) = \begin{pmatrix} 3\ 6\ 0\ 0 \\ 1\ 1\ 1\ 1 \\ 1\ 1\ 2\ 4 \\ 1\ 1\ 4\ 2 \end{pmatrix}, \qquad X = \begin{pmatrix} 3 & -1 & 0 & 0 \\ 1 & 1 & 1 & 1 \\ 1 & 1 & \zeta^2 & \zeta^4 \\ 1 & 1 & \zeta^4 & \zeta^2 \end{pmatrix},$$

where ζ is a primitive complex 6th root of unity, and we have replaced ζ^3 by -1. Of course, this character table would normally be printed with the unit character, which is $X[2]$, as the first row.   ⬜

With small examples like this, it is not really necessary to use the lemma to lift character values from $\mathbb{F}_p$ to $\mathbb{Z}[\zeta]$. The knowledge that χ_i^j is a sum of d_i $|g_j|$-th roots of unity allows us to guess the lifted value very quickly.

7.7.3 Further improvements

The algorithm as described so far is essentially that of Dixon [Dix67], except that we have followed Schneider in calculating characters as left row eigenvectors of the matrices M_j. Now we shall describe some of the improvements introduced in [Sch90b].

The bulk of the computing time goes into the calculation of the class matrices M_j, and so we aim to compute as few as these as possible and to avoid evaluating all columns of a particular M_j. (As explained earlier, we compute M_j a column at a time.) Notice that the first column of M_j has the single nonzero entry h_j in row j', so all first columns are known already.

The improvements to be discussed affect only Steps 3 and 4 of the algorithm. We do not compute all of the M_j in advance; rather, we compute them (completely or partially) one-by-one and, after each such computation, we attempt to complete Step 4 using only the M_j found so far. (A heuristic is suggested in [Sch90b] for deciding the order in which to compute the M_j.)

We maintain the notation of the previous subsection. So X is the matrix of character values, ζ is complex e-th root of unity with e the exponent of G, and $\Theta : \mathbb{Z}[\zeta] \to \mathbb{F}_p$ is a ring homomorphism for a suitably chosen prime p. We define a *character space* to be a subspace of $\mathbb{F}_p^r$ spanned by some of the rows of the matrix $\Theta(X)$.

We start by choosing some $j > 1$ and evaluating M_j. Since $\mathbb{F}_p^r$ is spanned by the eigenvectors of $\Theta(M_j)$, we have $\mathbb{F}_p^r = V_1 \oplus \cdots \oplus V_t$, where the V_i are the distinct eigenspaces of $\Theta(M_j)$. Notice that the V_i are all character spaces. They are the nullspaces of $\Theta(M_j) - \lambda_i I_r$ for the eigenvalues λ_i of $\Theta(M_j)$, and we compute echelonized bases of the V_i using NULLSPACE from Section 7.2.

If $\dim(V_i) = 1$ for all i, then we have completed Step 4 of the algorithm. In general, however, we will have $\dim(V_i) > 1$ for some V_i, and then we need to use a different M_j to split V_i into smaller dimensional character spaces.

The following lemma from [Sch90b] shows that we can decide whether a particular M_j will split V_i just from the first column of M_j, which is already known. So we need never evaluate an M_j unnecessarily.

LEMMA 7.2   *Let $b_1, \ldots, b_s$ be an echelonized basis of a character space V, and let W be the subspace of V spanned by $b_2, \ldots, b_s$. Then V is contained in a single eigenspace of $\Theta(M_j)$ if and only if W is fixed by $\Theta(M_j)$.*

PROOF   If W lies in a single eigenspace of $\Theta(M_j)$, then all nonzero vectors of W are eigenvectors of $\Theta(M_j)$, so clearly W is fixed by $\Theta(M_j)$.

Conversely, suppose that W is fixed by $\Theta(M_j)$ but that V is not contained in a single eigenspace of $\Theta(M_j)$. So V contains eigenvectors v_1 and v_2 for distinct eigenvalues λ_1 and λ_2 of $\Theta(M_j)$. Now the intersection of V with an eigenspace of $\Theta(M_j)$ is a character space and so it is spanned by rows of $\Theta(X)$, all of which have a nonzero first entry. Since b_1 is the only basis vector of V with nonzero first entry, we may assume that $v_1 = b_1 + w_1$ with $w_1 \in W_1$ and similarly $v_2 = b_1 + w_2$. Hence $v_1 \Theta(M_j) = \lambda_1 v_1 = \lambda_1 b_1 + \lambda_1 w_1$ and, since W is fixed by $\Theta(M_j)$, $b_1 \Theta(M_j)$ must have λ_1 as its first component. But, by the same argument applied to v_2, $b_1 \Theta(M_j)$ must have λ_2 as its first component, contradicting $\lambda_1 \neq \lambda_2$.   ∎

So the character space V_i lies in a single eigenspace of M_j if and only if each of its echelonized basis vectors $b_2, \ldots, b_s$ has zero in position j'. If not, then we can use M_j to decompose V_i into smaller character spaces. In fact, if the first nonzero entries of $b_1, \ldots, b_s$ are in columns $k_1, \ldots, k_s$ then, to calculate the induced action of $\Theta(M_j)$ on V_i, we only need columns $k_1, \ldots, k_s$ of M_j and so we need only evaluate these columns. Of course, if the same M_j is used to decompose more than one V_i, then the different V_i may require the evaluation of different columns of M_j.

Here is the modified version of Steps 3 and 4 of the algorithm in Subsection 7.7.2.

3. Calculate M_j for some j with $1 < j \leq r$.

4. Compute the set $\mathcal{V} := \{V_1, \ldots, V_t\}$ of eigenspaces of $\Theta(M_j)$ and an echelonized basis of each V_i.

While there exists i with $\dim(V_i) > 1$ repeat the following:

3'. Find a j such that V_i is not contained in a single eigenspace of $\Theta(M_j)$. Calculate the columns of M_j corresponding to the positions of the leading entries in the echelonized bases of each V_i that has this property.

4'. For each V_i not contained in a single eigenspace of $\Theta(M_j)$, compute the eigenspaces $V_{i1}, \ldots, V_{it_i}$ of $\Theta(M_j)$ on V_i together with their echelonized bases, and replace V_i by $V_{i1}, \ldots, V_{it_i}$ in $\mathcal{V}$.

Example 7.7
Let $G := \langle x, y \mid x^4 = y^3 = 1, x^{-1}yx = y^{-1} \rangle$. Then $|G| = 12$. There are 6 conjugacy classes, and we choose $g_1 = 1$, $g_2 = x^2$, $g_3 = y$, $g_4 = x$, $g_5 = x^{-1}$, $g_6 = x^2 y$, so $h_1 = h_2 = 1$, $h_3 = h_6 = 2$, $h_4 = h_5 = 3$. Note that $4' = 5$ and $j = j'$ for $j = 1, 2, 3, 6$. We have $e = 12$, so we can choose $p = 13$.

Let us start with
$$M_6 = \begin{pmatrix} 0\,0\,0\,0\,0\,1 \\ 0\,0\,1\,0\,0\,0 \\ 0\,2\,0\,0\,0\,1 \\ 0\,0\,0\,0\,2\,0 \\ 0\,0\,0\,2\,0\,0 \\ 2\,0\,1\,0\,0\,0 \end{pmatrix}.$$

The characteristic polynomial of $\Theta(M_6)$ factorizes over the ring $\mathbb{F}_{13}[x]$ as $(x + 1)(x + 2)(x + 11)(x + 12)$, and the matrices of the echelonized bases of the corresponding eigenspaces (which we also denote by V_i) are

$$V_1 = (1\,1\,6\,0\,0\,6), \qquad V_2 = \begin{pmatrix} 1 & 12 & 1 & 0 & 0 & 12 \\ 0 & 0 & 0 & 1 & 12 & 0 \end{pmatrix},$$

$$V_3 = \begin{pmatrix} 1 & 1 & 1 & 0 & 0 & 1 \\ 0 & 0 & 0 & 1 & 1 & 0 \end{pmatrix}, \qquad V_4 = (1\,12\,6\,0\,0\,7).$$

Now, for $i = 2$ and 3, we see that the second basis vector b_2 of V_i has a nonzero in position j' for $j = 4$ or 5, which means that we can use either of M_4 or M_5 to split these two subspaces. Let us use

$$M_4 = \begin{pmatrix} 0\,0\,0\,1\,0\,0 \\ 0\,0\,0\,0\,1\,0 \\ 0\,0\,0\,2\,0\,0 \\ 0\,3\,0\,0\,0\,3 \\ 3\,0\,3\,0\,0\,0 \\ 0\,0\,0\,0\,2\,0 \end{pmatrix}.$$

(Note that we need only the first and fourth columns of M_4 to compute its action on V_2 and V_3.) We find that the eigenspaces of M_4 on V_2 are spanned by vectors $(1, 12, 1, 8, 5, 12)$ and $(1, 12, 1, 5, 8, 12)$, whereas those on V_3 are spanned by $(1, 1, 1, 1, 1, 1)$ and $(1, 1, 1, 12, 12, 1)$. This completes the decomposition of $\mathbb{F}_{13}^6$ into one-dimensional character spaces. Using Step 5 of the original algorithm, we can calculate the degrees d_i (which are 2,1,1,1,1,2) and thereby find $\Theta(X)$. Then we can use Step 6 to compute X. Choosing 2 as our element ω of order 12 in $\mathbb{F}_p$, we get the result:

$$\Theta(X) = \begin{pmatrix} 2 & 2 & 12 & 0 & 0 & 12 \\ 1 & 12 & 1 & 8 & 5 & 12 \\ 1 & 12 & 1 & 5 & 8 & 12 \\ 1 & 1 & 1 & 1 & 1 & 1 \\ 1 & 1 & 1 & 12 & 12 & 1 \\ 2 & 11 & 12 & 0 & 0 & 1 \end{pmatrix}, \qquad X = \begin{pmatrix} 2 & 2 & -1 & 0 & 0 & -1 \\ 1 & -1 & 1 & \zeta^3 & \zeta^9 & -1 \\ 1 & -1 & 1 & \zeta^9 & \zeta^3 & -1 \\ 1 & 1 & 1 & 1 & 1 & 1 \\ 1 & 1 & 1 & -1 & -1 & 1 \\ 2 & -2 & -1 & 0 & 0 & 1 \end{pmatrix}.$$

☐

If we happen to know some of the irreducible characters of G before we start the computation, then we can use these to improve the algorithm. Define an inner product $\langle, \rangle$ on $\mathbb{F}_p^r$ by $\langle v, w \rangle = \sum_{j=1}^r h_j v[j] w[j]$. Then the matrix equation $XY = |G| I_d$ implies that the rows of $\Theta(X)$ are mutually orthogonal under this inner product. So we can compute the subspace V of $\mathbb{F}_p^r$ of vectors that are orthogonal to all of the known characters (or rather their images under Θ), and carry out our eigenspace computations in V rather than in $\mathbb{F}_p^r$.

In particular, it is generally worthwhile to compute the linear characters of G in advance, and then to apply the main algorithm within the orthogonal complement of the space that they span. The linear characters are just the homomorphisms from $G/[G,G]$ to the multiplicative group $\mathbb{C}^{\#}$ of the nonzero complex numbers, and these can be easily computed as follows. Express $G/[G,G]$ as a direct product of cyclic groups $C_{r_1} \otimes \cdots \otimes C_{r_s}$. Then

$$\mathrm{Hom}(G/[G,G], \mathbb{C}^{\#}) \cong \bigotimes_{i=1}^s \mathrm{Hom}(C_{r_i}, \mathbb{C}^{\#}),$$

where $\mathrm{Hom}(C_{r_i}, \mathbb{C}^{\#})$ is cyclic of order r_i, and is generated by the homomorphism mapping a generator of C_{r_i} to a primitive complex r_i-th root of unity.

Here is another trick to avoid having to calculate class matrices. If, after an intermediate pass of the algorithm, we have found one or more one-dimensional character spaces V_i, then we will have found some corresponding irreducible characters χ^i with $\dim(\chi^i) > 1$. Then the tensor product of two not necessarily distinct such characters $\chi^i \otimes \chi^{i'}$ is a (generally reducible) character of G given by $(\chi^i \otimes \chi^{i'})(g) = \chi^i(g) \chi^{i'}(g)$. By computing the inner product $\langle \chi^i \otimes \chi^{i'}, \chi \rangle$ for the already known irreducible characters χ of G, we can subtract from $\chi^i \otimes \chi^{i'}$ all of its irreducible components that are known characters, and hope that what remains will be a single new irreducible. Schneider reports in [Sch90b] that this does not happen sufficiently often for it to be worthwhile to do it by default, but it can occasionally be very effective in interactive implementations of the algorithm, which allow the user to choose a strategy after each pass.

Finally, we note that a character space of dimension two can very often be split without calculating a further class matrix; see [Sch90b] for details.

Exercise

1. Show that the dihedral and quaternion groups of order 8 both have 5 conjugacy classes. Calculate their class matrices, show that they are the same for D_8 and Q_8, and deduce that D_8 and Q_8 have the same character tables. Compute this character table.

7.8 Structural investigation of matrix groups

As we discussed in Chapter 3, there are three basic methods of representing groups on a computer: groups of permutations of a finite set, groups of matrices over a ring, and groups defined by a finite presentation. Chapter 4 is devoted to computing in permutation groups, whereas algorithms for computing with groups defined by a finite presentation are handled in Chapter 5 and also in several later chapters of the book. That leaves matrix groups, which are the topic of this section.

As we shall see in the first subsection, finite matrix groups of moderate order and degree can be handled satisfactorily either by converting them explicitly to permutation groups acting on vectors or subspaces of their underlying vector space or, better, by extending the notion of base and strong generating set from permutation groups to matrix groups. This approach does not work for larger matrix groups, and it is only during the past decade that attempts have been made to develop effective algorithms in this area. This is currently an active area of research in CGT, and a detailed discussion lies outside the scope of this book, but we shall provide a summary of the methods developed together with pointers to the literature.

7.8.1 Methods based on bases and strong generating sets

Recall from Chapter 4 that a base and strong generating set (BSGS) is a key concept in investigating the structure of a permutation group. See Section 4.4 for definitions, and for a descriptions of the Schreier-Sims and random Schreier-Sims algorithms for computing a BSGS in a finite permutation group.

We now outline how this concept of fundamental computational importance for permutation groups has been adapted to matrix groups defined over finite fields.

There is of course a natural action of a matrix group $G \leq \mathrm{GL}(d, q)$ on the underlying vector space. Let V denote the d-dimensional space of row vectors over the field $F = \mathbb{F}_q$. We define the action of $g \in G$ on $v \in V$ by $v^g = v \cdot g$. This action is clearly faithful, so G can be considered as a permutation group acting on the vectors of V.

We can now apply the Schreier-Sims algorithm to G and construct a BSGS for its action on the vectors of V, where we take the base points to be standard basis vectors for V.

A central component in constructing a BSGS is the computation of the *basic orbits* — namely, the orbits of the base points β_i under $G^{(i)}$ for $i = 1, \ldots, k$. The performance and range of application of the technique depends greatly on the sizes of these basic orbits.

Observe that the size of V is q^d and so grows exponentially with d. Hence the basic orbits obtained can be very large; with simple groups in particular, the first basic index is often the order of the group.

One technique for finding smaller basic orbits is to consider the action of G on a set other than the vectors of V; for example, the set of subspaces of V having a specified dimension. Often this action is not faithful, but we can ensure that it becomes so by adding some vectors as additional base points.

The first application of the Schreier-Sims algorithm to matrix groups defined over a finite field was by Butler [But76, But79]. In an attempt to reduce the size of the basic orbits, he considered the action of G on the one-dimensional subspaces of V. The use of these subspaces may reduce the size of the basic orbits by a factor of up to $q - 1$. Since the action of G on the subspaces need not be faithful, each subspace in the base is followed by a nonzero vector contained in it.

Murray and O'Brien [MO95] developed and investigated a more general strategy for selecting base points for matrix groups, which we expect *a priori* to have 'small' orbits under the action of a subgroup in the stabilizer chain of a matrix group. In summary, the strategy selects some common eigenvectors for a collection of elements of the group and then uses them and related spaces to obtain a base.

We now consider this strategy in more detail. Let A be a generator of G and let $g(x)$ be a factor of the characteristic polynomial of A that is irreducible over $\mathbb{F}_q$. We call $v \in V$ a *generalized eigenvector* if $v \cdot g(A) = 0$, but throughout this section we shall refer to such a v simply as an eigenvector. The subspace of all such eigenvectors is the *eigenspace* of A corresponding to $g(x)$. We do not require that $g(x)$ be linear. If $g(x)$ has degree m, the size of the corresponding orbit is bounded by $q^m - 1$. Hence we choose as base points eigenvectors corresponding to a factor that has as small a degree as possible. Now suppose that $g(x) = x^m - \lambda$, for some $\lambda \in \mathbb{F}_q$. Then the orbit $v^{\langle A \rangle}$ contains at most $m(q - 1)$ points, since $v^{A^m} = \lambda v$ and $\lambda^{q-1} = 1$. Investigations suggest that it is also useful to choose as base points eigenvectors corresponding to a factor that has as few nonzero terms as possible.

An eigenvector of one generator need not have a small orbit under the action of a member of the stabilizer-chain. However, we can improve the probability of finding a small orbit by using a base point that is an eigenvector of more than one group element. First we calculate the set of eigenspaces of the generators and add to it all of the nontrivial intersections of these spaces. We order the elements of this set by considering the polynomial factors corresponding to each subspace and giving precedence to those spaces whose factors have the lowest degree and the smallest number of nonzero terms. We now choose as our good base points one vector from each of the list of spaces. When we select a vector as a base point, we always precede it in the base by the corresponding one-dimensional subspace. An alternative is to choose the complete subspaces, rather than selecting a vector from each. Occasionally, this can result in a reduction in orbit size sufficient to compensate for the increased time taken to calculate images.

One problem with this strategy is that the generating set for G usually contains very few elements; for example, every sporadic simple group can be generated by two elements. We can greatly improve our chances of finding a small orbit by also calculating eigenvectors of a number of random elements of the group. This not only improves our chances of finding a 'good' eigenvector, but also allows us to choose base points that are simultaneously eigenvectors of a larger number of group elements.

An additional criterion that proved useful is the *sparseness* of a row-space, defined as the number of zeros in the row-reduced matrix of basis vectors divided by the dimension of the space. Hence, we used sparseness as one of the criteria for ordering the intersections.

The algorithm to select base points is summarized below. The eigenspaces and intersections are sorted according to five criteria. Only the first of these has theoretical justification; the inclusion and order of the remaining criteria is largely based on experimentation.

GOODBASEPOINTS(G)

    Input: Group $G \leq \mathrm{GL}(d, q)$
    Output: List $\mathcal{L}$ of base points of G for use with Schreier-Sims
1   Construct a number of random group elements;
2   Calculate eigenspaces of the generators and random elements;
3   Calculate all nontrivial intersections of the eigenspaces;
4   Let $\mathcal{L}$ be the resulting list of eigenspaces and intersections;
5   Order $\mathcal{L}$ by the following criteria:
6   Smallest degrees of corresponding factors,
7   Smallest numbers of nonzero terms in corresponding factors,
8   Largest number of eigenspaces intersected,
9   Greatest sparseness,
10   Smallest dimension;
11 Return $\mathcal{L}$;

In seeking to order the collection of subspaces, we first order the eigenspaces according to the first two criteria, and then choose the best from the relevant factors as the corresponding factor for an intersection. Now we can sort all of the subspaces according to the five criteria.

Example 7.8

We illustrate by a simple example that the choice of base points is significant.

Consider the first Janko simple group J_1 of order $175\,560$, which has an irreducible representation of degree 14 over $\mathbb{F}_{11}$. Two generating matrices for this representation of J_1 are displayed below.

Let V be the natural vector space of degree 14 over $\mathbb{F}_{11}$ on which G acts. If the first base point is chosen to be either the first standard basis vector $V.1$ or the one-dimensional subspace $\langle V.1 \rangle$, then the first basic orbit has length $175\,560$, the order of G.

$$
\begin{pmatrix}
0 & 1 & 0 & 0 & 0 & 0 & 0 & 0 & 0 & 0 & 0 & 0 & 0 & 0 \\
1 & 0 & 0 & 0 & 0 & 0 & 0 & 0 & 0 & 0 & 0 & 0 & 0 & 0 \\
0 & 0 & 0 & 1 & 0 & 0 & 0 & 0 & 0 & 0 & 0 & 0 & 0 & 0 \\
0 & 0 & 1 & 0 & 0 & 0 & 0 & 0 & 0 & 0 & 0 & 0 & 0 & 0 \\
0 & 0 & 0 & 0 & 0 & 0 & 1 & 0 & 0 & 0 & 0 & 0 & 0 & 0 \\
0 & 0 & 0 & 0 & 0 & 0 & 0 & 1 & 0 & 0 & 0 & 0 & 0 & 0 \\
0 & 0 & 0 & 0 & 1 & 0 & 0 & 0 & 0 & 0 & 0 & 0 & 0 & 0 \\
0 & 0 & 0 & 0 & 0 & 0 & 0 & 0 & 0 & 0 & 1 & 0 & 0 & 0 \\
0 & 0 & 0 & 0 & 0 & 1 & 0 & 0 & 0 & 0 & 0 & 0 & 0 & 0 \\
5 & 2 & 1 & 5 & 1 & 4 & 0 & 4 & 0 & 1 & 1 & 1 & 0 & 2 \\
1 & 4 & 8 & 4 & 7 & 8 & 8 & 5 & 1 & 9 & 9 & 8 & 0 & 9 \\
0 & 0 & 0 & 0 & 0 & 0 & 1 & 0 & 0 & 0 & 0 & 0 & 0 & 0 \\
0 & 10 & 6 & 0 & 9 & 7 & 5 & 4 & 2 & 8 & 4 & 3 & 10 & 8 \\
7 & 0 & 4 & 6 & 7 & 5 & 1 & 6 & 3 & 1 & 6 & 10 & 0 & 0
\end{pmatrix}
,
\begin{pmatrix}
8 & 10 & 0 & 0 & 0 & 0 & 0 & 0 & 0 & 0 & 0 & 0 & 0 & 0 \\
0 & 0 & 1 & 0 & 0 & 0 & 0 & 0 & 0 & 0 & 0 & 0 & 0 & 0 \\
0 & 0 & 0 & 0 & 1 & 0 & 0 & 0 & 0 & 0 & 0 & 0 & 0 & 0 \\
0 & 0 & 0 & 0 & 0 & 1 & 0 & 0 & 0 & 0 & 0 & 0 & 0 & 0 \\
0 & 0 & 0 & 0 & 0 & 0 & 1 & 0 & 0 & 0 & 0 & 0 & 0 & 0 \\
0 & 0 & 0 & 0 & 0 & 0 & 0 & 0 & 1 & 0 & 0 & 0 & 0 & 0 \\
0 & 0 & 0 & 0 & 0 & 0 & 0 & 0 & 0 & 1 & 0 & 0 & 0 & 0 \\
0 & 0 & 0 & 0 & 0 & 0 & 0 & 0 & 0 & 0 & 0 & 0 & 0 & 1 \\
8 & 8 & 1 & 6 & 8 & 7 & 10 & 0 & 4 & 6 & 4 & 8 & 3 & 10 \\
3 & 3 & 9 & 1 & 0 & 2 & 3 & 0 & 5 & 4 & 10 & 6 & 6 & 10 \\
2 & 6 & 3 & 6 & 9 & 7 & 2 & 9 & 6 & 8 & 5 & 1 & 6 & 0 \\
0 & 10 & 3 & 1 & 6 & 9 & 0 & 5 & 8 & 10 & 3 & 0 & 1 & 1 \\
0 & 5 & 4 & 5 & 2 & 7 & 1 & 9 & 5 & 5 & 2 & 1 & 6 & 7
\end{pmatrix}
.
$$

Let g be the following random element of G:

$$
\begin{pmatrix}
1 & 5 & 0 & 0 & 3 & 0 & 0 & 2 & 10 & 10 & 6 & 1 & 0 & 6 \\
2 & 10 & 7 & 5 & 0 & 3 & 1 & 6 & 1 & 3 & 6 & 6 & 6 & 6 \\
6 & 2 & 8 & 7 & 5 & 9 & 0 & 0 & 9 & 0 & 1 & 0 & 2 & 2 \\
5 & 7 & 2 & 6 & 6 & 2 & 6 & 9 & 2 & 7 & 5 & 3 & 5 & 10 \\
7 & 1 & 3 & 0 & 6 & 2 & 6 & 0 & 9 & 5 & 3 & 2 & 2 & 1 \\
9 & 10 & 5 & 3 & 2 & 8 & 9 & 4 & 0 & 1 & 9 & 5 & 5 & 1 \\
4 & 4 & 8 & 9 & 3 & 4 & 0 & 4 & 3 & 3 & 3 & 9 & 4 & 0 \\
5 & 4 & 3 & 1 & 2 & 1 & 2 & 7 & 6 & 6 & 4 & 7 & 0 & 3 \\
1 & 3 & 2 & 5 & 2 & 8 & 9 & 10 & 7 & 6 & 9 & 10 & 3 & 7 \\
9 & 10 & 7 & 6 & 9 & 6 & 7 & 4 & 8 & 5 & 8 & 6 & 5 & 10 \\
5 & 9 & 10 & 6 & 1 & 0 & 5 & 6 & 8 & 9 & 5 & 5 & 10 & 8 \\
2 & 3 & 5 & 3 & 5 & 9 & 7 & 4 & 10 & 10 & 9 & 8 & 10 & 10 \\
7 & 8 & 6 & 1 & 5 & 2 & 4 & 0 & 3 & 6 & 2 & 5 & 6 & 6 \\
10 & 9 & 5 & 9 & 6 & 1 & 6 & 4 & 7 & 8 & 7 & 7 & 10 & 8
\end{pmatrix}
.
$$

All nonzero elements of $\mathbb{F}_{11}$ occur as eigenvalues of g, where $1, 2, 6, 10$ have multiplicity 2. The eigenspace for 3 is the 1-dimensional space

$$E := \langle (1, 1, 4, 2, 6, 5, 9, 2, 6, 6, 6, 3, 4, 5) \rangle.$$

If we choose as the base points $[E, \langle V.1 \rangle]$, then the basic orbits have lengths $17\,556$ and 10. The eigenspace for 1 is the 2-dimensional space

$$E := \langle (1, 0, 2, 6, 7, 2, 5, 7, 2, 0, 2, 1, 1, 9), (0, 1, 10, 8, 7, 1, 8, 0, 0, 8, 1, 8, 3, 9) \rangle.$$

If we choose as the base points $[E, \langle V.1 \rangle]$ then the basic orbits have lengths 8778 and 20. Of course, we achieved a smaller basic orbit at the cost of computing and storing images of a 2-dimensional space. If we choose the other eigenspaces of g (or random vectors from these spaces) as the first base point for G, we observe identical behaviour.

The first generator of G has an eigenvalue 1 with multiplicity 6. Random vectors from its 6-dimensional eigenspace have orbits under the action of G of length $87\,780$; the eigenspace itself has an orbit of length only 1540. Again we must compare the relative cost of computing and storing images of spaces of dimension 1 or 6.   □

Of course, it is possible that all of the computed characteristic polynomials are irreducible and the resulting intersection is the complete space. If this occurs, our method will choose a base consisting of standard basis vectors: it defaults to Butler's strategy of choosing subspaces of dimension one, where each subspace is followed by a spanning vector.

By choosing base points that give shorter basic orbits, we extend significantly the range of application of the Schreier-Sims. These algorithms are the most important component of the current machinery for structural investigation of matrix groups. Implementations are available in GAP and MAGMA. While this model borrows heavily from permutation groups, algorithms in MAGMA do not rely on writing down an explicit permutation representation for the matrix group. Instead they use, as a central component, the (permutation group) concept of a stabilizer-chain, where one stabilizer has 'moderate' index in its predecessor. Some of these algorithms mimic their counterparts for permutation groups, in making use of the stabilizer-chain. For details of the algorithm to construct, for example, centralizers of elements of matrix groups, see the paper by Butler and Cannon [BC82].

7.8.2   Computing in large-degree matrix groups

In practice, many structural questions cannot be answered for an arbitrary matrix group, because a stabilizer-chain having 'moderate indices' simply does not exist (or, less commonly, cannot be found).

Critical to the successful application of the Schreier-Sims algorithm is the index $|G^{(i)} : G^{(i+1)}|$. Since S_n has a subgroup S_{n-1} of index n, we obtain a 'nice' subgroup chain, and hence the algorithm performs well even for large values of n. However, the 'optimal' subgroup chain for $GL(d, q)$ is

$$GL(d, q) \geq q^{d-1} . GL(d-1, q) \geq GL(d-1, q) \geq \ldots,$$

where the leading index is $q^d - 1$ and so grows exponentially with d.

In practice, many matrix groups may have no 'small-degree' permutation representation and so no useful stabilizer-chain. For example, the largest maximal subgroup of the sporadic simple group $J_4 \leq GL(112, 2)$ has index $173\,067\,389$.

Thus, in many cases, the stabilizer-chain approach is limited to matrix groups of degree up to about 10; however, any such claim is highly dependent on the defining field.

Hence, it is frequently impossible to study the structure of a matrix group of relatively small dimension. In particular it may be impossible to answer such a fundamental question as the order of the group. Indeed, it is only recently that the order of a *single* matrix can be found efficiently, using the algorithm of Celler and Leedham-Green [CLG97].

The paucity of effective tools for their study has prompted a major ongoing international research program to develop effective algorithms for structural investigations of matrix groups.

One of the competing approaches seeks to investigate instead whether a matrix group G satisfies certain natural and inherent properties in its action on the underlying vector space. If we can decide these questions, then we exploit this additional insight to assist us in analyzing the group structure.

A classification of the maximal subgroups of $GL(d, q)$ by Aschbacher [Asc84] underpins this approach. Let Z denote the group of scalar matrices of G. We say that G is *almost simple modulo scalars* if there is a nonabelian simple group T such that $T \leq G/Z \leq \text{Aut}(T)$, the automorphism group of T.

We can summarize Aschbacher's classification as follows: a matrix group preserves some natural linear structure in its action on the underlying space and has a normal subgroup related to this structure, or it is almost simple modulo scalars. More formally, we can paraphrase the theorem as follows.

THEOREM 7.3   *Let V be the vector space of row vectors on which $GL(d, q)$ acts, let G be a subgroup of $GL(d, q)$, and let Z be the subgroup of scalar matrices of G. Then one of the following is true:*

1. *G acts reducibly.*

2. *G acts imprimitively: G preserves a decomposition of V as a direct sum $V_1 \oplus V_2 \oplus \cdots \oplus V_r$ of $r > 1$ subspaces of dimension s, which are permuted transitively by G, and so $G \leq GL(s, q) \wr \text{Sym}(r)$.*

3. *G acts on V as a group of semilinear automorphisms of a space of dimensional d/e over the extension field $\mathbb{F}_{q^e}$ for some $e > 1$, and so G embeds in $\Gamma L(d/e, q^e)$. (This covers the class of 'absolutely reducible' matrix groups, where G embeds in $GL(d/e, q^e)$.)*

4. *G preserves a decomposition of V as a tensor product $U \otimes W$ of spaces of dimensions $d_1, d_2 > 1$ over F. Then G is a subgroup of a central product of $GL(d_1, q)$ and $GL(d_2, q)$.*

5. *G is definable modulo scalars over a subfield: for some proper subfield $\mathbb{F}_{q'}$ of $\mathbb{F}_q$, $G^g \leq GL(d, q').Z$ for some $g \in GL(d, q)$.*

6. *For some prime r, $d = r^m$ and G/Z is contained in the normalizer of an extraspecial group of order r^{2m+1}, or of a group of order 2^{2m+2} and symplectic type.*

7. *G is tensor-induced: it preserves a decomposition of V as $V_1 \otimes V_2 \otimes \cdots \otimes V_m$, where each V_i has dimension $r > 1$ and the set of V_i is permuted by G, and so $G/Z \leq PGL(r, q) \wr Sym(m)$.*

8. *G normalizes a classical group in its natural representation.*

9. *G is almost simple modulo scalars.*

Of course, the nine Aschbacher categories are not mutually exclusive.

In broad outline, this theorem suggests that a first step in investigating a matrix group G is to determine (at least one of) its categories in the Aschbacher classification. If a category of G can be recognized, then we can investigate its structure more completely using algorithms designed specifically for members of this category. Sometimes, this investigation will be greatly facilitated because we have reduced the size and nature of the problem. For example, if G acts imprimitively, then we obtain a permutation representation of degree at most d for G; if G is a tensor product, we now consider two matrix groups of smaller degree. Where a relevant normal subgroup N exists, (as it does for the first 7 cases), we recognize G/N recursively, ultimately obtaining a composition series for the group. If a composition series can be constructed for G, then we expect that many questions about the structure of G can be answered by solving first the problem for the composition factors of G.

What can we say about the almost simple groups? Liebeck [Lie85] proved that the maximal nonclassical subgroups of $GL(d, q)$ have order at most q^{3d}. Since these are small by comparison with $GL(d, q)$ (which has order $O(q^{d^2})$), it may be possible to identify these groups using variations of stabilizer-chain techniques.

We provide here only a top-level summary of the substantial work carried out on certain parts of this project.

1. The *Meataxe*, initially developed by Parker [Par84] and later generalized and analyzed by Holt and Rees [HR94] can decide if matrix groups, having degrees up to the high hundreds, are reducible. This was described in detail in Section 7.4.

2. An algorithm to decide whether a group acts imprimitively is presented in [HLGOR96b].

3. An algorithm to decide whether a group is 'absolutely reducible' is presented in [HR94]. The general case of semilinear groups is considered in [HLGOR96a].

4. An algorithm to decide whether a matrix group preserves a tensor-decomposition of the underlying vector space is presented by Leedham-Green and O'Brien in [LGO97b, LGO97a].

5. An algorithm to decide if a group is conjugate to one defined over a smaller field was developed by Glasby and Howlett in [GH97].

6. An algorithm to recognize whether a given subgroup G of $GL(r, q)$ contains a normal extraspecial r-group of order r^3 and odd prime exponent r is presented by Niemeyer in [Nie]. (This is the extraspecial normalizer case for $m = 1$.)

7. An algorithm to decide if a matrix group is tensor-induced is presented by Leedham-Green and O'Brien in [LGO02].

As part of the research program, two major types of algorithms have been developed.

- Nonconstructive recognition algorithms identify or 'name' a group.

- Constructive recognition algorithms 'name' the group G, provide a constructive homomorphism from the group to a 'useful' (or natural) representation H of G, and permit an arbitrary element of G to be written as a word in the generators of G.

A seminal paper by Neumann and Praeger [NP92], which presents a nonconstructive algorithm for the special linear group in its natural representation, provided much of the impetus for the Aschbacher-based research program. Algorithms to recognize other classical groups in their natural representation are developed by Niemeyer and Praeger in [NP98].

Constructive recognition algorithms for black-box groups isomorphic to the alternating and symmetric groups are presented by Bratus and Pak in [BP00] and by Beals, Leedham-Green, Niemeyer, Praeger, and Seress in [BLGN+02].

If a simple group is known to be sporadic, then it can be named readily, usually by considering element orders. Various constructive recognition algorithms have been developed for these groups. The Schreier-Sims algorithms can exploit known chains of subgroups described via the standard generators for such groups defined by Wilson in [Wil96]. An alternative approach reduces the problem of writing $g \in G$ as a word in its generators to writing related elements of G as words in the generators of centralizers of involutions; see [HLO+04].

The classical groups in their natural representation and the groups of Lie type in arbitrary representations constitute the most difficult outstanding cases.

If the group is a classical or exceptional group of Lie type in an arbitrary representation where the defining characteristic is known, then the polynomial-time Monte Carlo algorithm described in [BKPS02] names the group. More generally, it names the nonabelian composition factor of a quasi-simple group of Lie type.

A constructive recognition algorithm for $SL(d, q)$ in its natural representation is described by Celler and Leedham-Green in [CLG98]; algorithms for the other classical groups in their natural representation are developed by Brooksbank in [Bro03].

More generally, recognition algorithms for the classical groups in black-box representations have been developed by Kantor and Seress in [KS01]. The complexity of these algorithm involved a factor of q. In [CLGO], a *discrete log oracle* is used to obtain a constructive Las Vegas algorithm for a group isomorphic to $PSL(2, q)$ in any irreducible representation in characteristic dividing q, running in time that is polynomial in the input length. This 'base-case' solution has been further exploited in [BK01] to obtain better constructive recognition algorithms for certain classical families.

Constructive recognition algorithms for the exceptional groups in black-box representations are under development by Kantor and Magaard (2004).

Babai and Beals [BBR93, BB99] have developed an alternative and powerful theoretical framework for the study of matrix groups defined over finite fields. They focus on a characteristic series for a black-box group, whose smallest nontrivial term is the solvable radical; they use this series to develop randomized algorithms that solve many problems for linear groups in polynomial time.

Deterministic polynomial-time algorithms for nilpotent and solvable linear groups, which exploit their special structure. are developed by Luks in [Luk92].

Most of these algorithms assume that random elements of the group can be obtained efficiently; see Subsection 3.2.2. In many cases, the theoretical analysis of (the performance of) the algorithms employed in matrix recognition relies on detailed knowledge of the distributions of elements in the groups.

Chapter 8

Computation with Polycyclic Groups

In this chapter, we shall discuss methods for computing with polycyclic groups that are defined by a very particular type of group presentation, known as a *polycyclic presentation*. Unlike a general group presentation, a polycyclic presentation allows efficient structural computation within the group.

A group G is said to be *polycyclic* if it has a descending chain of subgroups $G = G_1 \geq G_2 \geq \cdots \geq G_{n+1} = 1$ in which each G_{i+1} is a normal subgroup of G_i, and the quotient group G_i/G_{i+1} is cyclic. Such a chain of subgroups is called a *polycyclic series*.

Every finitely generated nilpotent or abelian group is polycyclic and every polycyclic group is solvable. The polycyclic groups can also be defined as those solvable groups in which every subgroup is finitely generated. In particular, every polycyclic group is finitely generated. Further, the class of polycyclic groups is closed under taking subgroups, factor groups, extensions, and finite direct products. The classes of finite solvable and finite polycyclic groups are the same.

Polycyclic presentations are certain finite presentations of polycyclic groups, which reflect the polycyclic structure of the groups they define. Polycyclic presentations allow effective computations with their underlying groups, as we shall show in this chapter. For example, we shall observe that the word problem is effectively solvable for polycyclically presented groups.

In the first part of this chapter we shall introduce the basic background for computing with polycyclic groups and we shall recall the definition and the elementary properties of polycyclic presentations. Then we shall discuss some special classes of polycyclic groups and we shall provide some examples of polycyclic groups. The remainder of this chapter is then devoted to the introduction of various methods to compute with polycyclic presentations. This part of the chapter is restricted to finite polycyclic groups only, as this simplifies the outline of the methods significantly.

As we observed in Subsection 4.4.6, if we wish to perform detailed structural computations in a solvable finite permutation group, then it is usually worthwhile to use the methods described there to transfer to a polycyclic presentation of the group, and to use the methods to be described in this chapter to perform the computations in that context. The same applies to finite solvable matrix groups.

For background on the theory of polycyclic groups we refer the reader to Segal's book [Seg83]. An introduction to the fundamentals for computations with polycyclic groups can also be found in Sims' book [Sim94]. Further, extensions of the methods for finite polycyclic groups to arbitrary polycyclic groups are described by Eick in [Eic01a].

8.1 Polycyclic presentations

In this section, we shall introduce polycyclic sequences and polycyclic presentations together with an outline of their elementary features. This also includes a description of the *collection* algorithm, which can be used to solve the word problem for polycyclic presentations.

8.1.1 Polycyclic sequences

Let G be a polycyclic group with a polycyclic series $G = G_1 \geq G_2 \geq \cdots \geq G_{n+1} = 1$. As G_i/G_{i+1} is cyclic, there exist elements $x_i \in G$ with $\langle x_i G_{i+1} \rangle = G_i/G_{i+1}$ for every index i.

DEFINITION 8.1   A sequence of elements $X = [x_1, \ldots, x_n]$ such that $\langle x_i G_{i+1} \rangle = G_i/G_{i+1}$ for $1 \leq i \leq n$ is called a *polycyclic sequence* for G.

It is straightforward to show that every tail-sequence $X_i = [x_i, \ldots, x_n]$ is a polycyclic sequence for the subgroup G_i. This feature will be used frequently for induction arguments and algorithmic methods.

It is also easy to verify that the subgroups of the polycyclic series are exhibited by the polycyclic sequence X as $G_i = \langle x_i, \ldots, x_n \rangle$. In particular, the group G is generated by the elements in the sequence X. Moreover, this yields that the polycyclic series $G = G_1 \geq \cdots \geq G_{n+1} = 1$ is uniquely determined by X.

In the literature, there are various names that have been used to describe polycyclic sequences: for example, they are called polycyclic generating sequences in [Sim94] and they have been denoted as AG-systems in [LNS84].

DEFINITION 8.2   Let X be a polycyclic sequence for G. The sequence $R(X) := (r_1, \ldots, r_n)$ defined by $r_i := |G_i : G_{i+1}| \in \mathbb{N} \cup \{\infty\}$ is called the *sequence of relative orders* for X. The set $\{i \in \{1 .. n\} \mid r_i \text{ finite}\}$ is denoted by $I(X)$.

The sequence $R(X)$ and the set $I(X)$ depend on the underlying polycyclic series $G_1, \ldots, G_n$ of X only. Hence if Y is another polycyclic sequence for G defining the same polycyclic series as X, then $R(X) = R(Y)$ and $I(X) = I(Y)$.

The relative orders exhibit some basic information about the group G. For example, the group G is finite if and only if every entry in $R(X)$ is finite or, equivalently if and only if $I(X) = \{1 \ldots n\}$. If G is finite, then $|G| = r_1 \cdots r_n$, the product of the entries in $R(X)$.

Example 8.1
Let $G := \langle (1,2,3,4), (1,3) \rangle \cong D_8$ the dihedral group of order 8. We shall introduce a variety of polycyclic sequences for this group G. Using this example, we note that the length of a polycyclic sequence for G is not uniquely defined by G. Also, we observe that different polycyclic sequences of G may define the same polycyclic series.

a) Let $G_2 := \langle (1,2,3,4) \rangle \cong C_4$. Then the chain $G = G_1 \geq G_2 \geq G_3 = 1$ is a polycyclic series for G. This series allows several different polycyclic sequences. For example, the sequences $X := [(1,3), (1,2,3,4)]$ and $Y := [(2,4), (1,4,3,2)]$ are polycyclic sequences defining this series. They have the relative orders $R(X) = R(Y) = (2,4)$ and the finite orders set $I(X) = I(Y) = \{1,2\}$.

b) Let $G_2 := \langle (1,2)(3,4), (1,3)(2,4) \rangle \cong V_4$ and $G_3 := \langle (1,3)(2,4) \rangle \cong C_2$. Then the chain $G = G_1 \geq G_2 \geq G_3 \geq G_4 = 1$ is a polycyclic series for G. This series also allows several different polycyclic sequences. For example, the sequences $X := [(2,4), (1,2)(3,4), (1,3)(2,4)]$ and $Y := [(1,2,3,4), (1,2)(3,4), (1,3)(2,4)]$ are polycyclic sequences defining this series. They have the relative orders $R(X) = R(Y) = (2,2,2)$ and the finite orders $I(X) = I(Y) = \{1,2,3\}$.

We note that $8 = |G| = 2 \cdot 2 \cdot 2 = 2 \cdot 4$ is the product of the relative orders in all cases. ∎

Example 8.2
Let $G := \langle a, b \rangle$ with

$$a := \begin{pmatrix} -1 & 0 \\ 0 & 1 \end{pmatrix} \text{ and } b := \begin{pmatrix} -1 & -1 \\ 0 & 1 \end{pmatrix}.$$

Then $G \cong D_\infty$, the infinite dihedral group. As above, we shall describe a variety of polycyclic sequences for G. They will show that an infinite polycyclic group can have polycyclic sequences of unbounded length.

a) A shortest polycyclic sequence for G is the sequence $X := [a, ab]$ with the relative orders $R(X) = (2, \infty)$ and the finite orders $I(X) = \{1\}$. This sequence exhibits the structure of G as a semidirect product of an infinite cyclic group $\langle ab \rangle$ with a cyclic group $\langle a \rangle$ of order 2.

b) The group G has many other polycyclic sequences. In fact, the group G has infinitely many polycyclic sequences of almost arbitrary length and almost arbitrary relative orders. To observe this, let $l \in \mathbb{N}$ and let $n_i \in \mathbb{N}$ with $n_i \neq 1$ for $1 \leq i \leq l$. Then $Y := [a, ab, (ab)^{n_1}, \ldots, (ab)^{n_1 \cdots n_l}]$ is a polycyclic sequence with relative orders $R(Y) = (2, n_1, \ldots, n_l, \infty)$.

The number of infinite entries in the relative orders sequences is the same in both cases. It is known that this number of infinite entries in $R(X)$ is an invariant for G, which is called the *Hirsch length* or the rank of G; see [Rob82], 5.4.13. $\quad \square$

We shall now investigate an elementary, but important feature of polycyclic sequences.

LEMMA 8.3   Let $X = [x_1, \ldots, x_n]$ be a polycyclic sequence for G with the relative orders $R(X) = (r_1, \ldots, r_n)$. Then for every $g \in G$ there exists a unique sequence $(e_1, \ldots, e_n)$, with $e_i \in \mathbb{Z}$ for $1 \leq i \leq n$ and $0 \leq e_i < r_i$ if $i \in I(X)$, such that $g = x_1^{e_1} \cdots x_n^{e_n}$.

PROOF   Since $G_1/G_2 = \langle x_1 G_2 \rangle$, we find that $gG_2 = x_1^{e_1} G_2$ for some $e_1 \in \mathbb{Z}$. If $i \in I(X)$, then $r_1 < \infty$ and we can choose $e_i \in \{0 .. r_1 - 1\}$. With this additional property, we find that e_i is unique. Let $h = x_1^{-e_1} g \in G_2$. By induction on the length of a polycyclic sequence, we can assume that we have found an expression of the desired form for h; that is, $h = x_2^{e_2} \cdots x_n^{e_n}$. This yields $g = x_1^{e_1} x_2^{e_2} \cdots x_n^{e_n}$ as desired. $\blacksquare$

DEFINITION 8.4   The expression $g = x_1^{e_1} \cdots x_n^{e_n}$ of Lemma 8.3 is called the *normal form* of G with respect to X. The sequence $(e_1, \ldots, e_n)$ is the *exponent vector* of g with respect to X. We shall write $\exp_X(g) = (e_1, \ldots, e_n)$.

It is standard notation to neglect leading trivial entries in a normal form expression; that is, if $e_1 = \cdots = e_{i-1} = 0$, then it is standard to write $g = x_i^{e_i} \cdots x_n^{e_n}$ as a normal form for g. But the exponent vector is always written as a vector of length n.

Lemma 8.3 and Definition 8.4 yield an injective map $G \to \mathbb{Z}^n : g \mapsto \exp_X(g)$ from G into the additive group of $\mathbb{Z}^n$. It is important to note that this is not a group homomorphism and the structure of the polycyclic group G is in general not closely related to an abelian lattice! Nonetheless, the exponent vectors form the basic underlying structure for computations in polycyclic groups, and many algorithms for polycyclic groups generalize methods for abelian lattices in a noncommutative form.

DEFINITION 8.5 Let G be a polycyclic group with polycyclic sequence X and let $g \in G$ with $g \neq 1$ and $\exp_X(g) = (e_1, \ldots, e_n)$.

a) The *depth* of g is defined by:
$\text{dep}_X(g) = i$, if $e_1 = \cdots = e_{i-1} = 0$ and $e_i \neq 0$.

b) The *leading exponent* of g is defined by:
$\text{led}_X(g) = e_i$, if $\text{dep}_X(g) = i$

If $g \in G$ with $g = 1$, then $\exp_X(g) = (0, \ldots, 0)$. In this case we define $\text{dep}_X(g) = n + 1$ and we leave the leading exponent undefined.

The depth of an element $g \in G$ describes how far down in the polycyclic series the elements g lies. Thus $\text{dep}_X(g) = i$ is equivalent to $g \in G_i \setminus G_{i+1}$. In turn, this implies that the normal form for g is of the form $g = x_i^{e_i} \cdots x_n^{e_n}$.

Example 8.3
Consider the polycyclic sequence $X := [(1,3), (1,2,3,4)] = [x_1, x_2]$ for the dihedral group G. Let $g := (1,3)(2,4) \in G$. Then g has the normal form $g = x_1^0 x_2^2 = x_2^2$ and the exponent vector $\exp_X(g) = (0, 2)$. Thus $\text{dep}_X(g) = 2$ and $\text{led}_X(g) = 2$. ⬚

The exponent vectors of elements in a polycyclic group can be used to describe relations for G in its generators X. These relations are the first and fundamental step towards a polycyclic presentation for G.

LEMMA 8.6 *Let $X = [x_1, \ldots, x_n]$ be a polycyclic sequence for G with relative orders $R(X) = (r_1, \ldots, r_n)$.*

a) *Let $i \in I(X)$. Then the normal form of a power $x_i^{r_i}$ is of the form*
$x_i^{r_i} = x_{i+1}^{a_{i,i+1}} \cdots x_n^{a_{i,n}}$.

b) *Let $1 \leq j < i \leq n$. Then the normal form of a conjugate $x_j^{-1} x_i x_j$ is of the form $x_j^{-1} x_i x_j = x_{j+1}^{b_{i,j,j+1}} \cdots x_n^{b_{i,j,n}}$.*

c) *Let $1 \leq j < i \leq n$. Then the normal form of a conjugate $x_j x_i x_j^{-1}$ is of the form $x_j x_i x_j^{-1} = x_{j+1}^{c_{i,j,j+1}} \cdots x_n^{c_{i,j,n}}$.*

PROOF a) As $r_i = |G_i : G_{i+1}|$, we have that $x_i^{r_i} \in G_{i+1}$ and thus $\text{dep}_X(x_i^{r_i}) > i$. Thus the normal form expression for this power has the form described in a).

b+c) Since $j < i$, we have $x_i \in G_{j+1}$. As G_{j+1} is normal in G_j, this yields that the conjugates $x_j x_i x_j^{-1}$ and $x_j^{-1} x_i x_j$ are contained in G_{j+1}. Thus their depths are at least $j+1$ and their normal form expressions are of the form given in b) and c), respectively. ∎

8.1.2   Polycyclic presentations and consistency

DEFINITION 8.7   A presentation $\langle x_1, \ldots, x_n \mid R \rangle$ is called a *polycyclic presentation* if there exists a sequence $S = (s_1, \ldots, s_n)$ with $s_i \in \mathbb{N} \cup \{\infty\}$ and integers $a_{i,k}, b_{i,j,k}, c_{i,j,k}$ such that R consists of the following relations:

$$x_i^{s_i} = R_{i,i} \text{ with } R_{i,i} := x_{i+1}^{a_{i,i+1}} \cdots x_n^{a_{i,n}} \text{ for } 1 \leq i \leq n \text{ with } s_i < \infty,$$

$$x_j^{-1} x_i x_j = R_{i,j} \text{ with } R_{i,j} := x_{j+1}^{b_{i,j,j+1}} \cdots x_n^{b_{i,j,n}} \text{ for } 1 \leq j < i \leq n,$$

$$x_j x_i x_j^{-1} = R_{j,i} \text{ with } R_{j,i} := x_{j+1}^{c_{i,j,j+1}} \cdots x_n^{c_{i,j,n}} \text{ for } 1 \leq j < i \leq n.$$

These relations are called *polycyclic relations*. The first type are the *power relations* and the second and third types are the *conjugate relations*. And S is called the sequence of *power exponents* of the presentation.

Conjugate relations of the form $x_j^{-1} x_i x_j = x_i$ or $x_j x_i x_j^{-1} = x_i$ are called *trivial polycyclic relations*. They are often omitted from a polycyclic presentation to simplify the notation. This means that polycyclic presentations have to be distinguished from arbitrary finite presentations. For this purpose we denote them with

$$\text{Pc}\langle x_1, \ldots, x_n \mid R \rangle.$$

A group that is defined as and represented by a polycyclic presentation is known as a *PC-group*.

If a polycyclic presentation has n generators and k finite entries in S, then it has $k + n(n-1) = O(n^2)$ relations. This set of relations is usually highly redundant as a set of defining relations, but we shall observe in the following that it is useful for computational purposes.

We consider the special case of a polycyclic presentation on a single generator in the following example.

Example 8.4

Let $G = \text{Pc}\langle X \mid R \rangle$ be a group defined by a polycyclic presentation on one generator; that is, $X = [x_1]$. Then there are only two possibilities for R:

a) $R = \emptyset$. In this case G is the infinite cyclic group.

b) R contains a single power relation. In this case this relation is of the form $x_1^{s_1} = 1$ and thus G is a finite cyclic group of order s_1.

   ⬜

Every polycyclic group G has a polycyclic sequence X and every such sequence induces a complete set of polycyclic relations as outlined in Lemma 8.6. The power exponents S of the presentation equal the relative orders $R(X)$

in this case. It is straightforward to prove that the resulting set of polycyclic relations defines G as a group. This is summarized in the following.

THEOREM 8.8   *Every polycyclic sequence determines a (unique) polycyclic presentation. Thus every polycyclic group can be defined by a polycyclic presentation.*

Example 8.5
Let $G := \langle (1,3), (1,2,3,4) \rangle$ with polycyclic sequence $X := [(1,3), (1,2,3,4)]$ and relative orders $R(X) = (2,4)$. The polycyclic presentation defined by this sequence has the generators x_1, x_2, the power exponents $s_1 = 2$ and $s_2 = 4$ and its relations are $x_1^2 = 1$, $x_2^4 = 1$, $x_1 x_2 x_1^{-1} = x_2^3$ and $x_1^{-1} x_2 x_1 = x_2^3$.   ☐

Vice versa, we shall observe that every polycyclic presentation defines a polycyclic group.

THEOREM 8.9   *Let* $\mathrm{Pc}\langle x_1, \dots, x_n \mid R \rangle$ *be a polycyclic presentation and let G be the group defined by this presentation. Then G is polycyclic and $X = [x_1, \dots, x_n]$ is a polycyclic sequence for G. Its relative orders $R(X) = (r_1, \dots, r_n)$ fulfil $r_i \le s_i$ for $1 \le i \le n$.*

PROOF   We define $G_i := \langle x_i, \dots, x_n \rangle \le G$. Then the conjugate relations in R enforce that G_{i+1} is normal in G_i for $1 \le i \le n$. By construction, G_i/G_{i+1} is cyclic and hence G is polycyclic. As $G_i = \langle x_i G_{i+1} \rangle$ by definition, it follows that X is a polycyclic sequence for G. Finally, the power relations enforce that $r_i = |G_i : G_{i+1}| \le s_i$ for $1 \le i \le n$.   ∎

Example 8.6
We consider the group G defined by the following polycyclic presentation with power exponents $S = (3,2,\infty)$.

$$G := \mathrm{Pc}\langle x_1, x_2, x_3 \mid x_1^3 = x_3, \; x_2^2 = x_3,$$
$$x_1^{-1} x_2 x_1 = x_2 x_3, \; x_1 x_2 x_1^{-1} = x_2 x_3 \rangle.$$

By convention, this implies that the trivial polycyclic relations $x_1^{-1} x_3 x_1 = x_1 x_3 x_1^{-1} = x_2^{-1} x_3 x_2 = x_2 x_3 x_2^{-1} = x_3$ hold in G.

By Theorem 8.9, this yields that $X = [x_1, x_2, x_3]$ is a polycyclic sequence for G with relative orders $R(X) \le (3, 2, \infty)$. Using, for example, the methods of Chapter 5, it is straightforward to determine the precise relative orders as $R(X) = (3, 2, 1)$.

Hence this example shows that the power exponents in a polycyclic presentation give an upper bound for the relative orders only. It is not even possible to read off from the power exponents whether the given group is finite or infinite. ▯

Those polycyclic presentations for which the power exponents S and the relative orders $R(X)$ coincide play a central role in the algorithmic theory of polycyclic groups. In fact, computations with polycyclic groups are usually performed in such presentations.

DEFINITION 8.10 A polycyclic presentation $\mathrm{Pc}\langle X \mid R \rangle$ with power exponents S is called *consistent* (or *confluent*) if $R(X) = S$.

It is straightforward to observe that every polycyclic group has a consistent polycyclic presentation by using the same ideas as in Theorem 8.8. We summarize this as follows.

THEOREM 8.11 *Every polycyclic sequence determines a consistent polycyclic presentation. Thus every polycyclic group can be defined by a consistent polycyclic presentation.*

An effective method to check whether a given polycyclic presentation is consistent or to modify a given polycyclic presentation to an equivalent consistent one is outlined in Section 12.4.

8.1.3   The collection algorithm

Collection is a method that can be used to determine the normal form for an element in a group given by a consistent polycyclic presentation. Thus, in particular, collection can be used to solve the word problem in a group given by a consistent polycyclic presentation.

Collection can also be applied in the more general context of arbitrary polycyclic presentations and our outline in this section covers this more general case also. Hence we consider a group G defined by a polycyclic presentation $\mathrm{Pc}\langle X \mid R \rangle$ with power exponents S.

LEMMA 8.12 *Let $G = \mathrm{Pc}\langle X \mid R \rangle$ be a polycyclic presentation with power exponents S. For every element g in G there exists a word representing g of the form $w(g) = x_1^{e_1} \cdots x_n^{e_n}$ with $e_i \in \mathbb{Z}$ and $0 \leq e_i < s_i$ if $s_i \neq \infty$.*

PROOF This follows readily from Lemma 8.3 using that $R(X) \leq S$. ▮

DEFINITION 8.13 Words of the type considered in Lemma 8.12 are called *collected words*.

If the considered presentation $\text{Pc}\langle X \mid R \rangle$ is consistent, then $R(X) = S$ and the collected words coincide with the normal forms with respect to X. In this case there exists a unique collected word for every element in the given group G and thus we can solve the word problem.

The underlying group G is given as a finitely presented group (defined by a polycyclic presentation). Hence elements in G are usually described as words in the generators X. We use the following notation to deal with this situation.

DEFINITION 8.14 Let X be a set of abstract generators. In general, we write a (nonempty) word w in X as a string $w = x_{i_1}^{a_1} \cdots x_{i_r}^{a_r}$ with $a_j \in \mathbb{Z}$. We assume that $i_j \neq i_{j+1}$ for $1 \leq j \leq r-1$ and $a_j \neq 0$ for $1 \leq j \leq r$.

a) A word w is called *collected*, if $w = x_{i_1}^{a_1} \cdots x_{i_r}^{a_r}$ with $i_1 < i_2 < \cdots < i_r$ and $a_j \in \{1 .. s_j - 1\}$ if $s_j \neq \infty$. Otherwise w is *uncollected*.

b) A word v in X is a *minimal uncollected subword* of the word w if v is a subword of w and it has one of the following forms:

  i) $v = x_{i_j}^{a_j} \cdot x_{i_{j+1}}$ for $i_j > i_{j+1}$,

  ii) $v = x_{i_j}^{a_j} \cdot x_{i_{j+1}}^{-1}$ for $i_j > i_{j+1}$,

  iii) $v = x_{i_j}^{a_j}$ for $r_{i_j} \neq \infty$ and $a_j \notin \{1 .. s_{i_j} - 1\}$.

Note that a reduced word is collected if and only if it does not contain a minimal uncollected subword.

A central idea of the collection algorithm is to successively eliminate minimal uncollected subwords in a given word by applying the polycyclic relations. Before we introduce the formal outline of the collection algorithm, we give an example to illustrate the idea.

Example 8.7

Let $G := \text{Pc}\langle x_1, x_2 \mid x_1^2 = 1, x_1 x_2 x_1^{-1} = x_2^{-1}, x_1^{-1} x_2 x_1 = x_2^{-1} \rangle$ and let $w := x_2^2 x_1^7$, a word representing an element of G. We can find two different minimal uncollected subwords in w: x_1^7 and $x_2^2 x_1$.

a) We consider the minimal uncollected subword x_1^7 first. We modify w by applying the power relation $x_1^2 = 1$ and thus we obtain $w \equiv w' = x_2^2 x_1$. (Here '$\equiv$' denotes equality in the group G.) Now w' is a minimal uncollected subword itself. We modify this word by applying a conjugate relation in the following form:

$$w \equiv w' = x_2^2 x_1$$
$$= x_1 x_1^{-1} x_2^2 x_1$$
$$= x_1 (x_1^{-1} x_2 x_1)^2$$
$$\equiv x_1 (x_2^{-1})^2$$
$$= x_1 x_2^{-2}$$

b) We consider the minimal uncollected subword $x_2^2 x_1$ first. We modify w by applying a conjugate relation and thus we obtain $w \equiv w' = x_1(x_1^{-1}x_2^2 x_1)x_1^6 = x_1 x_2^{-2} x_1^6$. Iterating this process and applying a power relation as a final step yields

$$
\begin{aligned}
w \equiv w' &= x_1 x_2^{-2} x_1^6 \\
&\equiv x_1^2 x_2^2 x_1^5 \\
&\equiv x_1^3 x_2^{-2} x_1^4 \\
&\equiv x_1^4 x_2^2 x_1^3 \\
&\equiv x_1^5 x_2^{-2} x_1^2 \\
&\equiv x_1^6 x_2^2 x_1 \\
&\equiv x_1^7 x_2^{-2} \\
&\equiv x_1 x_2^{-2}
\end{aligned}
$$

The result in both cases is the same collected word (as it should be, because the presentation considered is consistent). We can also note at this stage already that in the first case we have to do significantly fewer applications of the polycyclic relations. ▯

We now give a more formal description of the collection process. The following contains a broad outline.

COLLECTEDWORD(X, R, w)

  Input: A polycyclic presentation $\mathrm{Pc}\langle\, X \mid R\,\rangle$, a word w in X
  Output: A collected word equivalent to w
1 **while** there exists a minimal uncollected subword in w
2   **do** choose a minimal uncollected subword v in w;
3   **if** $v = x_{i_j}^{a_j} x_{i_{j+1}}$ with $i_j > i_{j+1}$
4   **then** replace v by $x_{i_{j+1}}(R_{i_j, i_{j+1}})^{a_j}$ in w;
5   **if** $v = x_{i_j}^{a_j} x_{i_{j+1}}^{-1}$ with $i_j > i_{j+1}$
6   **then** replace v by $x_{i_{j+1}}^{-1}(R_{i_{j+1}, i_j})^{a_j}$ in w;
7   **if** $v = x_{i_j}^{a_j}$ with $a_j \notin \{1 \ldots s_{i_j} - 1\}$
8   **then** let $a_j = q s_{i_j} + r$ with $r \in \{0 \ldots s_{i_j} - 1\}$
9   replace v by $x_{i_j}^r (R_{i_j, i_j})^q$ in w;
10 **return** w;

In each step of this method we replace w by an equivalent word, since we only apply relations of R for the modifications of w. Hence, if the method terminates, then it returns a collected word equivalent to w. We shall now prove that this always happens.

THEOREM 8.15 *The collection algorithm terminates.*

PROOF We use induction on the number n of generators in X for this purpose. It is obvious that the method terminates if $n = 1$. (See Example 8.4 also.) Thus we assume by induction that the method terminates for all polycyclic presentations on $n-1$ generators. The given polycyclic presentation $\mathrm{Pc}\langle X \mid R \rangle$ contains a polycyclic presentation $\mathrm{Pc}\langle X_2 \mid R_2 \rangle$ where $X_2 = [x_2, \ldots, x_n]$ and R_2 consists of those relations in R that involve generators in X_2 only. By induction, the collection method terminates in this presentation $\mathrm{Pc}\langle X_2 \mid R_2 \rangle$. Hence the collection method terminates on every word that consists of generators in X_2 only.

Now let w be an arbitrary word in X. Then w contains only finitely many occurrences of x_1 and thus $w = x_1^{a_1} v_1 x_1^{a_2} v_2 \cdots x_1^{a_r} v_r$, where $v_1, \ldots, v_r$ are words in X_2. Since the collection method terminates on $v_1, \ldots, v_r$, we find that after finitely many steps, the collection method considers a minimal uncollected subword involving x_1. This minimal uncollected subword is eliminated by the algorithm and a word in X_2 is introduced for it. This process is iterated.

Since the method never introduces new occurrences of x_1, but only eliminates minimal uncollected subwords involving x_1, we find that after finitely many steps the word w is replaced by the equivalent word $x_1^{a_1 + \cdots + a_r \bmod s_1} v$, where v is a word in X_2. By induction, the collection algorithm terminates on v and hence it terminates on w also. ∎

The collection algorithm was implemented first by Neubüser [Neu61]. Since then, there have been several implementations and also various improvements and refinements of this algorithm. We shall briefly discuss some of these improvements and refinements here.

The most obvious refinement is in trying to find a good choice in Step 2 of the algorithm. This refinement has been considered in detail by Leedham-Green and Soicher in [LGS90]. As a result, the collection-from-the-left strategy is usually used for Step 2. This strategy chooses the first occurrence of a minimal uncollected subword by parsing through w from the left.

Another refinement of the general outline is to allow more complicated words as minimal uncollected subwords and to try to eliminate such words in a single collection step. For example, one can consider all subwords of the type $v = x_{i_j}^{a_j} x_{i_{j+1}}^{a_{j+1}}$ with $i_j > i_{j+1}$.

To collect such a subword, we rewrite it as

$$v = x_{i_{j+1}}^{a_{j+1}} (x_{i_{j+1}}^{-a_{j+1}} x_{i_j} x_{i_{j+1}}^{a_{j+1}})^{a_j} = x_{i_{j+1}}^{a_{j+1}} (x_{i_j}^{\varphi^{a_{j+1}}})^{a_j},$$

where φ is the isomorphism induced by the conjugation action of $x_{i_{j+1}}$ on $G_{i_{j+1}+1}$. It then remains to determine the image of x_{i_j} under the a_{j+1}-st power of the isomorphism φ. This image can be obtained effectively by evaluating the isomorphism φ.

Finally, it remains to mention that for special types of polycyclic presentations there are other methods available to determine collected words. The 'Deep-Thought' technique introduced by Leedham-Green and Soicher in [LGS98] is an example for such an alternative method. Deep-Thought applies to certain polycyclic presentations for nilpotent groups and it uses polynomial functions to determine collected words.

8.1.4 Changing the presentation

Usually, a polycyclic group G has several different polycyclic presentations. The number of generators and relations in these presentations can vary and some of these presentations may be more useful for effective computations than others. Hence we briefly consider the question how to modify a given polycyclic presentation to a computationally more useful one. The following lemma yields an obvious reduction.

LEMMA 8.16   Let $\mathrm{Pc}\langle X \mid R \rangle$ be a polycyclic presentation with power exponents S. If there exists an exponent $s_i = 1$, then the generator x_i is redundant in the polycyclic presentation and can be eliminated.

PROOF   If $s_i = 1$, then there exists a relation $x_i = R_{i,i}$ and $R_{i,i}$ is a word in the generators $x_{i+1}, \ldots, x_n$. Hence we can eliminate x_i from the list of generators and we replace each occurrence of x_i in a relation in R by $R_{i,i}$. Then we use the collection algorithm to determine collected words for all modified right-hand sides in relations in R.   ∎

Suppose that $G = \mathrm{Pc}\langle X \mid R \rangle$ is a finite group defined by a consistent polycyclic presentation with power exponents S. Computations with G are often easier if the entries in S are all primes. This can always be achieved as the following lemma shows.

LEMMA 8.17   Let $G = \mathrm{Pc}\langle X \mid R \rangle$ be a group defined by a consistent polycyclic presentation with power exponents S. Then there exists another polycyclic presentation $G = \mathrm{Pc}\langle X' \mid R' \rangle$ with power exponents S' such that each finite entry in S' is a prime.

PROOF   Let x_i be a generator with power exponent s_i. Suppose that s_i is finite and not a prime; that is, $s_i = ab$ with $a, b \in \mathbb{N}$ and $a, b \neq 1$. Let $x_i' := x_i^a$. Then we add this new generator to the sequence X. This yields a new polycyclic sequence X'. This new sequence defines a polycyclic presentation that expands the given one. Iterating this process eventually yields a polycyclic presentation of the desired type.   ∎

DEFINITION 8.18   We call a polycyclic presentation $\mathrm{Pc}\langle\, X \mid R \,\rangle$ whose power exponents S are all primes a *refined polycyclic presentation*.

The refined consistent polycyclic presentations will be used throughout for computations in finite polycyclic groups. The following lemma shows that these presentations yield a certain standardized form for finite groups.

LEMMA 8.19   *Let G be a finite polycyclic group. Then all refined consistent polycyclic presentations for G have the same number of generators and thus all these presentations have the same number of relations.*

PROOF   The number of generators in a refined consistent polycyclic presentation for a finite group G equals the number of prime factors of $|G|$. Hence this number is an invariant for G. The number of relations in a refined consistent polycyclic presentation depends on the number of generators only and hence it is also an invariant for G.   ∎

Finding good polycyclic presentations for infinite groups is significantly more difficult than for finite groups. For finite groups, we mostly work with refined consistent polycyclic presentations. There are various more advanced reductions and improvements available for these groups also. We refer to [CELG04] for a further discussion.

The relations in a polycyclic presentation depend on the generators and the power exponents only. Usually, the set of relations obtained is highly redundant. In the following lemma we observe that some of the polycyclic relations can always be omitted from a presentation, since they can be computed from the remaining relations.

LEMMA 8.20   *Let $\mathrm{Pc}\langle\, X \mid R \,\rangle$ be a consistent polycyclic presentation with power exponents S. If $s_j \neq \infty$. Then the relation $x_j x_i x_j^{-1} = R_{j,i}$ is a consequence of the relations $x_j^{s_j} = R_{j,j}$ and $x_j^{-1} x_k x_j = R_{k,j}$ for $j+1 \leq k \leq n$.*

PROOF   We note that $x_j^{-1} = x_j^{s_j-1} R_{j,j}^{-1}$ and $x_j = R_{j,j} x_j^{1-s_j}$. Thus

$$
\begin{aligned}
x_j x_i x_j^{-1} &= R_{j,j} x_j^{-(s_j-1)} x_i x_j^{s_j-1} R_{j,j}^{-1} \\
&= R_{j,j} x_j^{-(s_j-2)} R_{i,j} x_j^{s_j-2} R_{j,j}^{-1} \\
&= \ldots \\
&= R_{j,j} x_j^{-1} w(x_{j+1},\ldots,x_n) x_j R_{j,j}^{-1} \\
&= R_{j,j} v(x_{j+1},\ldots,x_n) R_{j,j}^{-1}
\end{aligned}
$$

∎

Exercises

1. Show that $X = [x_1, x_2, x_3, x_4] := [(1,2), (1,2,3), (1,2)(3,4), (1,3)(2,4)]$ is a polycyclic generating sequence for $G := \text{Sym}(4)$ and that, for $g \in G$, $g \in \text{Alt}(4) \Leftrightarrow \text{dep}_X(g) > 1$.

   Calculate the polycyclic presentation of G defined by X, and use the collection algorithm to write the element $x_4 x_3 x_2 x_1$ in normal form.

2. Let $G := \langle x, y \mid y^{-1}xy = x^2 \rangle$. Is G solvable? Is G polycyclic?

3. For which values of $m \in \mathbb{Z}$ is $\langle x, y \mid y^{-1}xy = x^m \rangle$ polycyclic?

8.2 Examples of polycyclic groups

In this section we shall outline some examples and discuss some special types of polycyclic presentations. Also, we include an application of the algorithmic theory of polycyclic groups to crystallographic groups.

8.2.1 Abelian, nilpotent, and supersolvable groups

Every finitely generated abelian group is polycyclic and hence has a polycyclic presentation. The following elementary lemma observes that the polycyclic presentations for abelian groups are easy to describe, and we can read off from a given polycyclic presentation whether the group defined by this presentation is abelian.

LEMMA 8.21 *Let $G = \text{Pc}\langle X \mid R \rangle$ be a group defined by a consistent polycyclic presentation. Then G is abelian if and only if all conjugate relations in R are trivial.*

Every finitely generated nilpotent group is polycyclic and hence has a polycyclic presentation. In fact, a finitely generated nilpotent group always has a polycyclic presentation of a special form, which we introduce in the following.

DEFINITION 8.22 *A polycyclic presentation $\text{Pc}\langle x_1, \ldots, x_n \mid R \rangle$ with power exponents S is called a nilpotent presentation if the conjugate relations R have the following form:*

$$x_j^{-1} x_i x_j = R_{i,j} := x_i x_{i+1}^{b_{i,j,i+1}} \cdots x_n^{b_{i,j,n}} \text{ for } 1 \le j < i \le n,$$
$$x_j x_i x_j^{-1} = R_{j,i} := x_i x_{i+1}^{c_{i,j,i+1}} \cdots x_n^{c_{i,j,n}} \text{ for } 1 \le j < i \le n.$$

Thus the right-hand sides of the conjugate relations $R_{i,j}$ and $R_{j,i}$ in a nilpotent presentation have depth i and leading exponent 1. In an arbitrary polycyclic presentation these relations can have smaller depth and arbitrary leading exponent.

LEMMA 8.23

a) *Every finitely generated nilpotent has a consistent nilpotent presentation defining it.*

b) *Every consistent nilpotent presentation defines a finitely generated nilpotent group.*

c) *If G is finitely generated nilpotent and torsion-free, then there exists a consistent nilpotent presentation defining G whose power exponents are all infinite. The number of generators in such a presentation is minimal for a polycyclic presentation of G.*

PROOF a) By Theorem 8.11 and Lemma 8.6, it is sufficient to show that G has a polycyclic sequence that induces the desired type of relations. This follows readily if we choose a polycyclic series $G = G_1 > \cdots > G_{n+1} = 1$ which refines a central series of G and we choose X as a polycyclic sequence defining this series.

b) The generators X of a consistent nilpotent presentation form a polycyclic sequence that can be seen, from the relations, to be a central series by the relations for the group.

c) If G is torsion-free, then the factors of the upper central series are also torsion-free (exercise). Hence there exists a polycyclic series for G which refines a central series and has infinite factors only. $\blacksquare$

In contrast to the abelian case, a finitely generated nilpotent group can also have consistent polycyclic presentations that are not nilpotent presentations.

Example 8.8

Let $G := \langle (1,2,3,4), (1,3) \rangle \cong D_8$. Then G is nilpotent and hence G has a nilpotent presentation. To compute such a presentation for G, we first determine a central series for G. For example, we can use the lower central series $G > \langle (1,3)(2,4) \rangle > \{1\}$ for this purpose. Then, we choose a polycyclic sequence whose polycyclic series refines the determined central series. For example, the sequence $X := [(1,3), (1,2,3,4), (1,3)(2,4)]$ is of this type. Finally, we compute the polycyclic presentation induced by X:

$$\text{Pc}\langle x_1, x_2, x_3 \mid x_1^2 = 1,\ x_2^2 = x_3,\ x_3^2 = 1,$$
$$x_1^{-1} x_2 x_1 = x_2 x_3,\ x_1 x_2 x_1^{-1} = x_2 x_3 \,\rangle.$$

However, the group G also has polycyclic presentations that are not nilpotent presentations. To obtain such a presentation, we choose a polycyclic sequence whose polycyclic series does not refine a central series. For example, $X := [(1,3),(1,2,3,4)]$ is of this type. The presentation induced by X reads:

$$\text{Pc}\langle\, x_1, x_2 \mid x_1^2 = 1,\ x_2^4 = 1,$$
$$x_1^{-1} x_2 x_1 = x_2^3,\ x_1 x_2 x_1^{-1} = x_2^3 \,\rangle.$$

☐

Yet another special case of polycyclic groups are the *supersolvable groups*, which are defined to be groups having a polycyclic series consisting of normal subgroups of G. A supersolvable group always has a polycyclic presentation of a special form.

DEFINITION 8.24   A polycyclic presentation $\text{Pc}\langle\, x_1, \dots, x_n \mid R\,\rangle$ with power exponents S is called a *supersolvable presentation*, if the relations R have the following special form:

$$x_j^{-1} x_i x_j = R_{i,j} := x_i^{b_{i,j,i}} \cdots x_n^{b_{i,j,n}} \text{ for } 1 \le j < i \le n,$$
$$x_j x_i x_j^{-1} = R_{j,i} := x_i^{c_{i,j,i}} \cdots x_n^{c_{i,j,n}} \text{ for } 1 \le j < i \le n.$$

Thus the right-hand sides of the conjugate relations $R_{i,j}$ and $R_{j,i}$ in a supersolvable presentation have depth i, while in an arbitrary polycyclic presentation they can have a smaller depth. The following lemma is proved in a similar way to Lemma 8.23.

LEMMA 8.25

   a) *Every supersolvable group has a supersolvable presentation defining it.*

   b) *Every supersolvable presentation defines a supersolvable group.*

8.2.2   Infinite polycyclic groups and number fields

Algebraic number fields can be used to construct interesting types of infinite polycyclic groups. We give a brief introduction to this construction here. First, we recall the underlying notations and definitions. We refer to [ST02] for background on algebraic number theory.

DEFINITION 8.26   An *algebraic number field* F is an extension of $\mathbb{Q}$ of finite degree $[F:\mathbb{Q}]$. Further, we have the following.

   a) The *maximal order* $O(F)$ of an algebraic number field F is defined as

$$O(F) := \{\, a \in F \mid \text{ there exists a monic } f_a(x) \in \mathbb{Z}[x] \text{ with } f_a(a) = 0 \,\}.$$

b) The *unit group* $U(F)$ of an algebraic number field F is defined as

$$U(F) := \{ a \in O(F) \mid a \neq 0 \text{ and } a^{-1} \in O(F) \}.$$

We collect a number of folklore results on algebraic number fields in the following.

THEOREM 8.27   *Let F be an algebraic number field and let $d := [F:\mathbb{Q}]$.*

a) *The maximal order $O(F)$ forms a ring whose additive group is isomorphic to $\mathbb{Z}^d$.*

b) *(Dirichlet) Let $d = s + 2t$, where s and $2t$ are the numbers of real and complex field monomorphisms $F \to \mathbb{C}$. Then the unit group $U(F)$ is a finitely generated abelian group of the form $U(F) \cong C_\infty^{s+t-1} \times C_m$ where $m \in 2\mathbb{N}$.*

The unit group $U(F)$ acts by multiplication from the right on the additive group of the maximal order $O(F)$. Thus we can form the split extension of these two groups and we obtain an interesting type of polycyclic group.

LEMMA 8.28   *Let F be an algebraic number field with degree $d \neq 1$. Then $O(F) \rtimes U(F)$ is an infinite polycyclic group.*

The resulting polycyclic groups are nonnilpotent (because $-1 \in U(F)$ inverts all elements of $O(F)$), and they have a complex structure. Computations with such groups translate directly to computations with algebraic number fields. These types of polycyclic groups are a major reason why computations with infinite polycyclic groups are much more complex than computations with finite polycyclic groups.

8.2.3   Application — crystallographic groups

Crystallographic groups are the symmetry groups of crystals. These symmetry groups have various applications in chemistry and physics, and it is in this context that the crystallographic groups were first studied.

Let C be a crystal with symmetry group G. What motions can we do with C without changing the crystal? There are two basic motions: the translation by an integer vector and the rotation of C in place. The symmetry group G contains all combinations of these two types of movements. The set of translations forms a normal subgroup T in G whose factor group corresponds to the motions in place and thus is finite. This leads to the following definition.

DEFINITION 8.29   An *(affine) crystallographic group G* is a subgroup of the group of all euclidean motions of d-dimensional space $\mathbb{R}^d$ such that the set T of all of its pure translations is a discrete normal subgroup of finite index. The factor group G/T is called the point group of G.

The translations T form a free abelian normal subgroup in a crystallographic group G and its rank is at most d. If the rank of T is d, then the group G is also called a *space group*.

Not all crystallographic groups are polycyclic, but many of them are. The following lemma indicates this. We refer to [BBN+78] for background.

LEMMA 8.30

a) *A crystallographic group is polycyclic if and only if its point group is polycyclic.*

b) *All space groups of dimension at most 3 are polycyclic.*

The algorithmic theory for polycyclic groups can be used to compute with polycyclic crystallographic groups. For this purpose we first need to determine a polycyclic presentation for a given crystallographic group. Since these groups usually arise naturally as affine subgroups of $\mathrm{GL}(d+1, \mathbb{Q})$, we use the affine structure for this purpose. We also note that various other methods to compute with crystallographic groups are outlined in [OPS98] and [EGN97].

We shall return to this topic in Section 11.5, where we shall be discussing the enumeration of space groups.

Exercises

1. Prove the claim in Lemma 8.23 that the factors in the upper central series of a torsion-free nilpotent group are torsion-free. Does the same property necessarily hold for the lower central series?

2. For the number field $\mathbb{Q}(\sqrt{2})$, we have $O(F) = \{a + b\sqrt{2} \mid a, b \in \mathbb{Z}\}$ and $U(F) = \langle 1+\sqrt{2} \rangle \times \langle -1 \rangle$. Find a polycyclic presentation of $O(F) \rtimes U(F)$, and show that this group has no nilpotent subgroup of finite index.

8.3 Subgroups and membership testing

Let G be a finite polycyclic group defined by the refined consistent polycyclic presentation $\mathrm{Pc}\langle X \mid R \rangle$. Suppose that $U \leq G$ is a subgroup described by a set of generators $U = \langle u_1, \dots, u_l \rangle$.

In the first part of this section we shall describe how to determine a polycyclic sequence for U. The sequence that is determined by the methods in this section is closely related to the sequence X, and thus it will be called an *induced* polycyclic sequence with respect to X.

The induced polycyclic sequences are the major tool for computing with subgroups in a polycyclically presented group G. For example, they allow us to compute a consistent refined polycyclic presentation for U and they facilitate an effective membership test for U.

In the second part of this section we describe a method to solve the subgroup problem for subgroups of G; that is, we introduce a method to check whether $U = V$ for two subgroups $U, V \leq G$. For this purpose we determine polycyclic sequences for U and for V, which are canonical with respect to X.

8.3.1   Induced polycyclic sequences

Let $G = G_1 > \cdots > G_{n+1} = 1$ be the polycyclic series defined by the polycyclic sequence X of G. Then, with $U_i := G_i \cap U$, we obtain the *induced polycyclic series* $U = U_1 \geq \cdots \geq U_{n+1} = 1$ for U. This induced series is a natural choice for a polycyclic series of U.

As G is defined by a refined presentation, the factors of the polycyclic series for G are nontrivial factors of prime order. The induced polycyclic series for a proper subgroup U contains trivial factors. Hence a polycyclic sequence for U defining the induced polycyclic series does not define a refined presentation. For computational purposes we want to remedy this situation.

First we investigate the situation in more detail. Since $|U_i : U_{i+1}|$ divides $|G_i : G_{i+1}| = r_i$ and r_i is a prime, we have $|U_i : U_{i+1}| = 1$ or $|U_i : U_{i+1}| = r_i$. Let $i_1 < \cdots < i_m$ denote those indices in $\{1 .. n\}$ with $|U_{i_j} : U_{i_j+1}| \neq 1$. Then $U = U_{i_1} > U_{i_2} > \cdots > U_{i_m} > U_{i_m+1} = 1$ is a polycyclic series for U with factors of prime order. This series leads to the following definition.

DEFINITION 8.31 $Y = [y_1, \ldots, y_m]$ is an *induced polycyclic sequence* for U with respect to X if Y defines the series $U = U_{i_1} > \cdots > U_{i_m} > 1$; that is, if $\langle y_j, U_{i_j+1} \rangle = U_{i_j}$ for $1 \leq j \leq m$.

Thus an induced polycyclic sequence for a subgroup U is a polycyclic sequence for U, which defines the natural induced polycyclic series of U. In the following we shall investigate induced polycyclic sequences in more detail.

LEMMA 8.32   *Let U be a subgroup of G and let $Y = [y_1, \ldots, y_m]$ be an induced polycyclic sequence for U defining the series $U = U_{i_1} > \cdots > U_{i_m} > 1$. Then*

a) Y *is a polycyclic sequence for U.*

b) $\mathrm{dep}_X(y_j) = i_j$ *and thus* $\mathrm{dep}_X(y_1) < \cdots < \mathrm{dep}_X(y_m)$.

c) $R(Y) = (r_{i_1}, \ldots, r_{i_m})$.

PROOF   a) is obvious from the definition.

b) We find that $y_j \in U_{i_j} \leq G_{i_j}$ and hence $\text{dep}_X(y_j) \geq i_j$. Suppose that $\text{dep}_X(y_j) > i_j$. Then $y_j \in G_{i_j+1}$ and thus $y_j \in U_{i_j+1} = U_{i_j+1}$ by construction. But $y_j U_{i_j+1}$ generates the nontrivial factor U_{i_j}/U_{i_j+1} and this yields a contradiction. Hence $\text{dep}_X(y_j) = i_j$.

c) The sequence Y defines the series $U = U_{i_1} > \cdots > U_{i_m} > U_n = 1$. Hence its relative orders are $r_{i_j} = |U_{i_j}:U_{i_j+1}|$ for $1 \leq j \leq m$. ∎

Let Y be an induced polycyclic sequence with respect to X and let B be the matrix whose rows consists of the exponent vectors of the elements in Y; that is,

$$B := \begin{pmatrix} \exp_X(y_1) \\ \vdots \\ \exp_X(y_m) \end{pmatrix} \in \mathbb{Z}^{m \times n}.$$

Then Lemma 8.32 b) says that B is a matrix in upper triangular form.

LEMMA 8.33   *Let U be a subgroup of G and let $Y = [y_1, \ldots, y_m]$ be an induced polycyclic sequence for U. Define $d := \{\text{dep}_X(y) \mid y \in Y\} \cup \{n+1\}$. Then*

a) *$\text{dep}_X(u) \in d$ for every $u \in U$.*

b) *$T := \{g \in G \mid \exp_X(g)_j = 0 \ \forall j \in d\}$ is a left transversal for U in G, where $\exp_X(g)_j$ is the j-th entry in the exponent vector of g.*

PROOF   a) Let $u \in U$. If $u = 1$, then $\text{dep}_X(u) = n+1 \in d$. If $u \neq 1$, then $u = y_j^{e_j} \cdots y_m^{e_m}$ such that $e_j \in \{1 \ldots r_{i_j}-1\}$ for some j. As $\text{dep}_X(y_k) > \text{dep}_X(y_j)$ for $k > j$, we have $\text{dep}_X(u) = \text{dep}_X(y_j) \in d$.

b) The set T contains $\prod_{j \notin d} r_j$ elements and thus $|T| = |G:U|$. It remains to show that two different elements in T lie in different left cosets. Thus let $s, t \in T$ and suppose that $sU = tU$. We assume w.l.o.g. that $\text{dep}_X(s) \geq \text{dep}_X(t)$. We use induction on $\text{dep}_X(t)$ to prove $s = t$. Suppose that $\text{dep}_X(t) = n+1$. In this case $t = 1$ and thus $s \in U$. By a) we have that $U \cap T = 1$ and thus $s = 1$. Hence $s = t$.

Now suppose that the claim has been proved for all $s, t \in T$ with $\text{dep}_X(s) \geq \text{dep}_X(t) \geq j+1$ for some $j \leq n$. Suppose that $\text{dep}_X(t) = j$. Then $s, t \neq 1$ and $\text{dep}_X(s), \text{dep}_X(t) \notin d$. Let $u \in U$ with $s = tu$. As $\text{dep}_X(u) \in d$ by a), we find that $\text{dep}_X(t) \neq \text{dep}_X(u)$. If $\text{dep}_X(u) < \text{dep}_X(t)$, then $\text{dep}_X(s) = \text{dep}_X(tu) = \text{dep}_X(u) \in d$ which is a contradiction. Thus we have that $\text{dep}_X(u) > \text{dep}_X(t)$ and $\text{dep}_X(s) = \text{dep}_X(tu) = \text{dep}_X(t)$. As $s = tu$, we obtain $\text{led}_X(s) = \text{led}_X(t)$ in this case. Hence $s = x_j^e s'$ and $t = x_j^e t'$ for $e = \text{led}_X(t)$ and $s', t' \in T$ with $\text{dep}_X(t') > j$ and $s'U = t'U$. By induction, we have $s' = t'$ and thus $s = t$. ∎

In the following we describe an alternative characterization of the induced polycyclic sequence. This characterization is more technical, but it will be more useful for computational purposes.

LEMMA 8.34 *A sequence $Y = [y_1, \ldots, y_m]$ is an induced polycyclic sequence for a subgroup U with respect to X if and only if:*

a) $\langle y_1, \ldots, y_m \rangle = U$.

b) *with $i_j := \operatorname{dep}_X(y_j)$, we have $i_1 < \cdots < i_m$.*

c) $y_j^{r_{i_j}} = y_{j+1}^{e_{j+1}} \cdots y_m^{e_m}$ *with $e_h \in \{0 \mathinner{.\,.} r_{i_h} - 1\}$ for $1 \le j \le m$.*

d) $[y_j, y_k] = y_{k+1}^{e_{k+1}} \cdots y_m^{e_m}$ *with $e_h \in \{0 \mathinner{.\,.} r_{i_h} - 1\}$ for $1 \le k < j \le m$.*

PROOF $\Rightarrow$: Since Y is an induced polycyclic sequence, it is also a polycyclic sequence for U and thus a), c) and d) follow from Lemma 8.6. Part b) follows from Lemma 8.32.

$\Leftarrow$: Put $i_0 := 0$ and $i_{m+1} := n+1$. Let $H_j := \langle y_j, \ldots, y_m \rangle$ for $1 \le j \le m+1$ and $U_i := G_i \cap U$ for $1 \le i \le n+1$. Our aim is to prove $H_j = U_{i_{j-1}+1} = \cdots = U_{i_j}$ for $1 \le j \le m+1$.

First, d) implies that H_{j+1} is normal in H_j for every j. By the construction of the subgroups H_j this yields that the series $H_1 > \cdots > H_{m+1} = 1$ is a polycyclic series for H_1 and Y is a polycyclic sequence defining this series. By a) we have that $H_1 = U$ and Y is a polycyclic sequence for U.

By a) and b) we have that $H_j \le U_{i_j}$. Let $u \in U_{i_j}$. As Y is a polycyclic sequence for U, we find that u can be written in the form $u = y_1^{e_1} \cdots y_m^{e_m}$. Let e_k be the first nonzero exponent. Then $\operatorname{dep}_X(u) = \operatorname{dep}_X(y_k) = i_k$. Since $u \in G_{i_j}$, we have $k \ge j$ and thus $u = y_j^{e_j} \cdots y_m^{e_m}$. This yields that $u \in H_j$. In summary, we now have $H_j = U_{i_j}$ for $1 \le j \le m+1$.

Now let $H = U_k$ with $i_j < k < i_{j+1}$. Then $H_j \le H \le H_{j+1}$. Note that $H_j \ne H_{j+1}$ by b) and thus $|H_j : H_{j+1}| = r_{i_j}$ by c), since r_{i_j} is a prime. Thus we have $H = H_j$ or $H = H_{j+1}$. Further, we find that $y_j \notin G_k$ by b) and hence $H \ne H_j$. This yields that $H = H_{j+1}$ as desired. ∎

Lemma 8.34 translates into an algorithm to compute an induced polycyclic sequence for a subgroup U. The basic idea of the algorithm is to complete a partial induced polycyclic sequence for U stepwise until it fulfils the criteria of Lemma 8.34. As a preparation for this algorithm, we introduce a method to sift an element through a partial induced sequence. For technical purposes, we define a *partial induced sequence* as a sequence of length n, which contains at place i either the trivial element of G or an element of depth i.

Sift(Z, g)

    Input: A sequence $Z = [z_1, \ldots, z_n]$ with $z_i = 1$ or $\mathrm{dep}_X(z_i) = i$,
        an element $g \in G$

    Output: An element $h \in G$ of maximal depth $d = \mathrm{dep}_X(h)$ such
        that $g = z_1^{e_1} \cdots z_{d-1}^{e_{d-1}} h$ with $0 \le e_i < r_i$

1   $h := g$;
2   $d := \mathrm{dep}_X(h)$;
3   **while** $d < n+1$ and $z_d \ne 1$
4   **do** $k := z_d$;
5          $e := \mathrm{led}_X(h) \cdot \mathrm{led}_X(k)^{-1} \bmod r_d$;
6          $h := k^{-e} \cdot h$;
7          $d := \mathrm{dep}_X(h)$;
8   **return** h;

We note that the depth $\mathrm{dep}_X(h)$ increases in every pass through the 'While' loop in the function Sift. Hence Sift terminates and the returned element h has at least the depth of g.

Now we can introduce an outline of the algorithm to compute an induced polycyclic sequence for a subgroup U that is given by generators $\{u_1, \ldots, u_t\}$. The algorithm is based on Lemma 8.34 and the method used in Sift.

InducedPolycyclicSequence($\{u_1, \ldots, u_t\}$)

    Input: Generators $\{u_1, \ldots, u_t\}$ for a subgroup U
    Output: An induced polycyclic sequence Y for U

1   Initialize Z as a sequence of length n with all entries 1;
2   Initialize $\mathcal{G} := \{u_1, \ldots, u_t\}$;
3   **while** $\mathcal{G} \ne \emptyset$
4   **do** select g from $\mathcal{G}$ and delete g from $\mathcal{G}$;
5          $h := \mathrm{Sift}(Z, g)$;
6          $d := \mathrm{dep}_X(h)$;
7   **if** $d < n + 1$
8   **then** add h^{r_d} to $\mathcal{G}$;
9   add $[h, z]$ to $\mathcal{G}$ for all $z \in Z$;
10               $Z[d] := h$;
11 Let Y be the sequence of the nontrivial entries of Z;
12 **return** Y;

LEMMA 8.35 *The algorithm* InducedPolycyclicSequence *terminates and it returns an induced polycyclic sequence for U.*

PROOF First we prove that the algorithm terminates. For this purpose we have to show that the set $\mathcal{G}$ is empty at some stage in the algorithm. In each pass through the 'While' loop of the algorithm, the set $\mathcal{G}$ is reduced by one element in line 4 and it is extended by a finite number of elements in lines 8 and 9. Whenever a new element u is added to $\mathcal{G}$ in lines 8 or 9, then we have that $\operatorname{dep}_X(u) > \operatorname{dep}_X(g)$ where g is the original element selected in line 4. Hence after finitely many steps there is at most the element of maximal depth $n+1$ left in the set $\mathcal{G}$. This element is eliminated at the next pass through the 'While' loop without further changes to $\mathcal{G}$. At this stage the algorithm terminates.

Next we prove that Y is an induced polycyclic sequence for U. For this purpose we show that Y fulfils the criteria of Lemma 8.34. Since any nontrivial elements $Z[d]$ always have depth d, condition b) is satisfied, and condition c) and d) follow from the fact that the powers and commutators of these nontrivial elements are adjoined to $\mathcal{G}$ for later processing in lines 8 and 9. It remains to prove condition a).

For this purpose we show that $U = \langle Z, \mathcal{G} \rangle$ at every step of the algorithm after the initialization. Clearly, $U = \langle Y, \mathcal{G} \rangle$ at line 3. Whenever we take an element from $\mathcal{G}$ and eliminate it from this set, then we sift it through Z and add the result to Z if it is nontrivial. Hence the property $U = \langle Z, \mathcal{G} \rangle$ remains unchanged through the algorithm. As $\mathcal{G}$ is empty at the end of the algorithm, it follows that the elements in Z and Y generate U at line 11. ∎

In the theory of vector spaces, a basis of a subspace can be computed by the Gauss algorithm. Due to this analogy, the algorithm INDUCEDPOLYCYCLIC-SEQUENCE has also been called the 'Noncommutative Gauss algorithm' by Laue, Neubüser, and Schoenwaelder in [LNS84].

The algorithm INDUCEDPOLYCYCLICSEQUENCE can be extended in such a form that it works for arbitrary (not necessarily refined) consistent polycyclic presentations of finite or infinite polycyclic groups. We refer to Eick's thesis [Eic01a] for further background.

Using a modification of the algorithm SIFT we can test membership of elements in a subgroup U and determine the exponent vector of an element $u \in U$ with respect to an induced polycyclic sequence Y for U. This is done in the function CONSTRUCTIVEMEMBERSHIPTEST.

Example 8.9

We consider the following polycyclic presentation (trivial polycyclic relations and conjugation relations with inverse generators are omitted):

$$\mathrm{Pc}\langle\, x_1, \ldots, x_4 \mid x_1^3 = 1, \; x_2^2 = x_4, \; x_3^2 = x_4, \; x_4^2 = 1,$$
$$x_1^{-1} x_2 x_1 = x_3, \; x_1^{-1} x_3 x_1 = x_2 x_3, \; x_2^{-1} x_3 x_2 = x_3 x_4 \,\rangle.$$

Let $U := \langle x_2, x_2 x_3 \rangle$. Applying the algorithm INDUCEDPOLYCYCLICSEQUENCE

CONSTRUCTIVEMEMBERSHIPTEST(Y, g)

> **Input**: An induced polycyclic sequence $Y = [y_1, \ldots, y_m]$ for a
> subgroup U, an element $g \in G$
> **Output**: The exponent vector $\exp_Y(g)$ if $g \in U$ and **false** otherwise

1   Let e be a list of length m containing 0's;
2   $h := g$;
3   $d := \dep_X(h)$;
4   **while** there exists j with $\dep_X(y_j) = d$
5   **do** $f := \led_X(h) \cdot \led_X(y_j)^{-1} \bmod r_d$;
6           $h := y_j^{-f} \cdot h$;
7           $e[j] := f$;
8           $d := \dep_X(h)$;
9   **if** $h = 1$
10   **then return** e;
11   **else return false**;

yields the following assignments in the passes through the 'While' loop:

$$Z = [1, 1, 1, 1] \quad \mathcal{G} = \{x_2, x_2 x_3\}$$
$$Z = [1, x_2, 1, 1] \quad \mathcal{G} = \{x_2 x_3, x_4\} \quad (x_4 = x_2^2)$$
$$Z = [1, x_2, x_3, 1] \quad \mathcal{G} = \{x_4\} \quad (x_3^2 = [x_2, x_3] = x_4)$$
$$Z = [1, x_2, x_3, x_4] \quad \mathcal{G} = \{1\}$$
$$Z = [1, x_2, x_3, x_4] \quad \mathcal{G} = \{\}$$

Hence U has the induced polycyclic sequence $Y = [x_2, x_3, x_4]$.   □

8.3.2   Canonical polycyclic sequences

In this subsection we outline a method to solve the subgroup problem in polycyclically presented groups. That is, for two given subgroups U and V of a polycyclic presented group G, we want to check whether $U = V$. For this purpose we introduce canonical polycyclic sequences for subgroups of the polycyclically presented group G. They will yield that two subgroups are equal if and only if they have the same canonical polycyclic sequence.

DEFINITION 8.36   Let $Y = [y_1, \ldots, y_m]$ be an induced polycyclic sequence with respect to X for $U \leq G$. Put $b_i := \exp_X(y_i) = (b_{i1}, \ldots, b_{in})$ and let $d_i := \dep_X(y_i)$ for $1 \leq i \leq m$. Then Y is called a *canonical polycyclic sequence* for U if

a) $b_{i,d_i} = 1$ for $1 \leq i \leq m$, and

b) $b_{i,d_j} = 0$ for $1 \leq i < j \leq m$.

Let Y be a canonical polycyclic sequence with respect to X. Then the $m \times n$-matrix B whose rows consist of the exponent vectors $\exp_X(y_i)$ for $1 \le i \le m$ is a matrix in upper triangular form with leading entries in each row equal to 1, and 0's above these leading entries. This is an alternative way to characterize the canonical polycyclic sequences.

It is straightforward to modify a given induced polycyclic sequence to a canonical polycyclic sequence by adjusting the elements in the sequence. We give an outline of the method in the following function.

CANONICALPOLYCYCLICSEQUENCE(Y)

    Input: An induced polycyclic sequence $Y = [y_1, \ldots, y_m]$

    Output: A modified sequence Y which is canonical

1   $d := [\operatorname{dep}_X(y_1), \ldots, \operatorname{dep}_X(y_m)]$;

2   **for** $i \in [m .. 1 \text{ by} -1]$

3   **do** $b := \exp_X(y_i)$; $\;f := b_{d[i]}^{-1} \bmod r_{d[i]}$; $\;y_i := y_i^f$;

4   **for** $j \in [i+1 .. m]$

5   **do** $b := \exp_X(y_i)$; $\;y_i := y_i y_j^{-b_{d[j]}}$;

6   **return** Y;

The following theorem shows that canonical polycyclic sequences can be used to solve the subgroup problem.

THEOREM 8.37   *Every subgroup of G has a unique canonical polycyclic sequence with respect to X.*

PROOF   Let U be a subgroup of G. It is straightforward to observe that U has a canonical polycyclic sequence; for example, the algorithm CANONICALPOLYCYCLICSEQUENCE can be used to determine such a sequence. It remains to prove that, given two canonical polycyclic sequences Y and Z for U, the equality $Y = Z$ holds.

By induction, we assume that every subgroup of G_2 has a unique canonical polycyclic sequence with respect to $X_2 := [x_2, \ldots, x_n]$. If $U \le G_2$, then Y and Z are two canonical polycyclic sequences with respect to X_2 and the theorem is proved. Thus we assume that $U \not\le G_2$. Then $Y_2 := Y \cap X_2$ and $Z_2 := Z \cap X_2$ are two canonical polycyclic sequences with respect to X_2 for $U_2 := U \cap G_2$. Hence $Y_2 = Z_2$ and it remains to prove $y_1 = z_1$. We observe that $\operatorname{dep}_X(y_1) = \operatorname{dep}_X(z_1) = 1$ and $\operatorname{led}_X(y_1) = \operatorname{led}_X(z_1) = 1$. Suppose that $y_1 \ne z_1$ and let j be the smallest position in which $\exp_X(y_1)$ and $\exp_X(z_1)$ differ. Then $j > 1$ and $j \notin \{\operatorname{dep}_X(y) \mid y \in Y_2\}$. Let $u = y_1 z_1^{-1} \in U$. Then $\operatorname{dep}_X(u) = j$. But Y is an induced polycyclic sequence of U and thus all the elements in U have depths in $\{\operatorname{dep}_X(y) \mid y \in Y\} \cup \{n+1\}$ by Lemma 8.33. This is a contradiction and so $y_1 = z_1$.   ∎

Exercises

1. Calculate the canonical polycyclic sequence of the subgroup $\langle (1,4), (2,3) \rangle$ of $G = \mathrm{Sym}(4)$ with respect to the polycyclic sequence of G defined in Exercise 1 of Section 8.1.

2. Show that the set T in Lemma 8.33 is not necessarily a right transversal of U in G. (*Hint*: Let $G := \mathrm{Sym}(4)$ and $U := \langle (1,3)(2,4) \rangle$.)

8.4   Factor groups and homomorphisms

Factor groups and homomorphisms of groups are closely connected to each other. In this section we shall describe algorithms to compute with them. For this purpose we assume that G is a finite polycyclic group defined by the refined consistent polycyclic presentation $\mathrm{Pc}\langle X \mid R \rangle$. Let $X = [x_1, \ldots, x_n]$ and $R(X) = (r_1, \ldots, r_n)$.

8.4.1   Factor groups

Let N be a normal subgroup of G. As usual, denote the polycyclic series defined by X by $G = G_1 > \cdots > G_{n+1} = 1$. Then the sequence $X/N := [x_1 N, \ldots, x_n N]$ is a polycyclic sequence for the quotient group G/N, which defines the induced polycyclic sequence $G/N = G_1 N/N \geq \cdots \geq G_{n+1} N/N = N/N$.

This induced series for G/N contains trivial factors if N is a nontrivial subgroup. Then X/N forms a polycyclic sequence for G/N, but this sequence does not define a refined consistent polycyclic presentation. The following lemma shows how to determine a polycyclic sequence for G/N that yields a refined consistent polycyclic presentation.

LEMMA 8.38   *Let Y be an induced polycyclic sequence for $N \trianglelefteq G$ with respect to X. Let $d := \{ \mathrm{dep}_X(y) \mid y \in Y \}$ and $Z := [x_i \mid i \notin d]$. Then Z/N is a polycyclic sequence for G/N that induces a refined consistent polycyclic presentation for G/N.*

PROOF   The normal subgroup N covers all the factors G_i / G_{i+1} with $i \in d$ by Lemma 8.33 and N avoids the remaining factors of the polycyclic series of G. Hence Z/N is a polycyclic sequence for G/N that defines a properly descending polycyclic series. Hence Z/N also defines a refined consistent polycyclic presentation for G/N. ∎

It is straightforward to determine the exponent vector of an element $gN \in G/N$ with respect to the sequence Z/N of Lemma 8.38. Using this, we can then also determine a refined consistent polycyclic presentation for the factor group G/N.

8.4.2 Homomorphisms

Let $H = \mathrm{Pc}\langle X' \mid R' \rangle$ be another polycyclic group defined by a refined consistent polycyclic presentation. A homomorphism from G to H can be defined by describing the images of a generating set $\{g_1, \ldots, g_l\}$ of G:

$$\varphi : G \to H : g_i \mapsto h_i$$

Given a homomorphism in this form, the following problems are of interest (cf Section 3.4):

(1) Compute the image of $g \in G$ under φ.

(2) Determine the image I of φ.

(3) Compute a preimage of $h \in I$ under φ.

(4) Determine the kernel K of φ.

All of these problems can be solved effectively in the given situation. As a first step, we introduce two precomputations for the solutions.

(I) *Compute images for the polycyclic sequence X:* We compute $Y :=$ INDUCEDPOLYCYCLICSEQUENCE($\{g_1, \ldots, g_l\}$) and then $Z :=$ CANONICALPOLYCYCLICSEQUENCE(Y). Additionally, in this process we perform all computations with $\{g_1, \ldots, g_l\}$ simultaneously also with $\{h_1, \ldots, h_l\}$. Since X is the unique canonical polycyclic sequence of G, we obtain $Z = X$ and we determine $[\varphi(x) \mid x \in X]$.

(II) *Compute preimages for an induced polycyclic sequence of I:* We compute $Y' :=$ INDUCEDPOLYCYCLICSEQUENCE($\{h_1, \ldots, h_l\}$) to determine an induced polycyclic sequence for I. Additionally, in this process we perform all computations with $\{h_1, \ldots, h_l\}$ simultaneously also with $\{g_1, \ldots, g_l\}$. This yields a sequence of preimages $[\varphi^{-1}(y) \mid y \in Y']$.

These two precomputations yield effective solutions to some of the above problems. First, problem (1) can be solved effectively, since all elements in G are given as collected words in X and hence it is straightforward to translate a collected word from X to $\varphi(X)$.

Then, an induced polycyclic sequence for I is determined by (II) and hence problem (2) is solved. Similarly, we can compute preimages using (II), since we can translate collected words in an induced polycyclic sequence Y' for I to their preimages under φ.

Finally, it remains to discuss problem (4). Normal subgroup generators for the kernel can be computed by evaluating the relations of the image I. Thus, using (II) we can determine normal subgroup generators for the kernel K readily. It then remains to determine the normal closure of the subgroup generated by the normal subgroup generators. Note that we can read off the order of the source G and the order of the image I from their polycyclic sequences. Hence we know *a priori* the order of the kernel K, which is useful as a stopping criterion for the kernel computation.

Exercises

1. Let X be the polycyclic sequence for $G := \mathrm{Sym}(4)$ defined in Exercise 1 of Section 8.1, and let $\varphi : G \to G$ be the homomorphism defined by $\varphi : (1,2) \mapsto (2,3), (2,3) \mapsto (3,4), (3,4) \mapsto (2,3)$. Use the methods described above to compute $\varphi(X)$, and to find generators of $\ker(\varphi)$.

2. Let G be a group defined by a refined polycyclic presentation, let H be a second (black-box) group, let $\{g_1, \ldots, g_l\}$ be a set of generators of G, and let $h_1, \ldots, h_l$ be elements of H. How would you test whether the map $\varphi : g_i \mapsto h_i$ extends to a homomorphism $G \to H$. (You might want to refer back to the general discussion in Subsection 3.4.1.)

8.5   Subgroup series

In this section we shall discuss the determination of various subgroup series for the finite polycyclic group G. First, we note that, if X is the polycyclic generating sequence associated with a refined consistent polycyclic presentation of G, then the polycyclic series $G = G_1 > \cdots > G_{n+1} = 1$ defined by X is a composition series for G, since $|G_i : G_{i+1}| = r_i$ is a prime for every i.

In many algorithmic methods for polycyclic groups, a normal series with elementary abelian factors is used for induction purposes. For some algorithms it is also of interest to have a characteristic series of this type or a chief series. In the remainder of this section we describe methods to compute such series in G.

A key to many of these series is the determination of a commutator subgroup $[U, V] = \langle [u, v] \mid u \in U, v \in V \rangle$; that is, we want to determine an induced polycyclic sequence for $[U, V]$ from induced polycyclic sequences for U and V. This is addressed in the following elementary lemma. Note that the first part follows from Proposition 2.35, and the second part is proved similarly.

LEMMA 8.39 *Let Y and Z be induced polycyclic sequences for U and V, respectively.*

a) *$[U, V]$ is generated as a normal subgroup of $\langle U, V \rangle$ by the set $[Y, Z] := \{ [y, z] \mid y \in Y, z \in Z \}$.*

b) *If U and V normalize each other, then $[U, V]$ is generated as a subgroup by $[Y, Z]$.*

By iterating the commutator subgroup computation, we can calculate induced polycyclic sequences for each of the subgroups in the derived series $G = G^{[1]} \rhd \cdots \rhd G^{[d+1]} = 1$ and in the lower central series $G = \lambda_1(G) \rhd \lambda_2(G) \rhd \cdots \rhd \lambda_c(G) \rhd 1$ of G. Hence we can also determine the lower nilpotent series $G = L_1 \rhd \cdots \rhd L_r \rhd L_{r+1} = 1$. where L_{i+1} is defined as the smallest normal subgroup in L_i with L_i/L_{i+1} nilpotent; that is, L_{i+1} is the last term in the lower central series of L_i. This series can be refined to a series with abelian factors using the lower central series of its factors. Thus we define $L_{ij}/L_{i+1} = \lambda_j(L_i/L_{i+1})$ and we obtain $L_i = L_{i1} \rhd \cdots \rhd L_{ic_i} \rhd L_{i+1}$, a series with central factors. Combining this with the lower nilpotent series yields the refined lower nilpotent series $G = L_1 = L_{11} \rhd \cdots \rhd L_{ij} \rhd \cdots \rhd L_{r+1} = 1$.

Some properties of this series are summarized in the following lemma. For a proof and for a more efficient algorithm to determine the refined lower nilpotent series of G, we refer to the recent article by Cannon, Eick, and Leedham-Green [CELG04].

LEMMA 8.40 *The refined lower nilpotent series of G is a characteristic series with abelian factors. Its factor $L_{i,j-1}/L_{i,j}$ is central in $L_i/L_{i,j}$ and its factor L/L_{i2} splits over L_i/L_{i2} .*

Suppose that $G = N_1 \rhd \cdots \rhd N_l \rhd N_{l+1} = 1$ is a normal (or characteristic) series with abelian factors. Our next aim is to refine this series to a normal (or characteristic) series with elementary abelian factors. For this purpose we use the methods of Section 8.4 to compute with the factors $A = N_i/N_{i+1}$. Each of these factors A is a finite abelian group and we can read off the order $|A|$ from a polycyclic sequence for A. For a prime p dividing $|A|$ we can then determine $A^p := \{ a^p \mid a \in A \}$ and thus obtain a characteristic subgroup with elementary abelian factor A/A^p. Iterating this approach yields a normal (or characteristic) series with elementary abelian factors through A. Combining all of these refinements we obtain a normal (or characteristic) Series with elementary abelian factors for G.

Suppose that $G = N_1 \rhd \cdots \rhd N_l \rhd N_{l+1} = 1$ is a normal series with elementary abelian factors. Our next aim is to refine this series to a chief series for G. Let $A := N_i/N_{i+1}$ be a factor in this series. Then A is an elementary abelian p-group of rank d, say. The group G acts on A by conjugation. We identify A with the additive group of the vector space $\mathbb{F}_p^d$ and switch to additive notation.

The conjugation action yields a homomorphism $G \to GL(d, \mathbb{F}_p)$ in this setting. Using the methods of Section 7.4 we can compute a composition series of $\mathbb{F}_p^d$ as a G-module. This translates to a series $A = A_1 > \cdots > A_r > 1$ such that each A_j is invariant under the action of G and irreducible as a $\mathbb{F}_p G$-module. If we refine every factor in the original series of G by such a module composition series, then we obtain a chief series for G.

8.6   Orbit-stabilizer methods

Let G be a finite group defined by a refined consistent polycyclic presentation and suppose that G acts on a finite set Ω. Then one can determine orbits and stabilizers of elements in Ω using the methods of Section 4.1. But in the special case that G is polycyclic, there is a more efficient method to determine orbits and stabilizers available. This method will be outlined in this section.

Let $G = \mathrm{Pc}\langle X \mid R \rangle$ be the defining presentation of G. Then $X = [x_1, \ldots, x_n]$ is a polycyclic sequence. Its relative orders $R(X) = (r_1, \ldots, r_n)$ are primes and they coincide with the power exponents of the presentation. For $\omega \in \Omega$, we want to determine the orbit $\omega^G = \{ \omega^g \mid g \in G \}$ and an induced polycyclic sequence for the stabilizer $\mathrm{Stab}_G(\omega) = \{ g \in G \mid \omega^g = \omega \}$.

We start by considering the special case when the polycyclic sequence X has length 1; so G is cyclic of prime order r_1. Then there are two possibilities:

(a) $\omega^{x_1} = \omega$. In this case $\mathrm{Stab}_G(\omega) = G$ and thus $\mathrm{Stab}_G(\omega)$ has the induced polycyclic sequence $[x_1]$. Further, we have $\omega^G = \{\omega\}$.

(b) $\omega^{x_1} \neq \omega$. In this case $\mathrm{Stab}_G(\omega) = \{1\}$ and $\mathrm{Stab}_G(\omega)$ has the induced polycyclic sequence $[]$. Further, we have $\omega^G = \{ \omega^{x_1^{e_1}} \mid 0 \leq e_1 < r_1 \}$.

Thus to determine the orbit and the stabilizer in this special case it is sufficient to check whether $\omega^{x_1} = \omega$ holds. The result can then be read off.

The central idea of the polycyclic orbit-stabilizer algorithm is to generalize this method for the cyclic group to polycyclic groups. We use induction on the length of X for this purpose. Recall that $G_2 = \langle x_2, \ldots, x_n \rangle$ has the polycyclic sequence $X_2 = [x_2, \ldots, x_n]$ of length $n-1$. We assume by induction that we have computed the orbit $\Delta := \omega^{G_2}$ and that we have an induced polycyclic sequence $Y = [y_1, \ldots, y_m]$ for the stabilizer $S := \mathrm{Stab}_{G_2}(\omega)$. The following lemma shows how to extend these results to G.

LEMMA 8.41   *Let $\Delta := \omega^{G_2}$ and $S := \mathrm{Stab}_{G_2}(\omega)$ with induced polycyclic sequence $Y = [y_1, \ldots, y_m]$ and relative orders $R(Y) = (z_1, \ldots, z_m)$. We distinguish the following two cases.*

a) $\omega^{x_1} \in \Delta$:

   In this case there exists an element $g \in G_2$ with $\omega^{x_1} = \omega^g$. Then $[x_1 g^{-1}, y_1, \ldots, y_m]$ is an induced polycyclic sequence for $\mathrm{Stab}_G(\omega)$ and its relative orders are $(r_1, z_1, \ldots, z_m)$. Further, we have $\omega^G = \Delta$.

b) $\omega^{x_1} \notin \Delta$:

   In this case we have $\mathrm{Stab}_G(\omega) = S$ and thus it has the induced polycyclic sequence Y with relative orders $R(Y)$. Further, we have

$$\omega^G = \Delta \cup \Delta^{x_1} \cup \cdots \cup \Delta^{x_1^{r_1 - 1}}.$$

PROOF   Let $T := \mathrm{Stab}_G(\omega)$ to shorten notation. First, we note that $|G : G_2| = r_1$ is a prime. As $T \cap G_2 = S$, either $|T : S| = r_1$ or $T = S$. Since $|\omega^G| = |G : T| = |G : S|/|T : S| = |G : G_2||G_2 : S|/|T : S| = r_1 |\Delta|/|T : S|$, we find that $|\omega^G| = |\Delta|$ if $|T : S| = r_1$ and $|\omega^G| = r_1 |\Delta|$ if $T = S$.

In case a), the element $g \in G_2$ exists by construction, since Δ is the orbit of ω under the action of G_2. This yields that $x_1 g^{-1} \in T$. Since $x_1 g^{-1} \notin G_2$, we have $x_1 g^{-1} \in T \setminus S$ and hence $T \neq S$. By the above argument, this yields that $|T : S| = r_1$ and $\omega^G = \Delta$. Further, we can extend the induced polycyclic sequence Y by an arbitrary element of $T \setminus S$ to obtain an induced polycyclic sequence for T. Hence $x_1 g^{-1}$ is suitable for this purpose.

In case b), $\omega^G \neq \Delta$. By the above argument, we have $T = S$ and $|\omega^G| = r_1 |\Delta|$. Since G_2 is normal in G, the orbit Δ is a block for the action of G; that is, $\Delta^g = \Delta$ or $\Delta^g \cap \Delta = \emptyset$ for every $g \in G$. As $\Delta^{x_1} \neq \Delta$, we have $\Delta^{x_1} \cap \Delta = \emptyset$. Similarly, we find that the sets $\Delta, \Delta^{x_1}, \ldots, \Delta^{x_1^{r_1-1}}$ are pairwise disjoint and thus their union is the orbit ω^G.   ∎

Lemma 8.41 allows us to extend orbits and stabilizers from G_2 to G readily and hence this lemma yields the inductive step in an orbit-stabilizer algorithm for polycyclic groups. It only remains to discuss how to find the element g in Lemma 8.41 a). An outline of the complete algorithm is presented in POLYCYCLICORBITSTABILIZER.

It remains to discuss line 6 of this algorithm: the determination of the element g. In the general orbit-stabilizer algorithm of Section 4.1, such an element is computed either by storing a transversal of the orbit, or by storing a Schreier vector for the orbit. In the polycyclic orbit-stabilizer algorithm a transversal is not computed or stored. However, by line 10 of the algorithm, the construction of Δ is set up in such a way that we know how each orbit point has been obtained without storing a transversal. This can be used to determine g efficiently. This method is similar to the Schreier vector method, but we do not store all of the powers x_i^j that arise in line 10; we just store the x_i, and then the values of j and the powers x_i^j are computed when required.

POLYCYCLICORBITSTABILIZER(ω)

> **Input**: $\omega \in \Omega$.
> **Output**: An induced polycyclic sequence Y for $\mathrm{Stab}_G(\omega)$, and ω^G

1   Initialize $\Delta := [\omega]$;
2   Initialize $Y := []$;
3   **for** $i \in [n .. 1$ by $-1]$
4   **do** $\alpha := \omega^{x_i}$;
5   **if** $\alpha \in \Delta$
6   **then** find $g \in G_2$ with $\alpha = \omega^g$;
7            $Y := [x_i g^{-1}] \cup Y$;
8   **else** $\Gamma := []$;
9   **for** $j \in [1 .. r_i - 1]$
10   **do** $\Gamma := \Gamma$ cat $[\delta^{x_i^j} \mid \delta \in \Delta]$;
11           $\Delta := \Delta$ cat Γ;
12   **return** Y and Δ;

8.7   Complements and extensions

In this section we shall describe effective methods to determine the conjugacy classes of complements of a normal subgroup and to compute the equivalence classes of extensions for finite polycyclic groups.

Complements are a frequently used tool in the theory of polycyclic groups. The complement method outlined here was described first by Celler, Neubüser, and Wright in [CNW90] and it is based on the algorithm to compute the first cohomology group presented in Subsection 7.6.1.

Extensions are an important tool for constructing and describing polycyclic groups. Here, we shall provide an effective method to compute the equivalence classes of extensions of an elementary abelian G-module by a polycyclic group G. This algorithm is based on the determination of the second cohomology group, which is also outlined below.

Throughout the section we shall assume that $G = \mathrm{Pc}\langle X \mid R \rangle$ is a finite group defined by a refined consistent polycyclic presentation with $X = [x_1, \ldots, x_n]$.

8.7.1   Complements and the first cohomology group

Let $N \trianglelefteq G$ be a normal subgroup of G. We shall describe a method to determine the conjugacy classes of complements of N in G. Every conjugacy class is described by a conjugacy class representative U and its normalizer $\mathbf{N}_G(U)$ in G. Note that we do not work with full lists of subgroups in a conjugacy class, as this would be too space consuming. We denote a conjugacy class with representative U by U^G.

As a first step towards this aim, we determine a G-invariant series with elementary abelian factors $N = N_1 \triangleright \cdots \triangleright N_l \triangleright N_{l+1} = 1$. For example, a characteristic series with elementary abelian factors of N would be suitable for this purpose; see Section 8.5. The next lemma observes that we can use induction down this series to determine the conjugacy classes of complements of N in G.

LEMMA 8.42   *Let $M \leq N \leq G$ with $M, N \trianglelefteq G$. Let $L_1/M, \ldots, L_r/M$ be a set of representatives for the conjugacy classes of complements of N/M in G/M and let $R_i/M := \mathbf{N}_{G/M}(L_i/M)$ be their normalizers. For $1 \leq i \leq r$ let $U_{i1}, \ldots, U_{il_i}$ be a set of representatives for the R_i-classes of complements of M in L_i and let $S_{ij} := \mathrm{Stab}_{R_i}(U_{ij}) = \mathbf{N}_{R_i}(U_{ij})$. Then*

   a) The set $\{ U_{ij} \mid 1 \leq i \leq r, 1 \leq j \leq l_i \}$ is a complete and irredundant set of representatives of the conjugacy classes of complements of N in G.

   b) The normalizer in G of a conjugacy class representative can be obtained as $S_{ij} := \mathbf{N}_G(U_{ij})$ for $1 \leq i \leq r$ and $1 \leq j \leq l_i$.

PROOF   a) Let U be a complement of N in G. Then UM/M is a complement of N/M in G/M. Thus U/M is conjugate to L_i/M for some i; that is, there exists $g \in G$ with $U^g/M = L_i/M$. Hence U^g is a complement of M in L_i and thus U^g is conjugate to U_{ij} for some j; that is, there exists $h \in R_i$ with $U^{gh} = U_{ij}$. So the set $\{ U_{ij} \mid i, j \}$ is complete.

Now suppose that two representatives U_{ij} and U_{kl} are conjugate. Then the quotients $U_{ij}M/M$ and $U_{kl}M/M$ are also conjugate. By construction, we have $U_{ij}M/M = L_i/M$ and $U_{kl}M/M = L_k/M$. As L_i/M and L_k/M are either equal or not conjugate, we have $U_{ij}M = L_i = L_k = U_{kl}M$, and U_{ij} and U_{kl} are two conjugate complements of M in L_i. Then they have to be conjugate under R_i, since M is normal in G. This yields $U_{ij} = U_{kl}$ by construction.

b) This can be proved in a similar way to a) and we omit the proof. ∎

Choose $M = N_l$ in Lemma 8.42 and assume by induction that we have determined the conjugacy classes of complements of N/M in G/M. Let L/M be a representative for such a class and let R/M be its normalizer. It remains to determine the R-conjugacy classes of complements of M in L.

First, using the methods of Section 7.6, we check whether there exists a complement to M in L. If such a complement exists, then we determine one such complement U explicitly and we compute the cohomology groups $Z^1(L/M, M)$ and $B^1(L/M, M)$. Then the elements of the factor group $H^1(L/M, M) = Z^1(L/M, M)/B^1(L/M, M)$ correspond one-to-one to the L-conjugacy classes of complements of M in L.

We digress at this point to introduce the concept of an *affine action* of a group on a vector space, which we shall need here and also later in the chapter, in Subsections 8.8.1 and 8.8.2.

DEFINITION 8.43   Let $V \cong K^n$ be a finite-dimensional vector space over a field K, let $\varphi : G \to \operatorname{Hom}_K(V, V)$ be a linear action of a group G on V, and let $\delta : G \to V$ be a derivation with respect to the action φ. Then the map $\alpha : G \to \operatorname{Sym}(V)$ defined by $v^{\alpha(g)} = v^{\phi(g)} + \delta(g)$ is called an *affine action* of G on V.

It is straightforward to check that α is a homomorphism, so an affine action is an action of G on the set V, but not a linear action unless $\delta = 0$.

If we identify V with K^n and $\operatorname{Hom}_K(V, V)$ with $\operatorname{GL}(n, K)$, then the affine action α corresponds to the homomorphism $\hat{\alpha} : G \to \operatorname{GL}(n+1, K)$, given by

$$
\hat{\alpha}(g) = \left(\begin{array}{c|c} \varphi(g) & \begin{matrix} 0 \\ \vdots \\ 0 \end{matrix} \\ \hline \delta(g) & 1 \end{array} \right),
$$

where, for $v \in K^n$, $(\, v^{\alpha(g)} \mid 1 \,) = (\, v \mid 1 \,) \cdot \hat{\alpha}(g)$.

The following lemma is the first step towards determining the R-conjugacy classes of complements of M in L.

LEMMA 8.44   *Let U be a complement of M in L and let $R := \mathbf{N}_G(L)$.*

a) *For $r \in R$ and $\gamma \in Z^1(L/M, M)$ we define*

$$
\gamma^{\varphi(r)} : L/M \to M : gM \mapsto ((g^{r^{-1}} M)^{\gamma}))^r.
$$

   Then $\gamma^{\varphi(r)} \in Z^1(L/M, M)$ and $\gamma \to \gamma^{\varphi(r)}$ defines a linear action φ of R on $Z^1(L/M, M)$.

b) *Every $g \in L$ can be written uniquely as $g = u(g)m(g)$ with $u(g) \in U$ and $m(g) \in M$. For $r \in R$ we define*

$$
\beta_r : L/M \to M : gM \mapsto u(g)^{-1}(u(g^{r^{-1}}))^r.
$$

   Then $\delta : R \to Z^1(L/M, M) : r \mapsto \beta_r$ is a derivation.

c) *By combining the linear action of a) and the derivation of b), we obtain an affine action α of R on $Z^1(L/M, M)$ given by $\zeta^{\alpha(r)} = \zeta^{\phi(r)} + \delta(r)$ for $r \in R$ and $\zeta \in Z^1(L/M, M)$.*

PROOF   All three parts of the lemma are straightforward to prove. We refer to [CNW90] or to [Eic02] for further information.   ∎

The setup provided by Lemma 8.44 can be used to determine the R-conjugacy classes of complements of M in L as follows.

THEOREM 8.45   *Let U be a complement of M in L. We consider the affine action α of $R := \mathbf{N}_G(L)$ on $Z^1(L/M, M)$ defined in Lemma 8.44 c).*

a) *The orbits of R on $Z^1(L/M, M)$ are in one-to-one correspondence with the R-conjugacy classes of complements of M in L.*

b) *If $\gamma \in Z^1(L/M, M)$, then $\mathrm{Stab}_R(\gamma) = \mathbf{N}_R(U)$.*

PROOF   If V is a complement corresponding to $\gamma \in Z^1(L/M, M)$, then the complement V^r corresponds to the cocycle γ^r where R acts affinely on $Z^1(L/M, M)$.   ∎

Hence it remains to determine the orbits and the stabilizers of the elements in the elementary abelian p-group $Z^1(L/M, M)$ under the affine action of R to determine the R-conjugacy classes of complements of M in L. We identify the elementary abelian p-group $Z^1(L/M, M)$ with the additive group of $\mathbb{F}_p^h$ and we switch to additive notation. As we remarked above, the affine action α of R corresponds to a homomorphism $\hat{\alpha} : R \to \mathrm{GL}(h+1, \mathbb{F}_p)$. Now it remains to calculate the orbits and stabilizers of the elements of the affine space under the action of R via $\hat{\alpha}$. This can be achieved effectively using the orbit-stabilizer methods of Section 8.6.

Finally, we note that, as a further reduction, we can use the fact that the group $B^1(L/M, M)$ is the orbit of the trivial cocycle under the affine action of L, and thus $B^1(L/M, M)$ is a block for the action of R on $Z^1(L/M, M)$. Hence it is sufficient to determine orbits and stabilizers under the action of R on $H^1(L/M, M)$ and then to lift the results to $Z^1(L/M, M)$.

8.7.2   Extensions and the second cohomology group

Let M be an elementary abelian group having the structure of a G-module. We shall describe an algorithm to determine the equivalence classes of extensions of M by G. Every extension of M by G is a polycyclic group and we want to determine a refined consistent polycyclic presentation for each equivalence class representative. It is well-known that the equivalence classes of extensions are in a one-to-one correspondence with the second cohomology group $H^2(G, M) = Z^2(G, M)/B^2(G, M)$. Hence our algorithm provides a method to determine this cohomology group also.

First, we introduce some notation. Let M be an elementary abelian p-group of rank d, and let $Y = [y_1, \ldots, y_d]$ be a polycyclic sequence for M. We describe the G-module structure of M by giving the action of every generator x_i on M. Since M can be considered as the additive group of the vector space $\mathbb{F}_p^d$, we describe the action of x_i on M by a matrix $X_i \in \mathrm{GL}(d, \mathbb{F}_p)$. We denote the inverse of X_i by $\overline{X_i}$.

Next, we analyze the refined consistent polycyclic presentations of an extension E of M by G. We identify M with its corresponding normal subgroup in E and we also identify G with the corresponding quotient E/M. Using this setup, it follows that E has a polycyclic sequence $[x_1, \ldots, x_n, y_1, \ldots, y_d]$. If $R(X) = (r_1, \ldots, r_n)$, then this polycyclic sequence has the relative orders $(r_1, \ldots, r_n, p, \ldots, p)$. Thus it induces a refined consistent polycyclic presentation and the polycyclic relations of this presentation have the following form:

$$(R_1) \quad \begin{aligned} x_j^{-1} x_i x_j &= R_{i,j}(x_{j+1}, \ldots, x_n) \, t_{i,j}(y_1, \ldots, y_d) \ \ (1 \leq j < i \leq n), \\ x_j x_i x_j^{-1} &= R_{j,i}(x_{j+1}, \ldots, x_n) \, t_{j,i}(y_1, \ldots, y_d) \ \ (1 \leq j < i \leq n), \\ x_i^{r_i} &= R_{i,i}(x_{i+1}, \ldots, x_n) \, t_{i,i}(y_1, \ldots, y_d) \ \ (1 \leq i \leq n), \end{aligned}$$

$$(R_2) \quad \begin{aligned} x_j^{-1} y_i x_j &= y_1^{X_{j,i,1}} \cdots y_d^{X_{j,i,d}} \ \ (1 \leq i \leq d, 1 \leq j \leq n), \\ x_j y_i x_j^{-1} &= y_1^{\overline{X}_{j,i,1}} \cdots y_d^{\overline{X}_{j,i,d}} \ \ (1 \leq i \leq d, 1 \leq j \leq n), \end{aligned}$$

$$(R_3) \quad y_i^p = 1 \ \ (1 \leq i \leq d).$$

The relations of type (R_3) are the polycyclic relations of M, and the trivial conjugation relations are omitted from them. The relations of type (R_2) reflect the action of E on M and thus are determined by the G-module structure of M only. The interesting relations are those of type (R_1). These relations exhibit the relations of the quotient group G of E and the tails $t_{i,j}$ of these relations determine the extension E. These tails are of the form

$$t_{i,j}(y_1, \ldots, y_d) = y_1^{t_{i,j,1}} \cdots y_d^{t_{i,j,d}} \quad \text{with } 0 \leq t_{i,j,k} < p \ \text{ for } 1 \leq i, j \leq n.$$

Hence every tail $t_{i,j}(y_1, \ldots, y_d)$ can be identified with an element of M. For $l := n^2$ we denote by $t := (t_{i,j} \mid 1 \leq i, j \leq n) \in M^l$ the tail-vector consisting of all tails $t_{i,j}$.

DEFINITION 8.46   The polycyclic presentation

$$P(t) := \mathrm{Pc}\langle\, x_1, \ldots, x_n, y_1, \ldots, y_d \mid R_1(t) \cup R_2 \cup R_3 \,\rangle$$

is called the polycyclic presentation for the tail-vector $t \in M^l$.

Let E be an extension of M by G that is defined by the cocycle $\gamma \in Z^2(G, M)$. Then E has the polycyclic sequence $[x_1, \ldots, x_n, y_1, \ldots, y_d]$ and this sequence induces a polycyclic presentation $P(t)$ for some $t \in M^l$. Hence we obtain a mapping

$$\varphi : Z^2(F, M) \to M^l : \gamma \mapsto t.$$

The following lemma investigates the mapping φ.

LEMMA 8.47   *The map φ is a homomorphism of abelian groups with $\ker(\varphi) \leq B^2(G, M)$. Thus $H^2(G, M) \cong Z^2(G, M)^\varphi / B^2(G, M)^\varphi$.*

PROOF Let $\gamma_1, \gamma_2 \in Z^2(G, M)$. Then, using additive notation for $Z^2(G, M)$ and M, we have $(\gamma_1 + \gamma_2)(g, h) = \gamma_1(g, h) + \gamma_2(g, h)$ for $g, h \in G$, and so $E_{\gamma_1 + \gamma_2}$ has a polycyclic presentation of the type $P(t_1 + t_2)$ with $t_i := \gamma_i^\varphi$ for $i = 1, 2$. Thus $(\gamma_1 + \gamma_2)^\varphi = t_1 + t_2 = \gamma_1^\varphi + \gamma_2^\varphi$ and φ is a homomorphism. Now let $\gamma \in \ker(\varphi)$. Then $t = \gamma^\varphi = 0$. Thus $P(t)$ defines a split extension of M by F. Hence $\gamma \in B^2(F, M)$. ∎

The group $H^2(G, M)$ describes the equivalence classes of extensions of M by G and our aim is to determine a basis for this elementary abelian group. For this purpose we determine bases for $Z^2(G, M)^\varphi$ and $B^2(G, M)^\varphi$. A basis for $B^2(G, M)^\varphi$ can be determined as the rows-pace of the matrix returned by Z1-MATRIX as described in Section 7.6. In the following, we consider the computation of $Z^2(G, M)^\varphi$. The following lemma yields a characterization of this group.

LEMMA 8.48 *We have $t \in Z^2(G, M)^\varphi$ if and only if $P(t)$ is a consistent polycyclic presentation.*

PROOF If $t \in Z^2(G, M)^\varphi$, then there exists $\gamma \in Z^2(G, M)$ with $\gamma^\varphi = t$. The extension E_γ induces the presentation $P(t)$ and thus $P(t)$ is consistent. Conversely, if $P(t)$ is consistent, then it defines an extension E of M by G. The extension E is defined by a cocycle $\gamma \in Z^2(G, M)$ and $\gamma^\varphi = t$. ∎

Let $T = (T_{i,j} \mid 1 \le i, j \le n)$ be a vector of n^2 different indeterminates that can take values in M. Then we can consider the (parameterized) polycyclic presentation $P(T)$. We determine the values for $t \in M^l$ with $P(t)$ consistent by performing a consistency check in $P(T)$.

DEFINITION 8.49 A word is collected with respect to $P(T)$ if it is of the form

$$x_1^{e_1} \cdots x_n^{e_m} \cdot y_1^{e_{m+1}} \cdots y_d^{e_{m+d}} \cdot T_{11}^{f_{11}} T_{12}^{f_{12}} \cdots T_{nn}^{f_{nn}}$$

where $0 \le e_i < r_i$ for $1 \le i \le n$ and $0 \le e_i < p$ for $m+1 \le i \le m+d$ and $f_{ij} \in \mathbb{F}_p G$ for $1 \le i, j \le n$.

The collection algorithm of Section 8.1.3 generalizes readily to the determination of collected words in $P(T)$. Hence we can assume that we can compute a collected word that is equivalent to an arbitrary word in $P(T)$ effectively. Using this collection algorithm we can perform the consistency check of Chapter 12 in a symbolic form in $P(T)$. This yields equations of the type

$$x_1^{e_1} \cdots x_n^{e_n} T_{11}^{f_{11}} \cdots T_{nn}^{f_{nn}} = x_1^{e_1'} \cdots x_n^{e_n'} T_{11}^{f_{11}'} \cdots T_{nn}^{f_{nn}'}$$

for $0 \le e_i, e_i' \le r_i$ and $f_{ij}, f_{ij}' \in \mathbb{F}_p G$. Since G is given by a consistent polycyclic presentation, we have $e_i = e_i'$ for $1 \le i \le n$. Hence the equation can be rewritten as

$$T_{11}^{f_{11}-f_{11}'} \cdots T_{nn}^{f_{nn}-f_{nn}'} = 1.$$

The elements $t \in M^l$ for which $P(t)$ is consistent are exactly the solutions of these resulting equations for T. It remains to solve this system of equations over the elementary abelian group M. For this purpose we identify $M \cong \mathbb{F}_p^d$ and switch to additive notation. Using the explicit G-module structure of M, we can translate each coefficient $f_{ij} - f_{ij}' \in \mathbb{F}_p G$ to a $d \times d$ matrix over $\mathbb{F}_p$. Thus the system of equations obtained translates into a homogeneous system of linear equations over $\mathbb{F}_p$. This can be solved readily.

Example 8.10
Let $G := \mathrm{Pc}\langle x_1 \mid x_1^3 = 1 \rangle$ and let $M := \langle y_1, y_2 \rangle \cong \mathbb{F}_3^2$ be a G-module via

$$X_1 := \begin{pmatrix} 0 & -1 \\ 1 & -1 \end{pmatrix}.$$

Then we obtain the parameterized polycyclic presentation (where conjugation with inverses is omitted):

$$P(T) = \mathrm{Pc}\langle x_1, y_1, y_2 \mid x_1^3 = T_{11},\ y_1^3 = 1,\ y_2^3 = 1,$$
$$x_1^{-1} y_1 x_1 = y_2^{-1},$$
$$x_1^{-1} y_2 x_1 = y_1 y_2^{-1} \rangle.$$

To evaluate the consistency relations in $P(T)$, we compute

$$x_1 T_{11} = T_{11} x_1 = x_1 x_1^{-1} T_{11} x_1 = x_1 T_{11}^{x_1}$$

and thus we obtain the relation

$$T_{11}^{x_1 - 1} = 1.$$

We have to find all possible solutions in M for this equation. To find these solutions we switch to additive notation and consider T_{11} as a vector in $\mathbb{F}_3^2$. Then x_1 acts by conjugation as X_1 and hence we have to find the nullspace of $X_1 - 1$. This nullspace can be computed readily as $\langle (1,1) \rangle$. This yields that $Z^2(G, M)^\varphi = \mathbb{F}_p$ and, translated back to the multiplicative setting, it yields that we have to choose $t_{11} \in \langle y_1 y_2 \rangle$.

Similarly, we can determine that $B^2(G, M)^\varphi = \{0\}$ in this example. Hence $H^2(G, M) \cong Z^2(G, M)^\varphi / B^2(G, M)^\varphi \cong \mathbb{F}_3$. So there are three equivalence classes of extensions of M by G. Their refined consistent polycyclic presentations can be obtained by choosing $t_{11} = 1$, or $t_{11} = y_1 y_2$, or $t_{11} = y_1^2 y_2^2$.

We note here (without proof) that the two nonsplit extensions are inequivalent as extensions, but they are isomorphic as groups. Thus the method outlined here does not provide isomorphism classes of extensions. $\quad\square$

8.8 Intersections, centralizers, and normalizers

In this section we shall outline three algorithms for finite polycyclic groups: methods for computing intersections of subgroups, centralizers of elements, and normalizers of subgroups. Also, we observe that similar methods can be used to solve the conjugacy problems for elements and subgroups and to determine all conjugacy classes of elements or subgroups.

The outlines in this section are overviews of the basic ideas of the algorithms considered. More detailed discussions and more background on the methods can be found in the papers by Mecky and Neubüser [MN89], Felsch and Neubüser [FN70], Glasby and Slattery [GS90], and by Celler, Neubüser, and Wright [CNW90]. For the larger class of infinite polycyclic groups we refer to papers of Eick and Ostheimer [EO03, Eic02, Eic01b, Eic01a].

The algorithmic problems discussed in this section could all be solved by straightforward applications of the orbit-stabilizer algorithm of Section 8.6. The methods introduced in this section provide a refinement for this orbit-stabilizer approach. Their basic idea is typical for algorithms in polycyclic groups: they proceed by induction downwards along a normal series with elementary abelian factors.

Throughout the section let $G = \mathrm{Pc}\langle X \mid R \rangle$ be defined by a refined consistent polycyclic presentation and let $G = N_1 \rhd \ldots \rhd N_l \rhd N_{l+1} = 1$ be a normal series with elementary abelian factors; see Section 8.5.

8.8.1 Intersections

We shall introduce a method to compute $U \cap V$ for two subgroups U and V of G. We assume that U and V are both given by induced polycyclic sequences and we want to determine an induced polycyclic sequence for $U \cap V$. The following elementary lemma shows that $U \cap V$ could be computed by the orbit-stabilizer algorithm of Section 8.6.

LEMMA 8.50 *The subgroup U acts by multiplication from the right on the right cosets $V \setminus G := \{Vg \mid g \in G\}$ and $U \cap V = \mathrm{Stab}_U(V)$.*

However, the orbit arising in the computation of $\mathrm{Stab}_U(V)$ has the length $[U : U \cap V]$ and hence it can be large. Therefore, the application of the orbit-stabilizer method can be time and space consuming. We want to investigate refinements of this approach.

First, we consider the special case of a normalized subgroup in the following elementary lemma. In this special case we can compute $U \cap V$ using the Kernel algorithm of Section 8.4.2. This is usually the most effective approach to determine an intersection if it applies.

LEMMA 8.51   If U is normalized by V, then $U \cap V$ is the kernel of the natural homomorphism $\varphi : V \to VU/U$.

Now, we shall introduce a method to compute $U \cap V$ in the general case. This method refines the orbit-stabilizer approach of Lemma 8.50 and it uses the special case of Lemma 8.51. The basic idea of the method is an induction downwards along a normal series of G with elementary abelian factors. Let $N = N_l$ be the last nontrivial subgroup in such a series and suppose that N is a p-group of rank d, say. By induction, we assume that the intersection in the factor G/N is already computed; that is, we are given an induced polycyclic sequence for $UN/N \cap VN/N = (UN \cap VN)/N$. Thus we can read off an induced polycyclic sequence for $UN \cap VN$ using the preimages method of Section 8.4.

LEMMA 8.52   Let $\varphi : U \to UN/N$ be the natural homomorphism. Then $(U \cap VN)^\varphi = (UN \cap VN)/N$.

PROOF   This follows directly from $(U \cap VN)N = UN \cap VN$.   ∎

Thus $K := U \cap VN$, and $L := UN \cap V$ can be computed effectively as preimages of the subgroup $(UN \cap VN)/N$ using the methods of Section 8.4.2. We note that $U \cap V = K \cap L$ holds, and it remains to determine the latter intersection.

LEMMA 8.53   Let $M := N \cap L$. Then:

a) M is normalized by K and thus K acts on N/M by conjugation.

b) Every $k \in K$ can be written as $k = v_k n_k$ for $v_k \in V$ and $n_k \in N$ and the coset $n_k M$ is uniquely defined by k.

c) $\delta : K \to N/M : k \mapsto n_k M$ is a well-defined derivation.

PROOF   a) First we note that $M := N \cap L = N \cap (UN \cap V) = N \cap V$. Thus M is normalized by V, since N is normal in G. Also, M is normalized by N, since N is abelian. Thus M is normalized by VN and hence by $K \leq VN$.

b) As $K \leq VN$, we can write every element in K as product in VN. Clearly, this factorization is unique modulo $M = N \cap V$.

c) By b) it follows that δ is a well-defined mapping. Clearly, we have that $\delta(1) = M$. Further, since

$$kh = v_k n_k v_h n_h = v_k v_h (n_k^{v_h} n_h) = v_k v_h (n_k^h n_h) \equiv v_{kh} n_{kh} \bmod M,$$

we have $\delta(kh) = \delta(k)^h \delta(h)$.   ∎

We note that the subgroup M of N can be computed by Lemma 8.51, since N is a normal subgroup of G. We also note that N and thus also N/M are elementary abelian p-groups and thus we can identify N/M with the additive group of $\mathbb{F}_p^e$ where e is the rank of N/M. Using this identification and switching to additive notation, we can write the conjugation action of K on N/M as a homomorphism $\varphi : K \to \mathrm{GL}(e, \mathbb{F}_p)$ and we consider δ as a derivation of the form $\delta : K \to \mathbb{F}_p^e$. This yields the following homomorphism corresponding to the affine action (see Definition 8.43) of φ and δ on N/M:

$$
\alpha : K \to \mathrm{GL}(e+1, \mathbb{F}_p) : k \mapsto \left(\begin{array}{c|c} \varphi(k) & \begin{matrix} 0 \\ \vdots \\ 0 \end{matrix} \\ \hline \delta(k) & 1 \end{array}\right).
$$

Note that this affine action can be computed effectively by its definition. It yields the following characterization of $U \cap V$.

LEMMA 8.54 $U \cap V = \mathrm{Stab}_K((0, \ldots, 0, 1))$ *where K acts via α on $\mathbb{F}_p^{e+1}$.*

PROOF   We noted already that $U \cap V = K \cap L$. By Lemma 8.50 we have that $K \cap L = \mathrm{Stab}_K(L)$ where K acts by multiplication on the right on the cosets $L \backslash G$. We prove that $\mathrm{Stab}_K(L) = \mathrm{Stab}_K(v)$ for $v = (0, \ldots, 0, 1) \in \mathbb{F}_p^{e+1}$. First, let $k \in \mathrm{Stab}_K(L)$. Then $k \in K \cap L$ and thus $k \in V$. Hence $n_k M = M$ and $\delta(k) = 0$. Therefore $k \in \mathrm{Stab}_K(v)$. Now, let $k \in \mathrm{Stab}_K(v)$. Then $\delta(k) = 0$ and thus $n_k \in M$. As $M = N \cap L$, we have $k \in L$. Thus $k \in K \cap L = \mathrm{Stab}_K(L)$.
∎

Hence we obtain an effective method to compute the intersection $U \cap V$. Similar to the method proposed by Lemma 8.50, this algorithm is also based on the orbit-stabilizer algorithm of Section 8.6. But instead of computing one large orbit of right cosets and its corresponding stabilizer, it determines several smaller orbits of vectors and their corresponding stabilizers. This is usually much more efficient.

8.8.2   Centralizers

We shall introduce a method of computing $\mathbf{C}_G(g) = \{h \in G \mid hg = gh\}$ for an element $g \in G$. We assume that g is given as a collected word in the generators of G and we aim to determine an induced polycyclic sequence for $\mathbf{C}_G(g)$.

The centralizer $\mathbf{C}_G(g)$ is the stabilizer of g under the conjugation action of G on its elements. Hence $\mathbf{C}_G(g)$ can be computed using the orbit-stabilizer algorithm of Section 8.6. However, the orbit of g^G needs to be computed and stored explicitly in this approach. This can be time and space consuming. We introduce a more effective approach in the following.

The basic idea of the method is an induction downwards along a normal series with elementary abelian factors. Let $N = N_l$ be the last nontrivial term in such a series and suppose that N is a p-group of rank d, say. By induction, we assume that the centralizer in the factor group G/N is already computed; that is, we have given an induced polycyclic sequence for $C/N = \mathbf{C}_{G/N}(gN)$. So we can read off an induced polycyclic sequence for C. The following lemma is elementary.

LEMMA 8.55 *The mapping $\delta : C \to N : c \mapsto [g, c] \ (= g^{-1}c^{-1}gc)$ is a derivation.*

Now we identify N with the additive group of $\mathbb{F}_p^d$ and we switch to additive notation. Then we can write the conjugation action of C on N as a homomorphism $\varphi : C \to \mathrm{GL}(d, \mathbb{F}_p)$ and we can consider δ as a derivation of the form $\delta : C \to \mathbb{F}_p^d$. These can be combined to give an affine action:

$$
\alpha : C \to \mathrm{GL}(d+1, \mathbb{F}_p) : c \mapsto \left(\begin{array}{c|c} \varphi(c) & \begin{matrix} 0 \\ \vdots \\ 0 \end{matrix} \\ \hline \delta(c) & 1 \end{array} \right).
$$

This affine action yields the following characterization of $\mathbf{C}_G(g)$.

LEMMA 8.56 $\mathbf{C}_G(g) = \mathrm{Stab}_C((0, \ldots, 0, 1))$ *where C acts on $\mathbb{F}_p^{d+1}$ via α.*

PROOF   The stabilizer in C of $(0, \ldots, 0, 1)$ under the action of α is the kernel of the derivation δ. By the definition of δ, we have $\ker(\delta) = \mathbf{C}_C(g)$. As $\mathbf{C}_G(g) \leq C$, we have that $\mathbf{C}_C(g) = \mathbf{C}_G(g)$ which completes the proof. ∎

Hence we obtain an effective method to compute $\mathbf{C}_G(g)$. Similar to the method using a single orbit-stabilizer application, this algorithm is also based on the orbit-stabilizer algorithm of Section 8.6. But this induction method computes several relatively small orbits of vectors instead of one relatively large orbit of elements in a polycyclic group. Thus the induction method is usually more effective than the single orbit-stabilizer application.

8.8.3   Normalizers

For a given subgroup $U \leq G$, we now introduce a method of computing its normalizer $\mathbf{N}_G(U) := \{ h \in G \mid hU = Uh \}$. We assume that U is given by an induced polycyclic sequence and we aim to compute an induced polycyclic sequence for $\mathbf{N}_G(U)$.

The normalizer $\mathbf{N}_G(U)$ is the stabilizer of U under the conjugation action of G on its subgroups. Hence $\mathbf{N}_G(U)$ can be computed using the orbit-stabilizer algorithm of Section 8.6. However, the orbit U^G needs to be computed and stored explicitly in this approach. This can be time and space consuming. Thus we introduce a refinement of this approach that is usually more efficient.

Once again, we use an induction downwards along a normal series with elementary abelian factors. Let $N = N_l$ be the last nontrivial term in such a series such that N is an elementary abelian p-group of rank d, say. By induction, we assume that the normalizer in the factor group G/N is already computed; that is, we know an induced polycyclic sequence for $R/N := \mathbf{N}_{G/N}(UN/N)$, and we can read off an induced polycyclic sequence for R. Now we proceed in several steps.

As a first step we determine $M := U \cap N$ using the idea of Lemma 8.51. This is a usually very effective preliminary computation. As a second step, we then compute $S := \mathrm{Stab}_R(M)$. For this purpose we identify N with the additive group of $\mathbb{F}_p^d$ and we switch to additive notation. Then the conjugation action of R on N translates to a homomorphism $\varphi : R \to \mathrm{GL}(d, \mathbb{F}_p)$ and we have to compute the stabilizer of the subspace of N corresponding to M under the matrix action of R on N.

The following lemma investigates M and S further.

LEMMA 8.57 *Let $S := \mathrm{Stab}_R(M)$ with $M := U \cap N$. Then:*

a) *M is normal in UN and thus $UN \leq S$. Further, U/M is a complement to N/M in UN/M.*

b) *$\mathbf{N}_G(U) = \mathrm{Stab}_S(U/M)$ where S acts on the set of complements to N/M in UN/M by conjugation.*

PROOF a) M is normal in U, since N is normal in G, and M is normal in N, since N is abelian. Thus M is normal in UN. As $M = U \cap N$, it follows that U/M is a complement to N/M in $(U/M)(N/M) = UN/M$.

b) $\mathbf{N}_G(U)$ normalizes UN and $U \cap N = M$, since N is normal in G. Thus $\mathbf{N}_G(U) \leq S$ and hence we find that $\mathbf{N}_G(U) = \mathbf{N}_S(U)$. As S normalizes M, we have $\mathbf{N}_S(U) = \mathbf{N}_S(U/M) = \mathrm{Stab}_S(U/M)$, where S acts by conjugation on the set of complements. ∎

Hence it remains to determine the stabilizer of U/M under the conjugation action of S on the complements to N/M in UN/M. This situation has already been discussed in Section 8.7.1. There we showed in Lemma 8.44 that S acts affinely on $Z^1(UN/N, N/M)$. Using this action, we can describe the S-conjugacy class of U/M in S/M as an orbit under this affine action; see Theorem 8.45.

Thus as a third step in our algorithm, we determine the elementary abelian p-group $Z^1(UN/N, N/M)$ and the affine action of S on this group. Then, we identify the elementary abelian p-group $Z^1(UN/N, N)$ with the additive group of $\mathbb{F}_p^h$ and we switch to additive notation. The affine action of S translates into an action homomorphism $\varphi : S \to \mathrm{GL}(h+1, \mathbb{F})$. Now it remains to calculate a stabilizer of the trivial affine vector under this action of S. This is summarized in the following lemma.

LEMMA 8.58 $\mathbf{N}_G(U) = \mathrm{Stab}_S((0, \dots, 0, 1))$ *where S acts on $\mathbb{F}_p^{h+1}$ via the affine action of Lemma 8.44.*

Hence we obtain an effective method to compute $\mathbf{N}_G(U)$. The induction downwards along the normal series with elementary abelian factors splits the computation into a sequence of induction steps. In each induction step we have to perform a variety of computations. In particular, each induction step requires two applications of the orbit-stabilizer algorithm. These two applications are usually the time and space consuming parts of the algorithm.

8.8.4   Conjugacy problems and conjugacy classes

There are various related problems to the determination of centralizers and normalizers. For example, with $g, h \in G$ and $U, V \leq G$ there are the following problems.

(1) Determine the conjugacy class g^G or U^G explicitly.

(2) Check whether there exists $x \in G$ with $g^x = h$ or $U^x = V$.

(3) Determine all conjugacy classes of elements or subgroups in G.

All these problems can be solved by minor modifications of the algorithms in the Sections 8.8.2 and 8.8.3. We discuss these problems for the elements of G here briefly and we show that they can be solved by variations of the centralizer algorithm. The corresponding problems for subgroups can be solved by similar variations of the normalizer algorithm.

Problem (1) can be solved directly by the centralizer algorithm. If an induced polycyclic sequence for $\mathbf{C}_G(g)$ is given, then a transversal T for $\mathbf{C}_G(g)$ in G can be read off from Lemma 8.33. This yields that $g^G = \{g^t \mid t \in T\}$.

Problem (2) requires a variation of the centralizer algorithm. In each induction step of this algorithm a stabilizer computation is performed. To solve Problem (2) we also compute its underlying orbit and check whether the element induced by h is contained in this orbit. If not, then g and h are not conjugate. If so, then we modify h to its conjugate h^y such that h^y induces the same element as g in the considered factor. Then we proceed to the next induction step.

Problem (3) also requires a variation of the centralizer algorithm. In each induction step of this algorithm a stabilizer computation of a given element is performed. To solve Problem (3) we have to compute the orbits and the stabilizer of all possible elements and then use all these elements and their orbits and stabilizer in the next induction steps.

On a historical note, computing centralizers and conjugacy classes in finite p-groups using the downwards induction method with polycyclic presentations was introduced by V. Felsch and J. Neubüser in [FN79]. They used an implementation of the algorithm to find a counterexample of order 2^{34} to the 'class-breadth' conjecture. The conjecture was that, for a finite p-group G, $c(G) \leq b(G) + 1$, where $c(G)$ is the nilpotency class and $b(G)$ is the size of the largest conjugacy class of G.

Exercises

1. Use the algorithm described above to compute the centralizer of $(2,3)$ in $G := \mathrm{Sym}(4)$, using the normal series with $N_1 = G$, $N_2 = \mathrm{Alt}(4) = \langle (1,2,3), (1,2)(3,4) \rangle$, $N_3 = \langle (1,2)(3,4), (1,3)(2,4) \rangle$, $N_4 = 1$.

2. Devise an alternative algorithm for computing the centralizer and conjugacy class of an element g in a polycyclic group G, which works by computing $\mathbf{C}_{G_i}(g)$ and g^{G_i} for $i = n, n-1, \ldots, 1$, where $G = G_1 \geq G_2 \geq \cdots \geq G_{n+1} = 1$ is a polycyclic sequence for G.

8.9 Automorphism groups

Let $G = \mathrm{Pc}\langle X \mid R \rangle$ be a polycyclic group defined by a refined consistent polycyclic presentation. In this section we shall outline an algorithm to compute the automorphism group $\mathrm{Aut}(G)$; that is, we want to determine generators for $\mathrm{Aut}(G)$ and the order $|\mathrm{Aut}(G)|$.

The method that is outlined here was first described by D.J.S. Robinson in [Rob81] and an implementation has been investigated and outlined in detail by M.J. Smith in [Smi94]. We note at this point that a dual approach can be used to check isomorphism between two polycyclically presented groups. We remark also that the methods for solving these problems in the special case of finite p-groups, which will be discussed later in Subsections 9.4.5 and 9.4.6, are generally more efficient than those to be discussed here.

The basic approach of this method is to use induction downwards along a characteristic series with elementary abelian factors; see Section 8.5. Thus let $G = N_1 \rhd \ldots \rhd N_l \rhd N_{l+1} = 1$ be such a series and let $N = N_l$ be the last nontrivial term in this series. Then N is an elementary abelian p-group of rank d, say. By induction, we assume that we know generators and the order of $\text{Aut}(G/N)$ and we aim to determine generators and the order of $\text{Aut}(G)$. As N is characteristic in G, there exists a mapping

$$\varphi : \text{Aut}(G) \to \text{Aut}(G/N) \times \text{Aut}(N) : \alpha \mapsto (\alpha_{G/N}, \alpha_N),$$

where $\alpha_{G/N}$ and α_N denote the induced actions of α on G/N and N. We determine $\text{Aut}(G)$ by computing the kernel and the image of φ. The kernel can be obtained from the following lemma.

LEMMA 8.59   *Let N be an elementary abelian p-group.*

  a) *Each element $\gamma \in Z^1(G/N, N)$ defines an automorphism $\alpha_\gamma \in \text{Aut}(G)$ by $g^{\alpha_\gamma} := g\gamma(gN)$ for $g \in G$.*

  b) *The mapping $\psi : Z^1(G/N, N) \to \text{Aut}(G) : \gamma \mapsto \alpha_\gamma$ is a monomorphism with $\text{im}(\psi) = \text{ker}(\varphi)$.*

  c) *We have $\text{ker}(\varphi) \cong Z^1(G/N, N)$.*

PROOF   a) is obvious and c) follows directly from b).

b) It is straightforward to prove that ψ is a monomorphism. It remains to show that $\text{im}(\psi) = \text{ker}(\varphi)$. Let $\alpha \in \text{im}(\psi)$. Then there exists $\gamma \in Z^1(G/N, N)$ such that $\alpha = \alpha_\gamma$. By the definition of α_γ, α induces the identity on G/N and N. Hence $\alpha \in \text{ker}(\varphi)$. Conversely, let $\alpha \in \text{ker}(\varphi)$. Then α induces the identity on G/N and hence for each $g \in G$ there exists $n_g \in N$ with $g^\alpha = gn_g$. As α induces the identity on N, the function $\delta : G \to N : g \mapsto n_g$ is constant on cosets of N and thus it is in fact a function of G/N. It is straightforward to check that δ is a derivation of G/N and thus $\delta \in Z^1(G/N, N)$. Hence $\alpha \in \text{im}(\psi)$ as desired.   ∎

Hence generators and the order of $\text{ker}(\varphi)$ can be computed readily using the methods of Section 7.6 and Lemma 8.59.

It remains to determine $\text{im}(\varphi)$. First, we note that generators and the order of $\text{Aut}(G/N) \times \text{Aut}(N)$ are available, since for $\text{Aut}(G/N)$ this information is known by induction and $\text{Aut}(N) \cong \text{GL}(d,p)$. Our aim is to determine generators and the order for $\text{im}(\varphi)$ from $\text{Aut}(G/N) \times \text{Aut}(N)$.

DEFINITION 8.60   Let $\psi : G/N \to \text{Aut}(N) : h \mapsto \overline{h}$ denote the action homomorphism corresponding to the conjugation action of G/N on N. Then we define the set of compatible pairs $\text{Comp}(G, N)$ as

$$\{ (\nu, \mu) \in \text{Aut}(G/N) \times \text{Aut}(N) \mid \overline{h^\nu} = \mu^{-1}\overline{h}\mu \ \text{ for all } h \in G/N \}.$$

The following lemma notes that $\text{Comp}(G, N)$ is a subgroup of $\text{Aut}(G/N) \times \text{Aut}(N)$ and it also exhibits the connection between $\text{Comp}(G, N)$ and the image $\text{im}(\varphi)$.

LEMMA 8.61 $\operatorname{im}(\varphi) \le \operatorname{Comp}(G, N) \le \operatorname{Aut}(G/N) \times \operatorname{Aut}(N)$.

PROOF   It is straightforward to check that $\operatorname{Comp}(G, N)$ is closed under multiplication; so it is a subgroup of $\operatorname{Aut}(G/N) \times \operatorname{Aut}(N)$. Let $(\nu, \mu) \in \operatorname{im}(\varphi)$. Then there exists $\alpha \in \operatorname{Aut}(G)$ with $\alpha_{G/N} = \nu$ and $\alpha_N = \mu$. For $h \in G/N$, we have $\overline{h^\nu} = \overline{h^\alpha} = \overline{h}^\alpha = \overline{h}^\mu$.   ∎

Next, we describe a method of computing $\operatorname{Comp}(G, N)$ from $\operatorname{Aut}(G/N)$ and $\operatorname{Aut}(N)$. Let $K/N := \ker(\psi)$ and $I := \operatorname{im}(\psi)$ be the kernel and the image of the conjugation action of G/N on N. Then $K/N = \mathbf{C}_{G/N}(N)$ and $I \cong (G/N)/(K/N) \cong G/K$. Let $S := \operatorname{Stab}_{\operatorname{Aut}(G/N)}(K/N)$ and $T := \mathbf{N}_{\operatorname{Aut}(N)}(I)$. Then $\operatorname{Comp}(G, N) \le S \times T$.

Generators and the orders of S and T can be computed from $\operatorname{Aut}(G/N)$ and $\operatorname{Aut}(N)$ using the methods of Section 4.1 and thus we can also obtain generators and the order of $S \times T$. The group $S \times T$ acts on the set $\operatorname{Hom}(G/N, \operatorname{Aut}(N))$ of all homomorphisms from G/N to $\operatorname{Aut}(N)$ via

$$\iota^{(\nu,\mu)} : G/N \to \operatorname{Aut}(N) : h \mapsto \mu^{-1}((h^{\nu^{-1}})^\iota)\mu$$

and the compatible pairs form the stabilizer of ψ under this action. Hence generators and the order of $\operatorname{Comp}(G, N)$ can be computed from $S \times T$ using the orbit-stabilizer method of Section 4.1.

DEFINITION 8.62   For $\gamma \in Z^2(G/N, N)$ and $(\nu, \mu) \in \operatorname{Comp}(G, N)$ we define

$$\gamma^{(\nu,\mu)}(h, k) = \gamma(h^{\nu^{-1}}, k^{\nu^{-1}})^\mu.$$

It is straightforward to observe that this induces an action of $\operatorname{Comp}(G, N)$ on the group of 2-cocycles $Z^2(G/N, N)$. The subgroup $B^2(G/N, N)$ is setwise invariant under this action. Thus we obtain an induced action of $\operatorname{Comp}(G, N)$ on the second cohomology group $H^2(G/N, N)$.

DEFINITION 8.63   Let $\gamma \in Z^2(G/N, N)$ be a cocycle defining the extension G of N by G/N. Then we define the set of inducible pairs as $\operatorname{Indu}(G, N) = \operatorname{Stab}_{\operatorname{Comp}(G,N)}(\gamma B^2(G/N, N))$.

By definition, the inducible pairs form a subgroup of the compatible pairs, and they can be computed using the orbit-stabilizer algorithm of Section 4.1. This yields an algorithm to compute the image $\operatorname{im}(\varphi)$ by the following theorem.

THEOREM 8.64 $\operatorname{im}(\varphi) = \operatorname{Indu}(G, N)$.

PROOF   The proof of this theorem is straightforward, but technical. We refer to [Rob81].   ∎

Theorem 8.64 yields that we can compute $\mathrm{im}(\varphi)$ as the stabilizer of a cocycle coset $\gamma B^2(G/N, N)$ under the action of the group of the compatible pairs $\mathrm{Comp}(G, N)$. Hence we can compute generators and the order of $\mathrm{im}(\varphi)$ using the methods of Section 4.1.

8.10   The structure of finite solvable groups

The structure of the finite solvable groups is well investigated and subgroups such as Sylow and Hall subgroups and maximal subgroups play a major rôle in this theory. The finite solvable groups are exactly the finite polycyclic groups and hence every finite solvable group can be defined by a refined consistent polycyclic presentation.

In this section we shall describe methods to compute such structure theoretic subgroups in a group G that is defined by a refined consistent polycyclic presentation $\mathrm{Pc}\langle X \mid R \rangle$. For further details and more advanced methods of a similar nature we refer to papers by Cannon, Eick, Leedham-Green, and Wright [CELG04, Eic93, Eic97, EW02]. For background on the underlying theory of finite solvable groups see [DH92].

8.10.1   Sylow and Hall subgroups

DEFINITION 8.65   Let p be a prime and let π be a set of primes. Write $|G| = p_1^{e_1} \cdots p_r^{e_r}$ for distinct primes $p_1, \ldots, p_r$.

   a) A *Sylow p-subgroup* S of G is a subgroup of p-power order such that $p \nmid |G\!:\!S|$; that is, $|G\!:\!S|$ is coprime to $|S|$.

   b) A *Hall π-subgroup* H of G is a subgroup such that all prime divisors of $|H|$ are contained in π and $|G\!:\!H|$ is coprime to $|H|$.

   c) A *Sylow system* of G is a set of Sylow subgroups $\{S_1, \ldots, S_r\}$ such that S_i is a Sylow p_i-subgroup and $S_i S_j = S_j S_i$ holds for each $i \neq j$.

   d) A *complement system* of G is a set of subgroups $\{C_1, \ldots, C_r\}$ such that $|G\!:\!C_i|$ is a p_i-power and $|G\!:\!C_i|$ is coprime to $|C_i|$ for $1 \leq i \leq r$. The group C_i is called a p_i-*complement*.

It is well-known that every finite solvable group has a Sylow system and a complement system. In fact, this existence can be used to characterize the

solvable groups among the finite groups. A complement system gives rise to
Hall subgroups for every possible set of primes π and it also gives rise to a
Sylow system, as the following lemma recalls. We refer to [Hup67], VI 1.5 and
VI 2.2 for a proof.

LEMMA 8.66   *Let* $\{C_1, \ldots, C_r\}$ *be a complement system of* G.

a) *Then* $H = \bigcap_{p_i \notin \pi} C_i$ *is a Hall* π-*subgroup in* G.

b) *The set* $\{S_1, \ldots, S_r\}$ *with* $S_i = \bigcap_{j \neq i} C_j$ *forms a Sylow system for* G.

It is the aim of this section to outline an algorithm for computing a complement system in the finite polycyclic group G. For this purpose we use
induction downwards along a normal series with elementary abelian factors
$G = N_1 \rhd \ldots \rhd N_l \rhd N_{l+1} = 1$. Thus let $N = N_l$ be the last nontrivial term
in this series. Then N is an elementary abelian p-group of rank d, say.

By induction, we assume that we have determined a q-complement C/N of
G/N for some prime q. We want to compute a q-complement of G. We have
to distinguish two cases on p and q as outlined in the following lemma.

LEMMA 8.67   *Let* C/N *be a* q-*complement in* G/N *for a* p-*group* N.

a) *If* $p \neq q$, *then* C *is a* q-*complement of* G.

b) *If* $p = q$, *then there exists a complement to* N *in* C *and every such
complement is a* q-*complement of* G.

PROOF   a) In this case we have that $|G : C|$ is a q-power and $|C| =
|C/N||N|$ is coprime to q. Thus C is a q-complement in G.

b) In this case we have that $|C/N|$ is coprime to $q = p$ and thus $|C/N|$ and
$|N|$ are coprime. Hence there exists a complement to N in C by the Schur-
Zassenhaus theorem [Rot94, Theorem 7.24]. Let K be such a complement.
Then $|K| = |C/N|$ is coprime to q and $|G:K| = |G:C||N|$ is a q-power. Thus
K is a q-complement in G.   ∎

Lemma 8.67 yields a method to determine a q-complement by the methods
of Section 7.6. However, in this special situation, we can determine complements with a more effective method. We give a brief outline of this improvement in the following. It makes use of the fact that, in case b) of the lemma,
every complement of N in a subgroup of C is contained in a complement of
N in C. This follows from the other part of the Schur-Zassenhaus theorem,
which says that, when $|C/N|$ and $|N|$ are coprime, then all complements of
N in C are conjugate.

Let $p = q$ and let $Y := [c_1, \ldots, c_r, n_1, \ldots, n_d]$ be a polycyclic sequence
for C such that $N = \langle n_1, \ldots, n_d \rangle$ and such that the relative orders $R(Y) =$

$(s_1, \ldots, s_r, p, \ldots, p)$ are all primes. Then every complement T of N in C has a polycyclic sequence of the form $[c_1 l_1, \ldots, c_r l_r]$ for certain elements $l_i \in N$. Our aim is to determine such elements $l_1, \ldots, l_r$. The following lemma yields a characterization of these elements.

LEMMA 8.68   *Let $Z := [c_1 l_1, \ldots, c_r l_r]$ for $l_1, \ldots, l_r \in N$. Then Z is an induced polycyclic sequence for a complement to N in C with respect to Y if and only if $(c_j l_j)^{s_j}$ and the commutators $[c_j l_j, c_k l_k]$ can be written as normal words in Z for $1 \le j < k \le r$.*

PROOF   This follows directly from Lemma 8.34.   ∎

By induction, we assume that the elements $l_{j+1}, \ldots, l_r$ have already been determined and we want to compute l_j. Let $b_k := c_k l_k$ for $j+1 \le k \le r$, and let $Y' := [c_1, \ldots, c_j, b_{j+1}, \ldots, b_r, n_1, \ldots, n_d]$. Then Y' is an induced polycyclic sequence for C. Therefore, by Lemma 8.34, we find that the power $c_j^{s_j}$ and the commutators $[b_k, c_j]$ for $j+1 \le k \le r$ can be written as normal words in Y' of the form

$$c_j^{s_j} = w_{j,j}(b_{j+1}, \ldots, b_r) v_{j,j}(n_1, \ldots, n_d)$$
$$[b_k, c_j] = w_{j,k}(b_{j+1}, \ldots, b_r) v_{j,k}(n_1, \ldots, n_d)$$

The following lemma yields a method to determine l_j from this setup.

LEMMA 8.69   *Let C be a group and $N \trianglelefteq C$ an abelian normal subgroup.*

a) *Let $c, b \in C$ with $[b, c] = wv$ for $v \in N$. Then, for $l \in N$, $[b, cl] = w$ if and only if $l^a = v$ with $a = bw - 1$.*

b) *Let $c \in C$ with $c^s = wv$ for $v \in N$. Then, for $l \in N$, $(cl)^s = w$ if and only if $l^b = v$ with $b = -\sum_{i=0}^{s-1} c^i$.*

PROOF   The proof is an elementary computation. We refer to [CELG04] for an outline.   ∎

8.10.2   Maximal subgroups

In this section we shall give an outline of an effective algorithm to compute the conjugacy classes of maximal subgroups of G.

DEFINITION 8.70

a) N/L is called a *normal factor* of G if $L, N \trianglelefteq G$ with $L \le N$.

b) K is called a complement to the normal factor N/L if $K \cap N = L$ and $KN = G$.

Thus K is a complement to the normal factor N/L if $L \leq K$ and K/L is a complement to N/L in G/L. Complements to chief factors can be used to determine maximal subgroups, as the following lemma shows.

LEMMA 8.71 *Let $G = N_1 \rhd \ldots \rhd N_l \rhd N_{l+1} = 1$ be a chief series of G. The conjugacy classes of maximal subgroups in G coincide with the conjugacy classes of complements of the normal factors N_i/N_{i+1} for $1 \leq i \leq l$.*

PROOF Let M be a maximal subgroup of G and let i be minimal with $N_{i+1} \leq M$. Then $i \geq 1$, since M is a proper subgroup of G and thus $N_1 \not\leq G$. And $i \leq l$, since $N_{l+1} = \{1\} \leq M$. Since M is maximal in G and $N_i \not\leq M$, we obtain that $MN_i = G$. Further, we find that $N_{i+1} \leq M \cap N_i$ and $M \cap N_i$ is normal in G. As N_i/N_{i+1} is a chief factor, this yields $M \cap N_i = N_{i+1}$ and M complements N_i/N_{i+1}. Using similar arguments, one can prove readily that every complement to a chief factor N_i/N_{i+1} is a maximal subgroup in G. ∎

Lemma 8.71 yields a method of computing the conjugacy classes of maximal subgroups of G. We first determine a chief series of G as described in Section 8.5. Then we consider every factor of this series in turn and compute its conjugacy classes of complements with the method outlined in Section 7.6.

When applying this method it often happens that many of the chief factors do not have any complement at all. Hence this method incorporates a lot of redundant work in this case. This can be avoided with the second approach, which is outlined briefly in the remainder of this section.

We consider the lower nilpotent series $G = L_1 \rhd \ldots \rhd L_{r+1} = 1$ and, as in Section 8.5, we define L_{i2} as the commutator subgroup by $L_{i2}/L_{i+1} := (L_i/L_{i+1})'$. As noted in Lemma 8.40, there exists a complement K_i to the normal factor L_i/L_{i2} in G. Also, we note that every maximal subgroup M of G covers all but one of the factors of the lower nilpotent series of G. This setup allows us to describe the conjugacy classes of maximal subgroups in G as follows.

THEOREM 8.72 *For $1 \leq i \leq r$, let K_i be a complement to L_i/L_{i2} in G and let $A_{i1}/L_{i2}, \ldots, A_{ik_i}/L_{i2}$ be a set of maximal G-normal subgroups in L_i/L_{i2}.*

a) *The set $\bigcup_{i=1}^{r} \{K_i A_{i1}, \ldots, K_i A_{ik_i}\}$ is a complete set of conjugacy class representatives for the maximal subgroups of G.*

b) *We have $\mathbf{N}_G(K_i A_{ij}) = G$ if $i = 1$, and $\mathbf{N}_G(K_i A_{ij}) = K_i A_{ij}$ otherwise.*

PROOF   We refer to [CELG04].   ∎

Hence we obtain an algorithm to compute the conjugacy classes of maximal subgroups. First, we determine the refined lower nilpotent series as in Section 8.5. Then we consider each factor L_i/L_{i2} in turn and determine its G-normal maximal subgroups. As L_i/L_{i2} is abelian, the G-normal maximal subgroups translate to maximal G-submodules of this factors and hence they can be computed by an application of the Meataxe; see Section 7.4. Finally, we need to determine a complement K_i to L_i/L_{i2} in G. This can be done with a straightforward application of the methods in Section 8.7.1. We note here that a more effective approach for this purpose is described in [CELG04] and it follows similar ideas to the methods in Section 8.10.1.

Exercise

1. Use the methods described in this section to compute the maximal subgroups of Sym(4).

Chapter 9

Computing Quotients of Finitely Presented Groups

If we have a finitely presented group G, what can we discover about the structure of G? Unfortunately, as we remarked at the beginning of Chapter 5, most of the natural questions that we might be inclined to ask are provably undecidable in general.

In Chapters 12 and 13, we shall study general algorithms, which approach such problems by attempting to compute a normal form for the elements of G. They are only successful for certain specific types of groups, but when they do succeed, they enable many questions about G to be answered, such as "is G finite?".

In this chapter, we shall investigate and describe algorithms that aim to compute quotients of G of specific types. In many cases, these are theoretically guaranteed to succeed given adequate computational resources. Specifically, we shall study algorithms that compute general finite quotients, abelian quotients, and quotients that are finite p-groups. Particularly when used in combination with each other, these methods can on occasion be used to prove that G is infinite; see Subsection 9.3.3. A summary of other quotient algorithms that have been proposed or implemented, including nilpotent, solvable, and polycyclic quotient algorithms, will be given in Subsection 9.4.3.

Many of the algorithms to be described in this chapter are implemented within the interactive graphics package QUOTPIC [HR93], by Holt and Rees. This enables the user to plot and visualize quotients computed so far, and then to decide which further calculations to undertake.

In some cases, the algorithms to be described lead on naturally to other algorithms of central importance in CGT. The finite quotient algorithm can be easily adapted to compute automorphism groups of finite groups and to test pairs of finite groups for isomorphism. The algorithm to compute finite p-quotients can be adapted and further developed to enumerate representatives of the isomorphism classes of all finite p-groups up to a specified order, and this in turn gives rise to algorithms to compute automorphism groups of finite p-groups and to test pairs of finite p-groups for isomorphism.

9.1 Finite quotients and automorphism groups of finite groups

The low-index subgroups algorithm discussed in Section 5.4 finds representatives H of the conjugacy classes of subgroups of a finitely presented group F up to a given index n. For each such subgroup H, there is a related action of F by multiplication on the right cosets of H, and the low-index subgroups algorithm in fact finds the coset table of H in F, which defines the generator images in this action. So, effectively, LowIndexSubgroups finds all homomorphisms $F \to \mathrm{Sym}(n)$, and it can be regarded as a method of finding the finite quotient groups of F.

The performance of LowIndexSubgroups deteriorates rapidly with increasing n (experiments suggest that the complexity is worse than exponential in n), however, and so it can only find finite quotients that have a permutation representation of low degree. There is an alternative approach to finding finite quotients, which we shall discuss in the this section, in which we fix a target group G and search specifically for homomorphisms (or, more usually, for epimorphisms) $F \to G$. Like LowIndexSubgroups, its performance deteriorates rapidly with the number of generators of F, but otherwise its running time is a polynomial function of $|G|$ and does not rely on G having a low-degree permutation representation.

This method also provides a reasonably efficient approach to computing automorphism groups of finite groups G and testing two finite groups G and H for isomorphism, provided that $\mathrm{Aut}(G)/\mathrm{Inn}(G)$ is not too large. The algorithm, together with the application to automorphism group computation is described by Hulpke in [Hul96, V.5]. It has been implemented in GAP by Hulpke and as a stand-alone program and in Magma by Holt.

Both LowIndexSubgroups and the Epimorphisms algorithm are indispensable tools for finding finite quotients of finitely presented groups. The former will be much more effective for quotients such as $\mathrm{Alt}(n)$ and $\mathrm{Sym}(n)$, but the latter is needed for relatively small target groups, such as the simple group J_1 of order 175560, which have no small-degree faithful permutation representation; the smallest such representation of J_1 has degree 266, which is usually much too large for LowIndexSubgroups.

9.1.1 Description of the algorithm

Let $F = \langle X \mid R \rangle$ be a group defined by a finite presentation, where $X = [x_1, \ldots, x_r]$, and let G be a finite group. The problem to be discussed in this section is that of finding homomorphisms from F to G.

Often we are mainly interested in epimorphisms rather than in all homomorphisms, so we shall make our main algorithm return a list of epimorphisms, but in fact it will only check the surjectivity of a homomorphism found at the final stage, and so it can be modified to find all homomorphisms simply by leaving out this final check.

Sometimes we are only interested in finding a single epimorphism, but let us assume that we require a complete list of all such epimorphisms. If $\varphi : F \to G$ is a homomorphism, and $\alpha : G \to G$ is an automorphism, then the composite $\varphi' := \alpha\varphi$ is another epimorphism, and we say that φ and φ' are *equivalent modulo* $\mathrm{Aut}(G)$. It is not hard to show that epimorphisms φ and φ' from F to G are equivalent modulo $\mathrm{Aut}(G)$ if and only if $\ker(\varphi) = \ker(\varphi')$, and so the distinct quotient groups of F that are isomorphic to G correspond exactly to the equivalence classes modulo $\mathrm{Aut}(G)$ of epimorphisms $F \to G$.

For this reason, and also in order to save time and space, it makes sense to ask for a list of representatives of these equivalence classes, rather than of all epimorphisms; that is, we would like a list of epimorphisms from F to G up to equivalence modulo $\mathrm{Aut}(G)$. However, we do not normally want to make the computation of $\mathrm{Aut}(G)$ an integral and necessary part of the algorithm, and so we shall design it to take three input parameters, F, G and A, where G must be a normal subgroup of A, and a list of epimorphisms $F \to G$ up to equivalence modulo conjugation by elements of A is returned. That is, epimorphisms φ and φ' are defined to be equivalent if one is the composite of the other and the conjugation action of an element $\alpha \in A$ on G.

The user will usually want to choose A such that A induces all of $\mathrm{Aut}(G)$ on G; for example, $G = \mathrm{PSL}(2, q)$ and $A = \mathrm{P\Gamma L}(2, q)$. But a lazy user can just use the default of $A = G$, and get a possibly longer list of epimorphisms that are equal up to equivalence modulo $\mathrm{Inn}(G)$.

The basic idea is simple-minded. We consider all r-tuples $(g_1, \ldots, g_r) \in G^r$, and use the criterion in Theorem 2.52 to test whether the map $\varphi : X \to G$ defined by $\varphi(x_i) = g_i$ for $1 \leq i \leq r$ extends to a homomorphism $\varphi : F \to G$. Let us refer to this process as *considering* $(g_1, \ldots, g_r)$ *as an image*.

In the case when $R = \phi$ and F is free, all such maps define homomorphisms, and so we cannot improve much on this. However, in general, considering $|G|^r$ such r-tuples as images will be hopelessly slow except for very small groups G and small values of r, so we should try and use backtrack search methods to prune the search tree.

According to the ideas presented in Section 4.6, to prune the search tree, we should try to find methods of ruling out k-tuples $(g_1, \ldots, g_k)$ as initial subsequences of possible images $(g_1, \ldots, g_r)$, for $k < r$. This can be done if some of the relators in R only involve the generators $x_1, \ldots, x_k$ and their inverses, because then we can apply the criterion of Theorem 2.52 to those relators, using only $g_1, \ldots, g_k$. If the criterion fails, then we can prune the search tree by ruling out the initial subsequence $(g_1, \ldots, g_k)$. If the criterion succeeds with $k < r$, however, then we need to descend to the next level and consider possible images g_{k+1}. If the criterion succeeds with $k = r$, then we have found a homomorphism $\varphi : F \to G$, and to test whether it is an epimorphism, we just check whether $\langle g_1, \ldots, g_r \rangle = G$.

To make these tests as effective as possible, we begin by reordering the generating sequence $[x_1, \ldots, x_r]$ of F so as to make the subsets R_k of R consisting of those relators that only involve $x_1, \ldots, x_k$ and their inverses as large as possible. We shall not attempt here to define precisely what we mean by that, or to present the algorithm for doing this. In unfavourable cases, we might find that all relators involve all of the generators, in which case we have no means of pruning the search tree.

Many, but by no means all, presentations that arise in practice contain relators of the form x^n for some $x \in X$, $n \in \mathbb{N}$, which provides an upper bound on the order of x in F. In this situation, x can only be mapped by a homomorphism $\varphi : F \to G$ to an element of G of order dividing n.

Guided partly by this idea, we have designed the algorithm such that it starts by finding representatives $h_1, \ldots, h_c$ of the conjugacy classes of A that lie G. For $1 \le i \le r$, we then define I_i to be the subset of $\{h_1, \ldots, h_c\}$ consisting of those h_j that are possible images of x_i in a homomorphism $\varphi : F \to G$; that is, those h_j for which $\varphi(x_i) = h_j$ is not ruled out by its order and a relator of F of the form x_i^n. Then we take each r-tuple of elements $(g_1, \ldots, g_r)$ with each $g_i \in I_i$ in turn, and consider the r-tuples $(g_1^{\alpha_1}, \ldots, g_r^{\alpha_r})$ for elements $\alpha_i \in A$ as images. This splits up the search into a number of subsearches, one for each such r-tuple $(g_1, \ldots, g_r)$ with each $g_i \in I_i$.

Experiments and experience indicate that doing that usually speeds up running times, particularly when there are relators of the form g^n. Unfortunately, in some examples, particularly those in which there are no relators of the form g^n and (even more so) those in which all relators involve all elements of X, it is quicker simply to run through all r-tuples of group elements in the naive manner. This makes it difficult to design a single algorithm that performs as well as possible on all examples.

In order to get each conjugate $g_i^{\alpha_i}$ only once in the r-tuples $(g_1^{\alpha_1}, \ldots, g_r^{\alpha_r})$, we restrict the α_i to lie in a transversal T_i of $\mathbf{C}_A(g_i)$ in A.

We have not yet discussed how we find epimorphisms only up equivalence modulo conjugation by elements of A. For this, we want to consider not all $(g_1^{\alpha_1}, \ldots, g_r^{\alpha_r})$ as images, but only orbit representatives of the action of A on the set of all such r-tuples defined by

$$(g_1^{\alpha_1}, \ldots, g_r^{\alpha_r}) \to (g_1^{\alpha_1 \alpha}, \ldots, g_r^{\alpha_r \alpha})$$

for $\alpha \in A$.

To find such representatives, we apply the following easy proposition and corollary, of which we leave the proofs to the reader.

PROPOSITION 9.1   *Suppose that a group G acts on sets Ω_1 and Ω_2. Let $\alpha_1, \ldots, \alpha_t$ be representatives of the orbits of G on Ω_1 and, for $1 \le i \le t$, let $\beta_{i1}, \ldots, \beta_{ii_t}$ be representatives of the orbits of G_{α_i} on Ω_2. Then*

$$\{\, (\alpha_i, \beta_{ij}) \mid 1 \le i \le t, \, 1 \le j \le i_t \,\}$$

is a set of orbit representatives of G in its induced action on $\Omega_1 \times \Omega_2$.

COROLLARY 9.2 *Suppose that G acts on sets Ω_i for $1 \leq i \leq r$. Then we can find a set $\mathcal{R}$ of representatives $(\alpha_1, \ldots, \alpha_r)$ of G in its induced action on $\Omega_1 \times \cdots \times \Omega_r$, such that the following is true for all $1 \leq k \leq r$. The initial subsequences $(\alpha_1, \ldots, \alpha_k)$ of elements $(\alpha_1, \ldots, \alpha_r) \in \mathcal{R}$ form a set of representatives of the induced action of G on $\Omega_1 \times \cdots \times \Omega_k$ and, for each such $(\alpha_1, \ldots, \alpha_k)$, the sequences $(\alpha_{r+1}, \ldots, \alpha_r)$ such that $(\alpha_1, \ldots, \alpha_r) \in \mathcal{R}$ form a set of orbit representatives in the induced action of $\cap_{i=1}^{k} G_{\alpha_i}$ on $\Omega_{k+1} \times \cdots \times \Omega_r$.*

In our application, the set Ω_i in Corollary 9.2 is the A-conjugacy class g_i^A, and the action is conjugation by A, which is transitive. So, by choosing $k = 1$ in this corollary, we see that we only need to consider r-tuples $(g_1, g_2^{\alpha_2}, \ldots, g_r^{\alpha_r})$ beginning with g_1.

More generally, for $2 \leq k \leq r$, suppose that we are considering r-tuples with initial subsequence $(g_1, g_2^{\alpha_2}, \ldots, g_{k-1}^{\alpha_{k-1}})$. Then the intersection of the stabilizers in A of the $g_i^{\alpha_i}$ is equal to $C := \cap_{i=1}^{k-1} \mathbf{C}_A(g_i^{\alpha_i})$ (where $\alpha_1 = 1$). So we need to compute orbit representatives of the action of C by conjugation on g_k^A. These are given by $g_k^{\beta_j}$, for $1 \leq j \leq l$, where $\beta_1, \ldots, \beta_l$ are representatives of the double cosets $\mathbf{C}_A(g_k)\beta_j C$ in A.

It is convenient to compute these double coset representatives during the backtrack procedure for considering images. We discussed the computation of double coset representatives in Subsection 4.6.8, and we shall assume the availability of a function DCR that does this.

The function for finding epimorphisms of F onto G up to conjugation in A is displayed in EPIMORPHISMS.

This is a tricky function to understand. Here is an explanation of some of the variables that are used in the function.

k is the current level in the search; that is, the component in the r-tuples $(g_1, g_2^{\alpha_2}, \ldots, g_k^{\alpha_k}, \ldots, g_r^{\alpha_r})$ that we are currently considering.

$\Lambda[j]$ stores the intersection of centralizers $\cap_{i=1}^{j} \mathbf{C}_A(g_i^{\alpha_i})$; recall that $\Lambda[k-1]$ was denoted by C in the explanation above.

$\delta[k]$ is a list of representatives of the double cosets $\mathbf{C}_A(g_k)g\Lambda[k-1]$ of $\mathbf{C}_A(g_k)$, $\Lambda[k-1]$ in A.

$\mu[k]$ is the index such that $\delta[k][\mu[k]]$ is the double coset representative in the list $\delta[k]$ currently being considered.

im$[k]$, which represents the component $g_k^{\alpha_k}$ in the r-tuple being currently considered, should be interpreted as being an alias for $g_k^{\delta[k][\mu[k]]}$; that is, the conjugate of g_k under the $\mu[k]$-th element in the list of double coset representatives discussed above.

EPIMORPHISMS(F, G, A)

    Input: $F = \langle\, X \mid R \,\rangle$, finite groups G, A with $G \trianglelefteq A$

    Output: List of epimorphisms $F \to G$ up to conjugation by A

    Remark: im$[k]$ is an alias for $g_k^{\delta[k][\mu[k]]}$.

1   Reorder generators $x_1, \ldots, x_r$ of F such that the subsets $R_k \subseteq R$ of relators that involve just $x_i^{\pm 1}$ for $1 \le i \le k$ are as large as possible.

2   Calculate representatives h_i $(1 \le i \le c)$ of the conjugacy classes of A that lie in G.

3   **for** $i \in [1 .. r]$

4   **do if** there is a relator $x_i^{\pm t}$ in R with $t > 0$

5   **then** $I_i := \{\, h_i \mid 1 \le i \le r,\ t \bmod |h_i| = 0 \,\}$;

6   **else** $I_i := \{\, h_i \mid 1 \le i \le r \,\}$;

7    $\mathcal{H} := [\,]$; (* the list of epimorphisms to be computed *)

8   **for each** $(g_1, \ldots, g_r) \in I_1 \times \cdots \times I_r$

9   **do** (* Consider images $x_i \to g_i^{h_{ij}}$ for suitable $h_{ij} \in A$ *)

10          $\Lambda[1] := \mathbf{C}_A(g_1)$; $f := \mathtt{false}$; $k := 2$;

11   **while** $k > 1$

12   **do if not** f **then** $\delta[k] := \mathrm{DCR}(A, \mathbf{C}_A(g_k), \Lambda[k-1])$;

13                $\mu[k] := 1$; $f := \mathtt{false}$; $b := \mathtt{true}$;

14   **while** b **and** $k > 1$

15   **do** $b := \mathtt{false}$;

16   **while** $\mu[k] \le |\delta[k]|$

17   **do if not** TESTRELS$(k,\text{im}[1],\ldots,\text{im}[k])$

18   **then** $\mu[k] := \mu[k] + 1$;

19   **if** $\mu[k] > |\delta[k]|$

20   **then** $b := \mathtt{true}$; $k := k-1$;

21   **if** $k > 1$ **then** $\mu[k] := \mu[k] + 1$;

                (* Relations are satisfied by images at level k *)

22   **if** $k > 1$

23   **then if** $k < r$

24   **then** $\Lambda[k] := \mathbf{C}_{\Lambda[k-1]}(\text{im}[k])$;

25                      $k := k + 1$;

26   **else** $f := \mathtt{true}$;

27   **if** $\langle \text{im}[1], \ldots, \text{im}[k] \rangle = G$

28   **then** APPEND$(\sim \mathcal{H}, \text{im})$;

29                  $\mu[k] := \mu[k] + 1$;

30   **return** $\mathcal{H}$;

TESTRELS$(k, \text{im}[1], \ldots, \text{im}[k])$ at line 17 checks whether the first k components of the current r-tuple being considered satisfy the relators in R_k. If so, then the code proceeds to line 23 and we move down to the next level, $k+1$, except when $k = r$, in which case we have found a homomorphism $F \to G$. If not, then this k-tuple is not part of a possible image, so we move on to the next conjugate of g_k at the same level.

We have not written out the details of TESTRELS, and the overall performance depends critically on an efficient implementation of it, so we shall discuss this matter later.

b (for backtrack) is set to be true at line 13 (the top of the main loop in the backtracking process), and also when we have just decreased k.

f (for found) is set to be true when we have just found a homomorphism from F to G.

Example 9.1

As an example, let $F := \langle\, x, y \mid x^2, y^3, (xy)^7, [x,y]^9 \,\rangle$, let $G := \mathrm{PSL}(2,8)$ and let $A := \mathrm{P\Gamma L}(2,8)$. Then $|A:G| = 3$ and $A \cong \mathrm{Aut}(G)$. We use the natural permutation representation of G and A on 9 points. Representatives of the five conjugacy classes of A that lie in G are

$$1_G,\ (1,6)(3,7)(4,5)(8,9),\ (1,3,2)(4,7,8)(5,6,9),\ (1,7,5,8,6,3,2),$$

$$(1,9,8,7,6,5,4,3,2).$$

From the relators x^2 and y^3, we find that the sets I_1 and I_2 of conjugacy class representatives of possible targets of x and y in a homomorphism $F \to G$ are $I_1 = \{\, 1_G, (1,6)(3,7)(4,5)(8,9) \,\}$ and $I_2 = \{\, 1_G, (1,3,2)(4,7,8)(5,6,9) \,\}$. So there are four possibilities for $(g_1, g_2) \in I_1 \times I_2$ in the outer 'For' loop beginning at line 8 of EPIMORPHISMS.

Of these, $g_1 = g_2 = 1$ defines a homomorphism but not an epimorphism. It is immediately clear to us that, if $g_1 = 1$ or $g_2 = 1$ but not both, then the image of the relator $(xy)^7$ cannot be satisfied in G, and so no homomorphism is possible with these images. The function itself is not so clever, and will have to eliminate these possibilities by running through the search, but this will not be unduly time-consuming.

So let us consider $g_1 = (1,6)(3,7)(4,5)(8,9)$, $g_2 = (1,3,2)(4,7,8)(5,6,9)$. At the top of the 'While' loop at line 11, we have $k = 2$ and $\Lambda[k{-}1] = \mathbf{C}_A(g_1)$. We then go on to calculate representatives of the double cosets $\mathbf{C}_A(g_2)g\mathbf{C}_A(g_1)$ in A. It turns out that $|A : \mathbf{C}_A(g_2)| = 56$, but there are just 3 orbits of the corresponding right coset action of $\mathbf{C}_A(g_1)$ (which has order 24), and so there are three double coset representatives:

$$\alpha_1 := 1_G,\ \alpha_2 := (1,3,8,6,4,9,2),\ \alpha_3 := (1,4,3,8,6,9)(2,5).$$

The function then applies TESTRELS with $k = 2$, $\mathrm{im}[1] = g_1$ and $\mathrm{im}[2] = g_2^{\alpha_i}$ for $i = 1, 2, 3$. We find that $|g_1 g_2^{\alpha_1}| = 9$ and $|g_1 g_2^{\alpha_2}| = 2$, so TESTRELS fails on the relators $(xy)^7$ and returns `false`. But $|g_1 g_2^{\alpha_3}| = 7$ and $|[g_1, g_2^{\alpha_3}]| = 9$, so TESTRELS returns `true` for α_3. We now have $k = 2 = r$, so Theorem 2.52 tells us that the map $x \to g_1$, $y \to g_2^{\alpha_3} = (1,2,9)(3,7,6)(4,8,5)$ extends to a homomorphism $F \to G$, which is in fact an epimorphism. The search is now complete, so this epimorphism is unique up to conjugation by A, which proves that F has a unique quotient group F/K that is isomorphic to $PSL(2,8)$. $\square$

The example above is very quick and easy, and the corresponding epimorphism could have been found equally quickly by running the low-index subgroups algorithm up to index 9. It does raise one interesting question, however: could we have done anything to avoid executing the searches in which one of g_1 and g_2 was 1, in view of the fact that it was so obvious to us that they were doomed to be fruitless.

There is a possibility of this nature, which has been tried in implementations, and would enable us to avoid the fruitless searches in the example. For each generator x_i of F and each order n_i of a potential image g_i of x_i in G, we append the new relator $x_i^{n_i}$ to F and run a coset enumeration program over the trivial subgroup of the modified group presentation $F' := \langle X \mid R \cup \{x_i^{n_i}\} \rangle$ for a very short time. If this enumeration completes and shows that $|F'| < |G|$, then there can be no epimorphism $\varphi : F \to G$ in which $\varphi(x_i)$ has order dividing n_i (proof left to the reader!). Hence we can remove any g_i whose order divides n_i from the set I_i of possible targets.

9.1.2   Performance issues

The performance, in terms of running time, and hence the overall usefulness of the above algorithm depends critically on its implementation. We shall briefly discuss some of the issues involved here.

It helps to start with an approximate idea of the feasible range of applicability. It is generally not the most appropriate algorithm to use when the target group G is abelian or even solvable. In that case, it is usually more sensible to employ the quotient algorithms to be discussed in later sections of this chapter.

EPIMORPHISMS is most often used when G is close, in some way, to being a nonabelian simple group, so we can expect the number of conjugacy classes of G to be significantly smaller than $|G|$. This means that we have relatively few targets to consider for the first generator x_1 of F. In that situation, for 2-generator groups F, we would hope to be able to use EPIMORPHISMS for groups G up to order about 10^7 or 10^8. For 3-generator groups F, this drops to about $|G| \leq 1000$ and it becomes rapidly unusable with still larger numbers of generators of F. The range tends to increase for presentations F with one or more relators of the form x_i^n, and for which the subsets R_k of R are large.

A consequence of the above situation is that for the groups G involved, the computations within G, such as conjugacy class representatives, and even the double coset representative calculations are likely to account for a very small proportion of the overall running time.

The double coset representative calculations can be made most efficient by calculating and storing the transversals T_i of $\mathbf{C}_A(g_i)$ together with the associated coset action homomorphisms $A \to \mathrm{Sym}(\Omega_i)$ (see Subsection 4.6.7) before entering the 'While' loop at line 11. For groups G within the practical range of the algorithm there is likely to be enough memory available to do this, and it avoids having to recompute them with every call of DCR. As we saw in Subsection 4.6.8, double coset representatives are computed as orbit representatives on these transversals under the coset actions.

A high proportion of the running time will typically be used within the calls of TESTRELS, and so it is vital to implement TESTRELS as efficiently as possible. For this, we need to check whether the condition in Theorem 2.52 is satisfied for the current generator images under consideration, which entails evaluating the group relators on the current images. If, as is generally the case, G is given as a finite permutation group, then this amounts to checking whether the associated permutations obtained by substituting the current generator images in the group relators evaluate to 1_G. If a base of G is known (see Subsection 4.4.1) then, to do this, we need only check whether the base points are fixed by the permutation word. So, we should endeavour to use a permutation representation of G having a small base, and then we can avoid multiplying out the complete permutations within the word.

It is unclear whether we should evaluate the generator images $\mathrm{im}[k] = g_k^{\delta[k][\mu[k]]}$ as permutations, or whether we should leave them as permutation words; this is one reason why we have chosen to use $\mathrm{im}[k]$ as an alias. Experiments suggest, but not completely conclusively, that we might obtain the fastest times by evaluating $\mathrm{im}[k]$ for $k < r$, and leaving them as words at the bottom level $k = r$.

It is also interesting to observe that the EPIMORPHISMS algorithm seems to be highly parallelizable, and running times could undoubtedly be reduced by using a parallel version spread across several machines, but to the author's knowledge this has not yet been attempted.

9.1.3 Automorphism groups of finite groups

Methods for computing $\mathrm{Aut}(G)$ and for testing two groups G and H for isomorphism for finite solvable groups and finite p-groups are described in Sections 8.9 and 9.4.5, respectively. These problems can be solved for finite permutation groups G (respectively G and H) by application of EPIMORPHISMS with $F = G$ for the automorphism group and $F = H$ for the isomorphism testing computation. A finite presentation is required for F, but such a presentation can easily be calculated if necessary by use of the methods described in Section 6.1. This application is described in more detail by Hulpke in [Hul96, V.5].

Before embarking upon such a computation, it is worthwhile to devote some effort to selecting a suitable generating set for F, for this can significantly affect the time taken by the main computation. It is vital that this set should

be as small as possible. The minimal number r_a of generators for the abelianized group $F/[F, F]$ is equal to the number of its invariant factors, which can
easily be computed by the abelian quotient algorithm, to be described in the
next section.

If $[F, F] = 1$ and $r_a = 1$, then F is cyclic, and we can find a single generator,
so suppose not. Then starting with r equal to $\min(2, r_a)$, we make some
effort to find a generating set of F of size r. If we fail, then we replace r by
$r + 1$ and try again. In [Hul96, V.5], Hulpke advocates using a short run of
EPIMORPHISMS itself, but with F equal to the free group of rank r and G equal
to the permutation group G or H, in an attempt to find such a generating set
of size r.

For a given list $[x_1, \ldots, x_r]$ of generators of F and an isomorphism $\varphi : F \to
G$, each $\varphi(x_i)$ must lie in a conjugacy class of G that has the same size as
the class x_i^F of x_i in F. This enables us to estimate in advance the number
of images of $[x_1, \ldots, x_r]$ that will need to be tested in EPIMORPHISMS and,
subject always to r being as small as possible, it is worthwhile to attempt to
select the x_i from conjugacy classes that minimize this search space.

In this particular application, we will generally not know $\mathrm{Aut}(G)$ in advance,
and so we run EPIMORPHISMS with $A = G$. (But of course, if we do happen
to know a group A with $G < A$ in which A induces outer automorphisms
of G by conjugation, then it could be worthwhile to use this A.) If we put
$A = G$, then EPIMORPHISMS will find an element of each coset of $\mathrm{Inn}(G)$ in
$\mathrm{Aut}(G)$; it has no possibility of using methods like those of Subsection 4.6.3 to
work relative to intermediate subgroups found during the search. This means
that the method will not work effectively unless $\mathrm{Aut}(G)/\mathrm{Inn}(G)$ is reasonably
small, and this is the principal disadvantage of the method.

Hulpke points out in [Hul96, V.5] that the most common examples of groups
for which $\mathrm{Aut}(G)/\mathrm{Inn}(G)$ is large are solvable groups, and for such groups
the methods of Section 8.9 should be used instead, after using the method
described in Subsection 4.4.6 to move from a permutation representation to
a presentation of G as a PC-group if necessary. There is some truth in this,
but of course one can conjure up awkward examples just by taking a direct
product of a small nonsolvable group, such as $\mathrm{Alt}(5)$, with a solvable group
with large automorphism group, such as an elementary abelian p-group.

An alternative method for computing $\mathrm{Aut}(G)$ for general groups G, which
is a generalization of the algorithm of Section 8.9 and is not so dependent
on $\mathrm{Aut}(G)/\mathrm{Inn}(G)$ being small, is described by Cannon and Holt in [CH03].
This is still inferior to the p-group method, however.

Exercises

1. Show that epimorphisms φ and φ' from F to G are equivalent modulo
   $\mathrm{Aut}(G)$ if and only if $\ker(\varphi) = \ker(\varphi')$.

2. Prove Proposition 9.1 and Corollary 9.2.

3. Prove that the use of coset enumeration, as described at the end of Subsection 9.1.1, to eliminate some possible generator images in EPIMORPHISMS, works as intended.

9.2 Abelian quotients

We saw in Proposition 2.68 that it is straightforward to write down an abelian presentation for the largest abelian quotient $G/[G,G]$ of a finitely presented group G. We simply write the relators of the presentation of G in additive notation.

In this section, we consider the problem of computing the structure of this abelianized presentation. The reader is probably already familiar with the fundamental theorem of finitely generated abelian groups, which states that each such group is isomorphic to a direct sum of cyclic groups. We shall prove this result at the same time as we describe an algorithm to compute such an isomorphism explicitly from an abelian group presentation.

We saw in Proposition 2.65 that a free abelian group of rank n is isomorphic to $\mathbb{Z}^n$, and it is obvious how to write down $G/[G,G]$ as a quotient of $\mathbb{Z}^n$.

Example 9.2
Let
$$G := \langle\, x, y, z \mid (xyz^{-1})^2, (x^{-1}y^2z)^2, (xy^{-2}z^{-1})^2 \,\rangle,$$
then
$$G/[G,G] = \mathrm{Ab}\langle\, x, y, z \mid 2x+2y-2z, -2x+4y+2z, 2x-4y-2z \,\rangle,$$
which is isomorphic to the quotient of $\mathbb{Z}^3$ by its subgroup generated by
$$\{\, (2,2,-2),\, (-2,4,2),\, (2,-4,-2) \,\}.$$

▯

So we need to study subgroups and quotients of the group $\mathbb{Z}^n$. Much of the theory here is very similar to basic linear algebra over a field, although there are some differences. We shall first present a rapid overview of this theory.

9.2.1 The linear algebra of a free abelian group

The free abelian group $\mathbb{Z}^n$ has the natural free basis $e_1, \ldots, e_n$, where e_i is the vector with 1 in the i-th component and 0 elsewhere.

A homomorphism from $\mathbb{Z}^m$ to $\mathbb{Z}^n$ can be represented by an $m \times n$ matrix M with integer entries, where the i-th row of M is the image of e_i. The image of a general element $v \in \mathbb{Z}^n$ is then given by $v \cdot M$.

If $m = n$, then M represents an automorphism of $\mathbb{Z}^n$ if and only if M is invertible over the integers; that is, if M^{-1} has integral entries. In this case $\det(M) \det(M^{-1}) = 1$, so $\det(M) = \pm 1$, and conversely, if $\det(M) = \pm 1$, then the standard adjoint matrix formula for M^{-1} shows that M^{-1} has integral entries. Such matrices are called *unimodular*. The set of all unimodular matrices forms a group $\mathrm{GL}(n, \mathbb{Z})$ under multiplication, and the subgroup $\mathrm{SL}(n, \mathbb{Z})$ of matrices with determinant 1 has index 2 in $\mathrm{GL}(n, \mathbb{Z})$.

Clearly an automorphism of $\mathbb{Z}^n$ maps one free basis of $\mathbb{Z}^n$ onto another, and the uniqueness up to isomorphism of a free abelian group shows that, for any two free bases of $\mathbb{Z}^n$, there is an automorphism of $\mathbb{Z}^n$ mapping one to the other. So the free bases of $\mathbb{Z}^n$ are given by the rows of the matrices in $\mathrm{GL}(n, \mathbb{Z})$.

A free basis of $\mathbb{Z}^n$ is also characterized by the property that it is linear independent over $\mathbb{Z}$ and spans (generates) $\mathbb{Z}^n$. Unlike in linear algebra over a field, a linear independent set of n elements of $\mathbb{Z}^n$ need not be a free basis of $\mathbb{Z}^n$, but a spanning set of size n is a free basis (exercise: this will be easily deducible from the structure theorem of finitely generated abelian groups).

We saw in Proposition 2.69 that any subgroup of $\mathbb{Z}^n$ is finitely generated, so we can represent any such subgroup by an $m \times n$ matrix M with integral entries in which the rows generate the subgroup; that is, the subgroup is the row-space of M.

This row-space is not changed if we replace M by AM for $A \in \mathrm{GL}(m, \mathbb{Z})$. If we replace M by MB with $B \in \mathrm{GL}(n, \mathbb{Z})$, however, then we are replacing the subgroup by its image under an automorphism of $\mathbb{Z}^n$.

9.2.2   Elementary row operations

There are three types of unimodular elementary row operations that we can perform on an integral $m \times n$ matrix M. These are

(i) Add an integral multiple of one row of M to another.

(ii) Interchange two rows on M.

(iii) Multiply a row of M by -1 .

These operations can also be achieved by replacing M be EM, where E is the corresponding $m \times m$ elementary matrix, which is obtained by applying the corresponding row operation to the identity matrix I_m. We have $\det(E) = 1$, -1, and -1 respectively in the three cases, so E is unimodular in each case.

Let us denote these $m \times m$ elementary matrices $E \in \mathrm{GL}(m, \mathbb{Z})$ in cases (i), (ii) and (iii), respectively, by $\rho_1(m, i, j, t)$ ($1 \le i, j \le m$, $i \ne j$, $t \in \mathbb{Z}$), $\rho_2(m, i, j)$ ($1 \le i, j \le m$, $i \ne j$), and $\rho_3(m, i)$ ($1 \le i \le m$), where:

(i) $\rho_1(m, i, j, t)_{ij} = t$;

(ii) $\rho_2(m, i, j)_{ij} = \rho_2(m, i, j)_{ji} = 1$ and $\rho_2(m, i, j)_{ii} = \rho_2(m, i, j)_{jj} = 0$;

(iii) $\rho_3(m, i)_{ii} = -1$;

and, in each case, all entries of the matrix not specified are the same as those in I_m. For example:

$$\rho_1(4, 3, 1, 7) = \begin{pmatrix} 1 & 0 & 0 & 0 \\ 0 & 1 & 0 & 0 \\ 7 & 0 & 1 & 0 \\ 0 & 0 & 0 & 1 \end{pmatrix}, \quad \rho_2(3, 1, 3) = \begin{pmatrix} 0 & 0 & 1 \\ 0 & 1 & 0 \\ 1 & 0 & 0 \end{pmatrix}, \quad \rho_3(3, 2) = \begin{pmatrix} 1 & 0 & 0 \\ 0 & -1 & 0 \\ 0 & 0 & 1 \end{pmatrix}.$$

Multiplying an $m \times n$ matrix M on the left by $\rho_1(m, i, j, k)$, $\rho_2(m, i, j)$, $\rho_3(m, i)$ has the effect of adding k times row j of M to row i of M, interchanging rows i and j of M, and multiplying row i of M by -1, respectively.

9.2.3 The Hermite normal form

DEFINITION 9.3 The integral $m \times n$ matrix M is said to be in *(row) Hermite normal form* (HNF) if the following conditions hold.

(i) There is an r with $0 \le r \le m$ such that the first r rows of M are nonzero and the remaining $m - r$ rows are zero.

(ii) For each i with $1 \le i \le r$, let M_{ij_i} be the first nonzero entry in the i-th row of M. Then $j_1 < j_2 < \cdots < j_r$.

(iii) $M_{ij_i} > 0$ for $1 \le i \le r$.

(iv) For $1 \le i \le r$, we have $0 \le M_{kj_i} < M_{ij_i}$ for $k < i$.

For example, the matrix:

$$\begin{pmatrix} 3 & -2 & 0 & 4 & -5 & 1 \\ 0 & 0 & 0 & 5 & 1 & 0 \\ 0 & 0 & 0 & 0 & 0 & 2 \\ 0 & 0 & 0 & 0 & 0 & 0 \end{pmatrix},$$

is in Hermite normal form with $j_1 = 1$, $j_2 = 4$, $j_3 = 6$.

There are various minor variants of this definition in the literature. The column HNF can be defined analogously and [Coh73], for example, has the zero columns of the matrix coming first rather than last.

THEOREM 9.4   *Let M be an $m \times n$ matrix over $\mathbb{Z}$. Then we can put M into HNF by applying a sequence of elementary unimodular row operations to M. Hence there exists an $A \in \mathrm{GL}(m, \mathbb{Z})$ such that AM is in HNF.*

We shall prove this theorem by presenting an algorithm that puts M into Hermite normal form. The algorithm below returns both the HNF N of M and $A \in \mathrm{GL}(m, \mathbb{Z})$ with $AM = N$.

HermiteForm(M)

> **Input**: $m \times n$ integral matrix M
> **Output**: Matrix N in HNF and $A \in \mathrm{GL}(m, \mathbb{Z})$ with $AM = N$

1    $m := |M|$; $n := |M[1]|$; $i := 1$; $j := 1$; $A := I_m$;
2   **while** $i \leq m$ and $j \leq n$
3   **do** (∗ Are there ≥ 2 nonzero entries M_{sj} with $s \in [i \mathinner{..} m]$? ∗)
4   **while** $\exists\, s, t$ with $i \leq s \neq t \leq m$ **and** $0 < |M_{sj}| \leq |M_{tj}|$
5   **do** (∗ Yes. Reduce $|M_{tj}|$ ∗)
6   Let $c \in \mathbb{Z}$ with $|M_{tj} + cM_{sj}| \leq |M_{sj}|/2$;
7                 $A := \rho_1(m, t, s, c){\cdot}A$; $M := \rho_1(m, t, s, c){\cdot}M$;
         (∗ Move a nonzero entry M_{sj} with $s \in [i \mathinner{..} m]$ to M_{ij} ∗)
8   **if** $M_{ij} = 0$ **and** $M_{sj} \neq 0$ with $i < s \leq m$
9   **then** $A := \rho_2(m, i, s){\cdot}A$; $M := \rho_2(m, i, s){\cdot}M$;
10   **if** $M_{ij} < 0$
11   **then** (∗ Make M_{ij} positive ∗)
12               $A := \rho_3(m, i){\cdot}A$; $M := \rho_3(m, i){\cdot}M$;
13   **if** $M_{ij} > 0$
14   **then** (∗ Put the entries M_{sj} with $s \in [1 \mathinner{..} i{-}1]$
                     into the range $[0 \mathinner{..} M_{ij}{-}1]$ ∗)
15   **for** $s \in [1 \mathinner{..} i{-}1]$
16   **do** Let $c \in \mathbb{Z}$ with $0 \leq M_{sj} + cM_{ij} < M_{ij}$;
17                    $A := \rho_1(m, s, i, c){\cdot}A$; $M := \rho_1(m, s, i, c){\cdot}M$;
18              $i := i + 1$;
19          $j := j + 1$;
20   **return** M, A;

We have not specified precisely how we choose s and t in line 4, but $|M_{tj}|$ is decreased in lines 6 and 7 for the chosen value of t. So eventually there will be at most one nonzero M_{sj} with $i \leq s \leq m$, and then we exit this loop. Since either i, j, or both is incremented with each pass of the main 'While' loop at line 2, the function must terminate.

The nonzero entries M_{ij} encountered at line 13 are the entries M_{ij_i} in the HNF. When we reach line 10, all entries M_{sj} with $s > i$ are zero. So Property (ii) of the HNF definition holds. Also, we only increment i when we have a nonzero entry M_{ij_i} at line 10 and, if r is maximal with the property that we encounter a nonzero entry M_{rj_r} at line 13 then the entries M_{kj} for $k > r$

and all columns j are zero, and hence Property (i) holds. The code following lines 10 and 13 ensures that Properties (iii) and (iv) hold, so the function performs as intended, and we have proved Theorem 9.4.

The most obvious way to choose s and t in line 4 is to choose s such that $|M_{sj}|$ is minimal, and then choose t such that M_{tj} is some other nonzero entry. In fact, it has been observed in practice that better performance is achieved by choosing s and t such that $|M_{tj}|$ and $|M_{sj}|$ are the largest and second largest possible entries, measured by absolute value.

Example 9.3

Let

$$
M = \begin{pmatrix}
1 & -2 & -1 & 1 & 1 & -3 \\
-1 & 2 & -3 & 1 & -3 & -9 \\
1 & -2 & -5 & 3 & -1 & -3 \\
1 & -2 & 1 & 0 & 3 & 8 \\
4 & -8 & -4 & 4 & 6 & 2
\end{pmatrix}.
$$

Starting at $i = j = 1$. Let us adopt the obvious strategy at line 4 of choosing s such that $|M_{sj}|$ is minimal. So we choose $s = 1$ and $t = 2, 3, 4, 5$ in turn, and we multiply M on the left by $\rho_1(5, 2, 1, 1)$, $\rho_1(5, 3, 1, -1)$, $\rho_1(5, 4, 1, -1)$, $\rho_1(5, 5, 1, -4)$ in turn. At this stage, we have:

$$
A = \begin{pmatrix}
1 & 0 & 0 & 0 & 0 \\
1 & 1 & 0 & 0 & 0 \\
-1 & 0 & 1 & 0 & 0 \\
-1 & 0 & 0 & 1 & 0 \\
-4 & 0 & 0 & 0 & 1
\end{pmatrix},
\quad
M = \begin{pmatrix}
1 & -2 & -1 & 1 & 1 & -3 \\
0 & 0 & -4 & 2 & -2 & -12 \\
0 & 0 & -4 & 2 & -2 & 0 \\
0 & 0 & 2 & -1 & 2 & 11 \\
0 & 0 & 0 & 0 & 2 & 14
\end{pmatrix}.
$$

Now we move on to $i = j = 2$. Since $M_{is} = 0$ for $2 \le i \le 5$, we move on to $j = 3$. The smallest M_{sj} occurs with $s = 4$ and taking $t = 2, 3$, we multiply A and M by $\rho_1(5, 2, 4, 2)$ and $\rho_1(5, 3, 4, 2)$. Then, at line 9, we multiply them by $\rho_2(5, 2, 4)$ to move M_{sj} to M_{ij} and, at line 17, we multiply them by $\rho_1(5, 1, 2, 1)$. Now we have:

$$
A = \begin{pmatrix}
0 & 0 & 0 & 1 & 0 \\
-1 & 0 & 0 & 1 & 0 \\
-3 & 0 & 1 & 2 & 0 \\
-1 & 1 & 0 & 2 & 0 \\
-4 & 0 & 0 & 0 & 1
\end{pmatrix},
\quad
M = \begin{pmatrix}
1 & -2 & 1 & 0 & 3 & 8 \\
0 & 0 & 2 & -1 & 2 & 11 \\
0 & 0 & 0 & 0 & 2 & 22 \\
0 & 0 & 0 & 0 & 2 & 10 \\
0 & 0 & 0 & 0 & 2 & 14
\end{pmatrix}.
$$

We then move on to $i = 3$, $j = 4$ and immediately to $j = 5$. Choosing $s = 3$, we multiply M and A on the left by $\rho_1(5, 4, 3, -1)$, $\rho_1(5, 5, 3, -1)$, $\rho_1(5, 1, 3, -1)$ and $\rho_1(5, 2, 3, -1)$ to give

$$A = \begin{pmatrix} 3 & 0 & -1 & -1 & 0 \\ 2 & 0 & -1 & -1 & 0 \\ -3 & 0 & 1 & 2 & 0 \\ 2 & 1 & -1 & 0 & 0 \\ -1 & 0 & -1 & -2 & 1 \end{pmatrix}, \quad M = \begin{pmatrix} 1 & -2 & 1 & 0 & 1 & -14 \\ 0 & 0 & 2 & -1 & 0 & -11 \\ 0 & 0 & 0 & 0 & 2 & 22 \\ 0 & 0 & 0 & 0 & 0 & -12 \\ 0 & 0 & 0 & 0 & 0 & -8 \end{pmatrix}.$$

Finally, we proceed to $i = 4$, $j = 6$, and multiply A, M on the left by $\rho_1(5, 4, 5, -2)$, $\rho_1(5, 5, 4, 2)$, $\rho_1(5, 1, 4, 4)$, $\rho_1(5, 2, 4, 3)$, $\rho_1(5, 3, 4, -5)$, to give:

$$A = \begin{pmatrix} 19 & 4 & 3 & 15 & -8 \\ 14 & 3 & 2 & 11 & -6 \\ -23 & -5 & -4 & -18 & 10 \\ 4 & 1 & 1 & 4 & -2 \\ 7 & 2 & 1 & 6 & -3 \end{pmatrix}, \quad M = \begin{pmatrix} 1 & -2 & 1 & 0 & 1 & 2 \\ 0 & 0 & 2 & -1 & 0 & 1 \\ 0 & 0 & 0 & 0 & 2 & 2 \\ 0 & 0 & 0 & 0 & 0 & 4 \\ 0 & 0 & 0 & 0 & 0 & 0 \end{pmatrix},$$

and M is now in HNF.   ▯

The r nonzero rows of a matrix in HNF are clearly linearly independent over $\mathbb{Z}$, and there at most n of them. Since replacing M by AM with $A \in \mathrm{GL}(m, \mathbb{Z})$ does not change the row-space of M, we have proved:

PROPOSITION 9.5   *Any subgroup of a free abelian group of rank n is free abelian of rank at most n.*

The HNF provides a canonical description of the image or row-space of an integral matrix M. The following proposition, of which we leave the proof as an exercise, provides a description of the kernel or nullspace.

PROPOSITION 9.6   *If AM is in HNF and AM has r nonzero rows, then the rows $r+1$ to m of A form a free basis of the nullspace of M.*

The next result proves that the HNF of an integral matrix is determined by its row-space. So, in particular, the HNF of a matrix is unique.

THEOREM 9.7   *Suppose that the $m \times n$ matrices M and N over $\mathbb{Z}$ have the same row-space, and that AM and BN are in Hermite normal form, with $A, B \in \mathrm{GL}(m, \mathbb{Z})$. Then $AM = BN$.*

PROOF   We proceed by induction on $m + n$, starting with the case $m = n = 0$, which is trivial. Let $C := AM$, $D := BN$ be in HNF with $A, B \in \mathrm{GL}(m, \mathbb{Z})$. Then the row-spaces of C and D are both equal to the row-space U of M and N. Let $e_1, \ldots, e_n$ be the natural free basis of $\mathbb{Z}^n$, and let $V \cong \mathbb{Z}^{n-1}$ be the subspace spanned by $e_2, \ldots, e_n$.

We have $U \leq V$ if and only if M and N have zero first column, in which case $C_{11} = D_{11} = 0$, and the result follows by induction applied to M and N with their first columns removed.

Otherwise C_{11} and D_{11} are nonzero, and $U \cap V$ is spanned by the rows of C or of D other than the first. But then the matrices obtained by removing the first rows from C and from D have the same row-space and are both in HNF so, by inductive hypothesis, they are equal.

So we just need to prove that C and D have the same first rows $C[1]$ and $D[1]$. Let $v := C[1] - D[1]$. Since C_{11} and D_{11} are positive integers, they must both be equal to $|\mathbb{Z}^n : U + V|$, so $v[1] = C_{11} - D_{11} = 0$. Hence $v \in U \cap V$, and v is in the space spanned by rows 2 to m of C.

Let C (and hence also D) have r nonzero rows, with first nonzero entries C_{ij_i} for $1 \leq i \leq r$. Then $v = a_2 C[2] + \cdots + a_r C[r]$ for some $a_2, \ldots, a_r \in \mathbb{Z}$, and hence $v[j_i] = a_i C_{ij_i}$ for $2 \leq i \leq r$. But, by condition (iv) in the definition of the HNF, we have $0 \leq C_{1j_i}, D_{1j_i} < C_{ij_i}$ and hence $|v[j_i]| < C_{ij_i}$ for $2 \leq i \leq r$. So we must have $a_i = 0$ for $2 \leq i \leq r$, and hence $v = 0$ and we are done. ∎

9.2.4 Elementary column matrices and the Smith normal form

The unimodular elementary column operations are defined analogously to the corresponding row operations. They can be achieved by replacing the $m \times n$ matrix M be ME, where E is the corresponding $n \times n$ elementary matrix, which is obtained by applying the corresponding column operation to the identity matrix I_n.

We denote these $n \times n$ matrices E by $\gamma_1(n, i, j, t)$ $(1 \leq i, j \leq n, i \neq j, t \in \mathbb{Z})$, $\gamma_2(n, i, j)$ $(1 \leq i, j \leq n, i \neq j)$, and $\gamma_3(n, i)$ $(1 \leq i \leq n)$, where multiplying an $m \times n$ matrix M on the right by $\gamma_1(n, i, j, t)$, $\gamma_2(n, i, j)$, or $\gamma_3(n, i)$ has the effect of adding t times column j of M to column i of M, interchanging columns i and j of M, or multiplying column i of M by -1, respectively. Their definition is:

(i) $\gamma_1(n, i, j, k)_{ji} = c$;

(ii) $\gamma_2(n, i, j)_{ij} = \gamma_2(n, i, j)_{ji} = 1$ and $\gamma_2(n, i, j)_{ii} = \gamma_2(n, i, j)_{jj} = 0$;

(iii) $\gamma_3(n, i)_{ii} = -1$.

and, in each case, all entries of the matrix not specified are the same as in I_n.

DEFINITION 9.8   The integral $m \times n$ matrix M is said to be in Smith normal form (SNF) if the following conditions hold.

(i) $M_{ij} = 0$ whenever $i \neq j$.

(ii) $M_{ii} \geq 0$ for $1 \leq i \leq \min(m, n)$.

(iii) For $1 \leq i \leq \min(m, n)$, we have $M_{ii} | M_{i+1,i+1}$.

For example, the following matrix is in Smith normal form:

$$\begin{pmatrix} 1 & 0 & 0 & 0 & 0 & 0 \\ 0 & 6 & 0 & 0 & 0 & 0 \\ 0 & 0 & 24 & 0 & 0 & 0 \\ 0 & 0 & 0 & 0 & 0 & 0 \end{pmatrix}.$$

THEOREM 9.9   *Let M be any $m \times n$ matrix over $\mathbb{Z}$. Then we can put M into SNF by applying a sequence of elementary unimodular row and column operations to M. Hence there exists an $A \in \mathrm{GL}(m, \mathbb{Z})$ and $B \in \mathrm{GL}(n, \mathbb{Z})$ such that AMB is in SNF.*

Again we shall prove the theorem by presenting an algorithm that puts M into Smith normal form. The function SMITHFORM returns both the SNF N of M and $A \in \mathrm{GL}(m, \mathbb{Z})$, $B \in \mathrm{GL}(n, \mathbb{Z})$ with $AMB = N$.

The entry M_{ij} that is selected at line 4 is called the *pivot entry* for that iteration of the main loop at line 2. It has not been specified exactly how the pivot entry is selected, and the algorithm is only guaranteed to terminate if we adopt a suitable policy for choosing it. The easiest policy that guarantees termination is to select the pivot entry M_{ij} such that $|M_{ij}|$ is as small as possible subject to being nonzero. This does not always lead to the best performance, and we shall consider other strategies later, but let us use this policy for the purposes of proving correctness and hence Theorem 9.9.

If the Boolean variable a is set to `false` at line 9 or 13, then $|M_{sj}|$ or $|M_{it}|$ will be nonzero, but smaller than $|M_{ij}|$, and so f will be unchanged. But, since we are choosing pivot entries with minimal absolute value, this absolute value will be smaller on the next iteration of the main loop. Hence a will eventually be `true` at line 14.

Then, if the 'If' condition holds at line 16, there will again be a smaller pivot entry with f unchanged on the next iteration of the main loop, and so this condition will eventually fail to hold, and then the code following line 21 will be executed.

At this stage, the pivot entry M_{ij} is the only nonzero entry in its row and in its column, and it divides all other entries in the submatrix $M[f \mathinner{..} m][f \mathinner{..} n]$ of M. The pivot entry is then moved to position M_{ff} and made positive if necessary. At this stage, all off-diagonal entries in rows and columns 1 to f will be zero, and the diagonal entries will be nonnegative, so conditions (i) and (ii) of the SNF will hold when the function terminates.

All subsequent row and column operations take place on rows $f+1$ to m or columns $f+1$ to n of M, and so M_{ff} is not altered again, and the property that it divides all other entries in $M[f \mathinner{..} m][f \mathinner{..} n]$ is maintained. Hence condition (iii) in the definition of the SNF holds at the end of the function. After moving the pivot entry to M_{ff}, f is incremented, so the function must eventually terminate.

SMITHFORM(M)

    Input: $m \times n$ integral matrix M
    Output: Matrix N in SNF,
                $A \in \mathrm{GL}(m, \mathbb{Z})$, $B \in \mathrm{GL}(n, \mathbb{Z})$ with $AMB = N$
1   $m := |M|$; $n := |M[1]|$; $f := 1$; $A := I_m$; $B := I_n$;
2 **while** $f \leq m$ **and** $f \leq n$
3   **do if** $M_{ij} = 0$ for all $i \in [f \mathrel{..} m]$ and $j \in [f \mathrel{..} n]$ **then break**;
            (* Select pivot entry M_{ij} *)
4   Choose i, j with $i \in [f \mathrel{..} m]$, $j \in [f \mathrel{..} n]$, $M_{ij} \neq 0$;
5           $a := \texttt{true}$;
            (* Reduce entries in column j using row operations *)
6   **for** $s \in [f \mathrel{..} m] \setminus [i]$
7   **do let** $c \in \mathbb{Z}$ with $|M_{sj} + cM_{ij}| \leq |M_{ij}|/2$;
8                   $A := \rho_1(m, s, i, c) \cdot A$; $M := \rho_1(m, s, i, c) \cdot M$;
9   **if** $M_{sj} \neq 0$ **then** $a := \texttt{false}$;
            (* Reduce entries in row i using column operations *)
10   **for** $t \in [f \mathrel{..} n] \setminus [j]$
11   **do let** $c \in \mathbb{Z}$ with $|M_{it} + cM_{ij}| \leq |M_{ij}|/2$;
12                  $B := B \cdot \gamma_1(m, t, j, c)$; $M := M \cdot \gamma_1(m, t, j, c)$;
13   **if** $M_{it} \neq 0$ **then** $a := \texttt{false}$;
14   **if** a
15   **then** (* All other entries in row i and column j are zero *)
16   **if** $\exists \, s \in [f \mathrel{..} m]$, $t \in [f \mathrel{..} n]$: $M_{st} \bmod M_{ij} \neq 0$
17   **then** (* $M_{ij} \nmid M_{st}$ –
                             make $M_{sj} := M_{st}$ and reduce it *)
18                   $B := B \cdot \gamma_1(n, j, s, 1)$; $M := M \cdot \gamma_1(n, j, s, 1)$;
19   Let $c \in \mathbb{Z}$ with $|M_{sj} + cM_{ij}| \leq |M_{ij}|/2$;
20                   $A := \rho_1(m, s, i, c) \cdot A$; $M := \rho_1(m, s, i, c) \cdot M$;
21   **else**
                  (* Move M_{ij} to M_{ff} and make it > 0 *)
22   **if** $f \neq i$
23   **then** $A := \rho_2(m, i, f) \cdot A$;
24                          $M := \rho_2(m, i, f) \cdot M$;
25   **if** $f \neq j$
26   **then** $B := B \cdot \gamma_2(n, j, f)$;
27                          $M := M \cdot \gamma_2(n, j, f)$;
28   **if** $M_{ff} < 0$
29   **then** $A := \rho_3(m, f) \cdot A$;
30                          $M := \rho_3(m, f) \cdot M$;
31                  $f := f + 1$;
32 **return** M, A, B;

Let us work through the SNF algorithm using Example 9.3. We start with the pivot entry $M_{11} = 1$. As before, we multiply M on the left by $\rho_1(5, 2, 1, 1)$, $\rho_1(5, 3, 1, -1)$, $\rho_1(5, 4, 1, -1)$, $\rho_1(5, 5, 1, -4)$, but now we also multiply on the right by $\gamma_1(6, 2, 1, 2)$, $\gamma_1(6, 3, 1, 1)$, $\gamma_1(6, 4, 1, -1)$, $\gamma_1(6, 5, 1, -1)$ and $\gamma_1(6, 6, 1, 3)$, to reduce the rest of row 1 to zero. At this stage, we have

$$
A = \begin{pmatrix} 1 & 0 & 0 & 0 & 0 \\ 1 & 1 & 0 & 0 & 0 \\ -1 & 0 & 1 & 0 & 0 \\ -1 & 0 & 0 & 1 & 0 \\ -4 & 0 & 0 & 0 & 1 \end{pmatrix}, \quad B = \begin{pmatrix} 1 & 2 & 1 & -1 & -1 & 3 \\ 0 & 1 & 0 & 0 & 0 & 0 \\ 0 & 0 & 1 & 0 & 0 & 0 \\ 0 & 0 & 0 & 1 & 0 & 0 \\ 0 & 0 & 0 & 0 & 1 & 0 \\ 0 & 0 & 0 & 0 & 0 & 1 \end{pmatrix}, \quad M = \begin{pmatrix} 1 & 0 & 0 & 0 & 0 & 0 \\ 0 & 0 & -4 & 2 & -2 & -12 \\ 0 & 0 & -4 & 2 & -2 & 0 \\ 0 & 0 & 2 & -1 & 2 & 11 \\ 0 & 0 & 0 & 0 & 2 & 14 \end{pmatrix}.
$$

We choose $M_{44} = -1$ as our next pivot. We multiply M on the left by $\rho_1(5, 2, 4, 2)$, $\rho_1(5, 3, 4, 2)$, and on the right by $\gamma_1(6, 3, 4, 2)$, $\gamma_1(6, 5, 4, 2)$, $\gamma_1(6, 6, 4, 11)$ to clear the rest of row 4 and column 4. Then we multiply on the left by $\rho_2(5, 2, 4)$ and on the right by $\gamma_2(6, 2, 4)$ to bring the pivot entry to M_{22}, and then we multiply on the left by $\rho_3(5, 2)$ to make it positive. We now have

$$
A = \begin{pmatrix} 1 & 0 & 0 & 0 & 0 \\ 1 & 0 & 0 & -1 & 0 \\ -3 & 0 & 1 & 2 & 0 \\ -1 & 1 & 0 & 2 & 0 \\ -4 & 0 & 0 & 0 & 1 \end{pmatrix}, \quad B = \begin{pmatrix} 1 & -1 & -1 & 2 & -3 & -8 \\ 0 & 0 & 0 & 1 & 0 & 0 \\ 0 & 0 & 1 & 0 & 0 & 0 \\ 0 & 1 & 2 & 0 & 2 & 11 \\ 0 & 0 & 0 & 0 & 1 & 0 \\ 0 & 0 & 0 & 0 & 0 & 1 \end{pmatrix}, \quad M = \begin{pmatrix} 1 & 0 & 0 & 0 & 0 & 0 \\ 0 & 1 & 0 & 0 & 0 & 0 \\ 0 & 0 & 0 & 0 & 2 & 22 \\ 0 & 0 & 0 & 0 & 2 & 10 \\ 0 & 0 & 0 & 0 & 2 & 14 \end{pmatrix}.
$$

The remaining operations are as follows. First we multiply on the left by $\rho_1(5, 4, 3, -1)$, $\rho_1(5, 5, 3, -1)$, then on the right by $\gamma_1(6, 6, 5, -11)$, $\gamma_2(6, 3, 5)$, then on the left by $\rho_1(5, 4, 5, -2)$, $\rho_1(5, 5, 4, 2)$, and finally on the right by $\gamma_2(6, 4, 6)$. We now have

$$
A = \begin{pmatrix} 1 & 0 & 0 & 0 & 0 \\ 1 & 0 & 0 & -1 & 0 \\ -3 & 0 & 1 & 2 & 0 \\ 4 & 1 & 1 & 4 & -2 \\ 7 & 2 & 1 & 6 & -3 \end{pmatrix}, \quad B = \begin{pmatrix} 1 & -1 & -3 & 25 & -1 & 2 \\ 0 & 0 & 0 & 0 & 0 & 1 \\ 0 & 0 & 0 & 0 & 1 & 0 \\ 0 & 1 & 2 & -11 & 2 & 0 \\ 0 & 0 & 1 & -11 & 0 & 0 \\ 0 & 0 & 0 & 1 & 0 & 0 \end{pmatrix}, \quad M = \begin{pmatrix} 1 & 0 & 0 & 0 & 0 & 0 \\ 0 & 1 & 0 & 0 & 0 & 0 \\ 0 & 0 & 2 & 0 & 0 & 0 \\ 0 & 0 & 0 & 4 & 0 & 0 \\ 0 & 0 & 0 & 0 & 0 & 0 \end{pmatrix},
$$

and M is in SNF.

Once again, we have a uniqueness result.

THEOREM 9.10   *Let M be an $m \times n$ matrix over $\mathbb{Z}$. If AMB and CMD are both in Smith normal form, with $A, C \in \mathrm{GL}(m, \mathbb{Z})$ and $B, D \in \mathrm{GL}(n, \mathbb{Z})$, then $AMB = CMD$.*

Rather than prove this theorem directly, we prefer to interpret the theory of Smith normal forms in terms of finitely generated abelian groups, and then derive the theorem from a result about abelian groups.

We saw in Section 2.6 that any finitely generated abelian group G is isomorphic to a quotient $\mathbb{Z}^n/K$ for some subgroup K of a free abelian group $\mathbb{Z}^n$ where, by Proposition 2.69, K is finitely generated. We can represent K by an $m \times n$ integral matrix M, where K is spanned by the rows of M and, conversely, we can associate the finitely generated abelian group $\mathbb{Z}^n/K$ with the $m \times n$ matrix M.

As we remarked at the end of Subsection 9.2.1, if we replace M by AM for $A \in \mathrm{GL}(m, \mathbb{Z})$, then we do not change the row-space K of M. If we replace M by MB for $B \in \mathrm{GL}(n, \mathbb{Z})$, however, then the row-space of M is replaced by $\varphi(K)$, where φ is the automorphism of $\mathbb{Z}^n$ defined by B. But $\mathbb{Z}^n/K \cong \mathbb{Z}^n/\varphi(K)$. Hence, for any $A \in \mathrm{GL}(m, \mathbb{Z})$ and $B \in \mathrm{GL}(n, \mathbb{Z})$, the finitely generated abelian groups associated with M and AMB are isomorphic.

Recall that, for an abelian group G and $n \in \mathbb{Z}$, nG is defined to be the subgroup $\{ng \mid g \in G\}$ of G. Suppose that M is in SNF with $M_{ii} = d_i$ for $1 \le i \le \min(m, n)$, and define $d_i = 0$ for $\min(m, n) < i \le n$. Then the row-space of M is the subgroup

$$K := d_1\mathbb{Z} \oplus d_2\mathbb{Z} \cdots \oplus d_n\mathbb{Z}$$

of $\mathbb{Z}^n$, and so

$$G := \mathbb{Z}^n/K \cong \mathbb{Z}/d_1\mathbb{Z} \oplus \mathbb{Z}/d_2\mathbb{Z} \cdots \oplus \mathbb{Z}/d_n\mathbb{Z}.$$

In general, some of the d_i may be zero, in which case the nonzero d_i will all precede the zero d_i. In any case, $\mathbb{Z}/d_i\mathbb{Z}$ is a finite cyclic group of order d_i if $d_i > 0$, and an infinite cyclic group if $d_i = 0$. Moreover, some of the d_i may equal 1, and they come first in the list; the corresponding quotients $\mathbb{Z}/d_i\mathbb{Z}$ are trivial and can be omitted from the direct sum decomposition of G.

DEFINITION 9.11   An abelian group G has type $(d_1, \ldots, d_n)$, for $d_i \in \mathbb{N}_0$ if it is isomorphic to the direct sum of cyclic groups $\mathbb{Z}/d_i\mathbb{Z}$. We say that $(d_1, \ldots, d_n)$ satisfies the *divisibility condition* if $d_i \ne 1$ for $1 \le i \le n$, and $d_i \mid d_{i+1}$ for $1 \le i < n$.

The above discussion and Theorem 9.9 together prove the existence part of the fundamental theorem of finitely generated abelian groups.

THEOREM 9.12   *A finitely generated abelian group has type $(d_1, \ldots, d_n)$ for some $d_i \in \mathbb{N}_0$ that satisfy the divisibility criterion.*

We shall now prove the uniqueness part of the fundamental theorem, which (exercise) also proves Theorem 9.10.

THEOREM 9.13 *Suppose that the abelian group G has type $(d_1, \ldots, d_n)$ and also has type $(c_1, \ldots, c_m)$, where $(d_1, \ldots, d_n)$ and $(c_1, \ldots, c_m)$ both satisfy the divisibility condition. Then $m = n$ and $d_i = c_i$ for $1 \le i \le n$.*

PROOF If p is prime and H is a cyclic group of order n, then $|H : pH| = p$ if $p|n$, and otherwise $|H : pH| = 1$. So $|G/pG| = p^n$ if $p|d_1$ and $|G/pG| < p^n$ otherwise. Since the same applies to c_1, this proves that $m = n$ and that the same primes divide d_1 as c_1.

If $d_1 = 0$ then all primes divide d_1, so $c_1 = 0$ and the result follows. Otherwise we use induction on n and, for a fixed n, on d_1. The result is true for $n = 1$, since $d_1 = c_1 = |G|$ in that case. If $n > 1$, then let p be a prime dividing d_1 and c_1. Then G/pG is abelian of type $(d_1/p, \ldots, d_n/p)$ and of type $(c_1/p, \ldots, c_n/p)$. These types both satisfy the divisibility condition after removing any leading 1's, and so by inductive hypothesis we have $d_i/p = c_i/p$ for $1 \le i \le n$, and the result follows immediately. ∎

Without the divisibility condition, the result no longer holds, and even the number n of factors is not uniquely determined by G. For example, abelian groups of types $(4, 3, 5)$, $(12, 5)$, $(4, 15)$, $(3, 20)$, and (60) are all isomorphic to each other, but only the last of these types satisfies the divisibility condition.

The principal use of the SNF algorithm in CGT is for finding the abelian invariants of $G/[G, G]$ in a finitely presented group G. So, in Example 9.2,

$$G := \langle \, x, y, z \mid (xyz^{-1})^2, (x^{-1}y^2z)^2, (xy^{-2}z^{-1})^2 \, \rangle,$$

the matrix M of the abelianized presentation is

$$M = \begin{pmatrix} 2 & 2 & -2 \\ -2 & 4 & 2 \\ 2 & -4 & -2 \end{pmatrix},$$

and we easily calculate the SNF N with $AMB = N$, where

$$N = \begin{pmatrix} 2 & 0 & 0 \\ 0 & 6 & 0 \\ 0 & 0 & 0 \end{pmatrix}, \quad A = \begin{pmatrix} 1 & 0 & 0 \\ 1 & 1 & 0 \\ 0 & 1 & 1 \end{pmatrix}, \quad B = \begin{pmatrix} 1 & -1 & 1 \\ 0 & 1 & 0 \\ 0 & 0 & 1 \end{pmatrix}.$$

So $G/[G, G]$ is an infinite group of type $(2, 6, 0)$.

The matrix A is of no particular interest in this regard, but the matrix B is important, since its rows define the images of the epimorphism μ from G to $H = \mathbb{Z}/2\mathbb{Z} \oplus \mathbb{Z}/6\mathbb{Z} \oplus \mathbb{Z}$. To be specific, if a, b, c denote the generators of the three direct summands of H, then the homomorphism is defined by $\mu(x) = a - b + c$, $\mu(y) = b$, $\mu(z) = c$.

Exercises

1. Prove Proposition 9.6.

2. Describe a method of solving a system $v = z \cdot M$ of linear equations over $\mathbb{Z}$, where M is an integral $m \times n$ matrix, $v \in \mathbb{Z}^n$ is a fixed vector, and $z \in \mathbb{Z}^m$ is an unknown vector. (*Hint*: First show how to do this when M is in HNF. In general, let AM be in HNF and solve the system $v = (z \cdot A^{-1}) \cdot (AM)$.)

3. Show that any set of n elements of $\mathbb{Z}^n$ that span $\mathbb{Z}^n$ form a free basis of $\mathbb{Z}^n$.

4. Show that the HNF of a matrix in $M \in \mathrm{GL}(m, \mathbb{Z})$ is equal to I_m. Deduce that M is equal to a product of unimodular elementary row matrices.

5. Prove that, if $a, b \in \mathbb{N}$, then abelian groups of types (a, b) and (ab) are isomorphic if and only if a and b are coprime.

6. Prove that Theorem 9.10 follows from Theorem 9.13.

9.3 Practical computation of the HNF and SNF

In Example 9.3, the largest absolute value of an entry in M is 9, and in the Hermite and Smith forms for M it is 4. However, the number 22 appeared in M during the course of the both of these calculations, and there is a 25 in the final transformation matrix B.

This might not seem particularly extreme or remarkable, but with larger matrices, the problem of integer entry explosion becomes a serious obstacle to satisfactory performance of the algorithms. In this section, we shall briefly discuss methods that can and have been used to alleviate this problem. We shall not go into much detail here, and we shall direct the reader to other sources for a more complete discussion. We shall concentrate more on the SNF calculation, because it is more important than the HNF in applications to CGT.

Some recent papers on the computation of the HNF and SNF, and on the closely related extended greatest common divisor problem for sets of integers, are those by Storjohann and Labahn [SL96] and [Sto98], by Havas, Majewski, and Matthews [HMM98, HMM99], and by Lübeck [Lüb02]. The HNF is more important in applications to computational algebraic number theory, and is discussed in Section 2.4 of [Coh73].

9.3.1 Modular techniques

Modular techniques involve performing all calculations modulo some fixed positive integer m. The advantage of this, of course, is that no integers larger than m will arise during the computations. The main ideas involved were introduced by Havas and Sterling in [HS79]. There is also a complete treatment in Section 8.4 of [Sim94]. This approach provides an efficient polynomial-time algorithm for finding the SNF of an integral matrix, and hence the type of a finitely presented abelian group. The more recent paper of Lübeck [Lüb02], which we shall not discuss here, also makes partial use of modular methods.

The main disadvantage of modular methods is that they do not, in general, allow us to find the transformation matrices A and B as matrices over $\mathbb{Z}$ and, in particular, for finite presentations of infinite abelian groups G, they do not allow us to compute an explicit isomorphism from G to the corresponding direct sum of cyclic groups. If the transformation matrices are not required explicitly, then modular techniques probably provide the most efficient method for finding the SNF.

Calculating the SNF modulo a positive integer m is done in almost exactly the same way as over the integers, but working modulo m. As transformation matrices, we can use any A and B that are invertible modulo m, and hence, for the third type of elementary row and column operation, we may multiply a row or a column by any integer (mod m) that is invertible modulo m.

By Theorem 9.9, we can use elementary row and column operations to reduce an $m \times n$ matrix M modulo m to a matrix in SNF. We can then write each nonzero diagonal entry M_{ii} as $d_i e_i$ where $d_i | m$ and $\gcd(e_i, m) = 1$. So, by multiplying row i by the inverse of e_i modulo m, we can replace M_{ii} by d_i. So, modulo m, we can reduce M to a matrix in SNF in which each M_{ii} divides m, and we then say that M in SNF modulo m. It is not hard to show, as in Theorem 9.7, that the SNF modulo m is uniquely determined by M.

The idea of the modular methods is to find an integer m that is divisible by and larger than each of the nonzero entries in the SNF of our integral matrix M. For such an m, it is clear that the SNF of M can be obtained from the SNF of M modulo m by replacing any entries equal to m by 0.

Let r be the rank of M, and let m be the gcd of the determinants of the $r \times r$ nonsingular submatrices of M. It can be shown (see, for example, Section 8.4 of [Sim94]), that m is not changed when we perform unimodular row or column operations on M, and so m is just the product of the nonzero entries in the SNF of M. Hence $2m$ satisfies the condition in the preceding paragraph but, in fact, assuming that we know r, we also know the number of nonzero entries in the SNF of M, and so it suffices to calculate the SNF modulo m.

The algorithm proceeds by first finding r, and then finding a small number of nonsingular $r \times r$ submatrices of M and using the gcd of their determinants for m. In practice, this will typically be a very small multiple of, and often equal to, the gcd of all nonsingular $r \times r$ submatrices of M.

Finding r is the most difficult part of this process. Notice that the associated finitely presented abelian group G is finite if and only if $d_n > 0$, where $(d_1, \ldots, d_n)$ is the type of G, and this is the case if and only if $r = n$. So a knowledge of r decides the finiteness of G which, in many applications, is exactly what we want to know!

If p is prime, then it is straightforward, by working modulo p in the finite field $\mathbb{F}_p$, to find the rank of M modulo p, which is a lower bound for r. The rank modulo p is equal to r for all primes p that do not divide any of the positive entries in the SNF of M; that is, for almost all primes p. So, in practice, we just need to calculate the rank modulo p for three or four moderately sized (within the range $[50..100]$ is usually adequate) random primes, and the value of r will be clear.

However, mathematicians understandably prefer to have mathematically substantiated proofs of the correctness of such calculations, even when the likelihood of error is vanishingly small. In fact, there is a well-known bound, the *Hadamard bound*, on the determinant of a square matrix, which enables us to choose primes $p_1, \ldots, p_s$ such that the maximum of the ranks of M modulo p_i for $1 \le i \le s$ is guaranteed to be equal to r: it is sufficient to choose them such that

$$p_1 p_2 \cdots p_s > H(M) := \prod_{i=1}^{m} \sqrt{\sum_{j=1}^{n} M_{ij}^2}.$$

See [Sim94] for further details.

Although it is too small and easy an example to warrant the use of modular techniques, let us consider Example 9.3. We have $H(M) \le 32409$, and M has rank 4 modulo primes 59, 61, 67, for example, of which the product exceeds $H(M)$, so we can conclude that M has rank 4. The 4×4 submatrices of M consisting of rows 1–4 and columns 3–6, and of rows 2–5 and columns 3–6, have determinants -24 and 56, respectively, where $\gcd(24, 56) = 8$, so we can find the SNF by calculating the SNF modulo 8.

9.3.2 The use of norms and row reduction techniques

The remaining methods to be discussed allow us to calculate the transforming matrices A and B as well as the SNF of M. The first polynomial-time algorithm for the SNF and the HNF was due to Kannan and Bachem [KB79]. We shall not discuss this method in detail here. We shall, however, briefly discuss some methods described by Havas, Holt, and Rees in [HHR93], which perform very well in practice, particularly during the earlier stages of the reduction of large sparse matrices, which arise frequently in the applications that we shall discuss in Subsection 9.3.3 below.

Further improvements, which use the lattice reduction (LLL) algorithm for square matrices and the modified lattice reduction (MLLL) algorithm for nonsquare matrices, were introduced by Havas and Majewski in [HM97]. The LLL and MLLL algorithms play a central rôle in many areas of computer algebra, but they are beyond the scope of this book. The reader could consult Section 8.6 of [Sim94] for further details of their application to CGT.

The idea of the norm-based methods studied in [HHR93] is to choose a pivot entry that is likely to result in changes to the other entries of the matrix that are as small as possible when we perform row and column reduction using this pivot. To achieve this, we choose the pivot such that the other entries in its row and column are small. The size of these other entries can be measured by using any one of a number of possible norms; for example, we could use the largest entry, or the average of the absolute values of all of these entries.

The best results in terms of restricting the size of intermediate entries, however, were obtained by choosing M_{ij} such that the product of $E(i,j)$ of the standard Euclidean lengths of row i and column j of M is as small as possible. To be precise,

$$E(i,j)^2 := (\sum_{k=1}^{n} M_{ik}^2)(\sum_{k=1}^{m} M_{kj}^2).$$

In order to ensure termination of SMITHFORM if a is `false` at line 9 or 13, or if the 'If' condition at line 14 holds, on the next iteration of the main loop, we choose our pivot entry to be either in the same row or in the same column as the current pivot. That ensures that the next pivot entry will be smaller in absolute value than the current one, and so after finitely many iterations of the main loop, we will get through to line 16, and then f will be incremented.

Example 9.4

Here is a 10×12 integral matrix. This was chosen at random, subject to all entries having absolute value at most 15.

$$M = \begin{pmatrix}
-3 & -8 & 0 & -4 & -4 & -13 & -4 & -3 & -7 & -2 & -7 & -9 \\
-6 & -5 & -1 & -2 & -2 & -8 & -8 & 5 & -3 & 1 & 2 & -6 \\
-9 & -2 & 3 & 3 & 0 & 3 & -5 & 1 & 4 & 1 & 7 & -7 \\
-8 & -7 & 1 & 0 & -9 & 6 & -3 & 7 & 0 & -3 & -1 & -9 \\
-2 & -7 & -3 & -1 & -1 & 1 & -2 & 3 & -6 & -3 & 0 & -6 \\
-3 & -2 & 4 & -1 & -7 & 2 & -1 & 1 & 5 & -6 & -5 & 0 \\
5 & 0 & 6 & -4 & -1 & -1 & 2 & -1 & -4 & 4 & -1 & 1 \\
-3 & -3 & -6 & -1 & -3 & -4 & 5 & -3 & -4 & 1 & -8 & -5 \\
1 & -12 & -4 & -5 & 2 & -4 & -4 & 1 & -12 & -4 & 0 & -9 \\
-13 & -13 & -6 & 0 & -10 & -4 & -9 & 10 & -5 & -6 & -3 & -15
\end{pmatrix}.$$

The rank is 10, and the nonzero entries in the SNF of M are 1 (8 times), 2 and 260. (Randomly chosen integral matrices typically have most of their diagonal entries equal to 1, and also have one such entry that is much larger than the others.)

If we use SMITHFORM and choose the pivot with minimal $|M_{ij}|$, then the largest entry in an intermediate M is 607590, whereas the largest entries in A and B are 2 864 915 944 860 487 and 101 232 331 673 348, respectively. With the choice of pivot based on minimal $E(i,j)$, as discussed above, the largest entry in an intermediate M is 47 838, whereas the largest entries in A and B are 4 338 759 921 705 589 and 111 961 267 948, respectively. So the matrix A has not improved but B has, as has the largest intermediate entry in M. ☐

The effectiveness of the norm-based method in the example above is perhaps not spectacular, but calculating and using this norm is very fast, so it adds virtually no time or space overheads to the algorithm, and its effect becomes more noticeable as the matrices get larger.

More dramatic improvements were obtained at the cost of a more significant increase in running time by the following simple technique, which we shall call *row simplification*. It can be thought of as a simple-minded but very effective alternative to the rather complicated MLLL-based methods.

From time to time, we interrupt the main algorithm, and attempt to reduce the size of the entries in M as follows. We consider each ordered pair of distinct rows $(M[i], M[j])$ of M and, if the sum of the absolute values of the entries in row $M[i]$ can be reduced by adding $M[j]$ or $-M[j]$ to $M[i]$, then we perform the corresponding row operation, and multiply M and A on the left by $\rho_1(m, i, j, \pm 1)$. We repeat this process until no further reductions can be obtained for any pair of rows of M.

The more often that we carry out such reductions, the greater will be the effectiveness in terms of reducing the size of the matrix entries but, since it is relatively expensive in terms of running time, we do not want to do it more than necessary. The function can usefully be given a parameter t, which means carry out the row simplification process once every t iterations of the main loop.

In the example above with $t = 1$, the largest entry in an intermediate M was 1300, whereas the largest entries in A and B were 1 696 131 and 448, respectively. Row simplification increased the running time by about 33%.

Row simplification can also be used to limit the size of matrix entries in HERMITEFORM. The HNF of the matrix M of Example 9.4 is displayed below.

Running HERMITEFORM on M with s and t chosen with the 'largest and second largest $|M_{sj}|$ and $|M_{tj}|$ in column j' strategy, resulted in largest entry in an intermediate M and largest entry in A equal to 10 915 109 187 628 210 751 and 1 151 887, respectively. If we use row simplification on every iteration of the main loop, then the largest entry in an intermediate M is reduced to 28 703 427 346 369, but at the cost of more than doubling the running time.

$$
\begin{pmatrix}
1 & 0 & 0 & 0 & 0 & 0 & 0 & 0 & 3 & 1043423 & 579606 & -978187 \\
0 & 1 & 0 & 0 & 0 & 0 & 0 & 0 & 3 & 615835 & 342087 & -577331 \\
0 & 0 & 1 & 0 & 0 & 0 & 0 & 1 & 0 & 113012 & 62777 & -105945 \\
0 & 0 & 0 & 1 & 0 & 0 & 0 & 1 & 4 & 581127 & 322809 & -544794 \\
0 & 0 & 0 & 0 & 1 & 0 & 0 & 0 & 4 & 33068 & 18371 & -31000 \\
0 & 0 & 0 & 0 & 0 & 1 & 0 & 1 & 0 & 688827 & 382633 & -645760 \\
0 & 0 & 0 & 0 & 0 & 0 & 1 & 0 & 3 & 1143452 & 635170 & -1071961 \\
0 & 0 & 0 & 0 & 0 & 0 & 0 & 2 & 0 & 85254 & 47358 & -79922 \\
0 & 0 & 0 & 0 & 0 & 0 & 0 & 0 & 5 & 203198 & 112875 & -190494 \\
0 & 0 & 0 & 0 & 0 & 0 & 0 & 0 & 0 & 1211808 & 673140 & -1136044
\end{pmatrix}.
$$

9.3.3   Applications

One of the oldest and most popular applications of CGT to other areas of mathematics relies on the following methodology. We are given one or more finitely presented groups G. First we look for suitable subgroups H of G of low finite index, using the methods described in Section 5.4 or 9.1. Next we compute presentations of these subgroups using the Reidemeister-Schreier algorithm described in Section 5.3. We then use the SNF algorithm to compute the abelian invariants of $H/[H,H]$.

When the index $|G:H|$ is large, the presentation of G has a large number of generators and relators. It is then advisable (and often essential) to use the Tietze transformation simplification techniques described in Subsection 5.3.3 to reduce the size of the presentation before constructing the matrix for the SNF computation. Experience shows that the best strategy is to simplify the presentation for as long as its total presentation length is decreasing.

Many of the applications based on these techniques are to topology, which is not surprising, given that finitely presented groups arise naturally in a variety of topological contexts, such as knot groups and fundamental groups. An early application of this nature is described by Havas and Kovács in [HK84], where the authors prove that the knot groups of 11 specific knots are not isomorphic; this immediately implies that the knots are all distinct.

Often, the principal aim is to find an H with infinite abelianization. This has long been a popular approach to proving that a finitely presented group is infinite, but it has a more specific topological significance. Under certain connectivity assumptions, the conjugacy classes of subgroups H of the fundamental group G of a topological space X are in one-one correspondence with the covering spaces $\tilde{X}$ of X, where the fundamental group of $\tilde{X}$ is isomorphic to the corresponding subgroup H, and subgroups of finite index correspond to finite covers. See, for example, Theorem 10.2 of Chapter V of [Mas91].

So the search for low-index subgroups H of G corresponds to the search for finite covers $\tilde{X}$ of X. It is proved in Theorem 7.1 of Chapter VIII of [Mas91] that the subgroup H has infinite abelianization, if and only if the homology group $H_1(\tilde{X})$ is infinite.

Specific applications are typically to manifold groups. We shall give only brief details here. A 3-manifold is defined to be *Haken* if it contains a topologically essential surface. The Virtual Haken Conjecture, due to F. Waldhausen, says that every irreducible 3-manifold with infinite fundamental group has a finite cover that is Haken. It turns out that a 3-manifold whose fundamental group has infinite abelianization is Haken, so a stronger form of the conjecture says that, for every irreducible 3-manifold with infinite fundamental group G, G has a finite index subgroup with infinite abelianization.

This conjecture was verified computationally by Dunfield and Thurston for a set of 10 986 small-volume closed hyperbolic 3-manifolds, using the techniques outlined above; see [DT03]. While many of the examples were settled using subgroups H of reasonably small index, a few were more difficult, and over a year of CPU-time, mainly using GAP and occasionally also MAGMA, was consumed in the complete exercise. Subgroups of index at most 6 worked for 40% of the examples, whereas index at most 200 worked 98% of the time.

There are applications of a similar nature in [Rei95], which describes the construction of the first explicit example of a nonHaken hyperbolic 3-manifold that has a finite cover which fibres over the circle, and in [CFJR01], as part of the authors' proof that the Weeks manifold is the smallest volume arithmetic hyperbolic 3-manifold.

An application in which the subgroup H is proved infinite by a different method, not involving $H/[H, H]$, will be described in Subsection 9.4.7.

Finally, we mention the paper of Havas, Holt, and Newman [HHN01], in which the use of these techniques on a small number of examples enabled the authors to notice a pattern, and thereby prove the infiniteness of all groups in an infinite family of examples. These groups had been originally investigated by Johnson, Kim, and O'Brien [JKO99].

Exercise

1. Prove that, for $m > 0$, the SNF modulo m of an integral matrix M is uniquely determined by M.

9.4 p-quotients of finitely presented groups

Substantial progress at an algorithmic and computational level has been made over the past thirty years in the study of various types of quotients of a finitely presented group. In particular, we can construct polycyclic presentations for those quotients of the group that have prime power order or are nilpotent or solvable.

Here we discuss in detail the commonly-used algorithm to compute such a presentation for a p-quotient; that is, a quotient that is a finite p-group for a prime p. We also study the p-group generation algorithm, used to generate descriptions of p-groups. Finally we report on some related algorithms, including computing automorphism groups of finite p-groups, and testing pairs of such groups for isomorphism.

9.4.1 Power-conjugate presentations

Recall from Chapter 8 that a group G is polycyclic if it has a sequence of subgroups

$$G = G_1 \geq \cdots \geq G_i \geq \cdots \geq G_{n+1} = 1$$

where $G_{i+1} \lhd G_i$ and G_i/G_{i+1} is cyclic.

Theorem 8.8 states that every polycyclic group G has a polycyclic presentation. In summary, we can associate with G a particular presentation as follows. We choose elements a_i where $G_i = \langle a_i, G_{i+1} \rangle$. The sequence of generators $A := [a_1, \ldots, a_n]$ is a *polycyclic generating sequence* for G. Let $I \subseteq \{1, \ldots, n\}$ denote the set of subscripts where G_i/G_{i+1} is finite and has order δ_i. Let R denote the set of defining relations

$$a_j a_i = a_i W_{ij} \text{ for } j > i$$
$$a_j a_i^{-1} = a_i^{-1} W_{ij}^* \text{ for } j > i$$
$$a_i^{\delta_i} = W_{ii} \text{ for } i \in I,$$

where the W_{ij}, W_{ij}^* are words involving only $a_{i+1}, \ldots, a_n$ and their inverses. Now G has a polycyclic presentation $\{A \mid R\}$.

These presentations have proved of central importance in allowing effective computation with such groups. In [Syl72], it is proved that every group of order p^n has such a presentation on n generators; Jürgensen first introduced them for finite solvable groups in [Jür70].

While every finitely generated nilpotent group and every finite solvable group is polycyclic, this need not be true for an arbitrary finitely generated solvable groups.

Some significant progress has been made in constructing polycyclic presentations for arbitrary solvable quotients of a finitely presented group. Assume we have a finite presentation for a group G and a polycyclic presentation for a quotient, G/N. In [BCM81b, BCM81a], Baumslag, Cannonito, and Miller describe a theoretical algorithm that can decide whether or not G/N' is polycyclic and, if so, obtain a polycyclic presentation for G/N'. In [Sim90b], Sims developed practical aspects of this algorithm to compute metabelian quotients of a finitely presented group.

This work is extended to the general case by Lo in [Lo98]. Let $G^{[n]}$ denote the nth term of the derived series of G. Then, given as input the finite presentation for G and $n > 1$, Lo's algorithm decides whether $G/G^{[n]}$ is polycyclic and, if so, returns a polycyclic presentation for $G/G^{[n]}$.

As we learned in Subsection 8.1.1, a critical feature of a polycyclic presentation is that every element of the presented group may be written in a *normal form* $a_1^{\alpha_1} a_2^{\alpha_2} \cdots a_n^{\alpha_n}$, where each α_i is an integer and $0 \le \alpha_i < \delta_i$ if $i \in I$. If this normal form is unique, then the presentation is *consistent*. See Subsection 8.1.3 for a discussion of collection algorithms – these rewrite an arbitrary word in the generators as a normal word equivalent to it using the power-conjugate presentation.

Most of the algorithms to construct a polycyclic presentation for a quotient of a finitely presented group G have a common structure. Each uses a chain of (sub)normal subgroups

$$G = G_1 \ge G_2 \ge \cdots \ge G_i \ge G_{i+1} \cdots \ge G_{n+1} = 1$$

and works down this chain, using the polycyclic presentation constructed for G/G_i to write down a presentation for G/G_{i+1}. We write down a presentation for a group that is a (downward) extension of G/G_i and has G/G_{i+1} as a quotient.

The input for one iteration of such an algorithm usually includes a finite presentation $\{X \mid S\}$ for a finitely presented group, G, and a polycyclic presentation for $H := G/G_i$, where G_i is the i-th term of the relevant series, and an epimorphism from G to H. The output includes a polycyclic presentation for $K := G/G_{i+1}$ and an epimorphism from G to K.

We now focus on the algorithm used to construct a polycyclic presentation for a finite p-group. In this context, we usually refer to *power-conjugate presentations* and list relations as commutators rather than conjugates.

9.4.2 The p-quotient algorithm

Recall from Subsection 2.3.4 that the *lower central series* of a group G is the sequence of subgroups

$$G = \gamma_1(G) \geq \cdots \geq \gamma_{i-1}(G) \geq \gamma_i(G) \geq \cdots$$

where $\gamma_i(G) = [\gamma_{i-1}(G), G]$ for $i > 1$.

The p-quotient algorithm uses a variation of the lower central series known as the *lower exponent-p central series*, where p is a prime. This is the descending sequence of subgroups

$$G = P_0(G) \geq \cdots \geq P_{i-1}(G) \geq P_i(G) \geq \cdots$$

where $P_i(G) = [P_{i-1}(G), G]P_{i-1}(G)^p$ for $i \geq 1$.

If G is a finite p-group then the series terminates in the identity. If $P_c(G) = 1$ and c is the smallest such integer then G has *exponent-p class c*. For the remainder of this section, *class* means *exponent-p class*.

We saw in Lemma 8.20 that the relations in a power-conjugate presentation of the type

$$a_j a_i^{-1} = a_i^{-1} W_{ij}^* \text{ for } j > i$$

are redundant in a finite group, and can be omitted from the presentation. We shall do this for the remainder of this section.

Example 9.5

Let G be the dihedral group of order 16 described as

$$\mathrm{Pc}\langle\, a_1, \ldots, a_4 \mid a_2^2 = a_3 a_4,\ a_3^2 = a_4,\ [a_2, a_1] = a_3,\ [a_3, a_1] = a_4 \,\rangle.$$

(Recall this notation from Subsection 8.1.2; trivial polycyclic relations are omitted from the presentation. In the context of polycyclic presentations of finite p-groups, we also omit trivial relations of the form $a_i^p = 1$.) The

lower exponent-p central series of G is the sequence of subgroups $P_0(G) = G$, $P_1(G) = \langle a_3, a_4 \rangle$, $P_2(G) = \langle a_4 \rangle$ and $P_3(G) = 1$, so G has exponent-2 class 3.
$\Box$

We first describe a number of basic but important properties of the lower exponent-p central series. The proofs of these results are similar to the analogous statements for the lower central series.

THEOREM 9.14   *If G is a d-generator group, then $G/P_1(G)$ is elementary abelian of order at most p^d. If G is a finite p-group then $P_1(G) = \Phi(G)$, the Frattini subgroup of G.*

Note that $P_1(G)$ is, by definition, the smallest normal subgroup of G having an elementary abelian quotient group, and so $P_1(G) = \Phi(G)$ by Proposition 2.45.

LEMMA 9.15   *If θ is a homomorphism of G then $\theta(P_i(G)) = P_i(\theta(G))$. Consequently each term of the lower exponent-p central series is a characteristic subgroup of G. In particular, if N is a normal subgroup of G then $P_i(G/N) = P_i(G)N/N$.*

LEMMA 9.16   *If $N \lhd G$ and G/N has class c then $P_c(G) \leq N$.*

LEMMA 9.17   *For positive integers i, j, $[P_i(G), P_j(G)] \leq P_{i+j+1}(G)$.*

This last lemma can be proved by using the three subgroups lemma; see Exercise 4 at the end of Section 2.3. The following result establishes that a finite p-group has a largest central elementary abelian extension.

THEOREM 9.18   *Let G be a d-generator finite p-group. Then there exists a d-generator finite p-group, G^*, with the property that every d-generator group H having a central elementary abelian p-subgroup, Z, such that H/Z is isomorphic to G, is a homomorphic image of G^*.*

PROOF   Let F be the free group of rank d freely generated by $a_1, \ldots, a_d$, and let R be the kernel of a homomorphism θ from F onto G. Define R^* to be $[R, F]R^p$ and G^* to be F/R^*. Then G^* has d generators and, since $R \lhd F$, we have $R^* \leq R$. Since H is a d-generator group and has a quotient, H/Z, which is isomorphic to G, the homomorphism, θ, may be factored through H and the resulting homomorphism, ψ, of F onto H maps R into Z. Since Z is elementary abelian and central, ψ maps both R^p and $[R, F]$ to the identity in H. Thus, $\psi(R^*) = 1$ and H is a homomorphic image of F/R^*. Since R/R^* is an elementary abelian p-group, G^* is a finite p-group.   ∎

We call G^* the *p-covering group* of G. It has some similarities to Schur covering groups, but critically its isomorphism type depends only on G and is independent of the chosen presentation.

LEMMA 9.19 *The isomorphism type of G^* depends only on G and not on R.*

PROOF Let R_1 and R_2 be normal subgroups of F; let $F/R_1 = G_1$ and $F/R_2 = G_2$ where $G_1 \cong G_2$. Following the notation of Theorem 9.18, we define R_1^*, G_1^*, R_2^*, and G_2^*. Then G_1^* is isomorphic to G_2^*, since each is a homomorphic image of the other. ∎

If F/R has class c, then G^* has class at most $c+1$. Further, if G is a class k quotient of F/R, then a quotient of F/R having class $k+1$ is isomorphic to a quotient of G^*.

Example 9.6

The 2-covering group G^* of the dihedral group G of order 16 defined in Example 9.5 has power-conjugate presentation

$$\text{Pc}\langle a_1, \ldots, a_4, a_5, a_6, a_7 \mid a_1^2 = a_6, a_2^2 = a_3 a_4 a_7, a_3^2 = a_4 a_5, a_4^2 = a_5,$$
$$[a_2, a_1] = a_3, [a_3, a_1] = a_4, [a_4, a_1] = a_5 \rangle.$$

Observe that $Z = \langle a_5, a_6, a_7 \rangle$ is elementary abelian and $G^*/Z \cong G$. ⬚

The next result describes generating sets for successive terms of the lower exponent-p central series. Its proof is similar to the corresponding statement for the lower central series.

THEOREM 9.20 *If $G/P_1(G)$ is generated by the images of $a_1, \ldots, a_d$, then $P_1(G)/P_2(G)$ is generated by the images of a_i^p where $1 \leq i \leq d$ and $[a_j, a_i]$ where $1 \leq i < j \leq d$. More generally, for $k > 0$, let S be a subset of G which generates G modulo $P_1(G)$ and let T generate $P_k(G)$ modulo $P_{k+1}(G)$. Then $P_{k+1}(G)$ is generated modulo $P_{k+2}(G)$ by $[s, t]$ for $s \in S$, $t \in T$ and t^p for $t \in T$.*

Let G be a d-generator p-group of order p^n and class c. Every refined normal series of G gives rise to a power-conjugate presentation and so usually every p-group has many power-conjugate presentations.

In practice, we construct a power-conjugate presentation for G arising from a normal series that refines the lower exponent-p central series of G. Hence, the power-conjugate presentation $\{A \mid R\}$ for G has additional structure.

Let $A := [a_1, \ldots, a_n]$. Then the assumption that the normal series associated with A refines the lower exponent-p central series of G together with Theorem 9.14 implies that $P_1(G) = \Phi(G) = \langle a_{d+1}, \ldots, a_n \rangle$. It follows from Proposition 2.44 that $\{a_1, \ldots, a_d\}$ is a generating set for G, and hence the remaining generators can be expressed as words in $\{a_1, \ldots, a_d\}$.

Define a function ω from A to $\{1..c\}$ that maps $g \in A$ to k if $g \in P_{k-1}(G) \setminus P_k(G)$. Then ω is a *weight* function and $\omega(g)$ is the weight of g. The assumption that the normal series associated with A refines the lower exponent-p central series of G means that $\omega(a_j) \geq \omega(a_i)$ for $j \geq i$.

Now the relations in R have the form:

$$a_i^p = \prod_{k=i+1}^{n} a_k^{\beta(i,k)}, \ 0 \leq \beta(i,k) < p, \ 1 \leq i \leq n,$$

$$a_j^{a_i} = a_j \prod_{k=j+1}^{n} a_k^{\beta(i,j,k)}, \ 0 \leq \beta(i,j,k) < p, \ 1 \leq i < j \leq n.$$

A power-conjugate presentation that satisfies the above conditions is called a *weighted* power-conjugate presentation. That a finite p-group G has a weighted power-conjugate presentation follows from Theorem 9.20.

For each a_k in $\{a_{d+1}, \ldots, a_n\}$, we require that there is at least one relation whose right-hand side is wa_k, where w is a (possibly empty) word in generators $a_1, \ldots, a_{k-1}$. One of these relations is designated the *definition* of a_k, and we say that the power-conjugate presentation has definitions.

Example 9.7

Consider once more the power-conjugate presentation for the dihedral group G of order 16:

$$\text{Pc}\langle a_1, \ldots, a_4 \mid a_2^2 = a_3 a_4, \ a_3^2 = a_4, \ [a_2, a_1] = a_3, \ [a_3, a_1] = a_4 \rangle.$$

Observe that $G = \langle a_1, a_2 \rangle$ and $[a_2, a_1] = a_3$, $[a_3, a_1] = a_4$. The weights of a_1, a_2, a_3, a_4 are $1, 1, 2, 3$, respectively. The definition of a_3 is $[a_2, a_1]$, and we choose $[a_3, a_1]$ as the definition of a_4. □

Now we can more precisely summarize: the p-quotient algorithm takes as input a finite presentation $\{X \mid S\}$, for a group G, a prime p, and a positive integer c. The output is a weighted consistent power-conjugate presentation with definitions for $G/P_c(G)$ and an epimorphism of G onto this quotient.

The algorithm works down the lower exponent-p central series, using the power-conjugate presentation constructed for $G/P_{t-1}(G)$ to write down a presentation for $G/P_t(G)$, where $t > 0$. All power-conjugate presentations in what follows are weighted and with definitions. A single iteration of the algorithm takes as input:

1. the finite presentation $\{X \mid S\}$ for G;

2. a consistent power-conjugate presentation for the factor group $H := G/P_{t-1}(G)$;

3. an epimorphism θ from G to H.

The output of this iteration is

1. a consistent power-conjugate presentation for the factor group $K := G/P_t(G)$;

2. an epimorphism φ from K to H;

3. an epimorphism τ from G to K where $\varphi\tau = \theta$.

The first iteration of the algorithm computes a consistent power-conjugate presentation for $G/P_1(G) = G/\Phi(G)$. If $G/\Phi(G)$ has order p^d (and so G has *Frattini rank d*), then this presentation is on d generators that commute pairwise, and each has order p. This is done by Gaussian elimination using the relators of G, where $G/P_1(G)$ is regarded as a vector space over $\mathrm{GF}(p)$.

Each subsequent iteration has two principal steps:

(i) Let $H := G/P_{t-1}(G)$ for some $t > 1$. Write down a consistent power-conjugate presentation for H^*, the p-covering group of H.

(ii) Evaluate the images of the defining relators of G in H^* to obtain the kernel of the homomorphism from H^* to K.

The first of these steps is performed by a *p-covering group algorithm*: it takes as input a consistent power-conjugate presentation $\{A \mid R\}$ for a group H and produces a power-conjugate presentation $\{A^* \mid R^*\}$ for its p-covering group, H^*. The generating set A^* contains A as a subset; the elements of $A^* \setminus A$ are *new generators*. Each relation in R^* ends in a (possibly empty) word in the new generators. If we make all new generators trivial, then the resulting presentation is just that input to the algorithm.

We shall now discuss step (i) in more detail. Consider the supplied power-conjugate presentation $\{A \mid R\}$ for H. Recall that for each a_k in $\{a_{d+1}, \ldots, a_n\}$, there is at least one relation whose right-hand side is wa_k, where w is a (possibly empty) word in generators $a_1, \ldots, a_{k-1}$. One of these relations is designated the *definition* of a_k. Hence there are precisely $d + \binom{n}{2}$ remaining (nondefinition) relations.

In order to write down a power-conjugate presentation for H^*, a total of $q := d + \binom{n}{2}$ new generators, $a_{n+1}, \ldots, a_{n+q}$, are introduced together with relations that make them central and of order p. We next add the *tails*: each of the remaining (nondefinition) relations is modified by inserting one of these new generators at the end of its right-hand side. The modified relation is now viewed as the definition of the new generator.

We now summarize an algorithm to construct a power-conjugate presentation for the p-covering group of H.

p-CoveringGroup(H)

    Input: Consistent power-conjugate presentation for p-group H

    Output: Power-conjugate presentation for its covering group H^*

1   Initialize A^* to be A;

2   Initialize R^* to consist of all relations from R which are definitions;

3   Modify each nondefining relation $a_i^p = w_{ii}$ or $[a_j, a_i] = w_{i,j}$ to be
    $a_i^p = w_{ii}a_r$ or $[a_j, a_i] = w_{i,j}a_r$ respectively for $r \in \{n+1 .. n+q\}$
    where different nondefining relations are modified by different a_r;

4   Add each a_r to A^*;

5   Add $a_r^p = 1$ to R^* for $r \in \{n+1 .. n+q\}$;

6   Add $[a_r, a_i] = 1$ to R^* for $r \in \{n+1 .. n+q\}$ and $i \in \{1 .. r-1\}$;

7   **return** $H^* = \langle A^* \mid R^* \rangle$;

This is a simplified version of the p-covering group algorithm; see the article by Newman, Nickel and Niemeyer [NNN98] for a detailed description and proof of the following result.

THEOREM 9.21   *The presentation $\{A^* \mid R^*\}$ produced by the covering group algorithm is a power-conjugate presentation for H^*.*

This power-conjugate presentation $\{A^* \mid R^*\}$ on $n + d + \binom{n}{2}$ generators is usually not consistent when $t > 2$.

The task of deciding whether a given presentation is consistent and, if not, its modification to produce a consistent one will be considered in Section 12.4 of Chapter 12. We briefly consider the topic here. In [Wam74], Wamsley proved that it is sufficient to establish that certain test words collected in two different ways evaluate to the same normal word to ensure that the presentation is consistent. Associativity of course dictates that the two normal words obtained must be identical; if they are not, then their quotient is a new relation that must hold. In [VL84], Vaughan-Lee significantly reduced the number of words that need to be evaluated. We summarize their results, which exploit the weights assigned to generators.

THEOREM 9.22   *A weighted power-conjugate presentation on $[a_1, \ldots, a_m]$ of a finite p-group of class c is consistent if the following words collect to the trivial word:*

$$((a_k a_j)a_i)(a_k(a_j a_i))^{-1} \qquad \begin{array}{l} 1 \le i < j < k \le m \text{ and } i \le d, \\ \omega(a_i) + \omega(a_j) + \omega(a_k) \le c; \end{array}$$

$$((a_j^{p-1}a_j)a_i)(a_j^{p-1}(a_j a_i))^{-1} \quad 1 \le i < j \le m \text{ and } i \le d, \ \omega(a_i) + \omega(a_j) < c;$$

$$((a_j a_i)a_i^{p-1})(a_j(a_i a_i^{p-1}))^{-1} \qquad 1 \le i < j \le m, \ \omega(a_i) + \omega(a_j) < c;$$

$$((a_i a_i^{p-1})a_i)(a_i(a_i^{p-1}a_i))^{-1} \qquad 1 \le i \le m, \ 2\omega(a_i) < c.$$

where words in inner parentheses are collected first.

For a proof of this theorem see Section 9.9 of [Sim94]. We apply this theorem in the p-covering group algorithm to the group H^*, where $m = n + q$ with $q = d + \binom{n}{2}$. The weight conditions in the theorem imply, in particular, that the tests need only be carried out on the generators $[a_1, \ldots, a_n]$. On the t-th iteration of the p-quotient algorithm we have $H = G/P_{t-1}(G)$, and H^* has class at most t, so we have $c = t$ in the theorem.

As we observed above, if the presentation is consistent, then associativity implies that all of these words evaluate to the identity. Hence the application of this theorem gives a set of new relations that are consequences of R^*. We add these to R^*. The words involve the new generators only, since $\{A^* \mid R^*\}$ extends the existing consistent presentation $\{A \mid R\}$.

We now consider briefly the second principal step of an iteration of the p-quotient algorithm. Let $M := \langle a_{n+1}, \ldots, a_{n+q} \rangle$ and let N be the kernel of the natural homomorphism from K onto H; then N is a homomorphic image M/L of M. In order to obtain a power-conjugate presentation for K, we must compute the kernel L of the map from M to N.

This is done by evaluating the images of the relators of G in M. To be more precise, we first lift θ to a homomorphism $\hat{\theta}$ from the free group on the generating set X of G to H^*, and then evaluate $\hat{\theta}(r)$ for each defining relator $r \in S$ of G. The elements $\hat{\theta}(r)$ lie in M and generate L. The epimorphism $\tau : G \to K$ to be returned by the algorithm is induced by $\hat{\theta}$, and $\varphi : K \to H$ is just the natural homomorphism with kernel $N = M/L$.

There is an additional technicality involved in lifting θ to $\hat{\theta}$, which we shall describe briefly. There will in practice be specific generators $x_1, \ldots, x_d$ of X such that $\theta(x_i) = a_i$ and, after the iteration, we will have $\tau(x_i) = a_i$, and so this property is maintained. For all other generators $x \in X$, we have $\hat{\theta}(x) = \theta(x)a_x$ where a_x is an unknown element of M. The elements a_x will be determined modulo L during the evaluation of $\hat{\theta}(r)$ for $r \in S$. In practice, they can be introduced as new temporary generators of M, which will be eliminated during the Gaussian elimination process described below. For further discussion of this aspect, see [HN80].

THEOREM 9.23 *The result of collecting the set of words in $\{a_1, \ldots, a_n\}$ listed in Theorem 9.22 in the power-conjugate presentation for H^* is a set U of elements of M; the result of evaluating the relators of G in the images of the generators of G under $\hat{\theta}$ in the power-conjugate presentation for H^* is a set V of elements of M; and N is isomorphic to $M/\langle U \cup V \rangle$.*

Critical to practical computation is the realization that M can be viewed as a vector space over the field of p elements. A basis $[n_1, \ldots, n_r]$ for $N = M/L$ can be computed using Gaussian elimination. Let $\hat{A}$ be $A \cup \{n_1, \ldots, n_r\}$. The image of each element a_{n+i} can be expressed in this basis of N, and we replace its occurrence in R^* by this image to obtain $\hat{R}$.

Hence, on the t-th iteration of the p-quotient algorithm, we obtain a consistent power-conjugate presentation $\{\hat{A} \mid \hat{R}\}$ for $G/P_t(G) = K$ and an epimorphism τ from G to K. We can now extend the weight function ω to $\{\hat{A} \mid \hat{R}\}$ by defining $\omega(n_i) := t$. This completes the construction of a consistent power-conjugate presentation $\{\hat{A} \mid \hat{R}\}$ for the class t p-quotient of G.

We now summarize the p-quotient algorithm.

P-QUOTIENT(Q, p, c)

    Input: Finitely presented group $Q = \langle X \mid S \rangle$, prime p, positive integer c

    Output: Consistent power-conjugate presentation for the largest class c
            p-quotient of Q

1   Calculate $G = \langle A \mid R \rangle$, the largest elementary abelian p-quotient of Q;

2   **for** $t \in [2 .. c]$

3   **do** $H := $ P-COVERINGGROUP (G);

4   Apply consistency tests to H;

5   Evaluate defining relations for G;

6   Eliminate redundant generators from H;

7   **if** $|H| = |G|$ **then return** G;

8               $G := H$;

9   **return** G;

Example 9.8

As an example, we construct the class 3 2-quotient of the finitely presented group $Q := \langle x, y \mid [y, x, x] = x^2, (xyx)^4 \rangle$ (where $[y, x, x] := [[y, x], x]$).

Its class 1 quotient is $G = C_2 \times C_2$ which has power-conjugate presentation

$$\langle a_1, a_2 \mid a_1^2 = 1, a_2^2 = 1, [a_2, a_1] = 1 \rangle,$$

where the epimorphism θ maps x to a_1 and y to a_2, and each generator has weight 1.

To write down its 2-covering group G^*, we introduce new generators as right-hand sides of the three nondefining relations. We also add relations to ensure that these new generators are central and of order 2. Hence G^* has power-conjugate presentation

$$\langle a_1, \ldots, a_5 \mid a_1^2 = a_4, a_2^2 = a_5, [a_2, a_1] = a_3, a_3^2 = a_4^2 = a_5^2 = 1,$$
$$[a_3, a_1] = [a_3, a_2] = 1 \rangle.$$

Is this presentation consistent? For example, is $a_1(a_1 a_1) = (a_1 a_1)a_1$? Observe that $a_1(a_1 a_1) = a_1 a_5$ and $(a_1 a_1)a_1 = a_5 a_1 = a_1 a_5$ and so we obtain only the trivial relation. If we enforce all of the consistency tests, we deduce no new relations and so the listed power-conjugate presentation for G^* is indeed consistent. (In fact, it follows from Theorem 9.22 that any weighted power-conjugate presentation of a p-group of class 2 is consistent so, in general, we do not need to carry out any consistency checks on the second iteration of the p-quotient algorithm.)

Now we enforce relations: $\theta([y, x, x]) = \theta(x^2)$ implies that $1 = a_4$ whereas $\theta((xyx)^4) = 1$ gives the trivial relation. Hence we factor G^* by $\langle a_4 \rangle$ and rename a_5 as a_4 to obtain the class 2 2-quotient G having consistent power-conjugate presentation

$$\langle a_1, \ldots, a_4 \mid a_2^2 = a_4, [a_2, a_1] = a_3, a_1^2, a_3^2, a_4^2, [a_3, a_1],$$
$$[a_3, a_2], [a_4, a_1], [a_4, a_2], [a_4, a_3] \rangle.$$

The unique relations having right-hand sides a_3, a_4 are chosen as the definitions of these generators, both having weight 2; the epimorphism θ from Q to G maps x to a_1 and y to a_2.

To construct the class 3 2-quotient of Q, we first construct a power-conjugate presentation for G^* by introducing new generators as right-hand sides of all 8 nondefining relations. We also add relations to ensure that the new generators are central and of order 2. Hence G^* is

$$\text{Pc}\langle a_1, \ldots, a_{12} \mid a_1^2 = a_{12}, a_2^2 = a_4, a_3^2 = a_{11}, a_4^2 = a_{10}, [a_2, a_1] = a_3, [a_3, a_1] = a_5,$$
$$[a_3, a_2] = a_6, [a_4, a_1] = a_7, [a_4, a_2] = a_8, [a_4, a_3] = a_9 \rangle.$$

We apply Theorem 9.22 to decide if this power-conjugate presentation is consistent.

1. Is $a_2(a_2 a_2) = (a_2 a_2) a_2$? Observe that $a_2(a_2 a_2) = a_2 a_4$ and $(a_2 a_2) a_2 = a_4 a_2 = a_2 a_4 a_8$. Hence a_8 is trivial.

2. Is $a_2(a_1 a_1) = (a_2 a_1) a_1$? Observe that $(a_2 a_1) a_1 = a_2 a_5 a_{11} a_{12}$ and $a_2(a_1 a_1) = a_2 a_{12}$. Hence $a_5 a_{11}$ is trivial.

3. Is $a_2(a_2 a_1) = (a_2 a_2) a_1$? Observe that $a_2(a_2 a_1) = a_1 a_4 a_6 a_{11}$ and that $(a_2 a_2) a_1 = a_1 a_4 a_7$. Hence $a_6 a_7 a_{11}$ is trivial.

4. Is $a_3(a_2 a_2) = (a_3 a_2) a_2$? Observe that $a_3(a_2 a_2) = a_3 a_4$ and $(a_3 a_2) a_2 = a_3 a_4 a_9$. Hence a_9 is trivial.

We leave as an exercise to verify that there are no other independent consequences of consistency.

Hence, we factor G^* by $\langle a_8, a_9, a_5 a_{11}, a_6 a_7 a_{11} \rangle$ and deduce that G^* has consistent power-conjugate presentation

$$\text{Pc}\langle a_1, \ldots, a_4, a_5, \ldots, a_8 \mid a_1^2 = a_8, a_2^2 = a_4, a_3^2 = a_7, a_4^2 = a_6, [a_2, a_1] = a_3,$$
$$[a_3, a_1] = a_7, [a_3, a_2] = a_5 a_7, [a_4, a_1] = a_5 \rangle.$$

Now we enforce relations: $\theta([y, x, x]) = \theta(x^2)$ implies that $a_7 = a_8$ and $\theta((xyx)^4) = 1$ implies that $a_6 = 1$. Hence we factor G^* by $\langle a_7 a_8, a_6 \rangle$ to obtain the class 3 2-quotient

$$\text{Pc}\langle a_1, \ldots, a_4, a_5, a_6 \mid a_1^2 = a_6, a_2^2 = a_4, a_3^2 = a_6, [a_2, a_1] = a_3,$$
$$[a_3, a_1] = a_6, [a_3, a_2] = a_5 a_6, [a_4, a_1] = a_5 \rangle.$$

☐

9.4.3 Other quotient algorithms

The first description of a p-quotient algorithm was by Macdonald in [Mac74]. Other algorithms are presented in [Mac87]. The version of the p-quotient algorithm that we described above was introduced by Newman in [New76]; for additional details and applications of the algorithm, see [HN80] and [NO96].

Sims, in [Sim87, Sim90b, Sim94], discusses computing certain quotients of finitely presented groups; his book [Sim94] is the most comprehensive source available on the topic. Nickel, in [Nic95], presents an algorithm to construct nilpotent quotients of a finitely presented group; his implementation is available in GAP and MAGMA.

A number of algorithms for computing finite solvable quotients of finitely presented groups have been proposed. They include that of Wamsley [Wam77], Leedham-Green [LG84], Plesken [Ple87], and Niemeyer [Nie93]. H. Brückner further developed and implemented a version of Plesken's algorithm; it is available in MAGMA. Niemeyer's algorithm is available in GAP V3.

In [Lo98], Lo describes an infinite polycyclic algorithm which constructs a polycyclic presentation for $G/G^{[n]}$, where $G^{[n]}$ is the nth term of the derived series of a finitely presented group G; it is available in GAP V3. A new infinite polycyclic quotient algorithm has recently been developed by Eick, Niemeyer and Panaia in [ENP]; it generalises the approach taken by Niemeyer in [Nie93]; an implementation is available in GAP.

9.4.4 Generating descriptions of p-groups

The p-group generation algorithm calculates power-conjugate presentations for particular extensions of a finite p-group.

Let G be a finite d-generator p-group of class c. A group H is a *descendant* of G if H is d-generator and the quotient $H/P_c(H)$ is isomorphic to G. A group is an *immediate descendant* of G if it is a descendant of G and has class $c + 1$.

In summary, the p-group generation algorithm takes as input a p-group, G, and produces as output a *complete and irredundant* list of the immediate descendants of G: all isomorphism types occur in the list and no two elements have the same isomorphism type.

We now describe the algorithm in more detail. Observe, from Theorem 9.18, that every immediate descendant of $G = F/R$ is isomorphic to a quotient of $G^* = F/R^*$. The factor group R/R^* is the *p-multiplicator* of G and the group $P_c(G^*)$ is called the *nucleus* of G. (Both the p-multiplicator and nucleus are elementary abelian p-groups.) An *allowable subgroup* is a subgroup of the p-multiplicator which is the kernel of a homomorphism from G^* onto an immediate descendant of G.

If we wish to construct the immediate descendants of G efficiently, then we must characterize the required quotients of G^*.

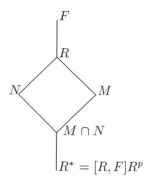

Figure 9.1: Various subgroups of the p-covering group.

THEOREM 9.24   *A subgroup is allowable if and only if it is a proper subgroup of the p-multiplicator of G which supplements the nucleus.*

PROOF   Let M/R^* be an allowable subgroup — that is, the kernel of a map from F/R^* onto an immediate descendant H of G. Since G has class c and H has class $c + 1$, it is clear that M is a proper subgroup of R. Lemma 9.16 shows that $P_c(F)$ is a subgroup of R and M is also a subgroup of R and, hence, $MP_c(F)$ is a subgroup of R. Following Theorem 9.18, $\psi(R)$ is a subgroup of $P_c(H)$. Since F/R has class c, $P_c(\psi(F))$ is also a subgroup of $\psi(R)$ showing that $\psi(R)$ equals $P_c(H)$. But $\psi(R)$ also equals R/M and $P_c(H) = P_c(F)M/M$. Therefore $R/M = P_c(F)M/M$ giving $R = P_c(F)M$. Hence, R^* can be factored out showing that $(M/R^*)P_c(F)R^*/R^* = R/R^*$. Lemma 9.15 now gives the required statement.

Conversely, let M/R^* be a proper subgroup of the p-multiplicator that supplements the nucleus. Then $P_c(F)M/R^* = R/R^*$ and hence $P_c(F)M/M = R/M$. Lemma 9.15 gives $P_c(F/M) = R/M$. Since F/M is a quotient of F/R^*, it has generator number d and the quotient $(F/M)/P_c(F/M)$ is isomorphic to G showing that F/M is a descendant of G. But $P_c(F/M) = R/M$ which is nontrivial; hence F/M has class $c+1$ and is an immediate descendant of G. ∎

Figure 9.1 illustrates the situation, where N/R^* represents the nucleus and M/R^* is an allowable subgroup.

Example 9.9
The 2-covering group G^* of the dihedral group G of order 16 has power-conjugate presentation

$$\mathrm{Pc}\langle a_1, \ldots, a_4, a_5, a_6, a_7 \mid a_1^2 = a_6,\ a_2^2 = a_3 a_4 a_7,\ a_3^2 = a_4 a_5,\ a_4^2 = a_5,$$
$$[a_2, a_1] = a_3,\ [a_3, a_1] = a_4,\ [a_4, a_1] = a_5\,\rangle.$$

The 2-multiplicator is $\langle a_5, a_6, a_7 \rangle$ and the nucleus is $\langle a_5 \rangle$. The subgroups $\langle a_6, a_7 \rangle$, $\langle a_5 a_6, a_7 \rangle$, $\langle a_6, a_5 a_7 \rangle$ are allowable and the corresponding immediate descendants have order 32. The subgroup $\langle a_5 a_6, a_5 a_7 \rangle$ is also allowable, but the resulting quotient is isomorphic to the quotient of G^* by $\langle a_6, a_5 a_7 \rangle$. ☐

If G has immediate descendants, it is called *capable*; otherwise, it is called *terminal*. Clearly, G is capable if and only if G^* has class exactly $c+1$. In particular, if G has order p^n and its nucleus has rank m, then G has immediate descendants of order p^{n+s} where $1 \le s \le m$.

If G is capable then, on taking factor groups of G^* by allowable subgroups, a complete list of immediate descendants is obtained; this list usually contains redundancies. We define an obvious equivalence relation: two allowable subgroups M_1/R^* and M_2/R^* are *equivalent* if and only if their quotients F/M_1 and F/M_2 are isomorphic.

A complete and irredundant set of immediate descendants of G can be obtained by factoring G^* by one representative of each equivalence class. In practice, this definition is useful only because the equivalence relation can be given a different characterization by using the automorphism group, $\mathrm{Aut}(G)$, of G.

In summary, we now describe an extension of each automorphism, α, of G to an automorphism, α^*, of G^*. The action of α^* when restricted to the p-multiplicator of G is uniquely determined by α, and α^* induces a permutation of the allowable subgroups. Finally we prove that the equivalence classes of allowable subgroups are exactly the orbits of the allowable subgroups under the action of these permutations.

THEOREM 9.25 Let M_1/R^* and M_2/R^* be allowable subgroups of F/R^* which are contained in R/R^*, and let φ be an isomorphism from F/M_1 to F/M_2. Then there exists an automorphism, α^*, of G^* which maps M_1/R^* to M_2/R^* and the map from F/M_1 to F/M_2 induced by α^* agrees with φ.

PROOF For each $i \in \{1 .. d\}$, let b_i be a word in F such that $\varphi(a_i M_1) = b_i M_2$. Lemma 9.15 implies that

$$\varphi(R/M_1) = \varphi(P_c(F/M_1)) = P_c(\varphi(F/M_1)) = P_c(F/M_2) = R/M_2.$$

Therefore φ induces an automorphism, α, on F/R.

For each automorphism, α, of F/R we shall now describe how to define an automorphism α^* of F/R^*. For each $i \in \{1 .. d\}$, choose a representative u_i in F of the coset $(a_i R)^\alpha$; then $(a_i R)^\alpha = u_i R$. Let $v(a_1, \ldots, a_d)$ be a word in F; then $(v(a_1, \ldots, a_d)R)^\alpha = v(u_1, \ldots, u_d)R$. If $v(a_1, \ldots, a_d)$ is an element of R, then $R^\alpha = v(u_1, \ldots, u_d)R$. But $R^\alpha = R$ and so $v(u_1, \ldots, u_d)$ is in R. Since $R^* = [R, F]R^p$, it follows that if $w(a_1, \ldots, a_d) \in R^*$ then $w(u_1, \ldots, u_d)$ is also an element of R^*.

Assume that $w_1(a_1, \ldots, a_d)R^* = w_2(a_1, \ldots, a_d)R^*$ where each w_i is a word in F. Then $w_2(a_1, \ldots, a_d)^{-1}w_1(a_1, \ldots, a_d) \in R^*$. Hence $w_1(u_1, \ldots, u_d)R^* = w_2(u_1, \ldots, u_d)R^*$.

We can now define the mapping α^* as follows: for each word $w(a_1, \ldots, a_d)$ in F, put

$$(w(a_1, \ldots, a_d)R^*)^{\alpha^*} = w(u_1, \ldots, u_d)R^*.$$

Clearly, α^* is a homomorphism and it remains to show that it is onto. But α is an automorphism of F/R and $(a_iR)^\alpha = u_iR$; therefore, F/R^* is generated by $\{u_1R^*, \ldots, u_dR^*, R/R^*\}$. Since $R/R^* \leq P_1(F/R^*)$, it follows that

$$F/R^* = \langle (a_1R^*)^{\alpha^*}, \ldots, (a_dR^*)^{\alpha^*} \rangle.$$

Hence α^* is an automorphism of F/R^*.

While α^* is not uniquely determined by α, its restriction to R/R^* is. Assume that $(a_iR)^\alpha = u_iR = u_ir_iR = v_iR$ for some $r_i \in R$. Then there are two automorphisms, α_1^* and α_2^*, where $(w(a_1, \ldots, a_d)R^*)^{\alpha_1^*} = w(u_1, \ldots, u_d)R^*$ and $(w(a_1, \ldots, a_d)R^*)^{\alpha_2^*} = w(v_1, \ldots, v_d)R^*$. Since each α_i^* is an automorphism, $(R/R^*)^{\alpha_i^*}$ equals R/R^*. Restricting both automorphisms to R/R^* shows that both $w(u_1, \ldots, u_d)$ and $w(v_1, \ldots, v_d)$ are elements of R. But words in R are products of pth powers and commutators; since $[v_j, v_i]R^* = [u_j, u_i]R^*$ and $v_i^pR^* = u_i^pR^*$, it follows that $w(u_1, \ldots, u_d)R^* = w(v_1, \ldots, v_d)R^*$. Hence the restriction of α^* to R/R^* is uniquely determined by α.

It remains to establish that $(M_1/R^*)^{\alpha^*}$ is equal to M_2/R^*. Let $w(a_1, \ldots, a_d)$ be an element of M_1 and let $\hat{\alpha}^*$ denote the restriction of α^* to R/R^*; then

$$(w(a_1, \ldots, a_d)R^*)^{\hat{\alpha}^*} = w(b_1, \ldots, b_d)R^*.$$

Further

$$\begin{aligned}
w(b_1, \ldots, b_d)M_2 &= w(b_1M_2, \ldots, b_dM_2) \\
&= w(\varphi(a_1M_1), \ldots, \varphi(a_dM_1)) \\
&= \varphi(w(a_1, \ldots, a_d))\varphi(M_1) \\
&= \varphi(M_1) \\
&= M_2.
\end{aligned}$$

It follows that $(M_1/R^*)^{\hat{\alpha}^*}$ is a subgroup of M_2/R^* and, since both have the same index in F/R^*, they are equal. ∎

We have the immediate corollary.

COROLLARY 9.26 *Every automorphism α of F/R extends to an automorphism α^* of F/R^* and the restriction of α^* to R/R^* is uniquely determined by α.*

PROOF   The results follows from Theorem 9.25, where both M_1/R^* and M_2/R^* are chosen to be R/R^*.   ∎

The automorphism α^* is an *extended automorphism*. We now show that it permutes the allowable subgroups.

LEMMA 9.27   *Each extended automorphism α^* induces a permutation of the allowable subgroups.*

PROOF   The nucleus, $P_c(F/R^*)$, of G is characteristic in G and the p-multiplicator, R/R^*, is fixed by α^*. Let M/R^* be an allowable subgroup. Then

$$(M/R^*)^{\alpha^*} P_c(F/R^*) = ((M/R^*)P_c(F/R^*))^{\alpha^*} = R/R^*$$

showing that $(M/R^*)^{\alpha^*}$ is an allowable subgroup.   Clearly the mapping is one-to-one and onto.   ∎

The permutation α' of the allowable subgroups induced by α^* depends only on the automorphism α of G. Let P be the permutation group generated by the α'. Then the mapping $\alpha \longmapsto \alpha'$ is a homomorphism from Aut G onto P.

We now establish the critical result about equivalence classes of the allowable subgroups.

THEOREM 9.28   *The orbits of the allowable subgroups under the action of P are exactly the equivalence classes of the allowable subgroups.*

PROOF   Let M_1/R^* and M_2/R^* be allowable subgroups in the same equivalence class; then F/M_1 and F/M_2 are isomorphic. By Theorem 9.25, there exists an automorphism, α^*, of F/R^* which maps M_1/R^* to M_2/R^* and α^* induces a permutation α' of the allowable subgroups. Thus, $(M_1/R^*)^{\alpha'} = M_2/R^*$ showing that M_1/R^* and M_2/R^* lie in the same orbit.

In order to establish the converse, it is simpler to use the following general result, which can be established easily. Let N be a normal subgroup of a group H and let γ be an automorphism of H; then $H/N \cong H/N^\gamma$. Now let M_1/R^* and M_2/R^* be allowable subgroups that are elements of the same orbit under P. Then there exists a permutation, α', of the allowable subgroups such that $(M_1/R^*)^{\alpha'} = M_2/R^*$. This permutation is induced by an automorphism, α^*, of F/R^* where $(M_1/R^*)^{\alpha^*} = M_2/R^*$. We use the general result to deduce that there is an isomorphism from F/M_1 to F/M_2.   ∎

In summary, the p-group generation algorithm takes as input a consistent power-conjugate presentation for a p-group and a generating set for its automorphism group. It produces consistent power-conjugate presentation for representatives of the isomorphism classes of immediate descendants of the group. (In practice, concurrently, we construct the automorphism groups of the immediate descendants, hence permitting recursion. See Section 9.4.6 for more details.) In applying the algorithm, there is a natural division of the calculation of the immediate descendants according to their order. A top-level outline of the algorithm is the following.

GENERATE-P-GROUPS(G, A, s)

> **Input**: Group G of order p^n, its automorphism group A, $s \in \mathbb{N}$
> **Output**: Immediate descendants of G having order p^{n+s}

1   Construct a consistent power-conjugate presentation for the
    p-covering group G^* of G;
2   Initialize L to be the empty list;
3   **if** the order of the nucleus of G is less than p^s **then return** L;
4   **for** each generator α of A
5   **do** Calculate its extended automorphism α^*;
6   Calculate the permutation α' of the allowable subgroups
            of index p^{n+s} in G^* induced by α^*;
7   Calculate orbits of the group generated by the permutations α';
8   **for** each orbit
9   **do** Choose a representative;
10   Factor G^* by the representative allowable subgroup
            to obtain an immediate descendant H;
11   Append H to L;
12 **return** L;

Example 9.10

We construct the immediate descendants of the elementary abelian group $G = C_2 \times C_2$ having power-conjugate presentation

$$\langle\, a_1, a_2 \mid a_1^2 = 1,\ a_2^2 = 1,\ [a_2, a_1] = 1 \,\rangle.$$

Its 2-covering group G^* is

$$\mathrm{Pc}\langle\, a_1, \ldots, a_5 \mid a_1^2 = a_4,\ a_2^2 = a_5,\ [a_2, a_1] = a_3 \,\rangle.$$

The 2-multiplicator $\langle a_3, a_4, a_5 \rangle$ is elementary abelian and it coincides with the nucleus. Hence every proper subgroup of the 2-multiplicator supplements the nucleus and so is allowable.

The automorphism group of G is of course isomorphic to $\mathrm{GL}(2,2)$ and we choose as its generators

$$\alpha_1 : \begin{aligned} a_1 &\longmapsto a_1 a_2, \\ a_2 &\longmapsto a_2 \end{aligned} \qquad \alpha_2 : \begin{aligned} a_1 &\longmapsto a_2 \\ a_2 &\longmapsto a_1. \end{aligned}$$

The extensions of these automorphisms to G^* are:

$$\alpha_1^* : \begin{aligned} a_3 &\longmapsto a_3, \\ a_4 &\longmapsto a_3 a_4 a_5 \\ a_5 &\longmapsto a_5 \end{aligned} \qquad \alpha_2^* : \begin{aligned} a_3 &\longmapsto a_3 \\ a_4 &\longmapsto a_5 \\ a_5 &\longmapsto a_4 . \end{aligned}$$

Assume we wish to construct the immediate descendants of order 8. Then the seven allowable subgroups each have rank 2 and are

$$\langle a_4, a_5 \rangle, \ \langle a_4, a_3 a_5 \rangle, \ \langle a_3 a_4, a_5 \rangle, \ \langle a_3, a_5 \rangle, \ \langle a_3, a_4 a_5 \rangle, \ \langle a_3, a_4 \rangle, \ \langle a_3 a_4, a_3 a_5 \rangle$$

The orbits of the allowable subgroups induced by α_1^* and α_2^* are

$$\{\langle a_4, a_5 \rangle, \langle a_4, a_3 a_5 \rangle, \langle a_3 a_4, a_5 \rangle\}, \ \{\langle a_3 a_4, a_3 a_5 \rangle\}, \ \{\langle a_3, a_5 \rangle, \langle a_3, a_4 a_5 \rangle, \langle a_3, a_4 \rangle\}.$$

We choose one representative from each orbit and factor it from G^* to obtain three immediate descendants:

$$\begin{aligned} &\text{Pc}\langle a_1, a_2, a_3 \mid [a_2, a_1] = a_3 \rangle \\ &\text{Pc}\langle a_1, a_2, a_3 \mid a_1^2 = a_3, \ a_2^2 = a_3, \ [a_2, a_1] = a_3 \rangle \\ &\text{Pc}\langle a_1, a_2, a_3 \mid a_1^2 = a_3 \rangle. \end{aligned}$$

These immediate descendants are, of course, well-known: the dihedral and quaternion groups of order 8, and the abelian group $C_2 \times C_4$, respectively.

Now we consider the construction of the immediate descendants of $C_2 \times C_2$ having order 16. Generators for the seven cyclic allowable subgroups are

$$a_3, \ a_3^\delta a_4^\gamma a_5, \ a_3^\zeta a_4,$$

where each of δ, γ, ζ is 0 or 1. The orbits of the allowable subgroups induced by α_1^* and α_2^* are

$$\{\langle a_3 \rangle\}, \ \{\langle a_5 \rangle, \langle a_3 a_4 a_5 \rangle, \langle a_4 \rangle\}, \ \{\langle a_4 a_5 \rangle, \langle a_3 a_5 \rangle, \langle a_3 a_4 \rangle\}.$$

We choose one representatives from each orbit to obtain 3 immediate descendants of order 16. If, for example, we factor G^* by a_3 then we obtain the immediate descendant

$$\text{Pc}\langle a_1, a_2, a_3, a_4 \mid a_1^2 = a_3, \ a_2^2 = a_4 \rangle,$$

which is $C_4 \times C_4$.

Observe that $C_2 \times C_2$ has just one immediate descendant of order 2^5, obtained by factoring G^* by the trivial allowable subgroup. □

A detailed description of the algorithm appears in the articles by Newman [New77] and O'Brien [O'B90].

9.4.5 Testing finite p-groups for isomorphism

The isomorphism problem of determining whether two given presentations present the same group was introduced by Tietze in [Tie08] and formulated by Dehn in [Deh11]. It is proved by Adian in [Adi58] and by Rabin in [Rab58] that the isomorphism problem for finitely presented groups is unsolvable, by exhibiting its unsolvability for a particular class of examples. However, Segal proves in [Seg90] that there is an algorithm to decide the isomorphism of two polycyclic-by-finite groups given by finite presentations.

In [HR92], Holt and Rees seek to establish isomorphism by running a Knuth-Bendix procedure on the supplied group presentations, in an attempt to generate a normal form algorithm for words in the generators. Concurrently, they attempt to establish nonisomorphism of the two groups by finding the number of finite quotients each has of a particular order.

In [O'B94], O'Brien describes an algorithm that solves the problem for finite p-groups. He defines a *standard presentation* for each p-group and provides an algorithm for its construction. Hence, given two p-groups presented by arbitrary finite presentations, the determination of their isomorphism is essentially the same problem as the construction of their standard presentations and the easy comparison of these presentations.

One view of the p-group generation algorithm is that it is a method for constructing a particular power-conjugate presentation for a given p-group, G. If G has Frattini rank d and class c, then $G/P_1(G)$ is the elementary abelian group of order p^d; further G is a descendant of this elementary abelian group and $G/P_{i+1}(G)$ is an immediate descendant of $G/P_i(G)$ for $i < c$.

Assume we now construct the immediate descendants of $G/P_1(G)$. Among these immediate descendants is the class two quotient, $G/P_2(G)$, of G. It is now possible to calculate the immediate descendants of $G/P_2(G)$ in order to obtain a power-conjugate presentation for the class three quotient of G. We may iterate this construction until we construct the class c quotient of G. Therefore it is possible to construct G by iterating a method for calculating immediate descendants, starting with the elementary abelian group of rank d. We designate the presentation obtained by constructing a power-conjugate presentation for a given p-group using the p-group generation algorithm in this way as the *standard presentation* for this group. For a detailed description of the algorithm, see [O'B94].

An alternative approach for p-groups, involving modular group algebra techniques, is described by Wursthorn in [Wur93].

9.4.6 Automorphism groups of finite p-groups

The automorphism group of a p-group can be constructed by induction down the lower exponent-p central series of a p-group G; that is, we successively compute $\mathrm{Aut}(G/P_i(G))$. For a detailed description of the algorithm, see the paper by Eick, Leedham-Green, and O'Brien [ELGO02].

9.4.7   Applications

The Burnside problem, first posed by W. Burnside in 1902, asks whether a finitely generated torsion group is necessarily finite and, more specifically, whether the *Burnside group*

$$B(d,n) := \langle\, x_1, \ldots, x_d \mid w^n = 1 \ \forall w \in F(\{x_1, \ldots, x_d\})\,\rangle$$

is finite. It is now known that $B(d,n)$ is finite for $n = 2, 3, 4, 6$ and is infinite for $d \geq 2$ and sufficiently large n. Indeed, this is the case for all $n \geq 8000$ and for odd $n \geq 665$; see [Adi79, Iva94, Lys96]. However, despite intensive investigation, much of it computational, it is not yet known if $B(2,5)$ is finite.

One early motivation for the development of the p-quotient algorithm was the restricted Burnside problem, which asks if the order of a finite group with d generators and exponent n is bounded by a function of d and n; that is, whether there is an upper bound on the order of the finite quotients of $B(d,n)$. Results of P. Hall and G. Higman [HH56] reduced this to the case when $n = p^e$ is a prime power. In [Zel91a, Zel91b], using Jordan algebra techniques, Zel'manov provided a positive answer to the question; Vaughan-Lee [VL93] obtains similar results using Lie algebras.

Hence, given d and n, there exists a finite group $R(d,n)$, the *restricted Burnside group* with d generators and exponent n, such that all other finite groups with at most d generators and exponent dividing n are homomorphic images of $R(d,n)$. A natural question is: what is the order (and structure) of this group? While this remains a very hard problem, we are able to construct consistent power-conjugate presentations for some of the groups and so deduce structural information in these limited cases. The property $x^n = 1$ for all $x \in R(d,n)$ can be enforced via a *finite* list of relations which the p-quotient at each successive class must satisfy. In summary, the Higman lemma (see [Hig59], or [Sim94, Sec. 11.7] for a practical realization) states that a group of class c described by a consistent power-conjugate presentation has exponent p^m if every normal word of weight at most c has order dividing p^m.

In Table 9.1 we record the orders of $R(d,n)$ for selected values of d and n. The order of $R(2,4) = B(2,4)$ was determined by S.J. Tobin in 1954; in 1902, Burnside proved in [Bur02] that its order divided 2^{12}. Early constructions of power-conjugate presentations by machine were for $R(3,4)$ in [BKW74] and $R(2,5)$ in [HWW74]. As one indicator of the difficulty involved in obtaining such descriptions, a consistent power-conjugate presentation for $R(2,7)$ was constructed by O'Brien and Vaughan-Lee [OVL02] only after using a special purpose implementation of the p-quotient algorithm and approximately one year of CPU time.

M.F. Newman has been the driving force in the development and application of the p-quotient algorithm to this problem. In work spanning more than 30 years, and with various collaborators, including Havas, O'Brien, and Vaughan-Lee, he established the remaining results cited in Table 9.1 and many others. For a survey of the known results, see [VLZ99].

d	n	Order
2	4	2^{12}
3	4	2^{69}
4	4	2^{422}
5	4	2^{2728}
3	5	5^{2282}
2	7	7^{20416}

Table 9.1: Orders of $R(d, e)$

Newman also used the p-quotient algorithm to prove that $F(2, 9)$ is infinite. The Fibonacci group $F(2, n)$ is generated by $\{a_1, a_2, \ldots, a_n\}$ with defining relations $a_i a_{i+1} = a_{i+2}$, $i \in \{1 .. n\}$, with subscripts read modulo n. These groups were introduced by J.H. Conway in 1965, who asked whether $F(2, 5)$ is cyclic of order 11. By 1988, the (in)finiteness of all $F(2, n)$ except $F(2, 9)$ had been decided. In [New90], Newman introduced a refinement of the famous Golod-Shafarevich bound (which states that the number of defining relations needed for a finite p-group defined minimally on d generators is larger than $d^2/4$) and combined this with the result of p-quotient calculations to deduce that $F(2, 9)$ is infinite.

In more detail, he applied the p-quotient with prime 5 to a subgroup of index 152 in $F(2, 9)$. As with the applications of the abelian quotient algorithm discussed in Subsection 9.3.3, the first parts of this calculation involved applying the low-index subgroups and Reidemeister-Schreier algorithms.

In their construction of a counterexample to the class-breadth conjecture (see [LGNW69]), Felsch and Neubüser [FN79] used the p-quotient algorithm extensively to construct presentations for finite 2-quotients of infinite finitely presented groups.

In contrast to (some of) the Burnside computations, most applications of the p-quotient algorithm are very routine and take almost no resources in time or space to compute a consistent power-conjugate presentation for a p-quotient of a finitely presented group. The Havas, Newman, and O'Brien implementation is available in MAGMA, GAP, and as a stand-alone. For more details of the current algorithm and its implementation, see [NO96] and [NNN98].

The p-group generation algorithm has been applied successfully to determine the p-groups of orders up to 2000. It has also played an important role in verifying (and usually correcting) results of various theoretical classifications. In Table 9.2 we report the number of 2-groups of orders 2^k for $k \in \{1 .. 10\}$; these were first obtained using the p-group generation algorithm. Further details on the classification of all groups of order up to 2000 will be given in Section 11.4.

The p-group generation algorithm has also been an important tool in ob-

Order	Number
2^1	1
2^2	2
2^3	5
2^4	14
2^5	51
2^6	267
2^7	2 328
2^8	56 092
2^9	10 494 213
2^{10}	49 487 365 422

Table 9.2: The number of groups for selected orders

taining insight into the the structure of p-groups, particularly those of fixed *coclass*. It played a fundamental role in shaping the conjectures of Leedham-Green and Newman in their seminal study [LGN80] of p-groups of given co-class; and it was also used by Newman and O'Brien [NO99] in their study of 2-groups of coclass 3.

The most significant limiting factor in applications of the p-group generation algorithm is the degree of the permutation group defined by the action of the automorphism group on the allowable subgroups: this grows very rapidly with the prime and Frattini rank. Various practical steps are taken to reduce this limitation: these include exploiting the characteristic subgroup structure of the p-multiplicator. For further details see the papers of O'Brien [O'B90, O'B91]. The algorithm is an important tool for investigating small Frattini-rank p-groups. Implementations of the p-group generation algorithm are available in **GAP** and MAGMA.

Certain variations of the p-group generation algorithm, which make use of the Cauchy-Frobenius formula (Lemma 2.17) to *count* the precise number up to isomorphism of p-groups of a given order, have also been developed; see the article by Eick and O'Brien [EO99] for further details.

Chapter 10

Advanced Computations in Finite Groups

In this chapter, we shall be introducing a particular approach to some more advanced and sophisticated types of computations in finite groups, such as finding all subgroups of a particular type, or calculating automorphism groups. While most of these applications, such as automorphism groups, lie beyond the scope of this book, we shall illustrate the techniques involved with a reasonably detailed description of a method for finding representatives of the conjugacy classes of subgroups in a finite group.

The approach does not yet have a standard name, but for the purpose of the discussion in this chapter, we shall call it the *solvable radical* method, since the first step is always to find the solvable radical of the group.

We shall, for the most part, restrict our attention to permutation groups, but it is a feature of the method that it is to a large extent independent of the representation type of the finite group, provided that the more basic facilities are available for computing within that representation type.

In summary, we compute the solvable radical $L := \mathbf{O}_\infty(G)$ of G, and a series

$$1 = N_0 \lhd N_1 \lhd \cdots \lhd N_r = L \lhd G$$

of subgroups N_i of L for which each $N_i \unlhd G$ and N_i/N_{i-1} is elementary abelian. We then solve our problem first in $G/L = G/N_r$, and then, successively, we solve it in G/N_i, for $i = r-1, \ldots, 1, 0$.

For finite solvable groups defined by PC-presentations, the method can generally be used, but with the added simplification that $L = G$, and so the solution of the problem in G/L is generally trivial! Indeed, some of the examples, such as computing maximal subgroups or automorphism groups, can be seen as direct generalizations of the methods already described for solving those problems in solvable groups in Chapter 8. It should be noted, however, that the methods described there have been more carefully designed and optimized in the PC-presentation context, and so they will generally perform more efficiently in that setting than the general-purpose methods to be described in this chapter.

Some applications of the solvable radical method involve two other characteristic subgroups of G, in addition to L. In the first section, we shall define these subgroups, and then describe how to compute them.

In the following section, we shall describe how these subgroups can be used as a step towards computing chief and composition series of a finite permutation group. Another important ingredient of the chief series algorithm is the 'Meataxe' algorithm that was introduced in Section 7.4 for finding a composition series of a module for a group over a finite field.

It is interesting to observe that our approach is slightly different from that described by Seress in Chapter 6 of [Ser03], who advocates computing a composition series first, before computing subgroups such as $\mathbf{O}_p(G)$.

We shall then give a brief overview of the current range of algorithms that use the solvable radical method, and a more detailed description of the conjugacy classes of subgroups application.

10.1   Some useful subgroups

10.1.1   Definition of the subgroups

The author has used the following nonstandard terminology in a number of his recent papers, and will continue to do so here!

DEFINITION 10.1   A group is called a *TF-group* if it has no solvable normal subgroup.

Equivalently, a TF-group is one with no nontrivial nilpotent normal subgroup, which is equivalent to its having trivial Fitting subgroup; hence the name.

For a finite group G, let $L := \mathbf{O}_\infty(G)$ be the largest normal solvable subgroup of G. Then G/L is a TF-group.

Let $M/L := \mathrm{soc}(G/L)$ be the socle of G/L (Definition 2.39). Then, since M/L is a TF-group, it has no abelian minimal normal subgroups, and Proposition 2.40 tells us that M/L is the direct product of the minimal normal subgroups M_1/L, $M_2/L, \ldots, M_m/L$ of G/L. Furthermore, each M_i/L is itself the direct product of isomorphic finite nonabelian simple groups S_{ij} $(1 \le j \le m_i)$ which, by Corollary 2.41, are permuted transitively under the conjugation action of G/L, and hence of G, on M_i/L.

In other words, if we let Δ be the set containing all of the groups S_{ij}, then conjugation by G defines an action $\psi : G \to \mathrm{Sym}(\Delta)$ of G on Δ in which the orbits are the sets $\{\, S_{ij} \mid 1 \le j \le i_m \,\}$ for $1 \le i \le m$.

Let $K := \mathrm{ker}(\psi)$. Since M/L is a direct product of nonabelian simple groups, we have $\mathbf{Z}(M/L) = 1$, and hence $\mathbf{C}_{G/L}(M/L) = 1$, because otherwise $\mathbf{C}_{G/L}(M/L)$ would contain minimal normal subgroups of G/L disjoint from

M/L. So conjugation by G also induces a monomorphism $\alpha : G \to \mathrm{Aut}(M/L)$, and $\alpha(K)$ is contained in the subgroup of $\mathrm{Aut}(M/L)$ that fixes all of the S_{ij}. Hence $\alpha(K)$ is contained in the direct product of the groups $\mathrm{Aut}(S_{ij})$, and $\alpha(K)/\alpha(M)$ is contained in the direct product of the groups $\mathrm{Out}(S_{ij})$.

Now the Schreier conjecture, which states that the outer automorphism group of any nonabelian finite simple group is solvable, was confirmed by the classification of finite simple groups, and so $K/M \cong \alpha(K)/\alpha(M)$ is solvable.

10.1.2 Computing the subgroups — initial reductions

In this and the following two sections, we shall describe an algorithm for computing the subgroups that we have just defined in the case when G is a permutation group on a finite set Ω. There are a number of slightly different approaches to this problem. We shall take the easiest possible route here, although the algorithm to be described runs in polynomial time and is reasonably efficient in practice. Some parts of it are similar to the method described by Beals in [Bea93b].

In this subsection, we reduce the problem to that of finding a nontrivial proper normal subgroup in a primitive TF-group. We use the algorithm described in Subsection 4.7.5 to compute $L = \mathbf{O}_\infty(G)$ and then, by Corollary 4.15, we can compute an epimorphism $\rho : G \to H$, where $H \cong G/L$ and H is a permutation group of degree at most $|\Omega|$. This effectively reduces the problem to the TF-case, so we shall assume for the rest of the section that $\mathbf{O}_\infty(G) = 1$.

Now, provided that we can find the set Δ of the direct factors S_{ij} of $\mathrm{soc}(G)$, we can immediately compute the group M that they generate, and it is straightforward to compute the homomorphism $\psi : G \to \mathrm{Sym}(\Delta)$ and its kernel K. We can also compute the minimal normal subgroups M_i of G since, by Corollary 2.41, these correspond to the orbits of the action of G on the socle factors.

So we have now reduced the problem of finding the subgroups defined in Subsection 10.1.1 to that of finding the simple factors of the socle of a TF-group G.

LEMMA 10.2 *Let G be a TF-group and let $N \trianglelefteq G$. Then $\mathrm{soc}(G) = \mathrm{soc}(N) \times \mathrm{soc}(\mathbf{C}_G(\mathrm{soc}(N)))$.*

PROOF By Corollary 2.42, the simple factors of $\mathrm{soc}(G)$ and of $\mathrm{soc}(N)$ are just the simple subnormal subgroups of G and of N, respectively. Hence $\mathrm{soc}(N) \leq \mathrm{soc}(G)$ and similarly $\mathrm{soc}(\mathbf{C}_G(\mathrm{soc}(N))) \leq G$. Any simple factors of $\mathrm{soc}(G)$ that are not in $\mathrm{soc}(N)$ lie in $\mathbf{C}_G(\mathrm{soc}(N))$ and, since $\mathrm{soc}(N) \cap \mathrm{soc}(\mathbf{C}_G(\mathrm{soc}(N))) = 1$, the result follows. ∎

So, if we can find any proper nontrivial normal subgroup of G, then we can apply our algorithm recursively to N to compute the simple factors of $soc(N)$ and then compute $\mathbf{C}_G(soc(N))$ and the simple factors of its socle, thereby solving the problem in G. (As we remarked in Subsection 4.6.5, the centralizer in a permutation group G of a normal subgroup H of G can be found in polynomial time; see [Ser03, Sec. 6.1.4].)

This enables us to reduce immediately first to the case when G^Ω is transitive, and then to the case when G^Ω is primitive. If, for example, G^Ω is intransitive and Γ is a nontrivial orbit, then either G^Γ is faithful and we can replace G by the transitive group G^Γ, or the kernel of G^Γ is a proper nontrivial normal subgroup of G. The reduction to a primitive action is similar.

Before presenting the algorithm in the primitive case, we digress to discuss an important theorem.

10.1.3   The O'Nan-Scott theorem

The O'Nan-Scott theorem is a classification of finite primitive permutation groups according to the type and action of their minimal normal subgroups. It was first stated and proved by Scott in [Sco80]. There are many versions, which differ slightly in the subdivision of the cases, and in how much detail they describe the permutation actions involved.

DEFINITION 10.3   A group G with $S \leq G \leq \mathrm{Aut}(S)$ for some finite nonabelian simple group S is called *almost simple*.

So case (ii)(a) with $d = 1$ in our statement of the theorem below occurs when G is almost simple, and this is often singled out as a separate case. Detailed discussions and proofs can be found, for example, in Chapter 4 of [DM96] and in Chapter 4 of [Cam99].

THEOREM 10.4   (the O'Nan-Scott theorem) *Let G be a finite primitive permutation group on Ω. Then one of the following holds.*

(i) *G has an elementary abelian regular normal subgroup M, and M is the only minimal normal subgroup of G.*

(ii) *G has a unique minimal normal subgroup M, and M is nonabelian and isomorphic to $S_1 \times \cdots \times S_d$ for some $d \geq 1$, where the S_i are isomorphic simple groups. Furthermore, one of the following holds for $\alpha \in \Omega$.*

   (a) *$M_\alpha = H_1 \times \cdots \times H_d$, where each $H_i = (S_i)_\alpha$, and the H_i are isomorphic to each other with $\mathbf{N}_G(H_i)$ maximal in $\mathbf{N}_G(S_i)$. We have $|\Omega| = |S_1 : H_1|^d$.*

   (b) *We have $|\Omega| = |S_1|^{(a-1)b}$, for integers a, b with $ab = m$, $a > 1$ and $b \geq 1$, $M_\alpha = D_1 \times \cdots \times D_b$ where D_i is a diagonal subgroup of*

$S_{(i-1)a+1} \times \cdots \times S_{ia}$, and G acts imprimitively by conjugation on $\{ S_i \mid 1 \le i \le d \}$ with minimal block system

$$\{ \{ S_{(i-1)a+1}, \ldots, S_{ia} \} \mid 1 \le i \le b \}.$$

(c) M acts regularly on Ω, so $|\Omega| = |M|$. We have $d \ge 6$ in this case.

(iii) G has exactly two minimal normal subgroups M_1 and M_2, which are nonabelian, isomorphic to each other, and act regularly on Ω.

This theorem has played an important rôle in the recent development of algorithms for analyzing the structure of permutation groups. Generally, one tries to reduce to the primitive case, and then to use the O'Nan-Scott theorem.

For our immediate purposes, we shall just need the following easy corollary. We leave the proof as an exercise.

COROLLARY 10.5 *If G is a finite primitive permutation group on Ω with no regular normal subgroup, then we are in case (ii)(a) or (ii)(b) of Theorem 10.4. Furthermore, if $\mathrm{soc}(G) = G$, then we are in case (ii)(a) with $d = 1$, and G is simple.*

10.1.4 Finding the socle factors – the primitive case

Our algorithm for finding the socle factors in a primitive group G depends on the O'Nan-Scott theorem and the following lemma, which is a simplified version of Lemma 6.2.22 of [Ser03].

LEMMA 10.6 *Let G^Ω be primitive with $\alpha \in \Omega$, let M be either a solvable normal subgroup or a minimal normal subgroup of G_α, and assume that $\langle M^G \rangle = G$. Then*

(i) *If M is solvable then $G/\mathrm{soc}(G)$ is solvable.*

(ii) *If M is not solvable then either $\mathrm{soc}(G) = G$ or $\mathrm{soc}(G)_\alpha = \mathbf{C}_{G_\alpha}(M)$.*

PROOF Since G^Ω is primitive, $\mathrm{soc}(G)^\Omega$ is transitive, so $M\,\mathrm{soc}(G)$ is normal in G, and hence $\langle M^G \rangle = G$ implies that $M\,\mathrm{soc}(G) = G$. So $G/\mathrm{soc}(G)$ is isomorphic to the quotient group $M/(M \cap \mathrm{soc}(G))$ of M, which proves (i). From $M\,\mathrm{soc}(G) = G$ we have $M\,\mathrm{soc}(G)_\alpha = G_\alpha$. If $M \le \mathrm{soc}(G)$, then $\mathrm{soc}(G) = G$. Otherwise, since M is minimal normal in G_α, $M \cap \mathrm{soc}(G)_\alpha = 1$, so $G_\alpha = M \times \mathrm{soc}(G)_\alpha$ in which case $\mathrm{soc}(G)_\alpha = \mathbf{C}_{G_\alpha}(M)$. $\blacksquare$

Now let G^Ω be primitive. We are assuming that $\mathbf{O}_\infty(G) = 1$, so we cannot be in case (i) of the O'Nan-Scott theorem. As explained earlier, it suffices to find any proper nontrivial normal subgroup of G, or to prove that G is simple.

Now G^Ω cannot be regular so, if we can find a regular normal subgroup of G, then we are done. We can use the method described in Subsection 4.7.6 to test for the existence of such a subgroup, and to find such a subgroup if it exists.

So we may assume that G has no regular normal subgroup, in which case Corollary 10.5 tells us that we are in case (ii)(a) or (ii)(b) of the O'Nan-Scott theorem. We now use the algorithm P-CORE in Subsection 4.7.4 to decide whether $\mathbf{O}_p(G_\alpha) \neq 1$ for any prime p.

If so, choose such a prime p, let $M := \mathbf{O}_p(G_\alpha)$, let $K := \langle M^G \rangle$, compute the final term $N := K^\infty$ in the derived series of K, and return N. To justify this step, we need to show that either $N < G$, or $N = G$ and G is simple. So assume that $N = G$. Then $K = G$ and, by the lemma above, $K/\operatorname{soc}(G)$ is solvable. Since $K = K^\infty$, K has no nontrivial solvable quotients, so $G = K = \operatorname{soc}(G)$, and then G is simple by Corollary 10.5.

So we may assume that $\mathbf{O}_\infty(G_\alpha) = 1$; that is, G_α is a TF-group. We observed in Subsection 10.1.2 that, once we have found the socle factors of a TF-group, it is straightforward to compute its minimal normal subgroups. So we can apply our socle factor algorithm recursively to find a minimal normal subgroup M of G_α.

Again we compute $K := \langle M^G \rangle$. If $K \neq G$, then we can just return K. Otherwise, by the lemma above, either $\operatorname{soc}(G) = G$ or $\operatorname{soc}(G) = \langle \mathbf{C}_{G_\alpha}(M)^G \rangle$. In any case, we return $\langle \mathbf{C}_{G_\alpha}(M)^G \rangle$. If this group is also equal to G then, as before, $G = \operatorname{soc}(G)$ is simple.

We shall now summarize the algorithm.

NORMALSUBGROUP(G)

> **Input**: Primitive permutation group $G \leq \operatorname{Sym}(\Omega)$ with $\mathbf{O}_\infty(G) = 1$
>
> **Output**: N with $1 \neq N \trianglelefteq G$, and $N < G$ unless G is simple

1   Using the algorithm described in Subsection 4.7.6, decide whether G has
        a regular normal subgroup N;
    If so, then **return** N;
2   Let $\alpha \in \Omega$; Using the algorithm pCore in Subsection 4.7.4 decide
        whether $\mathbf{O}_p(G_\alpha) \neq 1$ for any prime p;
    If so, then let $1 \neq M := \mathbf{O}_p(G_\alpha)$ and **return** $\langle M^G \rangle^{(\infty)}$;
3   Now $\mathbf{O}_\infty(G_\alpha) = 1$; Let M be a minimal normal subgroup of G_α;
4   **if** $\langle M^G \rangle < G$, **then return** $\langle M^G \rangle$;
5   **return** $\langle \mathbf{C}_{G_\alpha}(M)^G \rangle$;

In an implementation, the performance can be improved considerably by imposing a limit on the degree $|\Omega|$ of G and keeping a table of the degrees that can actually occur in the different cases of the O'Nan-Scott theorem. For example, the current MAGMA implementation, which works for permutation degrees up to 10^7, is described by Cannon and Holt in [CCH97a], and the current GAP implementation is described in Section 6.2.5 of [Ser03]. By using results of Kantor from [Kan91], a list of the degrees up to 10^7 that can occur in case (ii)(a) was computed. The smallest degree for which case (ii)(c) occurs is 60^6, so this can be ignored and, in cases (ii)(b) and (iii), the degree must be a power of the order of a finite simple group.

Exercise

1. Prove Corollary 10.5.

10.2 Computing composition and chief series

In this section, we address the question of how to compute composition and chief series of a finite permutation group $G \leq \mathrm{Sym}(\Omega)$. We shall assume that the subgroups L, M, M_i, S_{ij}, K of G that were defined in Subsection 10.1.1 have been computed already, together with their associated homomorphisms $\rho : G \to H$ and $\psi : G \to \mathrm{Sym}(\Delta)$, where $H \cong G/L$ is a permutation group of the same degree as G, and $\ker(\psi) = K$.

The groups L, M, and K are characteristic in G, and so the problem reduces to finding composition or chief series of the sections L, M/L, K/M and G/K of G. Now $\mathrm{im}(\psi) \cong G/K$ is a permutation group of degree $|\Delta|$, which is always much smaller than $|\Omega|$, so can handle G/K by recursion. As for M/L, the subgroups M_i and S_{ij} can be used in the obvious way to construct a chief or composition series, respectively.

The remaining sections L and K/M of G are abelian, so we need to refine these sections appropriately.

10.2.1 Refining abelian sections

Let $A < B$ be normal subgroups of a permutation group G with B/A abelian. Refining B/A to elementary abelian sections is straightforward. We just compute the derived series of B (cf. Subsection 3.3.3) and use this to compute the descending series

$$B \geq AB^{[2]} \geq AB^{[3]} \cdots .$$

We can stop when we reach A, and remove any duplicate terms from this series. We have then refined B/A to a series with abelian sections.

So we may assume that B/A is abelian. Now just choose a prime p dividing $|B/A|$ and compute $C := AB^p$, where $B^p := \langle g^p \mid g \in \mathrm{Generators}(B) \rangle$. Then B/C is elementary abelian and, by repeating this process with B replaced by C, we can refine B/A to elementary abelian sections.

So now we may assume that B/A is an elementary abelian p-group for some prime p. For some applications, it is important to observe that, since we have been using characteristic subgroups of G during the refinement process, B and A are still normal subgroups of G, and indeed they are characteristic subgroups if the original B and A were.

If our aim is to find a composition series of G, then it is now easy to refine B/A to factors of order p. We simply adjoin generators of B to A one at a time to produce the subnormal subgroups of G in the refinement. So we end up with $B = \langle A, x_1, x_2, \ldots, x_d \rangle$, where $|B/A| = p^d$.

If we want a chief series, then we can use the following result to convert our problem into a problem about $\mathbb{F}_p G$-modules. The proof is straightforward, and is left as an exercise.

PROPOSITION 10.7   *Let $A < B$ be normal subgroups of a group G, where B/A is an elementary abelian p-group of order p^d for some prime p. Then B/A can be regarded as a vector space of dimension d over $\mathbb{F}_p$, and the action of G on B/A induced by conjugations makes B/A into an $\mathbb{F}_p G$-module. Furthermore, if $A \leq C \leq B$, then C/A is an $\mathbb{F}_p G$-submodule of B/A if and only if $C \trianglelefteq G$.*

To make the correspondence between B/A and the associated $\mathbb{F}_p G$-module explicit, we let $V := \mathbb{F}_p^d$. We change generators of B to a set containing generators of A together with the elements $x_1, \ldots, x_d$ defined above, and define a homomorphism $\tau : B \to V$ from the multiplicative group B to the additive group V with $A = \ker(\tau)$, and $\tau(g_i)$ equal to the i-th standard basis vector of V.

To make V into an $\mathbb{F}_p G$-module, we need to compute the $d \times d$ matrices $\alpha_1, \ldots, \alpha_r$ that define the action of the generators $g_1, \ldots, g_r$ of G on V. This is straightforward since, for each g_i, the j-th row of α_i is given by $\tau(x_j^{g_i})$.

Since normal subgroups C/A of B/A correspond to $\mathbb{F}_p G$-submodules of V, to calculate a refinement of B/A for a chief series of G, we need to find a composition series

$$1 < V_1 < V_2 < \cdots < V_s = V$$

of the $\mathbb{F}_p G$-module V, and then the groups $\tau^{-1}(V_i)$ for $1 \leq i < s$ will be the required normal subgroups of G in the refinement.

The composition series of the module can be found by repeated use of the MEATAXE algorithm, as described in Section 7.4. Since the dimension d of the module is typically rather small when considered as a problem in module theory, this calculation will generally be very fast and efficient.

10.2.2   Identifying the composition factors

We have now described how to find composition and chief series of each of the sections L, M/L, K/M and G/K of G, so we have completed our description of an algorithm to find such series for any finite permutation group G.

We have made no use in this algorithm of the classification of finite simple groups although, as we remarked at the end of Section 10.1, the MAGMA and GAP implementations use lists of possible permutation degrees, which do depend upon the classification.

For most of the more advanced applications of these techniques, however, we want to make use of known properties, often stored in databases, of the various types of finite simple groups. In order to do this, we must first identify the isomorphism types of the nonabelian composition factors as abstract finite simple groups.

The result below makes this problem a great deal easier than one might have expected! It says that, with the exception of one infinite family of examples and one isolated example, a finite characteristically simple group is determined up to isomorphism by its order. This was proved for the classical finite simple groups by Artin in [Art55] and extended to all finite characteristically simple groups using the classification of finite simple groups by D. N. Teague (see Note (ii) at the end of Cameron's paper [Cam81]).

PROPOSITION 10.8   *Let G_1 and G_2 be finite characteristically simple groups with $|G_1| = |G_2|$. Then one of the following holds:*

(i) $G_i \cong \mathrm{Alt}(8) \cong \mathrm{PSL}(4,2)$, $G_j \cong \mathrm{PSL}(3,4)$, *with* $\{i,j\} = \{1,2\}$;

(ii) $G_i \cong \mathrm{PSp}(2m,q)$, $G_j \cong O(2m+1,q)$ *for some* $m \geq 3$ *and odd* q, *with* $\{i,j\} = \{1,2\}$;

(iii) $G_1 \cong G_2$.

The extension from simple to characteristically simple groups is not actually needed in the methods that we have been describing, because we find the nonabelian composition factors at the same time as we find the chief factors. But, in practice, it can be used to speed things up, since in some situations we find the socle of a primitive TF-group before its direct factors, and it can be helpful to know the orders of these direct factors in advance.

Exercise

1. Prove Proposition 10.7.

10.3   Applications of the solvable radical method

The solvable radical method has been used in algorithms for solving the following problems for a finite group G:

- Finding representatives of the conjugacy classes of elements of G: Cannon and Souvignier [CS97].

- Finding all normal subgroups of G: Cannon and Souvignier [CS98].

- Finding representatives of the conjugacy classes of all subgroups of G: Cannon, Cox, and Holt [CCH01] (for permutation groups) and Hulpke [Hul99a] (for PC-groups).

- Finding representatives of conjugacy classes of maximal subgroups of G: Cannon and Holt [CH04] and Eick and Hulpke [EH01a].

- Computing $\mathrm{Aut}(G)$: Cannon and Holt [CH03] (for permutation groups) and Smith [Smi94] (for PC-groups).

For the first two of these, it is not necessary to identify the isomorphism type of the TF-group G/L. As we shall see in the next section, the algorithm for finding representatives of the classes of all subgroups is only designed for groups in which $|G/L|$ is moderately small ($|G/L| < 216\,000$ in the current implementation in MAGMA). We first identify the isomorphism type of G/L, and then set up an isomorphism between G/L and a standard isomorphic copy stored in a database, and extract all of the necessary information about the subgroups of G/L from this database.

The applications for computing maximal subgroups, automorphism groups, and testing groups for isomorphism are rather more complicated, and are designed to work with much larger groups G/L. The details of these applications are beyond the scope of this book but, for the lower dimensional families of simple groups of Lie type, such as $\mathrm{PSL}(d, q)$ for $d \leq 5$, we aim to be able to handle all groups in that family that arise as composition factors of G/L in a uniform fashion. These applications generalize the corresponding algorithms for finite solvable groups described in Sections 8.10 and 8.9 respectively, in the sense that their 'lifting through the elementary layers' parts are very similar to those for solvable groups.

As we saw above, identifying the isomorphism types of simple groups is not a problem; the difficult part of the problem is to find an explicit isomorphism from the given simple group to a standard copy of the group.

Fortunately, there is some prospect of this problem being solved in a satisfactory fashion for all finite simple groups within the next few years. This will be as a particular case of the project to identify finite simple groups given as black-box groups (see Subsection 3.1.4). An advantage is that the methods will apply equally well to finite groups in representations other than permutation groups, such as matrix groups over finite fields.

Methods for the black-box recognition of symmetric and alternating groups are described by Bratus and Pak in [BP00] and for the classical groups by Kantor and Seress in [KS01]. Certain difficult cases, such as the unitary groups and exceptional groups of Lie types, are currently being studied in detail, and improved methods are being developed for them.

10.4 Computing the subgroups of a finite group

In this section, as an example of the solvable radical method, we shall describe, in some detail, an algorithm for computing representatives of the conjugacy classes of subgroups of a finite group G. This description is taken mainly from the paper by Cannon, Cox, and Holt [CCH01]. We shall describe the algorithm for a finite permutation group G. Essentially the identical method for PC-groups is described by Hulpke in [Hul99a], but of course the computations in G/L do not arise in the PC-group setting.

As another important ingredient, this application makes use of the cohomological algorithms described in Section 7.6.

Previously, the only genuinely feasible approach to finding all subgroups of a group was based on the *cyclic extension method*. This is one of the oldest existing group theoretical algorithms, first introduced by Neubüser in [Neu60]. The idea is to start by finding all cyclic subgroups of prime order, which we regard as forming the first layer in the lattice. The subgroups H in subsequent layers are formed from those K in the previous layer as $H := \langle K, g \rangle$, where $K \triangleleft H$. Clearly only the solvable subgroups will be found by this method. The insolvable subgroups can be constructed in a corresponding manner by using the perfect subgroups as the first layer. For this, a complete stored list of the isomorphism classes of finite perfect groups up to a given order n is required, and the subgroup lattice found by this method can only be guaranteed correct for groups of order up to n. Chapter 6 of Butler's book [But91] gives a detailed description of this approach and further references.

The cyclic extension method is fast for most groups of order up to 1000, and is tolerable for many groups up to order 10 000. One major barrier to extending the range of applicability of the cyclic extension method is the policy of storing all perfect groups, which becomes impractical with increasing order. There are 66 isomorphism classes of perfect groups of order up to 10 000. These were categorized by Sandlöbes in [San81], and they are stored in the latest implementations that use this approach. There are a further 163 perfect groups up to order 50 000, but at order 61 440, the exact number becomes unknown. (See Chapter 5 of the book by Holt and Plesken [HP89] for tables of finite perfect groups.)

In the method to be described here, we start by computing $L := \mathbf{O}_\infty(G)$ together with the homomorphism $\rho : G \to H$, where H is a TF-group isomorphic to G/L. As described in Subsection 10.2.1, we can then refine L to a series

$$1 = N_0 \triangleleft N_1 \triangleleft \cdots \triangleleft N_r = L \triangleleft G$$

of normal subgroups of G in which each N_{i+1}/N_i is elementary abelian. It is not absolutely necessary for the factors N_{i+1}/N_i to be chief factors of G, although the algorithm becomes impractical if they are too large (2^8 is about the limit for the prime 2, for example), so we may sometimes need to use the module-theoretical methods described in Subsection 10.2.1 to further refine elementary abelian sections.

For TF-groups, we either store representatives of the conjugacy classes of all subgroups (after computing them once and for all, usually using the cyclic extension method) or, for larger groups, where this would be expensive in storage space, we store this information just for the maximal subgroups. In the latter case, we can find representatives of all classes when required, by applying the algorithm recursively to the maximal subgroups, and then testing for conjugacy of the resulting subgroups in G. This information is all stored in a database of TF-groups.

In the current MAGMA implementation, all subgroups or just the maximal subgroups are stored for each of the 154 TF-groups of order less than 216 000. For larger groups, more sophisticated methods are now available for finding maximal subgroups, which can be used in a recursive computation of all subgroups.

Having found class representatives of subgroups of G/N_r, we lift these successively through the layers, finding conjugacy class representatives of subgroups of G/N_{i-1} from those of G/N_i.

The cohomological calculations involved in the lifting process require a finite presentation for each subgroup in the previous layer that is being lifted. These need to be stored along with the subgroups themselves for the TF-groups, and we need to be able to compute presentations of the lifted subgroups of G/N_{i-1} from those of G/N_i. The stored presentations of the subgroups of the TF-groups were computed using Cannon's algorithm, which was described in Section 6.1.

In the following two subsections, we shall describe how we identify the TF-factor G/L and extract its subgroups from the database, and how we find the subgroups of G/N_{i-1} from those of G/N_i.

10.4.1 Identifying the TF-factor

To use the information for the TF-group isomorphic to G/L that is stored in the database, we first need to identify the isomorphism type of $H \cong G/L$ and then to define an explicit isomorphism between our group H and the group in the database.

For the isomorphism type, knowledge of the order is sufficient in most cases. There are a few orders, however, for which there are many distinct TF-groups. For example, there are 10 of order 120 960. We found that the sum of the orders of the conjugacy classes (see paper by Cannon and Souvignier [CS97]) was a useful and easily calculated invariant, which successfully distinguishes between isomorphism types in all cases up to order 216 000.

Assuming we have located the correct group X in the database, we now need to construct an isomorphism $\varphi : X \to H$. For a conjugacy class x^X of X, let us call the order $|x|$ of x and the length $|x^X|$ of the class the *parameters* of that class. Each TF-group X in the database is stored as a finitely presented group, and the parameters of the classes of the generators of X are also recorded. (In fact, X has just two generators in all cases with $|X| < 216\,000$ except $\text{Sym}(5) \times \text{P}\Gamma\text{L}(2,9)$, for which three are necessary.) The first generator x_1 of X has always been selected such that there is a unique class of X having the parameters of x_1^X. This means that $\varphi(x_1)$ can be chosen as a representative of the unique class in H having the same parameters as x_1^X.

For the images of the remaining generators x_i $(i > 1)$ of X, we choose random members of those conjugacy classes in H that have the same parameters as x_i^X, and for each such choice we check to see whether the defining relators of X are satisfied by the selected images $\varphi(x_i)$ in H. If so, then we must also check that the $\varphi(x_i)$ generate H. If this is the case then, by Theorem 2.52, we have found the required isomorphism $\varphi : X \to H$. The generators x_i $(i > 1)$ of X have been selected in such a way as to minimize the expected searching time for an isomorphism.

For small and moderately large examples, the database contains generators and defining relators for a representative of each conjugacy class of subgroups of X. These generators are stored as words in the x_i and their inverses, and we apply the map φ to each of these words to give us elements that generate the corresponding subgroups of H. The inverse images of these subgroups in G, which are needed as the initial input to the lifting part of the algorithm to find the subgroups of G, can then be found by using the map $\rho : G \to H$ mentioned above.

For larger examples, space is saved by storing only the representatives of the conjugacy classes of maximal subgroups Y of X. The main algorithm can then be applied recursively to the subgroups $\varphi(Y)$ of L. After doing this for each such Y, we have a collection of subgroups S of H which includes conjugacy class representatives of all subgroups of H. Some of these may be conjugate to each other, so we have to test each such subgroup for conjugacy with the earlier subgroups in the list of the same order.

10.4.2 Lifting subgroups to the next layer

In this subsection, we assume that G is a group with normal subgroups N and M, such that $N < M$ and M/N is an elementary abelian p-group of order p^d for some prime p and some integer $d > 0$. We assume that conjugacy class representatives of the subgroups of G/M are known, together with finite presentations for each of these subgroups on suitable generating sets.

More precisely, for each such subgroup S/M of G/M, we have elements $g_i \in G$ such that $S/M = \langle g_i M \rangle$, together with a set of words in the g_i and their inverses which, taken modulo M, form a set of defining relators for S/M. These words evaluate in G to elements of M. We shall describe our algorithm for computing the same information for G/N.

As described in Subsection 10.2.1, we can regard $V := M/N$ as the d-dimensional vector space $\mathbb{F}_p^d$ over $\mathbb{F}_p$, and then make V into an $\mathbb{F}_p G$-module via the conjugation action of G on M. Since M acts trivially on V, we can equivalently regard V as an $\mathbb{F}_p G/M$-module.

In fact, it is convenient to start by numbering all subgroups of M/N (or, equivalently, subspaces of V) in some standard fashion, and setting up a permutation action of G acting by conjugation on these subgroups.

Clearly, if two subgroups T_1/N and T_2/N of G/N are conjugate in G/N, then $T_1 M/M$ and $T_2 M/M$ are conjugate in G/M. It therefore suffices to take each class representative S/M of G/M in turn, and to compute class representatives of those subgroups T/N of G/N for which $TM/N = S/M$. From now on, we shall therefore take S/M to be a fixed subgroup of G/M, where $S = \langle M, g_1, \ldots, g_m \rangle$.

Although the computations relate to the quotient group S/M, they are all carried out using the subgroup S of G, and there is no need to construct the quotient group explicitly.

We first compute $R := \mathbf{N}_G(S)$; so R/M is the normalizer of S/M in G/M. The idea then is to find all possible intersections $J := T \cap M$ for subgroups T/N with $TM = S$, and then to take each such J in turn, and to find all T that have that particular intersection. Of course J/N must be normalized by S/N and, since we are only looking for conjugacy class representatives of subgroups, we must only consider one such group J from each orbit of the action of R/M on the S/N-invariant subgroups of M/N.

To find the groups J to be considered, we use the permutation action of G on the subgroups of M/N computed earlier, and then take our subgroups J/N to be orbit representatives of the orbits of R on the fixed points of S. So, from now on, we can fix our attention on a particular subgroup J/N of M/N normalized by S/N, and we shall look for those subgroups T/N of G/N such that $TM = S$ and $T \cap M = J$. We denote the stabilizer of J in R (that is, $\mathbf{N}_R(J)$) by Q.

The required subgroups T are precisely the complete inverse images in G of the complements T/J of M/J in S/J (if any). These are found using the cohomology calculations described in Subsections 7.6.1 and 7.6.2. Of course, we must first set up the section $W := M/J$ as an $\mathbb{F}_p S/M$-module, but we have already seen how to do that.

In fact, these calculations give us conjugacy class representatives of the complements under conjugation by M/J. If the subgroups T_1/N and T_2/N of G/N resulting from two such representatives T_1/J and T_2/J are conjugate in G/N by an element gN, then g must normalize S and J, and so $g \in Q$. We therefore need to compute the conjugation action of Q on the set of these complements.

We experimented with two approaches to this calculation. The first, and simpler, is to test all such pairs T_1 and T_2 for conjugacy within the permutation group Q, by using the standard default backtrack search method. The second is to use a cohomological calculation.

The chief disadvantage of this second method is that it appears to necessitate doing calculations within the regular representation of the quotient group S/M, and so we are forced to form this representation explicitly. (Of course, this is only necessary when Q is strictly larger than S. If, as often happens, $Q = S$, then we omit this part of the calculation.) However, our current experience indicates that this second approach is superior in general, mainly because the direct conjugacy test occasionally takes a very long time indeed.

The cohomological method of testing for conjugacy is rather technical, and the following brief description could be safely omitted by the reader.

We start by finding a set of elements h_j of Q such that the elements h_jS generate Q/S. This set should be as small as possible. For each such h_j and each of the generators g_k ($1 \le k \le m$) of S modulo M, we need to calculate a word $\sigma_{jk}(g_i)$ in the elements g_i and their inverses, such that $h_j^{-1}g_kh_jM = \sigma_{jk}(g_i)M$. It is to calculate these words that we need to construct the regular permutation representation of S/M.

Note that, if we conjugate back again by h_j^{-1}, then we get equations of the form $h_j\sigma_{jk}(g_i)h_j^{-1} = g_ky_{jk}$ ($1 \le k \le m$) for elements y_{jk} of M that we can calculate.

We now take each of our complements T/J in turn. Our aim is to compute and identify the complement $h_jTh_j^{-1}/J$ for each h_j. The complement T/J is defined by its corresponding one-cocycle in $Z^1(S/M, M/J)$, which consists essentially of vectors $w_i \in W := M/J$ for $1 \le i \le m$, such that T/J is generated by g_iw_i. Since the elements $\sigma_{jk}(g_i)M$ ($1 \le k \le m$) generate S modulo M, if we substitute g_iw_i for g_i in the words $\sigma_{jk}(g_i)$ for $1 \le k \le m$, then we obtain another set of generators for T/J, which have the form $\sigma_{jk}(g_i)w_{jk}$, for certain vectors $w_{jk} \in W$ that we can calculate.

If we now conjugate this new set of generators by h_j^{-1}, then we obtain generators for $h_jTh_j^{-1}$ modulo J, in the form $g_ky'_{jk}w'_{jk}$ ($1 \le k \le m$), where $y'_{jk}, w'_{jk} \in W$ and, modulo J, $h_j\sigma_{jk}(g_i)h_j^{-1} = g_ky'_{jk}$ and $h_jw_{jk}h_j^{-1} = w'_{jk}$. Note that the vectors y'_{jk} are obtained simply by taking the elements y_{jk} defined in the last subsection modulo J, and do not depend on the particular complement T/J, whereas the vectors w'_{jk} are obtained by applying the matrix for the action of h_j^{-1} on W to the vectors w_{jk}.

We now have generators for the conjugated complement $T'/J := h_jTh_j^{-1}/J$ in the form $g_iw'_i$, for vectors $w'_i \in W$. Then, as described in Subsection 7.6.1, the vectors w'_i ($1 \le i \le m$) can be combined to form a single row vector in $\mathbb{F}_p^{md}$, where $|W| = p^d$, and this vector is used in the cohomology algorithms to represent the element of $Z^1(T/M, W)$ corresponding to the complement T'/J. Hence we can identify the element of $H^1(T/M, W)$ defined by this complement.

These calculations enable us to set up the permutation action, induced by conjugation, of the group Q/S on the set of conjugacy class representatives under M of complements of M/J in S/J. Orbit representatives of this permutation action provide us with the required conjugacy class representatives in G of those subgroups T/N of G/N with $TM = S$ and $T \cap M = J$.

As the final part of the lifting process, we need to calculate presentations of the lifted subgroups T/N of S/M. As before, let g_i be the generators of S modulo M. Then the generators of T will consist of $g_i w_i$ and v_j, where w_i are elements of M that are computed in the cohomology calculation, and v_j generate J modulo N. (So the v_j form a basis of J/N regarded as a subspace of $V = M/N$.)

The defining relators of T/N are constructed as specified in Proposition 2.55, where T/N is regarded as an extension of J/N by S/M. They fall into three classes. The first consist of commutators and p-th powers of the v_j, to force them to generate an elementary abelian subgroup. The second specify the conjugation action of the g_i on the v_j; these are routine to calculate. For the third, we take each defining relator r of S/M and evaluate r in G, using $g_i w_i$ in place of $g_i M$ in r. The resulting element lies in J, and so it can be written (modulo N) as a word $v(r)$ in the generators v_j. Then we take $rv(r)^{-1}$ as the corresponding defining relator of T/N.

Exercise

1. Use the method described above to compute the conjugacy classes of subgroups of the symmetric group Sym(4).

10.5 Application – enumerating finite unlabelled structures

In Subsection 4.8.2, we discussed some techniques for using permutation groups G to construct examples of combinatorial structures that contain G in their automorphism groups. Here we shall briefly discuss the more general problem of counting and enumerating all such structures of a given type. We shall only have space to scratch the surface of this active and important area of research. Fortunately, there is an up-to-date book by Adalbert Kerber [Ker99] to which we can refer the reader for full details.

This area of research has important applications outside of mathematics, principally to the systematic enumeration of chemical isomers; that is, chemical compounds with the same formula, such as C_3H_7OH, but with different molecular structures. Within mathematics, the techniques have been used to construct many examples of 7–designs, and even some 8–designs (t–(n, k, λ) designs were defined in Subsection 4.8.2).

The fundamental result on which much of the theory is based is the Cauchy-Frobenius lemma, Lemma 2.17. Here is a simple and well-known example of how it is applied. Consider necklaces, consisting of n coloured beads spaced at regular intervals around a circular band. A natural question to ask, is how many different types of necklace can be made using n beads with r possible colours. Two necklaces are the same if one can be transformed into another by rotating it or turning it over (reflecting it).

We approach the problem as follows. Before considering the effect of rotations and reflections, we can make r^n different necklaces. Let Ω be a set containing exactly these r^n different necklaces. Now we define an action of the dihedral group D_{2n} on Ω, which is induced in the obvious way from its natural action as the symmetry group of a regular n-sided polygon. So, for example, a rotation ρ through an angle $2\pi/n$ will map a necklace in $\alpha \in \Omega$ to the necklace obtained by applying ρ to α. Then the distinct necklaces correspond exactly to the orbits of D_{2n} on Ω, and we can apply Lemma 2.17 to count the number of such orbits. This calculation still requires some combinatorial analysis, which we omit. The number is

$$\frac{1}{2n}\sum_{d|n}\Phi(d)r^{n/d} + \frac{1}{2}r^{\frac{n+1}{2}} \quad \text{or} \quad \frac{1}{2n}\sum_{d|n}\Phi(d)r^{n/d} + \frac{1}{4}(1+r)r^{\frac{n}{2}},$$

when n is odd and even, respectively, where Φ is the Euler Φ-function.

Let us now focus on a specific aspect of this topic in which computational group theory plays a significant rôle. In general, if the group G acts on the set Ω then, for $\alpha \in \Omega$ and $g \in G$, we have $G_{\alpha^g} = g^{-1}G_\alpha g$ (Exercise 1, Section 2.2). Hence the stabilizers of the points in a particular orbit of G form a conjugacy class of subgroups of G. In many applications, we wish to know not only the total number of orbits, but the total number of orbits for which the point stabilizers lie in some specified conjugacy class of subgroups of G. This information is related to the size of the orbit, and the automorphism group of the members of these orbits.

It can be computed as follows. This method was introduced originally by Burnside. Let the conjugacy classes of subgroups of G be $\{\tilde{U}_1, \ldots, \tilde{U}_d\}$, where $\tilde{U}_i$ has a representative subgroup $U_i \leq G$ and, for $1 \leq i, j \leq d$, let $\zeta(U_i, U_j)$ denote the total number of conjugates U_j^g of U_j for which $U_i \leq U_j^g$. Define $m_{ij} := |\mathbf{N}_G(U_j) : U_j|\zeta(U_i, U_j)$. The $d \times d$ matrix $M(G)$ with entries m_{ij} was called the *table of marks* of G by Burnside. Note that its entries are nonnegative integers. It has also been called the *super character table* of G.

For an action of G on Ω and a subgroup $H \leq G$, define

$$\Omega_H := \{\alpha \in \Omega \mid \alpha^h = \alpha \ \forall h \in H\}.$$

Let v_O be the $d \times 1$ column vector in which the i-th component is the number of orbits of G^Ω with stabilizers in $\tilde{U}_i$, and let v_F be the $d \times 1$ column vector in which the i-th component is $|\Omega_{U_i}|$. Then Burnside's lemma, which is proved in detail Section 4.1 of [Ker99], states that $M(G) \cdot v_O = v_F$, so v_O can be computed as $M(G)^{-1} \cdot v_F$.

To compute the table of marks, we first need to find the conjugacy classes of subgroups of G, which was the topic of Section 10.4.

Computations with characters of the symmetric group is another area of CGT that is important in this topic. We should also mention here some specialized computer algebra packages that are available.

- SYMMETRICA is devoted to the representation theory and combinatorics of finite symmetric groups.

- DISCRETA is for constructing t–(n, k, λ) designs with prescribed automorphism group.

- MOLGEN is for computing structural formulae of chemicals.

Exercise

1. Prove the formula displayed above for the number of necklaces that can be made with n beads of r possible colours. If this is too difficult, try it first when the n is prime.

Chapter 11

Libraries and Databases

The classification of groups with the aid of computers and the electronic publication of the resulting libraries and databases form an important part of computational group theory. 'Classification' of groups here means that the groups with certain properties or parameters are determined up to conjugacy or isomorphism. The algorithms used for these purposes frequently use a large part of the available machinery from CGT.

In this chapter, we shall discuss four such databases in some detail, which currently contain:

- all primitive permutation groups up to degree 1000;

- all transitive permutation groups up degree 30;

- all perfect groups of order up to 10^6, but with seven missing orders;

- all finite groups up to order 2000, excluding the order 1024, but including the 408 641 062 groups of order 1536.

In the final section of this chapter, we shall briefly describe and discuss the vast quantity of data that has been collected together or computed by the authors of the "ATLAS of Finite Groups" [CCN+85] and in its sequel, "An Atlas of Brauer Characters" [JLPW95].

The four databases listed above, and also much of the data printed in the "ATLAS" volumes, are available from both **GAP** and MAGMA. As one would expect from databases, some facilities are also provided to help the user search for groups in the lists with specified properties, and to locate the isomorphic or conjugate copy in the library of a corresponding user-supplied group.

Among the resources that have been provided by the CGT community in recent years, these databases, particularly the list of small groups, probably constitute the facility that is most frequently used by the general mathematical and scientific community. Of course, it is difficult to know exactly how often and by whom they are used, given that many such applications remain unrecorded. The lists are almost certain to be extended in the future; for example, work is already in progress to find the primitive permutation groups up to degree 2500.

The history of group classifications is long, and many publications in this area have mistakes. Of course, the earlier classifications and list compilations, many of which date back to the late 19th century, were undertaken by hand rather than by machine. We refer the reader to the extensive bibliography compiled by Short in [Sho92, Appendix A] for more details on the history of primitive and transitive group classification, and to the article by Besche, Eick, and O'Brien [BEO02] for a history of classifications of groups of small order.

11.1   Primitive permutation groups

The primary aim in the classification of primitive groups is to determine a complete and irredundant list of conjugacy class representatives for the primitive subgroups of $\mathrm{Sym}(n)$ for a given degree n. The generally used approach towards this aim is based on the O'Nan-Scott theorem, which we stated earlier as Theorem 10.4. This theorem divides the primitive permutation groups into two classes: the affine (case (i) of Theorem 10.4) and the nonaffine groups (cases (ii) and (iii) of Theorem 10.4). This is the primary invariant used to classify the primitive permutation groups of a given degree; see Sections 11.1.1 and 11.1.2.

Currently, the primitive permutation groups of degree less than 1000 and the solvable primitive permutation groups of degree less than 6561 have been classified. We shall give a very brief overview of the history of the determination of primitive groups in the following. We refer the reader to [Sho92] for details.

In [Sim70], a list of all primitive permutation groups up to degree 20 is published and Sims also circulated a correct list of groups up to degree 50.

The solvable primitive permutation groups of degree less than 256 were listed in [Sho92], with two omissions. This list has been corrected and extended to the solvable primitive permutations groups of degree less than 6561 in [EH03]. The solvable primitive permutation groups are affine and an overview of a method to determine these groups is included in Section 11.1.1.

There is a description of examples in [DM88] and also in Appendix B of [DM96] with one or two omissions, and without detailed classification of affine examples with EARNS. This uses O'Nan-Scott and the classification of finite simple groups. A list in electronic form was described by Theissen in [The97] and made for **GAP**. This includes the examples from Dixon and Mortimer's list, and the affine examples of degree less than 256.

The remaining affine groups of degrees 256, 625, and 729 were classified by Roney-Dougal and Unger in [RDU03], who also rechecked the entire list of primitive permutation groups of degree less than 1000.

11.1.1 Affine primitive permutation groups

First, we recall the definition and some elementary results on the affine primitive groups.

DEFINITION 11.1 Let G be a primitive permutation group and let S be its socle. If S is solvable, then G is called an affine primitive group.

The following theorem is folklore for the affine primitive groups. See the paper by Liebeck, Praeger, and Saxl [LPS88] for background.

THEOREM 11.2 *Let G be an affine primitive permutation group with socle S and degree n.*

a) *n is a prime power; that is, $n = p^m$ for some prime p and $m \in \mathbb{N}$.*

b) *S is an elementary abelian p-group of order p^m.*

c) *The point stabilizer $K = G_1$ is a complement to S in G.*

d) *The conjugation action of K on S induces an embedding $K \to \mathrm{GL}(m, p)$ with irreducible image.*

Theorem 11.2 d) implies that every affine primitive permutation group of degree p^m yields an irreducible subgroup of the matrix group $\mathrm{GL}(m, p)$. Conversely, every irreducible subgroup of $\mathrm{GL}(m, p)$ determines an affine primitive permutation group in this form and the conjugacy classes of affine primitive subgroups in $\mathrm{Sym}(p^m)$ correspond one-to-one to the conjugacy classes of irreducible subgroups of $\mathrm{GL}(m, p)$. Hence the classification of the affine primitive permutation groups translates into the classification of the irreducible subgroups of $\mathrm{GL}(m, p)$.

An approach to classifying the irreducible subgroups of $\mathrm{GL}(m, p)$ is provided by Aschbacher's theorem [Asc84]; see Subsection 7.8.2 for a statement of the theorem. In particular, an irreducible subgroup G of $\mathrm{GL}(m, p)$ falls into one of the classes (2) to (9) of Aschbacher's theorem. The primary approach for constructing all irreducible subgroups of $\mathrm{GL}(m, p)$ is to determine the conjugacy classes of maximal subgroups for each of the classes (2) to (9). Then each conjugacy class representative is considered in turn and its conjugacy classes of subgroups are determined recursively.

This approach lists each conjugacy class representative of the irreducible subgroups of $\mathrm{GL}(m, p)$ at least once. But, since the classes in Aschbacher's theorem are not disjoint, it may happen that an irreducible subgroup of $\mathrm{GL}(m, p)$ lies in several of these classes and is therefore constructed several times.

Hence it remains to reduce the constructed list of subgroups to conjugacy class representatives. For this purpose we need an effective conjugacy test for subgroups in $\mathrm{GL}(m,p)$. Clearly one can consider $\mathrm{GL}(m,p)$ as a permutation group on $p^m - 1$ points and then use a backtrack approach to check conjugacy; see Section 4.6. However, this approach has turned out to be practical for small prime powers p^m only. For larger p^m there are two methods available for the conjugacy class test: we refer to the papers by Eick and Höfling [EH03] and by Roney-Dougal [RD04] for details.

11.1.2 Nonaffine primitive permutation groups

As mentioned earlier, the nonaffine primitive permutation groups of degree less than 1000 were determined and listed by hand by Dixon and Mortimer in [DM88]; their list is also available as Appendix B of [DM96]. A few errors were later discovered by various people, and the results described in [RDU03] are now generally believed to be correct. (Or rather, almost correct! It has very recently been discovered that $\mathrm{PSL}(2,41)$ in a primitive representation of degree 574 on the cosets of a maximal subgroup isomorphic to $\mathrm{Alt}(5)$ was omitted from the table of the total numbers of primitive groups.)

Dixon and Mortimer used the classification of finite simple groups, the known structure of the automorphism groups of the finite simple groups, the O'Nan-Scott theorem, and information about the maximal subgroups of almost simple groups, which was available principally from results of Cooperstein, Kantor, and Liebeck. In fact nearly all of the information they needed could be found in the "ATLAS" [CCN+85].

They proceeded by considering all possible socles, which are direct products of isomorphic finite simple groups, and collecting together the groups with a common socle into *cohorts*. Most, but by no means all, of the examples have simple socle; this is because other types of examples that occur in Theorem 10.4 tend to have larger degrees. For this case, one just needs to know the finite almost simple groups G that have maximal subgroups H of index less than 1000, and a knowledge of which of the automorphisms of G fix the conjugacy class of H in G. It is important to be aware of the fact that $H \cap \mathrm{soc}(G)$ is not necessarily a maximal subgroup of $\mathrm{soc}(G)$. Such intersections are listed in the "ATLAS", and called *novelties*. We refer the reader to [DM88] for further details.

Exercises

1. Prove Theorem 11.2.

2. Find examples of almost simple groups G having a maximal subgroup H for which $H \cap \mathrm{soc}(G)$ is a proper nonmaximal subgroup of $\mathrm{soc}(G)$.

3. Let G be a primitive subgroup of $\mathrm{Sym}(n)$ for some n with G almost simple, let $S = \mathrm{soc}(G)$, and let N be the normalizer in $\mathrm{Sym}(n)$ of G. Prove that N is isomorphic to a subgroup of the normalizer in $\mathrm{Aut}(S)$ of G, and that

an element α in $\mathbf{N}_{\mathrm{Aut}(S)}(G)$ lies in N if and only if H^{α} is conjugate in N to H, where $H := G_1$ is a point stabilizer.

11.2 Transitive permutation groups

Many attempts to classify transitive groups of small degree (up to 15) were carried out in the late 19th and early 20th centuries. See [Sho92] for details and references. More recently, with assistance from computers, transitive groups up to degree 11 were listed by Butler and McKay [BM83], degree 12 by Royle [Roy87], and degrees 14 and 15 by Butler [But93]. Finally, transitive groups of degree less than 32 were listed in [Hul96] by Hulpke. These are available as libraries in GAP and MAGMA up to degree 30.

There is general confidence that the latest lists are correct, because they have been checked using a variety of different methods. To give the reader an idea of the number of classes of groups involved, the largest number is in degree 24, with (exactly) 25000 groups, and then there are 5712, 2392, 1954, 1854, 1117, and 983 groups of degrees 30, 27, 16, 28, 20, and 18, respectively.

11.2.1 Summary of the method

Here, we shall summarize the method of classification described in full detail by Hulpke in [Hul96] or [Hul]. The aim is to list all transitive permutation groups of a given degree n up to conjugation in $\mathrm{Sym}(n)$. As was observed in Subsection 4.6.5, testing subgroups of $\mathrm{Sym}(n)$ for conjugacy is one of the most difficult computations to perform efficiently, and so it is essential to avoid having to do this wherever possible. Fortunately, the methods to be summarized below can be carried out without resort to direct conjugacy testing except in a few exceptional cases.

Since primitive groups are listed up to much higher degrees, we can assume that these are known, so we need to find the transitive imprimitive groups. We can also assume that the transitive groups of degree less than n are known.

We consider each nontrivial factorization $n = lm$ of n in turn, and list those groups for which the smallest block of imprimitivity has size l. Let G be such a group, let B be a block of imprimitivity of size l in a block system $\mathcal{B}$ of size m with $1 \in B$, let $U = G_1$ and $V = G_B$. Then, by minimality of B, V^B is primitive and so, by Proposition 2.30, U is a maximal subgroup of V. Let $\varphi : G \to \mathrm{Sym}(m) \cong \mathrm{Sym}(\mathcal{B})$ be the induced action of G on $\mathcal{B}$ and let $M := \ker(\varphi)$.

Now we split into two cases, $M = 1$ and $M \neq 1$. The case $M = 1$ is easier, and the associated imprimitive groups are called *inflations* of the isomorphic transitive groups $G^{\mathcal{B}}$. We take each transitive group G of degree m in turn and then find the maximal subgroups U of its point stabilizer V. The image of the action of G on the right cosets of U is a (necessarily faithful) inflation of G of degree n. It can be proved that two such inflations for the same G are conjugate in $\mathrm{Sym}(n)$ if and only if the two corresponding maximal subgroups U_1 and U_2 of V are in the same orbit under the action of $\mathrm{Aut}(G)$. This means that $\mathrm{Aut}(G)$ needs to be computed. For this, the methods of Section 8.9 were used for solvable G and of Subsection 9.1.3 for nonsolvable G.

So suppose that $M \neq 1$. Clearly $M \leq V$. Since V^B is primitive, M^B is transitive by Proposition 2.29, and the maximality of U in V implies that $V = MU$. Also the transitivity of G^B implies that the group actions M^{B_i} are equivalent for all blocks $B_i \in \mathcal{B}$. Finally, $G \leq \mathbf{N}_{\mathrm{Sym}(n)}(M)$ implies that $\mathbf{N}_{\mathrm{Sym}(n)}(M)$ is transitive.

The first and most time-consuming part of the classification procedure in this case is to list all possible groups $M \leq \mathrm{Sym}(n)$ that have all of the properties listed in the preceding paragraph. That is, they have m orbits, $B_1, \ldots, B_m$ each of length l, the actions M^{B_i} are all permutation equivalent, and $\mathbf{N}_{\mathrm{Sym}(n)}(M)$ is transitive. The theory of how this is achieved is technical, and involves the theory of subdirect products. We shall not give any further details here, except to observe that we only need to consider those transitive groups H of degree l for which $\mathbf{N}_{\mathrm{Sym}(l)}(H)$ is primitive as candidates for M^{B_i}.

Now we process each individual candidate group M in turn. First, we compute $N := \mathbf{N}_{\mathrm{Sym}(n)}(M)$, which is transitive by assumption. Furthermore, by Proposition 4.9, N permutes the set $\mathcal{B}$ of orbits of M, and so $\mathcal{B}$ is a system of imprimitivity for N of the required type. As before, we let $\varphi : N \to \mathrm{Sym}(m) \cong \mathrm{Sym}(\mathcal{B})$ be the induced action on $\mathcal{B}$, and compute $R := \varphi(N)$ and $K := \ker(\varphi)$. So R is transitive.

For each group G that we are seeking, $\varphi(G)$ is a transitive subgroup of R, so the next step is to compute representatives T of the conjugacy classes of transitive subgroups of R, and consider each of these in turn as candidates for $\varphi(G)$. This is potentially a difficult computation, but it turns out that, most of the time, R is either small enough for it to create no difficulties, or R is a subgroup of low index in either $\mathrm{Sym}(m)$ or a wreath product $\mathrm{Sym}(r) \wr \mathrm{Sym}(s)$ with $m = rs$ (see Subsection 2.2.5). If $|\mathrm{Sym}(m) : R|$ is small, then we can use the already available list of transitive subgroups of $\mathrm{Sym}(m)$, and efficient methods can also be developed to handle the wreath product case.

For each candidate group T, we compute the inverse image $S := \varphi^{-1}(T)$ of T in N, so we are looking for $G \leq S$ with $\varphi(G) = \varphi(S)$. Since, by assumption, $\ker(\varphi) \cap G = M$, this means that G/M is a complement of K/M in S/M and, conversely, all such complements satisfy the required conditions for G. So we need to find representatives of these complements up to conjugation in $\mathbf{N}_N(S)$. This is a cohomological calculation, and can be carried out in a similar way to the computations of complements described in Subsection 10.4.2 in connection with finding the subgroups of a finite group.

The final problem in the classification of transitive groups of degree n arises from the fact that such a group G can have more than one block system of minimal size, in which case it will be found several times, once for each such block system. To eliminate such repetitions, all block systems with blocks of size l are computed for each G, and considerable effort is devoted to ordering these block systems in such a way that there is a unique 'minimal' block system under this ordering; in that case G is kept on the list if and only if the block system for which it arose is the minimal one. There were, however, a few cases in which no such unique canonical system could be defined, and it was occasionally necessary to resort to direct conjugacy tests in $\mathrm{Sym}(n)$ to decide equivalence of some of the groups G.

11.2.2 Applications

The most important application of the lists of transitive groups is to the calculation of Galois groups $\mathrm{Gal}(f)$ of irreducible polynomials f over the rationals. The most effective practical methods in current use require a complete list of such subgroups of degree equal to $\deg(f)$; so, at the present time, they can only be applied when $\deg(f) \leq 31$.

We can assume that $f \in \mathbb{Z}(x)$ with f monic. Let $n := \deg(f)$. We consider $\mathrm{Gal}(f)$ as a subgroup of $\mathrm{Sym}(n)$ in its action on the n roots $\alpha_1, \ldots, \alpha_n$ of f. We have an action of $\mathrm{Sym}(n)$ on the polynomial ring $\mathbb{Z}[x_1, \ldots, x_n]$ where permutations act by permuting the variables x_i; i.e. $x_i^g = x_{i^g}$ for $g \in \mathrm{Sym}(n)$. Very roughly, for each transitive group G of degree n, we need to find a polynomial in n variables $x_1, \ldots, x_n$ which is invariant under all $g \in G$, but not under all elements of any overgroup of G in $\mathrm{Sym}(n)$. For example,

$$\prod_{1 \leq i < j \leq n} (x_j - x_i)$$

is invariant under $g \in \mathrm{Alt}(n)$ but not under $g \in \mathrm{Sym}(n) \setminus \mathrm{Alt}(n)$.

By computing complex approximations of the roots α_i, we can decide whether this polynomial evaluated with α_i in place of x_i yields a result in $\mathbb{Z}$. If so then, under certain additional hypotheses, we can conclude that $\mathrm{Gal}(f) \leq G$. So, for example, we have $\mathrm{Gal}(f) \leq \mathrm{Alt}(n)$ if and only if the *discriminant*

$$\prod_{1 \leq i < j \leq n} (\alpha_j - \alpha_i)^2$$

is a square in $\mathbb{Z}$.

The algorithm proceeds by working downwards through a descending chain of subgroups of $\mathrm{Sym}(n)$ and halts when we find a subgroup G with $\mathrm{Gal}(f) \leq G$ but $\mathrm{Gal}(f) \not\leq H$ for all maximal subgroup H of G. So, in addition to the list of transitive groups of the given degree n, we require information on containment between the subgroups on the list, which is not completely straightforward because the list only contains subgroups up to conjugacy in $\mathrm{Sym}(n)$.

The above description is unfortunately too brief and approximate to be of much use to readers who are seriously interested! A detailed treatment of degrees $n \leq 7$ can be found in [Coh73, Section 6.3]. Another useful reference for practical computations of Galois groups is the article by Soicher and McKay [SM85]. In [Hul99b], Hulpke introduces some new ideas in Galois group computation, which aim to avoid the necessity of a complete list of transitive subgroups of $\mathrm{Sym}(n)$.

The lists of transitive groups of degree n have also been used by Royle and others in [RP89] and [RM90] for enumerating graphs of degree n with transitive automorphism group. The enumerations proceed essentially by considering all candidates for this automorphism group.

11.3 Perfect groups

We recall from Definition 2.36 that a nontrivial group is said to be *perfect* if it is equal to its own derived subgroup. A list of representatives of the isomorphism classes of finite perfect groups of order at most 10000 was compiled by Sandlöbes, and published in [San81]. This list was recomputed and extended up to the order $1\,000\,000$ by Holt and Plesken in [HP89], but there are seven orders for which their list is incomplete, the smallest of these being $2^{10} \cdot 60 = 61\,440$. Their list agrees with that of Sandlöbes up to order 10000.

The perfect groups in the library are stored by means of finite presentations. This provides a compact method of storage, but is not the most useful for subsequent computations with the groups. The library therefore contains information that enables the user to construct faithful permutation representations of the groups.

In most cases, this information consists of generators for a core-free subgroup of reasonably small index, from which a permutation representation can be found by coset enumeration. For some examples, however, there are faithful intransitive permutation representations of significantly smaller degree than the smallest-degree transitive representation; in these cases, generators for the point-stabilizers of each orbit in these intransitive representations are stored.

If a finite group G has $M := \mathbf{O}_p(G) \neq 1$ and $N := \mathbf{O}_q(G) \neq 1$ for two distinct primes p, q, then G is a subdirect product of G/M and G/N (see Exercise 1 below). From a knowledge of G/M, G/N, and $\mathrm{Aut}(G/MN)$, it is straightforward to reconstruct the possible groups G. In general, there could be more than one isomorphism class of such groups, although this does not happen with perfect groups of order up to a million. It therefore suffices to find the perfect groups in which $\mathbf{O}_p(G) \neq 1$ for at most one prime p.

The first step is to write down all of the perfect groups up to the required order n for which $\mathbf{O}_p(G) = 1$ for *all* primes p; that is, the perfect groups with trivial solvable radical. From the known list of simple groups, it is easy to show that the smallest such group that is not a direct product of finite nonabelian simple groups has order 60^6, which is considerably larger than the orders with which we are concerned. So, this step is straightforward.

We now, recursively, take each perfect group F in turn and, for each prime p with $p|F| \leq n$ and $\mathbf{O}_p(F) = 1$, we construct those perfect groups G for which $G/\mathbf{O}_p(G) \cong F$. For such a group G, the terms in the lower central series of $\mathbf{O}_p(G)$ are characteristic in $\mathbf{O}_p(G)$ and hence normal in G. It follows that the chief factors M/N of G for which $N < M \leq \mathbf{O}_p(G)$ are all centralized under the induced action of $\mathbf{O}_p(G)$, and so they can be regarded as $\mathbb{F}_p F$-modules, where the module action is induced by conjugation in G.

So we can find the groups G as follows. First, we use the methods of Subsection 7.5.5 to find all irreducible $\mathbb{F}_p F$-modules for which the dimension d satisfies $|F|p^d \leq n$. We then initialize the list of groups G to $[F]$ and recursively consider each group G in the list in turn. Then, for each irreducible $\mathbb{F}_p F$-module M of dimension d for which $|G|p^d \leq n$, we regard M as an $\mathbb{F}_p G$-module in which $\mathbf{O}_p(G)$ acts trivially, and compute the second cohomology group $H^2(G, M)$. The corresponding extensions provide us with a complete list of extensions of M by G up to equivalence of extensions. We use the action of the compatible pairs in $\mathrm{Aut}(G) \times \mathrm{Aut}(M)$ on $H^2(G, M)$, as discussed in Section 8.9 in the context of soluble groups, to refine this to a collection of extensions that are mutually nonisomorphic as groups. These can be added to the list of groups G with $G/\mathbf{O}_p(G) \cong F$, provided that they are not isomorphic to any group that is already on the list.

In fact, we know that each isomorphism type of group G will arise once for each of its minimal normal subgroups contained in $\mathbf{O}_p(G)$, and in practice we can use this fact to avoid getting repetitions in the list. We do not need to resort to explicit isomorphism testing between groups.

The principal difficulty that we encountered in carrying out the above program was that, in order to use the author's cohomology programs to compute $H^2(G, M)$, we needed to find a reasonably small-degree permutation representation of the groups G. Unfortunately, the extensions computed by the cohomology programs are output as presentations, and so these permutation representations needed to be found individually for each group. Although this was done with the aid of a coset enumeration program, it required human expertise to find the appropriate subgroups for the coset enumerations.

For two of the classes of extensions, namely $F = \mathrm{Alt}(5)$ and $F = \mathrm{PSL}(3, 2)$, both with $p = 2$, there were simply too many groups arising for it to be practical to complete the lists up to order a million. The lists were successfully computed up to a million for all other classes.

In fact, as the orders grow larger, finding extensions of p-groups by a relatively small perfect group like $\mathrm{Alt}(5)$ starts to acquire similar characteristics to finding all p-groups. As we saw in Subsection 9.4.4, finding all p-groups can be undertaken completely mechanically, using the p-group generation program, and it might be possible to generalize these methods so that they can be applied to extensions of p-groups by a small fixed finite group.

Exercises

1. A *subdirect product* of two groups is defined to be a subgroup of their direct product that projects surjectively onto both factors. Show that, if G is a group with normal subgroups M and N with $M \cap N = 1$, then G is isomorphic to a subdirect product of G/N and G/M.

   Find an example of two nonisomorphic groups G_i with normal nonintersecting subgroups M_i and N_i ($i = 1, 2$), such that $G_1/M_1 \cong G_2/M_2$ and $G_1/N_1 \cong G_2/N_2$. (*Hint*: There is an example of order 273.)

2. Show that the smallest degree of a faithful permutation representation of the group $\mathrm{SL}(2,5) = \langle\, a, b \mid a^4, b^3, (ab)^5, a^2b = ba^2 \,\rangle$ is 24.

3. Show that the group defined by the presentation

   $$\langle\, a, b, w, x, y, z \mid a^4,\, b^3,\, (ab)^5,\, a^2b = ba^2,$$
   $$w^a = y,\, x^a = z,\, y^a = w,\, z^a = x,\, w^b = xz,\, x^b = wxyz,\, y^b = yz,\, z^b = y \,\rangle,$$

   which is perfect of order 1920 and is a split extension of an irreducible module of order 16 by $\mathrm{SL}(2,5)$, has a faithful intransitive permutation representation of degree 40.

11.4 The small groups library

The small groups library, which was compiled principally by Besche, Eick, and O'Brien, provides a classification of all groups of certain 'small' orders. The groups in this library are listed up to isomorphism; that is, for each of the available orders a complete and irredundant list of isomorphism type representatives of the groups is given. At the present time, the library contains all groups of orders:

- at most 2000 except 1024 (423 164 062 groups);

- p^n where p is an odd prime and $n \le 6$;

- $p^n q$ where p^n divides 2^8, 3^6, 5^5 or 7^4, and $p \ne q$ are primes;

- orders that factorize into at most 3 primes.

The first of these covers an explicit range of orders. The last three provide access to infinite families of groups having orders of certain types; see [BE01, NOVL04] for background. In this section we shall discuss the algorithmic methods that were used to determine the groups of order at most 2000 except 1024; see also the papers by O'Brien, Besche, and Eick: [O'B89, BE99b] for first partial results, [BEO01] for an overview, and [BEO02] for more details.

The vast majority of groups of order up to 2000 are solvable, and are stored in the library by means of polycyclic presentations. In fact, given the very large number of groups involved, it was necessary to store them in a highly compressed format. The insolvable groups are stored as permutation groups.

We shall outline an algorithm to determine up to isomorphism the groups of order n for some given n. As a first step, we split the groups of order n into three natural classes: the nilpotent groups, the solvable nonnilpotent groups, and the nonsolvable groups.

We saw in Theorem 2.38 that a finite group is nilpotent if and only if it is the direct product of p-groups for primes p. Thus the construction of the nilpotent groups is based on the determination up to isomorphism of the p-groups of order p^e. In turn, the p-groups of order p^e can be constructed by the p-group generation algorithm; see Subsection 9.4.4.

Effective methods to construct up to isomorphism the solvable nonnilpotent groups of a given order n were introduced by Besche and Eick [BE99a, BE01]. There are two different methods involved: the Frattini extension method and the coprime split extension method. We shall give an outline of the Frattini extension method below. This method is the more general of the two algorithms, as it applies to all possible orders, while the coprime split extension method applies to orders of the type $p^n q$ only, but is more efficient for this particular type of order.

It remains to discuss the construction up to isomorphism of the nonsolvable groups of a given order n. We refer to [BE99a] for an outline and to the thesis of Archer [Arc02] for a more efficient approach to this problem.

If G is a nonsolvable group of order n, and we put $N := G^{[\infty]}$, the last term in the derived series of G as defined in Subsection 2.3.3, then N is perfect and $H := G/N$ is solvable. We use the list of nontrivial perfect groups of order m dividing n as described in Section 11.3 above, and the list of solvable groups of order n/m as constructed with the methods described in this section, to determine the candidates for N and H. Then, given a perfect group N of order m and a solvable group H of order n/m, our aim is to determine up to isomorphism the groups G that are extensions of N by H.

Since H is solvable, it has a subnormal series in which the factors are cyclic of prime order. This enables the problem to be reduced to the so-called *cyclic extension problem*: given a finite group N and a prime p, determine up to isomorphism all groups G having a normal subgroup N with $|G/N| = p$.

To solve this, we first need to determine $\mathrm{Aut}(N)$ (see Section 9.1 or 10.3). Then we find all pairs (α, n) such that $\alpha \in \mathrm{Aut}(N)$, $n \in N$, and α^p is equal to the inner automorphism of N defined by conjugation by n. It can be proved that, for any such pair, there is a cyclic extension G of N with $|G/N| = p$ in which $G = \langle N, \alpha \rangle$ and $\alpha^p = n$, and that all such extensions arise in this fashion.

Finally, we need to determine which of the cyclic extensions constructed in this fashion are isomorphic to each other. As with other enumeration problems, this is the lengthiest part of the computation. It can be done using the general-purpose isomorphism tests discussed in Section 9.1 or 10.3, but there are various methods that can be used to reduce the number of such tests that are necessary in this particular situation; we omit the details.

11.4.1 The Frattini extension method

The Frattini extension method can be used to determine all or certain solvable groups of a given order n. First, we recall from Subsection 2.3.6 that the Frattini subgroup $\Phi(G)$ is the intersection of all maximal subgroups of the finite group G. Further, a group H is called a *Frattini extension* of G if there exists $N \lhd H$ with $N \leq \Phi(H)$ such that $H/N \cong G$. Thus each group G is a Frattini extension of its Frattini factor $G/\Phi(G)$.

We use the following general approach to construct groups of order n:

(1) Determine up to isomorphism candidates F for the Frattini factors of the groups of order n.

(2) For each candidate F:

   (a) Compute the Frattini extensions of order n of F.

   (b) Reduce the resulting list of extensions to isomorphism type representatives.

For any finite group G, if $N \unlhd G$ and $N \leq \Phi(G)$, then $\Phi(G/N) = \Phi(G)/N$ (exercise), and in particular $\Phi(G/\Phi(G))$ is trivial. So a candidate F for a Frattini factor will always have a trivial Frattini subgroup. Hence the Frattini extensions computed in 2 (a) will have F as their Frattini factor.

It follows from the following theorem that the solvability or nilpotence of a group G can be read off from its Frattini factor (see exercise below for proof). Hence the above outline can easily be modified to determine the solvable or the solvable nonnilpotent groups of a given order n only.

THEOREM 11.3   *Let G be a finite group. Then:*

a) G is solvable if and only if $G/\Phi(G)$ is solvable;

b) G is nilpotent if and only if $G/\Phi(G)$ is abelian.

In the following we discuss the steps of the above approach in more detail.

Step (1): Finding the candidates for the solvable Frattini factors F relies on the work of Gaschütz [Gas53]. Let G be a solvable group of order n and $F := G/\Phi(G)$. Then $|F|$ divides n and each prime divisor of n divides $|F|$. Further, the socle $\mathrm{soc}(F)$ is a direct product of elementary abelian groups and has a complement K in F. The socle complement K acts faithfully on $\mathrm{soc}(F)$ and each Sylow p-subgroup of $\mathrm{soc}(F)$ is a semisimple $\mathbb{F}_p K$-module.

We determine candidates for the Frattini factors F of the desired solvable groups by considering all direct products of elementary abelian groups of order dividing n as possible socles S for F. We then construct up to conjugacy all subgroups K of $\mathrm{Aut}(S)$ that have suitable order and act semisimply on S. Finally, we obtain the desired candidates F as $S \rtimes K$. This yields a very effective solution for Step (1).

Step (2a): We compute the Frattini extensions of a candidate F using a recursive approach. Let H be a Frattini extension of F, where $|H|$ divides n. Then every Frattini extension of H is also a Frattini extension of F. Let p be a prime dividing $n/|H|$. Then we compute the irreducible $\mathbb{F}_p H$-modules M up to equivalence; see Plesken's paper [Ple87] for an effective algorithm. For each module M we calculate $H^2(H, M)$ as described in Section 8.7.2. Since M is irreducible, the nontrivial elements of $H^2(H, M)$ correspond one-to-one with the equivalence classes of Frattini extensions of H by M.

This approach yields every Frattini extension of F of order n at least once. However, the iterated computation of equivalence classes of extensions usually produces redundancy, since equivalence of extensions is weaker than group isomorphism. We use the action of the compatible pairs in $\mathrm{Aut}(H) \times \mathrm{Aut}(M)$ on $H^2(H, M)$ as discussed in Section 8.9 to eliminate some of this redundancy.

Step (2b): We reduce the list of groups computed by the above steps to representatives of distinct isomorphism types. One could use a general-purpose isomorphism test at this stage; for example, the automorphism group method of Section 8.9 can be modified to an isomorphism test for two solvable groups. However, in this particular setting it turned out that such an approach would not be efficient enough for the desired applications. We shall now introduce a more effective method, designed for this particular application only.

11.4.2 A random isomorphism test

Suppose that a list $G_1, \ldots, G_r$ of groups of order n is given and we want to determine a subset of this list such that every group G_i in the list is isomorphic to exactly one of the groups in the subset. In the application to the determination of groups of small order, the number of groups r is often large, say several thousand, and their order n is small, say at most 2000. Also, there are typically only relatively few different isomorphism types contained in the list $G_1, \ldots, G_r$. Our approach for this purpose uses two steps:

(1) 'Fingerprinting': For every group $G_1, \ldots, G_r$ evaluate various isomorphism-invariant properties. Split the list up into several sublists such that every sublist contains the groups with certain properties only.

(2) 'Random reduction': Take one of the sublists of (1) and search randomly for isomorphisms between the groups in the list. Whenever an isomorphism is found, discard one of the isomorphic copies from the list.

The hope is that the fingerprinting in Step (1) splits the given list up into sublists that contain few (hopefully just one) isomorphisms types only; that is, Step (1) should get close to finding all isomorphism types. The invariants need to be chosen suitably for this purpose. Step (2) is then a verification of the result of Step (1) and the hope is that in Step (2) almost all groups are discarded from the given list. If Step (2) reduces a given list to a list of length one, then an isomorphism type representative has been obtained.

Step (1): For this step it is most important to find isomorphism-invariant properties of groups that can be evaluated extremely fast in a small group. The invariants used in the construction of the small groups library are based on unions of certain conjugacy classes and power maps. For a more detailed description we refer to [BE99a].

Step (2): This step is the time-critical step in our algorithm. We include a brief and simplified outline here only; the actual implementation of this algorithm is technical and incorporates many heuristic optimizations, so it would go beyond the scope of this book to describe it in detail.

As a precomputation, we determine for each group in the given list a collection of minimal generating sets. These collections of minimal generating sets are chosen in such a form that if two groups G and H in the list are isomorphic, then G and H both have a minimal generating set X_G and X_H in the collection such that X_G maps onto X_H under an isomorphism. Also, the collections should be as short as possible with this property.

Then, we iterated over the given list of groups. In each pass through the loop we select randomly one of the minimal generating sets for each of the groups. The chosen minimal generating set is then enlarged to a polycyclic sequence. As observed in Section 8.1, such a polycyclic sequence defines a unique polycyclic presentation and we encode the relations of this polycyclic presentation as a single integer and store this code with the group. Whenever a new code is determined, then we compare the new code with all known codes. There are two interesting cases:

(1) The new code has already been determined for a different group in the list. Then these two groups have the same polycyclic presentation and hence they are isomorphic. We discard one of the groups and add its codes to the stored codes of the other group.

(2) The new code has already been determined for the same group. Then the group has two minimal generating sets that define the same presentation. Thus we obtain an automorphism of the considered group. We store this automorphism.

We run this loop until either only one group remains or a known code has been determined for every group in the remaining list at least l times for a given value of l. At this stage we use the counting argument based on the automorphism groups of the considered groups as described in [BE01], 5.3, and thus split the remaining list up further. Finally, if lists of length greater than one still remain, then we compute more expensive invariants or we try to prove isomorphisms between the groups using a deterministic isomorphism test for finite solvable groups; see Section 8.9.

We conclude the section with a few final remarks on the method. First, we note that the parameter l controls the probability of finding all possible isomorphisms in the given list. If $l = 0$, then no isomorphism is found. If l tends to infinity, then all isomorphisms are found. For a more detailed analysis of the probabilistic nature of the method we refer to [BE01]. Secondly, we note that the random reduction algorithm performs well when many of the groups in the given list are isomorphic, since in this case the limit l has to be reached for only a few groups. Thus the preceding computation of invariants in Step (1) is important for the performance of the method. Also, it is critical for the performance of this algorithm that the groups are constructed as Frattini extensions, since this ensures that minimal generating sets are exhibited.

Exercises

1. Let N be a finite group, $p \in \mathbb{N}$ and $(\alpha, n) \in (\mathrm{Aut}(N), N)$ such that α^p is the inner automorphism of N induced by n. Show that there is a cyclic extension G of N with $|G/N| = p$ in which $G = \langle N, \alpha \rangle$ and $\alpha^p = n$.

2. For any finite group G, verify that, if $N \trianglelefteq G$ and $N \leq \Phi(G)$, then $\Phi(G/N) = \Phi(G)/N$.

3. Prove Theorem 11.3. (*Hint*: For part a), use the Frattini lemma (Proposition 2.21) to show that, if M/N is a nonsolvable chief factor of G, then $M \nleq \Phi(G)$. For part b), use Theorem 2.38.)

11.5   Crystallographic groups

Crystallographic groups are the symmetry groups of crystals. They are used in the study of crystals in chemistry and physics. Classifications and enumerations of these groups are particularly interesting in this context.

The basic theory of crystallographic groups was introduced in Section 8.2.3. They can be defined as the subgroups G of the group of all Euclidean motions of d-dimensional space $\mathbb{R}^d$ such that the set of translations T in G is a discrete normal subgroup of G of finite index. Thus the translations T form a free abelian normal subgroup in a crystallographic group G. The corresponding factor group G/T is called the *point group* of G.

We call a crystallographic group G a *space group* if the rank of its translation subgroup T is maximal; that is, if $\text{rank}(T) = d$. The parameter d is also known as the *dimension* of a space group G. The following theorem analyzes the structure of a space group.

THEOREM 11.4

 a) *A group G is a space group if and only if there exists a free abelian normal subgroup T in G with G/T finite and $\mathbf{C}_G(T) = T$.*

 b) *If G is a space group of dimension d, then its point group G/T embeds into $\text{GL}(d, \mathbb{Z})$.*

 c) *If G is a space group of dimension d, then G embeds as an affine group into $\text{GL}(d+1, \mathbb{Q})$.*

By Bieberbach's famous theorems, there exist only finitely many conjugacy classes of space groups of dimension d, and two space groups are conjugate if and only if they are isomorphic. Hence one can ask for a classification of the space groups of dimension d, at least for small dimensions d. Such classifications are available for $d \leq 4$ in [BBN+78].

There is a well-known algorithm due to Zassenhaus [Zas48], implemented by H. Brown [Bro69], which can be used to classify the d-dimensional space groups. This algorithm proceeds in the following two steps:

 (1) Determine up to conjugacy the finite subgroups P of $\text{GL}(d, \mathbb{Z})$ and their normalizers $N := \mathbf{N}_{\text{GL}(d,\mathbb{Z})}(P)$.

 (2) For each such finite subgroup P, determine up to isomorphism the extensions of $T \cong \mathbb{Z}^d$ by P.

Step (1) requires a detailed analysis of the finite subgroups of $\text{GL}(d, \mathbb{Z})$ using the notion of Bravais groups and automorphism groups of lattices. Currently, the irreducible maximal finite subgroups of $\text{GL}(d, \mathbb{Z})$ have been determined up to dimension $d \leq 31$; see Nebe [Neb95], and the article by Nebe and Plesken [NP95] for further information. Step (2) can be achieved using a dimension-shifting process, as described in the following theorem.

THEOREM 11.5   Let $P \leq \text{GL}(d, \mathbb{Z})$ with P finite, $N := \mathbf{N}_{\text{GL}(d,\mathbb{Z})}(P)$, $T := \mathbb{Z}^d$, and $V := \mathbb{Q}^d$. Then:

 a) *P acts naturally on T and thus also on V and on V/T. Further, there exists a dimension-shifting isomorphism $H^2(P,T) \cong H^1(P, V/T)$.*

 b) *N acts on $H^1(P, V/T)$ via $\delta^n : P \to V/T : g \mapsto ((g^{n^{-1}})^{\delta})^n$. The isomorphism types of extensions of T by P correspond one-to-one with the N-orbits on the elements of the finite group $H^1(P, V/T)$.*

We also mention an interesting generalization of the crystallographic groups: the *almost crystallographic* groups. These are groups in which the translation subgroup may be any torsion-free nilpotent group, and they are significantly more difficult to classify. An approach towards such a classification is described by Dekimpe in [Dek96]. A library of such groups, maintained by Eick and Dekimpe, is available as the ACLIB package in GAP.

11.6 The "ATLAS of Finite Groups"

The "ATLAS of Finite Groups" [CCN+85], by Conway, Curtis, Norton, Parker, and Wilson, has become one of the most widely cited sources for information about the finite simple groups and their close relatives. It was published in 1985, but the enormous amount of work involved in collecting together and computing the data contained therein had been going on for more than 10 years prior to that time.

Its sequel, "An Atlas of Brauer Characters" [JLPW95], by Jansen, Lux, Parker and Wilson, was published 10 years later and, in addition to the Brauer character tables, contains amendments and additions to the original "ATLAS".

Although these are printed volumes, the vast majority of the data that they contain was computed by machine, using many of the techniques that have been discussed in this book, but also many one-off special-purpose programs written to help complete the computation of the large character tables.

The principal groups covered by the "ATLAS" include all of the 26 sporadic simple groups, and also the smaller members in the infinite families of finite simple groups, the aim of the authors being to include sufficiently many of the small groups to indicate the generic behaviour of groups in that family.

The information on these groups includes a variety of constructions of the groups with information about isomorphism and containment between them, presentations of the groups using generators and relations, their automorphism groups and Schur multipliers, their maximal subgroups (where known), and their character tables. The character tables include information on conjugacy classes, power maps, and the fusion of the classes and characters under group automorphisms. All of this information is printed not only for the simple groups themselves, but also for closely related groups, such as extensions of the simple groups by automorphisms, and cyclic perfect central extensions of the simple groups.

Virtually all of the data in the two "ATLAS" volumes, and much more besides, is now freely available to the mathematical community in electronic form by means of a number of convenient interfaces, including GAP and MAGMA.

Thomas Breuer maintains a library of ordinary and modular character tables of finite groups, which is accessed via GAP. It contains all character tables in the "ATLAS", and many others, including the maximal subgroups of the groups in the "ATLAS"; see the Web site [GAP]. A library of *table of marks* (see Section 10.5) of various finite groups is also available in GAP.

Wilson, Parker, and Bray provide an extensive online library of permutation and matrix representations of the finite simple groups and their variations, featuring explicit matrices and permutations for the group generators; see the Web site [ATL]. Information is also provided on standard generators for the groups, which were discussed in Subsection 7.8.2 in connection with setting up isomorphisms between a standard copy of the group G in question and an arbitrary group known to be isomorphic to G. This database is now also available via MAGMA.

As was mentioned in Section 10.3, MAGMA also has an internal library containing information about the finite simple groups and their variations of order up to about 10^7, including their (maximal) subgroups and outer automorphisms.

Chapter 12

Rewriting Systems and the Knuth-Bendix Completion Process

Coset enumeration, which was studied in detail in Chapter 5, is designed to provide information about the finite quotients of a finitely presented group. It can be used to prove that such a group is finite, but rarely to prove that it is infinite. The algorithms described in Chapter 9 can tell us about other types of quotients of finitely presented groups, such as abelian, nilpotent, or polycyclic quotients, from which we can, on occasion, deduce that the group is infinite.

The methods to be discussed in the final two chapters of this book provide a completely different approach to computing with finitely presented groups. Their principal aim is to find a normal form for group elements, together with an algorithm for putting words in the group generators into normal form. When successful, they enable us to determine the finiteness or infiniteness of the group (and often the finiteness or infiniteness of the orders of elements of the group), since we can generally count the number of distinct normal forms. They may also enable us to compute neighborhoods of the origin in the Cayley graph of the group, and possibly to compute the growth function of the group.

Of course, we can only achieve these aims if the group has a solvable word problem and, as we noted at the beginning of Chapter 5, this is not the case in general. Fortunately, many of the interesting classes of finitely presented groups that arise in practice, particularly those, such as knot groups, that arise from topology or geometry, do turn out to have a solvable word problem. Furthermore, as we shall see in Chapter 13, many of these are automatic groups and are consequently amenable to the computational methods to be described there.

In this chapter we shall present the basic theory of rewriting systems over a fixed finite alphabet, together with its applications to CGT. We shall come to that in Section 12.2. First we need to generalize the basic theory of finitely presented groups to semigroups and monoids.

411

12.1   Monoid presentations

The basic theory of monoid presentations follows the same lines as that of group presentations, which we dealt with in Section 2.4 and, as we shall see, a group presentation can be regarded as a special case of a monoid presentation. However, the method of defining a group presentation using the normal closure of a set of relators in a free group does not carry over directly to monoids; instead, we need to use the idea of a congruence on a monoid. Some of the material in this and later sections of this chapter is based on the presentation in Sims' book [Sim94].

12.1.1   Monoids and semigroups

Roughly, a monoid is a group without inverses, and a semigroup is a group without inverses or an identity. More formally:

DEFINITION 12.1   A *semigroup* is a set S equipped with an associative binary operation $\circ$; that is, $x \circ (y \circ z) = (x \circ y) \circ z$ for all $x, y, z \in S$.

DEFINITION 12.2   A *monoid* is a set M equipped with an associative binary operation $\circ$, together with an identity element $e \in M$ that satisfies $e \circ x = x \circ e = x$ for all $x \in M$.

As we did for groups, we shall generally omit the $\circ$ symbol, and write xy instead of $x \circ y$, and we shall write 1 rather than e for the identity element, or 1_M for the identity of M. For additive examples, we use 0 for the identity.

In fact, we shall not be very much concerned with semigroups here, although the theory of finitely presented semigroups is not really very different from that of monoids. From any semigroup we can immediately define a monoid simply by adjoining an identity element.

A *subsemigroup* of a semigroup S is a subset of S that is itself closed under the binary operation of S, and a submonoid of a monoid M is a subset that is closed under the binary operation and contains the identity element of M.

Here are some examples of semigroups and monoids. The nonnegative integers $\mathbb{N}_0$ form a monoid under addition with identity 0. The nonzero natural numbers $\mathbb{N}$ form a subsemigroup but not a submonoid of $\mathbb{N}$ under addition.

The elements of any ring form a semigroup under multiplication in the ring and, if the ring has a unit element 1, then they form a monoid.

Recall that, if X and Y are sets, then a *relation* from X to Y is simply a subset of $X \times Y$, and the relation R is called a *function* if, for each $x \in X$, there is a unique $y \in Y$ with $(x, y) \in R$. The composition RS of a relation R

from X to Y and a relation S from Y to Z is defined by

$$RS := \{\, (x, z) \mid \exists y \in Y : (x, y) \in R, (y, z) \in S \,\}.$$

If X is a fixed set, then the set $\mathrm{Rel}(X)$ of all relations from X to X forms a monoid under composition, and the set $\mathrm{Fun}(X)$ of all functions from X to X forms a submonoid of $\mathrm{Rel}(X)$. The identity element is of course the identity function $\{\, (x, x) \mid x \in X \,\}$. The set $\mathrm{Sym}(X)$ of all bijections from X to X is a submonoid, and indeed a subgroup, of $\mathrm{Fun}(X)$.

From now on we shall discuss monoids only. The reader should have no difficulty in formulating analogous results for semigroups.

If X is a subset of a monoid M, then the submonoid of M generated by X is defined to be the intersection of all submonoids of M that contain X, and this can easily be shown to be equal to the set of all elements of M that can be written as strings $x_1 \cdots x_r$ with each $x_i \in X$. We allow the empty string ε, which is defined to represent the identity 1_M. A monoid M is said to be *finitely generated* if there exists a finite set X generating M.

Monoid homomorphisms are defined as you might expect. For monoids M and N, a monoid homomorphism from M to N is a function $f : M \to N$ satisfying $f(xy) = f(x)f(y)$ for all $x, y \in M$ and $f(1_M) = 1_N$.

Notice that there is a difference here from group theory in that, for monoids, the property $f(1_M) = 1_N$ does not follow from $f(xy) = f(x)f(y)$ for all $x, y \in M$, and so we can have a map $f : M \to N$ that is a semigroup homomorphism but not a monoid homomorphism. For example, let $M = N = \{0, 1\}$ under multiplication, and let $f(0) = f(1) = 0$. It is easily checked that the image $\mathrm{im}(f)$ of any monoid homomorphism $f : M \to N$ is a submonoid of N.

Now we come to a significant difference from group theory. The kernel of a group homomorphism φ is a particular type of subgroup, from which we can form a quotient group that is isomorphic to $\mathrm{im}(\varphi)$.

We can do something similar for monoids, but not by using a special type of submonoid. The problem is that, although the set of elements $x \in M$ with $f(x) = 1_N$ forms a submonoid of M, unlike in the group case, the pairs $x, y \in M$ for which $f(x) = f(y)$ are not uniquely determined by this subgroup. We need to replace the idea of a special type of subgroup by a certain type of equivalence relation, known as a *congruence* on the monoid.

DEFINITION 12.3   A *congruence* on a monoid M is an equivalence relation $\sim$ on M with the property that $x \sim y$ implies $xz \sim yz$ and $zx \sim zy$ for all $x, y, z \in M$.

PROPOSITION 12.4   *Let $f : M \to N$ be a monoid homomorphism and define $\sim$ on M by $x \sim y \iff f(x) = f(y)$. Then $\sim$ is a congruence on M.*

The proof is straightforward. The congruence $\sim$ defined in this proposition is called the *kernel* $\ker(f)$ of the homomorphism f. From a congruence $\sim$ on

M, we can form the quotient structure $M/\sim$ in which the elements are the equivalence classes under $\sim$, with the multiplication $[x][y] = [xy]$, where $[x]$ denoted the equivalence class of $x \in M$. Notice that, if $x \sim x'$ and $y \sim y'$ then, by definition of a congruence, $xy \sim x'y \sim x'y'$, so this multiplication is well-defined, and $M/\sim$ clearly has the identity element $[1_M]$. The map $M \to M/\sim$ defined by $x \mapsto [x]$ is a monoid homomorphism.

Monomorphisms, epimorphisms and isomorphisms of monoids are defined in the same way as for groups. The analogue of the first isomorphism theorem for monoids is:

THEOREM 12.5   *Let* $f : M \to N$ *be a monoid homomorphism with kernel* $\sim$. *Then the map* $[x] \mapsto f(x)$ *defines a monoid isomorphism from* $M/\sim$ *to* $\mathrm{im}(f)$.

PROOF   It is routine to verify that this map is a well-defined monoid homomorphism that is injective and surjective.   ∎

Of course, a group is just a special case of a monoid, so we would hope that this theory of equivalences and quotient monoids would correspond properly when applied to a monoid that happens to be a group. This is indeed the case. When M and N are groups, and $f : M \to N$ is a homomorphism, and $\sim$ is the kernel of f as we have defined it for monoid homomorphisms, then $x \sim y$ if and only if x and y are in the same coset of K, where K is the kernel as it would be defined for a group homomorphism. So, although we have two distinct definitions for the kernel of f in this situation, each of them is derived easily from the other.

The normal closure of a subset X of a group G was defined in Definition 2.7 to be the smallest normal subgroup of G containing X. There is an analogous idea for monoids. An equivalence relation on any set M is, by definition, a subset of $M \times M$, and so we can talk about the intersection of a collection of equivalence relations, and it is routine to check that the result is itself an equivalence relation. (In the case of an empty collection, the intersection is defined to be $M \times M$.) It is equally routine to check that the intersection of a collection of congruences on a monoid M is itself a congruence. So for any subset $\mathcal{R}$ of $M \times M$, we can define the congruence on M generated by $\mathcal{R}$ to be the intersection of all congruences that contain $\mathcal{R}$.

PROPOSITION 12.6   *Let* M *be a monoid, let* $\mathcal{R} \subseteq M \times M$, *and let* $\sim$ *be the congruence generated by* $\mathcal{R}$. *Let us call* $m, n \in M$ *directly equivalent if we have* $m = uvw$, $n = uv'w$, *for some* $u, v, v', w \in M$ *with* $(v, v') \in \mathcal{R}$ *or* $(v', v) \in \mathcal{R}$. *Then, for* $m, n \in M$, $(m, n) \in \sim$ *if and only if there is a sequence* $m = m_0, m_1, \ldots, m_r = n$ *of elements of* M, *with* $r \geq 0$, *such that* m_i *and* m_{i+1} *are directly equivalent for* $0 \leq i < r$.

PROOF   Define $\sim'$ on M by $(m, n) \in \sim'$ if and only if the condition defined in the proposition holds. It is routine to check that $\sim'$ is a congruence on M that contains $\mathcal{R}$. Further, by definition of a congruence, any congruence on M that contains $\sim$ must contain $\sim'$, so we have $\sim = \sim'$.   ∎

12.1.2   Free monoids and monoid presentations

DEFINITION 12.7   A monoid F is *free* on the subset X of F if, for any monoid M and any map $\theta : X \to M$, there exists a unique monoid homomorphism $\theta' : F \to M$ with $\theta'(x) = \theta(x)$ for all $x \in X$.

The proof of the following result is identical to that of the equivalent result for free groups, Proposition 2.47.

PROPOSITION 12.8   *(i) Two free monoids on the same set X are isomorphic.*
(ii) Free monoids on X_1 and X_2 are isomorphic if and only if $|X_1| = |X_2|$.

Existence is much easier to prove than is the case for free groups. Recall from Definition 2.48 that, for any set X, X^* is defined to be the set of all strings or words over X. It is obvious that X^* forms a monoid with identity element ε_X under the operation of concatenation of strings.

PROPOSITION 12.9   *For any set X, the monoid X^* is free on X.*

PROOF   For a monoid M and a map $\theta : X \to M$, the unique monoid homomorphism $X^* \to M$ that restricts to θ on X is the map sending $x_1 \cdots x_r$ to $\theta(x_1) \cdots \theta(x_r)$.   ∎

We can now define a monoid presentation.

DEFINITION 12.10   Let X be a set, and let $\mathcal{R}$ be a subset of $X^* \times X^*$. Then $\mathrm{Mon}\langle X \mid \mathcal{R} \rangle$ is defined to be the quotient monoid X^*/σ, where σ is the congruence on X^* generated by $\mathcal{R}$. This is also called the monoid with generating set X and defining relations $\mathcal{R}$.

The elements of $\mathcal{R}$ are called the *defining relations* of the presentation. A relation $(u, v) \in \mathcal{R}$ is usually written as $u = v$. A monoid M is said to be *finitely presented* (or sometimes, more accurately, *finitely presentable*) if M is isomorphic to $\mathrm{Mon}\langle X \mid \mathcal{R} \rangle$ with X and $\mathcal{R}$ both finite. In specific examples, if $X = \{x_1, \ldots, x_n\}$ and $\mathcal{R} = \{(u_1, v_1), \ldots, (u_m, v_m)\}$, then we shall often write

$\mathrm{Mon}\langle\, x_1,\ldots,x_n \mid u_1 = v_1,\ldots u_m = v_m \,\rangle$ rather than $\mathrm{Mon}\langle\, X \mid \mathcal{R} \,\rangle$.

For a set X, let $A := X \cup X^{-1}$ as in Subsection 2.4.1. Let $\mathcal{I}_X$ be the subset $\{\,(xx^{-1}, \varepsilon_A) \mid x \in A\,\}$ of $A^* \times A^*$, and let $\sim_X$ be the congruence generated by $\mathcal{I}_X$. Then the reader can check that the free group $F(X)$ defined in Subsection 2.4.1 is equal to the quotient monoid $A^*/\sim_X = \mathrm{Mon}\langle\, A \mid \mathcal{I}_X \,\rangle$.

We have the following analogue of Theorem 2.52.

THEOREM 12.11   *Let X be a set, let $\mathcal{R}$ be a subset of $X^* \times X^*$, let M be a monoid, and let $\theta : X \to M$ be a map with the property that, for all $(u,v) \in \mathcal{R}$, we have $\theta(x_1)\cdots\theta(x_r) = \theta(y_1)\cdots\theta(y_s)$, where $u = x_1\cdots x_r$, $v = y_1\cdots y_s$, and $x_i, y_j \in X$. Then there exists a unique monoid homomorphism $\theta' : \mathrm{Mon}\langle\, X \mid \mathcal{R} \,\rangle \to M$ for which $\theta'([x]) = \theta(x)$ for all $x \in X$, where $[x]$ represents the equivalence class of x in $\mathrm{Mon}\langle\, X \mid \mathcal{R} \,\rangle$.*

PROOF   If θ' is to be a homomorphism, then it must satisfy $\theta([u]) = \theta(x_1)\cdots\theta(x_r)$ for all $u = x_1\cdots x_r \in X^*$, and so we have uniqueness.

By definition of a free monoid and Proposition 12.9, there exists a homomorphism $\theta'' : X^* \to M$ with $\theta''(x) = \theta(x)$ for all $x \in X$. The assumption on θ says precisely that each element of $\mathcal{R}$ is in the kernel of θ'', and so $\ker(\theta'')$ is a congruence on X^* that contains $\mathcal{R}$. Hence it also contains the congruence $\sim$ on X^* generated by $\mathcal{R}$, and so $\theta''(u) = \theta''(v)$ whenever $u \sim v$. But this says precisely that the map $\theta' : \mathrm{Mon}\langle\, X \mid \mathcal{R} \,\rangle \to M$ defined by $\theta'([u]) = \theta(x_1)\cdots\theta(x_r)$ for all $u = x_1\cdots x_r \in X^*$ is well-defined, and it is clearly a monoid homomorphism satisfying $\theta'([x]) = \theta(x)$ for all $x \in X$.   ∎

Example 12.1
Let $M := \mathrm{Mon}\langle\, x \mid x^2 = x^6 \,\rangle$. The reader can verify that the congruence generated by $\mathcal{R}$ consists of (x^r, x^{r+4s}) and (x^{r+4s}, x^r) for all $r, s \in \mathbb{N}_0$ with $r \geq 2$, and that $M = \{[\varepsilon], [x], [x^2], [x^3], [x^4], [x^5]\}$.   ⬜

As we did for groups, we shall generally denote an element $[w]$ in a finitely presented monoid simply by w, and use $v =_M w$ to mean $[v] = [w]$. The following theorem shows that, as an alternative to Definition 2.4, we could define a group presentation as a special case of a monoid presentation.

THEOREM 12.12   *Let X be a set, let $A := X \cup X^{-1}$, and let $\mathcal{R}$ be a subset of $A^* \times A^*$. Then the group defined by the presentation $\langle\, X \mid \mathcal{R} \,\rangle$ is equal to the monoid defined by the presentation $\mathrm{Mon}\langle\, A \mid \mathcal{I}_X \cup \mathcal{R} \,\rangle$.*

PROOF   The elements of $\langle\, X \mid \mathcal{R} \,\rangle$ and of $\mathrm{Mon}\langle\, A \mid \mathcal{I}_X \cup \mathcal{R} \,\rangle$ are equivalence classes of A^* under congruences $\sim_1$ and $\sim_2$, respectively, and we have to show that $\sim_1 = \sim_2$. Let R be the set of words uv^{-1} corresponding to elements

$(u, v) \in \mathcal{R}$ and let $N := \langle R^F \rangle$ with F the free group on X. Then, by definition, for $w_1, w_2 \in A^*$, we have $w_1 \sim_1 w_2$ if and only if $w_1 N = w_2 N$. This is certainly the case if $(w_1, w_2) \in \mathcal{I}_X \cup \mathcal{R}$ and, since $\sim_2$ is the congruence generated by $\mathcal{I}_X \cup \mathcal{R}$, it follows that $w_1 \sim_2 w_2$ implies $w_1 \sim_1 w_2$. Conversely, since N is generated as a subgroup of F by the elements $g^{-1}uv^{-1}g$ with $(u, v) \in \mathcal{R}$, $w_1 \sim_1 w_2$ implies that $w_1 =_F w_2 w$, where w is a product of some of these generators of N and their inverses. Since $\mathcal{I}_X \subseteq \sim_2$, this implies $w_1 \sim_2 w_2 w$. But if $(u, v) \in \mathcal{R}$, then $u \sim_2 v$ and hence $uv^{-1} \sim_2 \varepsilon$ and $g^{-1}uv^{-1}g \sim_2 \varepsilon$. Since $\sim_2$ is a congruence, we have $w_2 w \sim_2 w_2$ and hence $w_1 \sim_2 w_2$. ∎

Exercises

1. Prove that, if the finitely generated monoid M is generated by X, then X has a finite subset that generates M.

2. Find an example of monoids M, N and a submonoid of M that can occur as $\{ x \in M \mid f(x) = 1_N \}$ for more than one monoid homomorphism $f : M \to N$.

3. Complete the details of the proof that the intersection of the congruences on M containing a given subset X of $M \times M$ is itself a congruence on M that contains X.

4. Let M be a group, let $\mathcal{R}$ be a subset of $M \times M$, and let $\sim$ be the congruence generated by $\mathcal{R}$. Show that $x \sim y$ if and only if xy^{-1} is in the normal closure in M of the set $\{ uv^{-1} \mid (u, v) \in \mathcal{R} \}$.

5. Prove directly from the definition that, if a monoid M is free on X, then X generates M.

6. Prove that any monoid that is generated by a single element is defined by a presentation with a single defining relation.

12.2   Rewriting systems

We turn now to the theory of rewriting systems over a fixed finite alphabet. There is a large body of literature on this topic, which has applications to all branches of algebra. For a historical account, with an extensive bibliography, we refer the reader to the article by Buchberger [Buc87]. One of the earliest papers describing the use of critical pairs and the completion process in a very general setting is that of Knuth and Bendix [KB70], after whom this process is usually named. Applications to group theory are described in the two papers of Gilman: [Gil79] and [Gil84]. There is also an extensive treatment of this topic, with details of algorithms and their implementation in [Sim94].

We let A be a finite set, known as the alphabet, and let A^* be the free monoid of strings over A. A *rewriting system* on A^* is a set S of ordered pairs (w_1, w_2), with $w_1, w_2 \in A^*$. The elements of S are called *rewrite-rules*, and w_1 and w_2 are called, respectively, the left- and right-hand sides of the rule. The idea is that we can replace occurrences of w_1 in a string by w_2. We shall assume throughout that no two distinct rules have the same left-hand sides.

For $u, v \in A^*$, we write $u \to_S v$ if there exist strings $x, y, w_1, w_2 \in A^*$ such that $u = x w_1 y$, $v = x w_2 y$ and $(w_1, w_2) \in S$; in other words, if v is obtained from u by making a single substitution using a rewrite-rule. We shall omit the subscript S when there is no danger of ambiguity.

DEFINITION 12.13   A string $u \in A^*$ is said to be *(S-)irreducible* or *(S-)reduced* if there is no string $v \in A^*$ with $u \to v$.

We denote the reflexive, transitive closure of $\to$ by $\to^*$. Hence $u \to^* v$ if and only if, for some $n \geq 0$, there exist $u = u_0, u_1, \ldots, u_n = v \in A^*$ with $u_i \to u_{i+1}$ for $0 \leq i < n$. The symmetric closure of $\to^*$ is denoted by $\leftrightarrow^*$. Thus $u \leftrightarrow^* v$ if and only if, for some $n \geq 0$, there exist $u = u_0, u_1, \ldots, u_n = v$ with

$$
\begin{array}{ccccccc}
& u_1 & & u_3 & & u_{n-1} & \\
{}^*\!\swarrow & & \searrow^* & {}^*\!\swarrow & & \searrow^* \cdots {}^*\!\swarrow & & \searrow^* \\
u_0 & & u_2 & & & & u_n
\end{array} \qquad (\dagger)
$$

The system S is called *Noetherian* (or terminating) if there is no infinite chain of strings u_i $(i > 0)$ with $u_i \to u_{i+1}$ for all i. This implies that for any $u \in A^*$ there exists an irreducible $v \in A^*$ with $u \to^* v$.

We call S *confluent* if, whenever $u, v_1, v_2 \in A^*$ with $u \to^* v_1$ and $u \to^* v_2$, there exists $w \in A^*$ with $v_1 \to^* w$ and $v_2 \to^* w$, and we call S *complete* if it is Noetherian and confluent.

It is called *locally confluent* if whenever $u, v_1, v_2 \in A^*$ with $u \to v_1$ and $u \to v_2$, there exists $w \in A^*$ with $v_1 \to^* w$ and $v_2 \to^* w$.

The following sequence of lemmas demonstrates that completeness is a highly desirable property for rewriting-systems to possess. For $u \in A^*$, we define the set $\mathrm{desc}(u)$ of descendants of u to be $\{\, w \in A^* \setminus \{u\} \mid u \to^* w \,\}$. So u is irreducible if and only if $\mathrm{desc}(u)$ is empty.

LEMMA 12.14   *Suppose that S is Noetherian. Then* $\mathrm{desc}(u)$ *is finite for all $u \in A^*$.*

PROOF   Suppose that $\mathrm{desc}(u)$ is infinite. Since no two rules in S have the same left-hand side, there are only finitely many w with $u \to w$. Thus there exists $u_2 \in A^*$ with $u \to u_2$ and $\mathrm{desc}(u_2)$ infinite. Similarly, there exists

u_3 with $u_2 \to u_3$ and $\mathrm{desc}(u_3)$ infinite, and we get an infinite chain $\{u_i\}$ with $u_i \to u_{i+1}$, contradicting the assumption that S is Noetherian. ∎

LEMMA 12.15   *Suppose that S is Noetherian and locally confluent. Then S is complete.*

PROOF   We shall prove that, under the given hypotheses, for all $u \in A^*$ there is a unique irreducible w with $u \to^* w$. The result clearly follows from this. The proof is by induction on $|\mathrm{desc}(u)|$, the result being clear for $|\mathrm{desc}(u)| = 0$. Suppose that $u = u_0 \to u_1 \to \ldots \to u_m$ and $u = u_0 \to u_1' \to \ldots \to u_n'$, with u_m and u_n' irreducible. By local confluence, there exists v with $u_1 \to^* v$ and $u_1' \to^* v$. Let $v \to^* w$ with w irreducible. Since $|\mathrm{desc}(u_1)| < |\mathrm{desc}(u)|$ and $|\mathrm{desc}(u_1')| < |\mathrm{desc}(u)|$, it follows by inductive hypothesis that $u_m = w$ and $u_n' = w$, so $u_m = u_n'$, which completes the proof. ∎

LEMMA 12.16   *If S is Noetherian and locally confluent, then each equivalence class under $\leftrightarrow^*$ contains a unique S-irreducible element.*

PROOF   Let u and v be irreducible strings with $u \leftrightarrow^* v$. Then there exist $u = u_0, u_1, \ldots, u_n = v \in A^*$ satisfying the equation (†) above. We have to prove that $u = v$. This is clear if $n = 0$, so we use induction on n. By Lemma 12.15 S is confluent, so there exists $w \in A^*$ with $u \to^* w$ and $u_2 \to^* w$. Since u is irreducible, we must have $u = w$, and so we can replace u_2 by u in the diagram, and the result follows by induction. ∎

The systems that we shall be considering will all be Noetherian, but they will not always be confluent. The following lemma provides a method of checking confluence (and hence completeness) constructively, at least when S is finite.

LEMMA 12.17   *The system S is locally confluent if and only if the following conditions are satisfied for all pairs of rules $(u_1, t_1), (u_2, t_2) \in S$.*

(i) *If $u_1 = rs$ and $u_2 = st$ with $r, s, t \in A^*$ and $s \neq \varepsilon$, then there exists $w \in A^*$ with $t_1 t \to^* w$ and $r t_2 \to^* w$.*

(ii) *If $u_1 = rst$ and $u_2 = s$ with $r, s, t \in A^*$ and $s \neq \varepsilon$, then there exists $w \in A^*$ with $t_1 \to^* w$ and $r t_2 t \to^* w$.*

PROOF   First assume that S is locally confluent. Then in both cases, the existence of w follows by applying the definition of local confluence to the string rst. Conversely, assume that the conditions (i) and (ii) are valid, and let u be a string with $u \to v_1$ and $u \to v_2$. Then u must have two substrings u_1 and u_2 such that there exist rules $(u_1, t_1), (u_2, t_2) \in S$ that are used in the reductions to v_1 and v_2. If u_1 and u_2 do not overlap in u, then we have $u = r u_1 s u_2 t$ for some strings r, s, t and $v_1 = r t_1 s u_2 t, v_2 = r u_1 s t_2 t$. Thus $v_1 \to w$ and $v_2 \to w$, with $w = r t_1 s t_2 t$. Otherwise, u_1 and u_2 overlap in S, and (after interchanging the rules if necessary) we must have one of the two situations postulated in conditions (i) and (ii) of the lemma. It then follows by assumption that there exists $w \in A^*$ with $v_1 \to^* w$ and $v_2 \to^* w$, so S is locally confluent. ∎

A pair of rules satisfying either of the two conditions of the lemma is called a *critical pair*. If one of these conditions fail (so S is not confluent), we end up with two distinct irreducible strings w_1 and w_2 that are irreducible and equivalent under $\leftrightarrow^*$. We can resolve this instance of incompleteness by adjoining either (w_1, w_2) or (w_2, w_1) to S as a new rule. We can then continue with the check for local confluence.

We shall assume from now on that S is finite unless explicitly stated otherwise.

The procedure of examining critical pairs and adjoining new rules to S if necessary is known as the *Knuth-Bendix completion process*. It may happen that this process eventually terminates with a finite complete set S, which is what we would like. In many cases, however, the process does not complete, but it does, in principal, generate an infinite complete set of rules. In a few examples, as we shall see, such an infinite set can have a very transparent structure, which can make it almost as useful as a finite complete set.

In general, it can be difficult to decide which of (w_1, w_2) or (w_2, w_1) to adjoin to S in order to resolve a critical pair. In the applications that we shall be considering, we resolve this problem by imposing a well-ordering $\leq$ on A^* that has the property that $u \leq v$ implies $uw \leq vw$ and $wu \leq wv$ for all $w \in A^*$. Such an ordering is known as a *reduction ordering*. Then we make the larger of w_1 and w_2 the left-hand side of the new rule.

The assumption on the ordering ensures that $u \to^* v$ implies $u \geq v$ and, since it is a well-ordering, S will necessarily be Noetherian. Thus, if S is complete, then the irreducible elements correspond to the least representatives in the $\leftrightarrow^*$ equivalence classes.

One of the most commonly used examples of a reduction ordering is a *shortlex* ordering, as defined in Definition 2.60. For example, if $A = \{a, b, c\}$ and (just to be awkward) we decide to order A by $b < c < a$, then the first few strings in the associated shortlex order are

$$\varepsilon, b, c, a, bb, bc, ba, cb, cc, ca, ab, ac, aa, bbb, bbc, bba, \ldots$$

From now on, we shall assume that the rules are ordered according to some such well-ordering, although not necessarily a shortlex ordering. (A completely different well-ordering, appropriate for polycyclic groups, will be considered in Section 12.4.)

DEFINITION 12.18   A string $u \in A^*$ is called *(S-)minimal* if it is the least element in its $\leftrightarrow^*$ equivalence class

Then clearly minimal strings are irreducible, and Lemma 12.16 says that S is complete if and only if all irreducible strings are minimal.

A rewrite-rule (w_1, w_2) in S is called *(S-)irreducible* if w_2 and all proper substrings of w_1 are S-irreducible. If all rules are irreducible, then we say that S is *irreducible* or *reduced*.

If a rule (w_1, w_2) is not irreducible, then we can use the other rules in S to simplify it, without changing the relation $\rightarrow^*$. The rule may either turn out to be redundant, or we may replace it by an irreducible rule (w_1', w_2') (or (w_2', w_1')) in which $w_1' < w_1$ or $w_2' < w_2$. (Of course, (w_1', w_2') then becomes a new rule of S.)

The new rules that we adjoin to S during the Knuth-Bendix completion process will always be irreducible, but they may sometimes destroy the irreducibility of some existing rules. It is generally efficient to simplify S by replacing all rules by irreducible rules at regular intervals. This simplification process does not change the relation $\rightarrow^*$.

We shall not go further here into the algorithmic details of the Knuth-Bendix procedure, but we shall assume that, for any two rewrite-rules that are in S at any stage, either one or both of them will get removed or replaced during simplification, or the pair of rules will be checked for being a critical pair (bearing in mind that the same pair can be a critical pair in more than one way). We shall also assume that any rule that is not S-reduced will at some stage be removed during rule simplification.

DEFINITION 12.19   A pair of strings (w_1, w_2) is called an *(S-)essential rule* if $w_1 \leftrightarrow^* w_2$, and w_2 and all proper substrings of w_1 are *(S-)minimal*.

So any essential rule in S is reduced and, if S is complete, then the reduced rules in S are essential. We have not said anything about essential rules being rules of S, but the following proposition says that they will be if we run Knuth-Bendix on S for long enough.

PROPOSITION 12.20   *(i) Let $w \in A^*$ and suppose that $v \in A^*$ is minimal with $w \leftrightarrow^* v$. After running the Knuth-Bendix completion process on S for sufficiently long, we will have $w \rightarrow_S^* v$.*

(ii) Let (w_1, w_2) be an S-essential rule. After running the Knuth-Bendix completion process on S for sufficiently long, (w_1, w_2) will be a rewrite-rule of S.

PROOF   Let $\tilde{S}$ be the (possibly infinite) set of all rewrite-rules that are in S at some stage during the Knuth-Bendix completion process, and are not subsequently removed as a result of simplification of the existing rule set. We shall show first that the rules in $\tilde{S}$ form a complete rewriting system.

By the assumption that we made above about the Knuth-Bendix procedure, any pair of rules from $\tilde{S}$ will at some stage be checked for being a critical pair, after which the hypotheses of Lemma 12.17 will hold in S for that pair of rules. Indeed, since simplification of the rule set does not change the relation $\to^*$, the hypotheses of Lemma 12.17 will continue to hold for that pair of rules, and since eventually all rules not in $\tilde{S}$ that are involved in the associated word reductions will be removed from S, the hypotheses of Lemma 12.17 hold in $\tilde{S}$. It follows from Lemma 12.17 that $\tilde{S}$ is locally confluent, and hence complete.

So, if w, v are as in (i), then $w \to^*_{\tilde{S}} v$. Since all of the rules in $\tilde{S}$ are eventually in S, this proves (i). If (w_1, w_2) is an essential rule then, by (i), we will eventually have $w_1 \to^*_{\tilde{S}} w_2$. Since all proper substrings of w_1 are minimal, w_1 must then be the left-hand side of some rule (w_1, w_2') in S. Then $w_2' \to^*_{\tilde{S}} w_2$ and so, by (i) again, eventually we will have $w_2' \to^*_S w_2$. Then, the next time that the rules are simplified, (w_1, w_2') will be replaced by (w_1, w_2), which proves (ii). ∎

COROLLARY 12.21 *If there are only finite many $\leftrightarrow^*$ equivalence classes, then the Knuth-Bendix completion process will halt with a finite complete set of rewrite-rules.*

PROOF   If there are only finitely many $\leftrightarrow^*$-classes, then there are only finitely many S-minimal words, and hence only finitely many essential rules. By the proposition, these will eventually all be in S, at which stage S will be complete. ∎

Implementing the Knuth-Bendix completion process efficiently on a computer is difficult. To be seriously useful it needs to be able to handle very large rewriting systems, possibly with millions of rewrite-rules. So the implementation needs to be efficient in terms of both time and space. The two critical components of an implementation are the search for critical pairs of S, and the reduction of words to S-irreducible words. Experiments indicate that in typical implementations, most of the time is taken up with word reduction.

We shall not go into implementation details here. There is much discussion and analysis of the various issues involved in Chapters 2 and 3 of [Sim94]. The word-reduction process can be made to run fast by using a certain type of finite state automaton, and we shall describe that method later in Subsection 13.1.3.

There are also a number of algorithmic issues that need to be decided, such as how often do we carry out rule simplification, and in what order do we consider pairs of rewrite-rules in the search for critical pairs. For example, we might consider them in the order in which they are found, or in order of increasing length. A flexible implementation would allow the user to choose from a range of strategies of this kind.

Exercises

1. Show that the set $\tilde{S}$ defined in the proof of Proposition 12.20 is equal to the set of all S-essential rules.

2. Let $A = \{c, o, z, e\}$ and $S = \{\, ze \to ce, \; zc \to oz, \; oc \to cz \,\}$.

   (i) Show that S is confluent with respect to the shortlex ordering of A^* with $c < o < z < e$.

   (ii) Show that, for $n \geq 0$, $z^n e \to^* c^n e$, and that this reduction involves $3.2^n - 2n - 3$ applications of rules in S.

12.3   Rewriting systems in monoids and groups

We turn now to the applications of rewriting systems to finitely presented monoids and groups. The *word problem* in a monoid M defined by a presentation with generating set A is the problem of deciding whether two words $u, v \in A^*$ represent the same element of M; that is, whether $u =_M v$. If M is a group, then this is equivalent to deciding whether $uv^{-1} =_M 1_M$. The word problem is known to be undecidable in some finitely presented monoids and groups. However, we can attempt to use the Knuth-Bendix completion process to solve this problem in specific examples.

If S is a rewriting system over A, then $\leftrightarrow^*$ is precisely the congruence on A^* generated by S, and so $A^* / \leftrightarrow^*$ is equal to the monoid defined by the presentation $\mathrm{Mon}\langle A \mid S \rangle$. Lemmas 12.15 and 12.16 say that, if S is complete, then the S-irreducible words are the S-minimal words, and form a set of representatives for the $\leftrightarrow^*$-classes on A^*. Thus, if S is finite and complete, then we can solve the word problem in the monoid by using the rules in S to reduce the two words u and v to irreducible words, and then we just check whether these reductions are equal words. When M is finite, Corollary 12.21 tells us that the completion process is guaranteed to terminate eventually with a finite complete set of rules. This also happens occasionally when M is infinite, but not typically. It can also happen that there is a complete set of rules that is infinite but has a very regular pattern, and so it can still be used to solve the word problem.

If $G = \langle X \mid R \rangle$ or $\langle X \mid \mathcal{R} \rangle$ is a group presentation, where R or $\mathcal{R}$ is, respectively, a set of relators or relations then, as we described in Theorem 12.12, we can construct a monoid presentation $M_G := \mathrm{Mon}\langle A \mid \mathcal{I} \cup \mathcal{R} \rangle$ for the group, where $A := X \cup X^{-1}$. Then we can apply the completion process to this monoid presentation, which might enable us to solve the word problem in G.

In particular, if R (or $\mathcal{R}$) is empty, then G is a free group on X and $S = \mathcal{I}$. Then the only critical pairs are those resulting from pairs of rules (xx^{-1}, ε)

and $(x^{-1}x, \varepsilon)$ for $x \in A$, where we have the overlap x^{-1} between the two left-hand sides. But both of the possible reductions of $xx^{-1}x$ are equal to x, so it follows from Lemma 12.17 that $\mathcal{S}$ is already confluent. Hence the $\mathcal{S}$-reduced words are precisely those that contain no adjacent mutually-inverse generators; that is, the reduced words as defined in Subsection 2.4.1. This is precisely what Proposition 2.50 asserts, and in fact the proof that we gave of that result in Chapter 2 was essentially the same as the proof that we have given here, but applied to that particular situation.

By default, we use a shortlex ordering on A^* to decide on the left- and right-hand sides of the rules but, as we shall see in the next section, in particular situations other orderings may be more effective.

In examples, we shall frequently use the so-called 'inversion by case-change' convention, which means that the inverses of generators x, y, etc., are denoted by X, Y. This has proved to be very convenient on many occasions. Unfortunately, there is a conflict with the use of X for the generating set, which we hope will not cause too much confusion!

Example 12.2

Let $G := \langle x, y \mid x^3 = 1, y^2 = 1, (xy)^2 = 1 \rangle$ be the dihedral group of order 6. Then $A = \{x, X, y, Y\}$, and (using the order $x < X < y < Y$ on A) we can start with

$$\mathcal{S} := \{ (xX, \varepsilon), (Xx, \varepsilon), (yY, \varepsilon), (Yy, \varepsilon), (xx, X), (Y, y), (yX, xy) \}.$$

(When each generator has an inverse generator and we are using a shortlex ordering, then we can always rewrite our rules so that the length of the left-hand side is greater than the right-hand side by at most 2, and it is generally efficient to do this.)

One critical pair, for example, is $(Xx, \varepsilon), (xx, X)$ with the overlap x. This leads to the word Xxx reducing in two different ways to x and XX, and to resolve this we adjoin the rule (XX, x) to $\mathcal{S}$.

It is extremely tedious to examine all such critical pairs by hand, but easy for a computer, and we end up with the complete set of rules:

$$\mathcal{S} = \{ (xX, \varepsilon), (Xx, \varepsilon), (yY, \varepsilon), (Yy, \varepsilon),$$
$$(xx, X), (XX, x), (Y, y), (yy, \varepsilon), (yX, xy), (yx, Xy) \}.$$

We now observe that the rules (yY, ε) and (Yy, ε) are not irreducible, since the word Y is reducible. In fact they are both redundant, and if we remove them, we are left with the set of all essential rules.   □

Most of the time, if G is an infinite group, then the Knuth-Bendix procedure will not terminate, but there are some examples in which it does. There does not seem to be any satisfactory theory that will predict when this will happen; in many examples, termination depends on the correct choice of ordering of A^*, and even of A when a shortlex ordering is used.

Example 12.3

Let $G := \langle\, x, y \mid yx = xy \,\rangle$ be the free abelian group on two generators.

If we use the ordering $x < X < y < Y$ of A, then we end up with the finite complete set of rules:

$$\{\, (xX, \varepsilon), (Xx, \varepsilon), (yY, \varepsilon), (Yy, \varepsilon), (yx, xy), (yX, Xy), (Yx, xY), (YX, XY) \,\}.$$

If we use the ordering $x < y < X < Y$, however, then we generate the infinite complete set of rules:

$$\{(xX, \varepsilon), (Xx, \varepsilon), (yY, \varepsilon), (Yy, \varepsilon), (yx, xy), (Xy, yX), (Yx, xY), (YX, XY),$$
$$(xy^n X, y^n), (yX^n Y, X^n) \mid \text{for all } n > 0 \,\}.$$

$\Box$

In Example 12.3, the infinite complete set is regular (in the sense of regular languages) and can be used effectively to reduce words to normal form.

The infinite examples for which finite complete rewriting systems have been found for some choice of generators and ordering include polycyclic groups, the two-dimensional surface groups, the von Dyck groups $\langle\, x, y \mid x^p, y^q, (xy)^r \,\rangle$ (by I. Wanless), many Coxeter groups (by Hermiller [Her94]), and a number of isolated examples, such as $\langle\, x, y \mid xy = yx^2 \,\rangle$. Many of these finite complete rewriting systems are catalogued by LeChenadec in [LeC86].

Finally, it is worth noting that there are certain similarities between the Knuth-Bendix procedure and Todd-Coxeter coset enumeration over the trivial subgroup. Both are in some sense looking systematically for consequences of the defining relations of the group. Of course the Todd-Coxeter algorithm can only complete for finite groups, but it appears to be more efficient than Knuth-Bendix for most straightforward examples.

There are exceptions, however. Knuth-Bendix works very well for certain Coxeter groups, such as the groups $\mathrm{Sym}(n)$, which have a large order and relatively small complete rewriting systems (about $(n-1)^2$ equations).

For certain pathological examples, such as the presentation

$$G = \langle\, r, s, t \mid ssr^{-1}s^{-1}rs^{-1}tst^{-1}t^{-1}r^{-1}srs^{-1}s^{-1}tts^{-1}t^{-1}stts^{-1}t^{-1}s$$
$$= tts^{-1}t^{-1}st^{-1}rtr^{-1}r^{-1}s^{-1}tst^{-1}t^{-1}rrt^{-1}r^{-1}trrt^{-1}r^{-1}t$$
$$= rrt^{-1}r^{-1}tr^{-1}srs^{-1}s^{-1}t^{-1}rtr^{-1}r^{-1}ssr^{-1}s^{-1}rssr^{-1}s^{-1}r = 1 \,\rangle$$

of the trivial group, due to B.H. Neumann, Knuth-Bendix succeeds eventually, whereas no implementation of coset enumeration has done so to date.

It is also possible to find a string-rewriting analogue of coset enumeration over a nontrivial subgroup H of G. We shall not go into details here. This topic is covered in Sections 2.8 and 3.10 of [Sim94].

Efficient implementations of the Knuth-Bendix algorithm for groups and monoids are available in the Rutgers Knuth-Bendix Package (rkbp) written by Charles Sims and available from him, and the KBMAG package by Holt. Both offer a wide choice of controlling parameters and word orderings.

Exercises

1. Show that the dihedral group D_{2n} has a complete rewriting system with three rules on a monoid generating set containing two elements of order two.

2. Find a complete rewriting system for the symmetric group $\mathrm{Sym}(4)$ on three monoid generators of order two.

3. Let $M := \mathrm{Mon}\langle\, a, b \mid bab = aba \,\rangle$, and use the shortlex ordering of A^* with $a < b$. Show that there is an infinite complete rewriting system consisting of the rules $bab \to aba$ and $ba^{n+1}ba \to aba^2b^n$ for all $n \geq 1$.

12.4   Rewriting systems for polycyclic groups

In this section, we shall give a brief exposition of the theory of polycyclic groups in the context of string-rewriting and Knuth-Bendix completion.

Let us first quickly recall the basic definitions from Section 8.1. A group G is polycyclic if it has a series of subgroups $G = G_1 > G_2 > \cdots > G_{n+1} = 1$ in which each $G_{i+1} \trianglelefteq G_i$ and G_i/G_{i+1} is cyclic.

Any such group has a polycyclic generating sequence set $X = [x_1, \ldots, x_n]$, where $x_i \in G_i$ and $x_i G_{i+1}$ generates G_i/G_{i+1}. As usual, we define our alphabet to be the set $A := \{\, x_i, x_i^{-1} \mid 1 \leq i \leq n \,\}$, which generates G as a monoid.

We let $I(X) = I$ be the subset of $\{1, 2, \ldots, n\}$ consisting of those i for which G_i/G_{i+1} is finite, and let $r_i := |G_i/G_{i+1}|$ for $i \in I$.

We may, if we wish, use the smaller set of monoid generators

$$A' := \{\, x_i \mid 1 \leq i \leq n \,\} \cup \{\, x_i^{-1} \mid i \notin I \,\}.$$

It is easy to show (by induction on $n-i$) that the elements x_i^{-1} for $i \in I$ can be expressed, in G, as products of the generators in A', and so A' generates G as a monoid. In particular, if G is finite, then we may dispose of the inverse generators altogether.

By Lemma 8.3, each $g \in G$ has a unique normal form word in $(A')^*$ of the form $x_1^{e_1} x_2^{e_2} \cdots x_n^{e_n}$, where each $e_i \in \mathbb{Z}$ and $0 \leq e_i < r_i$ if $i \in I$.

We should like to find a reduction ordering of the words in A^* for which the normal form for each group element g is the least word representing g. Since the normal form may be far from being the shortest word for g, orderings based on length are not suitable. In fact, the appropriate ordering is an instance of a *wreath-product ordering*, as defined on page 48 of [Sim94].

A wreath-product ordering on B^*, for an alphabet $B = \{z_1, \ldots z_m\}$ is defined as follows. First we assign *levels*, which must be positive integers, to the letters z_i. The idea is that a single occurrence of a generator at a higher level renders a word larger than any number of generators at lower levels. For words involving generators of the same level only, we use some other standard ordering, which may, for example, be a shortlex ordering.

To be precise, suppose, inductively, that for some $r > 0$, we know how to order words involving only generators of level less than r. Then distinct words v, w involving generators of level at most r are ordered as follows.

First remove all generators of level less than r from v and w, and compare the two resulting words v', w' in the level r generators. If they are distinct and, say, $v' > w'$, then we define $v > w$.

On the other hand, if $v' = w' = z_{i_1} z_{i_2} \cdots z_{i_l}$, say, then we have $v = t_1 z_{i_1} t_2 z_{i_2} \cdots t_l z_{i_l} t_{l+1}$ and $w = u_1 z_{i_1} u_2 z_{i_2} \cdots u_l z_{i_l} u_{l+1}$ where, for $1 \le k \le l+1$, t_k and u_k are words involving only generators of level less than r. We define $v < w$ if $t_1 < u_1$ or, for some $k \le l$, $t_i = u_i$ for $1 \le i \le k$, but $t_{k+1} < u_{k+1}$.

It is routine to check that this is a well-ordering and that $u \le v$ implies $uw \le vw$ and $wu \le wv$ for all $u, v, w \in A^*$, which are the required conditions for a reduction ordering.

In the application to polycyclic groups, we give each of the monoid generators distinct levels, with x_1^{-1} having the highest level, followed by x_1, then x_2^{-1} then x_2, and so on, with x_n having the lowest level. (These orderings are also referred to as *recursive path orderings* in the literature.)

Now G has a monoid presentation in which the relations have the form:

$$
\begin{aligned}
x_i x_i^{-1} &= x_i^{-1} x_i = \varepsilon \quad (i \notin I), \\
x_i^{-1} &= w_i' \quad (i \in I), \\
x_i^{r_i} &= w_i \quad (i \in I), \\
x_j x_i &= x_i w_{ij} \quad (1 \le i < j \le n), \\
x_j x_i^{-1} &= x_i^{-1} w_{ij}' \quad (1 \le i < j \le n, i \notin I), \\
x_j^{-1} x_i &= x_i \overline{w_{ij}} \quad (1 \le i < j \le n), \\
x_j^{-1} x_i^{-1} &= x_i^{-1} \overline{w_{ij}}' \quad (1 \le i < j \le n, i \notin I),
\end{aligned}
$$

where, for all i and j, w_i, w_{ij}, $\overline{w_{ij}}$, w_{ij}', and $\overline{w_{ij}}'$ are words in normal form in $(A' \cap \{x_k, x_k^{-1} \mid k > i\})^*$, and $w_i' \in (A' \cap \{x_i, x_k, x_k^{-1} \mid k > i\})^*$. (Note that $\overline{w_{ij}}$ is equal in G to the inverse of w_{ij}.)

We call such a presentation a *polycyclic monoid presentation* of G, and the reader can verify that this is almost identical to a polycyclic group presentation of G as defined in Subsection 8.1.2.

Notice that, in each of these equations, the left-hand side is greater than the right-hand side in the wreath-product ordering just defined, and so we can define a corresponding rewriting system S in which the left- and right-hand sides of the rules are given by the left- and right-hand sides of the relations in the presentation.

It can be seen that the words in normal form are precisely those that do not have the left-hand side of any of these equations as a subword; that is, they are the irreducible words with respect to the ordering. Since distinct irreducible words represent distinct elements of G, they cannot be equivalent in $\leftrightarrow^*$, and so S is complete, and it is clear that all of its rules are minimal and hence essential. The process of using the rewrite-rules to reduce a word to normal form is exactly the collection process as defined in Subsection 8.1.3.

Conversely, suppose that we are given an arbitrary monoid presentation of the above form, over the alphabet A. Then we can define the rewriting system S as in the preceding paragraph.

We should start by checking that the words $x_i w_i'$ and $w_i' x_i$ reduce to the identity in S for all $i \in I$. If this is the case, then the presentation defines a group G, in which the inverse generators are what they seem.

It is not difficult to see that if, for $1 \leq i \leq n+1$, we define G_i to be the subgroup of G generated by $\{ x_j \mid j \geq i \}$, then G_{i+1} is normal in G_i with G_i/G_{i+1} cyclic for all i, so G is certainly polycyclic. Furthermore, the rewriting system S is Noetherian, and the set of reduced words is as before.

However, the order r_i' of G_i/G_{i+1} may be less than the number r_i given in the presentation when $i \in I$, or r_i' may be finite when $i \notin I$. In fact, this occurs if and only if some group element has more than one irreducible representative word; that is, if and only if S is not confluent. The confluence (and hence also the completeness) of S is equivalent to its consistency, as defined in Definition 8.10.

Lemma 12.17 provides us with a method for testing for confluence. This method was first used and justified by Wamsley for the case of nilpotent groups in [Wam70] (although not in the context of critical pair completion). The equations arising from the critical pairs that have to be checked include:

$$
\begin{aligned}
x_i^{-1} w_{ij}' x_i &= x_j & (j > i, i \notin I), \\
x_i &= x_j^{-1} x_i w_{ij} & (j > i, j \notin I), \\
x_i^{-1} &= x_j^{-1} x_i^{-1} w_{ij}' & (j > i, j \notin I, i \notin I), \\
x_i w_i &= w_i x_i & (i \in I), \\
x_j^{r_i - 1} x_i w_{ij} &= w_j x_i & (j \in I, j > i), \\
x_i w_{ij} x_i^{r_i - 1} &= x_j w_i & (i \in I, j > i), \\
x_j w_{jk} x_i &= x_k x_i w_{ij} & (k > j > i).
\end{aligned}
$$

(This is only a proper subset of all such equations, but it is proved by Sims in Proposition 8.3, Chapter 9 of [Sim94] that it suffices to check the listed equations only in order to check consistency.)

For large n, the set of equations in the final line of this list is the largest, and contains up to $n(n-1)(n-2)/6$ equations. In the nilpotent case, many more of these checks can be shown to be redundant by assigning weights to the generators according to how far down the lower central series of G they lie. In the case of finite p-groups, it is proved by Vaughan-Lee in [VL84] that it is sufficient to test approximately $n^2 d/2$ equations, where d is the minimal number of generators of the group. This result was stated as Theorem 9.22 in Subsection 9.4.1.

Exercises

1. The construction described in this exercise is the rewriting system analogue of Proposition 2.55. The complete presentations of polycyclic groups described above come from repeated applications of this construction.

   Let G be a group with normal subgroup N, and suppose that G/N and N have complete rewriting systems $\overline{\mathcal{R}}$, $\mathcal{S}$ on finite monoid generating sets $\overline{A}$ and B, respectively. Let $A \subseteq G$ with $|A| = |\overline{A}|$ and each $a \in A$ mapping onto some $\overline{a} \in \overline{A}$.

   By using a wreath-product ordering in which the generators in A have higher weights than those in B, show that G has a complete rewriting system on $A \cup B$ with rules $\mathcal{R} \cup \mathcal{S} \cup \mathcal{T}$, where rules in $\mathcal{R}$ have the form $u \to vw$ for a rule $\overline{u} \to \overline{v}$ of $\overline{\mathcal{R}}$ and an $\mathcal{S}$-reduced word $w \in B^*$, and those in $\mathcal{T}$ have the form $ba \to aw$ for $a \in A$, $b \in B$ and an $\mathcal{S}$-reduced word $w \in B^*$.

2. The *Baumslag-Solitar groups* are defined by their presentations $B(m, n) := \langle\, x, y \mid y^{-1} x^m y = x^n \,\rangle$. Use a wreath-product ordering on $A = \{x, X, y, Y\}$, where $X = x^{-1}$, $Y = y^{-1}$ and x, X, y, Y have levels 1,2,3,4, to show that $B(m, n)$ has a complete rewriting system consisting of the four inverse rules together with the four rules:

$$x^n Y \to Y x^m, \quad x^m y \to y x^n, \quad Xy \to x^{m-1} y X^n, \quad XY \to x^{n-1} Y X^m.$$

12.5 Verifying nilpotency

It is of course undecidable whether a given presentation defines a polycyclic group but, if it does, then it is possible to verify this fact. The method of doing this to be described here, which makes use of a computer implementation of the Knuth-Bendix completion algorithm, was first proposed by Sims, and is described in Chapter 11 of [Sim94]. It requires the use of a polycyclic quotient algorithm (PQA); see Subsection 9.4.3. But if we are trying to verify nilpotency, then a nilpotent quotient algorithm (NQA) suffices.

Let $G := \langle Y \mid R \rangle$ with $Y = \{y_1, \ldots, y_m\}$ be the given finitely presented group. If G is polycyclic, then it certainly has a largest polycyclic quotient G/K, and so we can use the PQA (or NQA, if we want to verify nilpotency) to compute G/K. Then G is polycyclic if and only if $K = 1$, and the objective is to verify that $K = 1$.

The output of the PQA consists of a consistent polycyclic presentation of G/K on a generating set $X = \{x_1, \ldots, x_n\}$, together with the images of the y_j in G/K, given as words u_j in normal form in the generators x_i and x_i^{-1}.

We can then use the relations in this polycyclic presentation to express each x_i as a word v_i in $Y \cup \{x_j \mid j < i\}$ and inverses. (In the existing implementations of the NQA, at least, the output is in such a form as to make this process straightforward.)

Let $D := \{\, v_i x_i^{-1} \mid 1 \leq i \leq n\,\}$. Then the presentation $\hat{G} := \langle\, Y \cup X \mid R \cup D\,\rangle$ defines a group isomorphic to G. (This is because the relators in D can be used to eliminate the generators in X using Tietze transformations.)

Now let $A := \{\, y_i, y_i^{-1}, x_j, x_j^{-1} \mid 1 \leq i \leq m,\, 1 \leq j \leq n\,\}$ be the complete set of monoid generators of $\hat{G}$, and define a wreath-product ordering on A^* (as defined in Section 12.4), in which the list of generators, in decreasing order of levels, is

$$y_1^{-1}, y_1, y_2^{-1}, \ldots, y_m, x_1^{-1}, x_1, \ldots, x_n^{-1}, x_n.$$

We run the Knuth-Bendix completion algorithm on the presentation $\hat{G}$ using this ordering.

If $K = 1$ and $G/K = G$, then the relations $y_j =_G u_j$, $y_j^{-1} =_G u_j^{-1}$ for $1 \leq j \leq m$, together with each of the relations in the x_i and their inverses, which constitute the consistent polycyclic presentation of G/K, will all hold in $\hat{G}$, and it is easily seen that they form a complete rewriting system for $\hat{G}$ with respect to the specified ordering of A^*.

These relations are all essential (at least after writing u_j^{-1} in normal form), and will therefore inevitably be generated by the Knuth-Bendix process provided that we run it for long enough. On the other hand, if these relations are generated by the Knuth-Bendix process, then it follows immediately that G is polycyclic. Hence we can, in principal, verify the polycyclicity of G.

Example 12.4

Let us now work through a moderately straightforward example to illustrate this process. For group elements g, h, the commutator $[g, h]$ is defined to be $g^{-1}h^{-1}gh$, and we define $[g, h, k] := [[g, h], k]$, etc. The following group was proved nilpotent theoretically by Heineken in [Hei61].

$$G := \langle\, a, b \mid [b, a, a, a],\ [b^{-1}, a, a, a],\ [a, b, b, b],\ [a^{-1}, b, b, b],$$
$$[a, ab, ab, ab],\ [a^{-1}, ab, ab, ab]\,\rangle.$$

Running the NQA (using the implementation by Nickel described in [Nic93]) reveals that G has a largest nilpotent quotient of class 4, with 6 polycyclic generators $x_1, \ldots, x_6$, $I = \{6\}$ and $r_6 = 2$.

Since a maps to x_1 and b to x_2 we have $u_1 = x_1$, $u_2 = x_2$, and so we can take $v_1 = a$, $v_2 = b$. In fact, in an example like this, we can simplify matters by identifying a and b with x_1 and x_2, and we shall rename the generators $x_1, \ldots, x_6$ as a, b, c, d, e, f, respectively.

The relations in the full polycyclic presentation are

$$f^2 = 1, \ ba = abc, \ ba^{-1} = a^{-1}bc^{-1}d, \ ca = acd, \ ca^{-1} = a^{-1}cd^{-1}, \ cb = bce,$$
$$cb^{-1} = b^{-1}ce^{-1}, \ db = bdf, \ db^{-1} = b^{-1}df, \ ea = aef, \ ea^{-1} = a^{-1}ef,$$

where we have omitted relations between pairs of commuting generators.

Of these, we can take $c = [b, a]$, $d = [c, a]$, $e = [c, b]$ and $f = [d, b]$ as definitions of d, e and f, respectively, and so we have

$$\hat{G} = \langle \, a, b, c, d, e, f \mid [b, a, a, a], \ [b^{-1}, a, a, a], \ [a, b, b, b], \ [a^{-1}, b, b, b],$$
$$[a, ab, ab, ab], \ [a^{-1}, ab, ab, ab], \ c = [b, a], \ d = [c, a], \ e = [c, b], \ f = [d, b] \, \rangle,$$

and use this as input to the Knuth-Bendix algorithm, with the wreath-product ordering defined above. In this example, the relations in the polycyclic presentation were produced very rapidly, thereby verifying nilpotency. $\quad \square$

Although this particular example was very easy, with more challenging examples the Knuth-Bendix process can struggle and get itself tied up with very long words if the implementation is too simplistic. There is a lengthy discussion of possible techniques for improving the performance in Chapters 2 and 3 of [Sim94], together with implementation details.

For example, it can be a good idea only to store relatively short reduction equations in the early stages of the process. Although this means that the resulting set of stored equations may not be complete at the end, it is relatively easy to verify completeness later by restarting on the enlarged set of rules, once the polycyclic relations have all been deduced.

A slightly more difficult example is

$$G := \langle \, a, b \mid [b, a, b], \ [b, a, a, a, a], \ [b, a, a, a, b, a, a] \, \rangle,$$

which turned out to be nilpotent of class 7, with 8 polycyclic generators.

12.6 Applications

Applications of the Knuth-Bendix completion procedure within group theory include verification of nilpotency, as described in the preceding section, and a recent, more spectacular application by Charles Sims, using his rkbp package, in which he shows that the Burnside group $B(2, 6) = R(2, 6)$ (cf Subsection 9.4.7) can be defined with a presentation on 2 generators and about 60000 relators, all of which are sixth powers. The previous upper bound on the number of sixth-power relators had been 2^{124}, which was derived from a careful analysis of the original finiteness proof of $B(2, 6)$ by Marshall Hall.

The Knuth-Bendix procedure can also be used to help prove that two finitely presented groups G_1 and G_2 are isomorphic, and this has led to some applications outside of group theory. By Theorem 2.52, if we are given a map θ from the generators of G_1 to G_2 then, if we can prove that the images in G_2 of the defining relators of G_1 are equal to the identity in G_2, then we will have proved that θ extends to a homomorphism from G_1 to G_2. By running Knuth-Bendix on the presentation of G_2, we may be able to collect enough reduction equations in G_2 to carry this out. It is not necessary to obtain a complete rewriting system for G_2 for this purpose. Indeed, if θ really does extend to a homomorphism then, if we run Knuth-Bendix for long enough, we will be able to verify this fact. If θ happens to extend to an isomorphism from G_1 to G_2, then we can also verify this fact, by constructing its inverse map from G_2 to G_1 and checking that this is also a homomorphism.

The above process is described in more detail by Holt and Rees in [HR92], together with the authors' implementation of a procedure that attempts to prove that G_1 and G_2 are isomorphic by systematically searching for an isomorphism θ. Because of the huge search-space involved, this process can only succeed if the images of the generators of G_1 under θ are short words in the generators of G_2 but, even so, it has on occasion been successfully applied.

One such application occurred in the classification of flat 4-manifold groups by Hillman in Section 8.4 of [Hil02]. For example, for

$$G_1 := \langle\, s, t, z \mid st^2 s^{-1} = t^{-2},\ szs^{-1} = z^{-1},\ ts^2 t^{-1} = z,\ tzt^{-1} = s^2 \,\rangle$$

and

$$G_2 := \langle\, x, y, t \mid yx^2 y^{-1} = x^{-2},\ ty^2 t^{-1} = (xy)^2,\ t(xy)^2 t^{-1} = y^{-2},\ x = t^2 \,\rangle,$$

it was proved that

$$\theta : s \mapsto y,\ t \mapsto yt^{-1},\ z \mapsto ty^2 t^{-1}$$

extends to an isomorphism $G_1 \to G_2$ with

$$\theta^{-1} : y \mapsto s,\ t \mapsto t^{-1}s,\ x \mapsto (t^{-1}s)^2.$$

Chapter 13

Finite State Automata and Automatic Groups

In this final chapter, we shall describe the principal algorithms for computing automatic structures associated with automatic groups. For some types of infinite finitely presented group, they provide the only computational tool that is currently capable of deciding important structural properties, such as the finiteness or infiniteness of the groups concerned.

The algorithms have been implemented as the major component of Holt's KBMAG package. An earlier version was programmed by Holt, in collaboration with David Epstein and Sarah Rees. The papers [EHR91] and [Hol95] describe the procedure and this earlier implementation.

These methods involve *finite state automata*, which we shall abbreviate, in both singular and plural, as fsa. For background information on the theory of fsa and the related theory of *regular languages*, the reader should consult a textbook on formal language theory, such as [HMU01]. Here, we shall be concerned with algorithms for constructing and manipulating them, and with their applications to CGT, of which there are several. For example, they can be used to reduce words to their S-reduced form with respect to a finite rewriting system S.

The definition of the class of *automatic groups* evolved in the mid-1980s following the appearance of a paper by Jim Cannon on groups acting discretely and cocompactly on a hyperbolic space [Can84]. Thurston noticed that some of the properties that were proved about these groups could be reformulated in terms of finite state automata, which gave rise to the definition that is currently in use. The most comprehensive reference on automatic groups is by D.B.A. Epstein and coauthors in the six-author book [ECH+92], and we refer the reader to that book for further information and references.

From the viewpoint of CGT, what is important is that there are procedures, which take a finite group presentation as input, and attempt to construct a collection of finite state automata that form an *automatic structure* for the presentation. These procedures involve the use of the Knuth-Bendix completion process described in the previous chapter, but they are capable of succeeding even when Knuth-Bendix alone does not terminate with a complete set of rewrite rules, and they also involve the construction and manipulation of various fsa.

If successful, then they prove that the group is automatic, and they enable words in the group generators to be reduced to a normal form. This in turn, allows various important properties of the group to be decided or calculated. For example, the finiteness or infiniteness of the group can be decided, and its growth function can be computed. So the methods provide a new computational tool for handling infinite finitely presented groups which, in several examples, has been successful where all other known methods, both theoretical and computational, have failed.

In this chapter, we shall give the definition of an automatic group, and summarize the known properties and examples of these groups. Then, we shall provide a moderately detailed description (but without going down to the level of precise pseudocode) of the procedures mentioned above. These descriptions are taken and adapted from [EHR91] and [Hol95].

13.1 Finite state automata

We start with an account of the algorithms for manipulating finite state automata that we shall require for the automatic group programs.

13.1.1 Definitions and examples

Let us first collect all of the basic definitions together.

DEFINITION 13.1 For a finite set A is a finite set, we define $A' := A \cup \{\varepsilon\}$, where it is assumed that $\varepsilon \notin A$. A *finite state automaton* (fsa) M is a quintuple $(\Sigma, A, \tau, S, F) = (\Sigma_M, A_M, \tau_M, S_M, F_M)$ where Σ (the set of states) and A (the alphabet) are finite sets, S (the set of *start states*) and F (the set of *final* or *accept states*) are subsets of A, and τ (the set of *transitions*) is a subset of $\Sigma \times A' \times \Sigma$.

A transition $(\sigma_1, a, \sigma_2) \in \tau$ can also be thought of as an arrow labelled a from state σ_1 to state σ_2. It is called an *ε-transition* if $a = \varepsilon$.

For $w = a_1 a_2 \cdots a_r \in (A')^*$, a *path of arrows* labelled w from state σ_0 to state σ_r is a sequence of transitions $(\sigma_{i-1}, a_i, \sigma_i)$ for $1 \leq i \leq r$.

For $w \in (A')^*$ let us denote the word $w \in A^*$ obtained by deleting all occurrences of ε from w by $\rho(w)$.

The *language* $L(M) \subseteq A^*$ of M is defined as follows. For $w \in A^*$, $w \in L(M)$ if and only if there exists $w' \in (A')^*$ with $\rho(w') = w$, and a path of arrows labelled w' from a start state of A to a final state of A.

Two fsa M_1 and M_2 with the same alphabet A are called *equivalent* if $L(M_1) = L(M_2)$.

The fsa M is called (partial) *deterministic* if $|S| \leq 1$, there are no ε-transitions and, for each $\sigma_1 \in S$ and $a \in A$, there is at most one transition $(\sigma_1, a, \sigma_2) \in \tau$. We shall abbreviate "deterministic finite state automaton (automata)" as dfa.

It is called *complete deterministic* if $|S| = 1$, there are no ε-transitions and, for each $\sigma_1 \in S$ and $a \in A$, there is exactly one transition $(\sigma_1, a, \sigma_2) \in \tau$.

Notice that we can convert a partial dfa into an equivalent complete dfa by adjoining an extra state σ' to Σ with $\sigma' \notin F$, and adjoining (σ, a, σ') to τ for all $\sigma \in \Sigma \cup \{\sigma'\}$ and $a \in A$ such that there is no transition $(\sigma, a, \sigma'') \in \tau$ already. We put $\sigma' \in S$ if and only if S is empty to begin with.

We shall generally assume throughout this section that $|S| = 1$ for a dfa, and leave the reader to contemplate any changes needed for the trivial case when S is empty, in which case $L(M)$ is also empty.

For a dfa M, $\sigma \in \Sigma$ and $w \in A^*$, there is at most one state σ' such that there is a path of arrows from σ to σ' labelled w, and we define $\sigma^w = \sigma'$ if σ' exists and say that σ^w is undefined otherwise.

The new state σ' used to complete a partial dfa has the property that there is no word $w \in A^*$ with $(\sigma')^w \in F$; that is, we cannot reach an accept state from σ'. Such a state is called a *dead state*. Dead states can be removed from a dfa, yielding an equivalent partial dfa.

Finite state automata are normally studied as part of formal language theory. The fundamental result in this area about fsa is that a subset of A^* is the language of an fsa (or, equivalently, of a dfa) if and only if it is a *regular language*, which means that it can be defined by means of a *regular expression*. We refer the reader to Chapter 1 of [ECH+92] or, for a detailed treatment, to Chapter 3 of [HMU01] for the definitions and a proof of this theorem. In fact our treatment of fsa and of automatic groups in this chapter will not involve regular languages.

Example 13.1

A coset table, as defined in Chapter 5, would become a (partial or complete) dfa with $\Sigma = \Omega$ if we were to specify the sets S and F. The obvious choice for S is $\{1\}$. If we put $F = \{i\}$ for some $i \in \Omega$, then the language is a subset of the coset Hg corresponding to i, and is the whole of that coset if the table is complete. ∎

For demonstration purposes, an fsa can be represented by a diagram in which the states are circles (single circles for nonaccept states and double circles for accept states), where a start state σ is represented by a small unlabelled arrow with no source, and target σ. An example with $A = \{x, X, y, Y\}$ is displayed below. Notice that there are no arrows labelled Y. The language $L(M)$ of M is finite in this example, and is equal to $\{\varepsilon, x, X, y, xy, Xy\}$.

Example 13.2

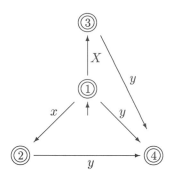

⧫

An alternative method of describing M (which is the one that we used for coset tables in Chapter 5), is to list the transitions as a table, known as the *transition table* of M, where rows are labelled by states, columns are labelled by alphabet members, and the entry in row σ and column a is σ^a if it is defined. If it is undefined, we can either leave the entry blank or put a 0 there. (This is assuming that M is a dfa. For a general fsa, we would need a column for ε, and table entries would be the subset of all targets of arrows with source σ and label a.)

In Example 13.2, the table is:

	x	X	y	Y
1	2	3	4	0
2	0	0	4	0
3	0	0	4	0
4	0	0	0	0

DEFINITION 13.2   Let $M = (\Sigma, A, \tau, \{\sigma_0\}, F)$ be a dfa. A state $\sigma \in \Sigma$ is said to be *accessible* if it can be reached from the start state; that is, if there exists $w \in A^*$ with $\sigma_0^w = \sigma$.

Dually, $\sigma \in \Sigma$ is said to be *coaccessible* if there exists $w \in A^*$ such that $\sigma^w \in F$.

The dfa M is said to accessible (respectively coaccessible) if all of its states are accessible (respectively coaccessible).

We say that M is *trim* if it is both accessible and coaccessible.

Finding the accessible states is easy and can be done using a method similar to that used in ORBIT in Section 4.1. Inaccessible states can be removed, together with all of their transitions, to yield an equivalent accessible dfa, which will be complete if the original dfa was complete.

We can also remove states that are not coaccessible without changing the language, but, if we do that, then we may end up with a partial dfa. Furthermore, locating these states is not as straightforward as for accessible states so, if we require an equivalent trim automaton, then it is probably easiest to use the dfa minimization algorithm, which will be discussed in Subsection 13.1.6. The function FSA-COUNT described in the next subsection requires a trim dfa as input.

Note also that, if $L(M)$ is empty, then there are no coaccessible states, and so the only trim fsa that accepts the empty language is one with no states. (That is the reason why we allowed S to be empty in the definition of a dfa; many books stipulate that $|S| = 1$ for a dfa.)

13.1.2 Enumerating and counting the language of a dfa

One of the standard operations that one can perform on a dfa M is to enumerate its language $L(M)$. The following function does this, but lists only those words $w \in L(M)$ up to a specified length n. The code can easily be modified to remove this restriction on length, but then it would run forever whenever $L(M)$ was infinite. It works using a straightforward backtrack search through the words in A^*. With the alphabet A numbered as $A = [a_1, \ldots, a_r]$, the current word during the search is $a_{g[1]} a_{g[2]} \cdots a_{g[l]}$, and the states $\varsigma[i]$ for $1 \le i \le l+1$ reached when M reads the prefix of this word of length $i - 1$ are also stored. We interpret $\sigma^a = 0$ to mean that the transition σ^a is undefined.

FSA-ENUMERATE(M, n)

    Input: dfa $M = (\Sigma, A = [a_1, \ldots, a_r], \tau, \{\alpha\}, F)$, $n \in \mathbb{N}_0$
    Output: List $\mathcal{L}$ of words $w \in L(M)$ with $|w| \le n$
1    $\mathcal{L} := []$; $l := 0$; $g := []$; $\varsigma[1] := \alpha$;
2   **repeat**
3   **if** $\varsigma[l+1] \in F$ **then** APPEND$(\sim \mathcal{L}, a_{g[1]} a_{g[2]} \cdots a_{g[l]})$;
4            $\beta := (l = n)$;
5   **if not** β
6   **then** $l := l+1$; $g[l] := 0$; $\beta := $ **true**;
7   **while** β **and** $l > 0$
8   **do** $\beta := $ **false**;
9                $g[l] := g[l] + 1$;
10   **while** $g[l] \le r$ **and** $\varsigma[l]^{a_{g[l]}} = 0$ **do** $g[l] := g[l] + 1$;
11   **if** $g[l] \le r$
12   **then** $\varsigma[l+1] := \varsigma[l]^{a_{g[l]}}$;
13   **else** $l := l-1$; $\beta := $ **true**;
14   **until** $l = 0$;

The search through A^* undertaken by FSA-ENUMERATE is known as a *depth first search*, which means that the words in A^* are considered in increasing order using the 'lexicographical order' (see Definition 2.59) with respect to the ordering $a_1 < a_2 < \cdots < a_r$ of A. Hence the words occur in this order in the list returned.

We may prefer to have the words ordered according to the shortlex ordering (corresponding to a *breadth first search* through A^*), in order to get all of the shorter words occurring before longer words. We can achieve this by modifying FSA-ENUMERATE such that it outputs only those words that have length exactly n, and then running FSA-ENUMERATE(M, k) for $k = 0, 1, 2, \ldots, n$. This will take a little longer than for the 'lexicographical order', but since the running time of FSA-ENUMERATE(M, k) is usually exponential in k, the increase in time is rarely critical.

If we are only interested in the cardinality $|L(M)|$ of the accepted language, then we can compute this much more quickly than by enumerating $L(M)$, although the function below requires the input dfa M to be trim.

FSA-COUNT(M, n)
    Input: Trim dfa $M = (\Sigma, A, \tau, \{\alpha\}, F)$
    Output: $|L(M)|$
1   **for** $\sigma \in \Sigma$ **do** $\delta_\sigma := |\{ (\sigma', a, \sigma) \in \tau \mid \sigma' \in \Sigma, a \in A \}|$;
2   $\varsigma[1] := \alpha$;
3   **if** $\delta_{\varsigma[1]} > 0$ **then return** ∞;
4   $k := 1$; $c := 1$;
5   **while** $k \leq c$
6   **do for** $a \in A$
7   **do if** $\sigma := \varsigma[k]^a$ is defined
8   **then** $\delta_\sigma := \delta_\sigma - 1$;
9   **if** $\delta_\sigma = 0$ **then** $c := c+1$; $\varsigma[c] := \sigma$;
10       $k := k + 1$;
11 **if** $c < |\Sigma|$ **then return** ∞;
12 **for** $k \in [c \, . \, . \, 1 \text{ by } -1]$
13   **do** $\sigma := \varsigma[c]$;
14   **if** $\sigma \in F$ **then** $\gamma_\sigma := 1$; **else** $\gamma_\sigma := 0$;
15   **for** $a \in A$
16   **do if** $\sigma' := \sigma^a$ is defined **then** $\gamma_\sigma := \gamma_\sigma + \gamma_{\sigma'}$;
17 **return** $\gamma_{\varsigma[1]}$;

The above function requires some explanation! Since M is trim, if M contains a circuit of arrows — that is, a nonempty word w and a state σ with $\sigma^w = \sigma$ — then $L(M)$ is infinite, because $w_1 w^n w_2 \in L(M)$ for all $n \geq 0$, where w_1 is a word sending the start state to σ and w_2 is a word sending σ to a final state.

The algorithm consists of three main steps. In the first step, at line 1, for each $\sigma \in \Sigma$ we calculate the number δ_σ of arrows of M with target σ. If this is nonzero for the start state, then, since M is trim, M has a circuit based at the start state and $L(M)$ is infinite.

In the second step, starting at line 4, we attempt to order the states in Σ as $\varsigma[1], \varsigma[2], \ldots$ such that $\varsigma[1]$ is the start state, and $i < j$ for all arrows $(\varsigma[i], a, \varsigma[j])$ of M. Initially, only $\varsigma[1]$ is defined. We then consider $\varsigma[i]$, for $i = 1, 2, \ldots$ in turn, and for each arrow $(\varsigma[i], a, \sigma)$ with source $\varsigma[i]$, we subtract one from δ_σ. If this reduces δ_σ to 0, then we know that all arrows with target σ have been accounted for, and their sources must all be in the currently defined list ς. Hence we can append σ to the end of the list of defined states $\varsigma[i]$, and the required property will hold for ς.

After the 'While' loop at line 5 exits, if the list ς does not contain all states of Σ, and Γ is the set of states not in ς, then $\delta_\sigma > 0$ for each $\sigma \in \Gamma$ (since otherwise it would have been appended to ς), and so there must be at least one arrow with target σ and with its source in Γ. (All arrows with source in the list ς were subtracted off from δ_σ during the 'While' loop.) This implies that there is a circuit of arrows within Γ, and so $L(M)$ is infinite. (To see that there must be a circuit within Γ, choose $\sigma \in \Gamma$ and trace a path of arrows backwards, starting from σ. We must eventually reach a state that we have already visited.)

So suppose that ς contains all states of Σ after completing the 'While' loop at line 5. Then we have succeeded in ordering the states of Σ with the property described above. In the third main step, starting at line 12, for each state σ, we count the number γ_σ of words $w \in A^*$ such that $\sigma^w \in F$. This is clearly equal to the sum of the numbers $\gamma_{\sigma'}$ over all arrows (σ, a, σ') with source σ plus an extra one, for $w = \varepsilon$, if $\sigma \in F$. Since all arrows go from $\varsigma[i]$ to $\varsigma[j]$ with $i < j$, we can calculate $\gamma_{\varsigma[i]}$ for all i by running through the list ς in reverse order, and then $|L(M)| = \gamma_{\varsigma[1]}$, which is finite in this situation.

13.1.3 The use of fsa in rewriting systems

In practice, in implementations of the Knuth-Bendix completion process on finite rewriting systems S over A^*, most of the time is taken up in reducing words to equivalent irreducible words using the rules in S. This process can be rendered much more efficient by the use of a certain dfa M_S, which we shall describe briefly in this section.

The alphabet of M_S is A and $\Sigma = \Sigma_S$ is defined to be the set of all prefixes of all left-hand sides of rules in S. The start state is ε. The set F of final states is the set of proper prefixes of left-hand sides of rules. For $\sigma \in F$ and $a \in A$, σ^a is defined to be the longest suffix of σa that lies in Σ. For $\sigma \in \Sigma \setminus F$, σ^a is undefined for all $a \in A$. So the nonaccept states, all of which are complete left-hand sides of rules, are dead states.

It is straightforward to show by induction on $|w|$ that, when M reads a word $w \in A^*$, then ε^w is either undefined or is equal to the longest suffix v of w with $v \in \Sigma$.

Let us assume that S is reduced, and hence that no left-hand side of any rule of S is a proper substring of any other left-hand side. Then the nonfinal states of M_S are exactly the left-hand sides of rules. Furthermore, if $w = w_1 w_2 w_3$, where w_2 is the left-hand side of a rule, and is the first occurrence of the left-hand side of a rule in w, then after reading the prefix $w_1 w_2$ of w, M will be in the dead state w_2. So $L(M_S)$ is equal to the set of S-irreducible words.

To use S to reduce words, we read w into M_S, and if we find $\varepsilon^{w_1 w_2} = w_2$ is a dead state for a prefix $w_1 w_2$ of w, then we locate the rule $(w_2, v_2) \in S$, replace w_2 by v_2 in w, and restart the process. (Or, for greater efficiency, if we remember the 'state history' of M reading $w_1 w_2$, then we can restart reading the modified word w into M after w_1 and before v_2.)

For example, consider the complete rewriting system for the dihedral group of order 6 with $A = \{x, X, y, Y\}$ that we considered in Example 12.2. Let us number the states Σ of M_S as follows:

$$1 = \varepsilon, \quad 2 = x, \quad 3 = X, \quad 4 = y, \quad -1 = Y, \quad -2 = xX,$$
$$-3 = Xx, \; -4 = xx, \; -5 = XX, \; -6 = yy, \; -7 = yX, \; -8 = yx$$

So $F = \{1, 2, 3, 4\}$ and the dead states are those with negative numbers. The transitions are as follows:

	x	X	y	Y
1	2	3	4	−1
2	−4	−2	4	−1
3	−3	−5	4	−1
4	−8	−7	−6	−1

Notice that this is the same as Example 13.2, except that we have replaced the undefined transitions by transitions to numbered dead states. This type of fsa is sometimes called an *index automaton* because the dead states provides an index to the reduction rules of S.

For example, let us use this fsa to reduce the word $w = yxYX$. On reading the first letter y of W, M moves from state 1 to $1^y = 4$, and then 4^x is the dead state -8, which represents the left-hand side of the rule $(yx, Xy) \in S$, so Ww is replaced by $XyYX$. We then get $1^{Xy} = 4$, $4^Y = -1$, so we use $(Y, y) \in S$ to replace Y by y, giving $w = XyyX$. Then $1^{Xy} = 4$, $4^y = -6$, and we use $(yy, \varepsilon) \in S$ and replace w by XX. Finally, $1^X = 3$, $3^X = -5$, and we use $(XX, x) \in S$, and replace w by x. Now $1^w = 1^x = 2$, an accept state, so w is irreducible.

If S is not reduced, then M_S may sometimes fail to notice a substring w_2 of w which is a left-hand side of a rule but is also a proper substring of the left-hand side of another rule. Hence it may accept and fail to reduce some S-reducible words. This is not disastrous provided that S is reduced at regular intervals during the Knuth-Bendix completion process.

We shall not go into the details of the algorithm for constructing M_S from S here, and refer the reader to Section 3.5 of Sims' book [Sim94] for the code for doing this. During the Knuth-Bendix completion process, the set S of rules is continually changing, and so M_S needs to be reconstructed very frequently. It is therefore vital to use an efficient implementation of the construction of

M_S. The fsa M_S can also be used to speed up the search for critical pairs; this usage is also discussed in detail in Section 3.5 of [Sim94].

A disadvantage of using M_S for word reduction is that M_S will often use more memory than S itself. There are other methods of string reduction, such as one using the Rabin-Karp string-matching algorithm, which are slower but use much less memory.

13.1.4 Word-acceptors

DEFINITION 13.3   If $G = \langle X \mid R \rangle$ is a finitely presented group and $A := X \cup X^{-1}$, then a finite state automaton W with input alphabet A is called a *word-acceptor* for G if it accepts at least one word in A^* for each group element. It is called a *unique word-acceptor* for G if it accepts a unique word for each group element.

We could define (unique) word-acceptors in the same way for finite monoid presentations, but we shall be concerned almost exclusively with groups in this book.

Suppose that S is a rewriting system for a finitely presented group, as described in Section 12.3. The rules in S all have the form (w_1, w_2) where $w_1 =_G w_2$ and $w_1 > w_2$ in the reduction ordering on A^* used to define S. Let us say that a word w is $<$-minimal (for the ordering $<$) if there is no $v \in A^*$ with $v < w$ and $v =_G w$. So the $<$-minimal words are precisely the minimal elements of the classes of the equivalence relation $\leftrightarrow^*$ defined on A^* by S.

The $<$-minimal words are certainly S-irreducible, and so they lie in $L(M_S)$, where the fsa M_S is defined as in the preceding subsection. Hence M_S is word-acceptor for G. If S is complete, then $L(M_S)$ contains only the $<$-minimal words, and then M_S is a unique word-acceptor for G.

If W is a unique word-acceptor for a group, such as the fsa M_S for a finite complete rewriting system S for the group, and we apply FSA-ENUMERATE to W, then we get exactly one representative for each group element. In that case, we can calculate the order of G, and hence decide whether G is finite or infinite, by applying FSA-COUNT to W.

If W is a word-acceptor but not a unique one, then we get more than one representative for some group elements but, in the case of $W = M_S$, if we have run the Knuth-Bendix process for a long time, then it may well happen that there is no repetition of group elements on short words. Some of the most important applications that this subject has found have been outside of group theory, and in fields such as hyperbolic geometry and ergodic theory. In these applications, all that is required is to enumerate the group elements with as little repetition as possible.

A further application worth mentioning is to the calculation of growth functions. For an fsa M, let $\alpha(n)$ denote the number of accepted words of M of length n. Then the function $f(t) := \sum_{i=0}^{\infty} \alpha(n)t^n$ is called the *growth function* of M, and it turns out that this is a rational function of t that can be calculated from M. We shall not discuss this algorithm here, but refer the interested reader to [EIFZ96].

If W is a unique word-acceptor for a group, and all accepted words are shortest possible for the group element that they represent, then the growth function for W is the same as that for G. This can be calculated if we can find a finite complete set S using a shortlex ordering on A^*.

Unfortunately, it is only rarely that we can find a complete rewriting system for an infinite group. In Section 13.2 below, we shall study a different possibility for constructing a unique word-acceptor for G.

13.1.5 2-variable fsa

The theory of automatic groups involves fsa that read two words $w_1, w_2 \in A^*$ simultaneously. We can almost do this by using a fsa with alphabet $A \times A$, but this does not work if $|w_1| \neq |w_2|$. To deal with words of unequal length, we introduce an extra alphabet symbol \$, known as a *padding symbol*, which is assumed not to be in A, and use it to pad the shorter of the two words at the end, so that the resulting padded word has the same length as the longer word.

To be precise, let $A^+ := A \cup \{\$\}$, and let $w_1, w_2 \in A^*$ with $w_1 = a_1 a_2 \cdots a_l$ and $w_2 = b_1 b_2 \cdots b_m$ with each $a_i, b_i \in A$. Then we define $(w_1, w_2)^+ \in (A^+ \times A^+)^*$ to be the word $(\alpha_1, \beta_1)(\alpha_2, \beta_2) \cdots (\alpha_n, \beta_n)$, where $n := \max(l, m)$, and

(i) $\alpha_i = a_i$ for $1 \leq i \leq l$ and $\alpha_i = \$$ for $l < i \leq n$;

(ii) $\beta_i = b_i$ for $1 \leq i \leq m$ and $\beta_i = \$$ for $m < i \leq n$.

For example, if $w_1 = abc$ and $w_2 = d$, then $(w_1, w_2)^+ = (a, d)(b, \$)(c, \$)$.

We define a 2-variable fsa M over the alphabet A to be a fsa with alphabet $A^+ \times A^+$ with the property that all words in $L(M)$ are of the form $(w_1, w_2)^+$ for words $w_1, w_2 \in A^*$. Notice that $(\$, \$)$ never occurs in an accepted word of M, so we could, if we wished, take the alphabet to be $A^+ \times A^+ \setminus \{(\$, \$)\}$, and some authors do this.

We can define an n-variable fsa over A for any $n > 1$ in a similar way. In fact, the construction of the composite of two 2-variable fsa, which we shall discuss below in Subsection 13.1.7, involves a 3-variable fsa.

13.1.6 Operations on finite state automata

The procedures for computing the fsa for automatic structures of automatic groups that we shall be describing in Section 13.2 involve two main components. The first is the Knuth-Bendix completion process, which we have already discussed, and the second is the use of various operations on fsa, which we shall describe in this and the following subsection.

We shall assume that all fsa arising in this subsection have the same fixed alphabet A. The general idea is that we are given one or more fsa as input, from which we want to construct another fsa. The first two algorithms construct a dfa equivalent to the input fsa. Given a general fsa we describe how to construct an equivalent dfa, and given a dfa, we describe how to construct an equivalent dfa with the smallest possible number of states. This minimization algorithm also enables us to test two dfa for equivalence.

The other algorithms all take one or two dfa M_1 (and M_2) as input. They output a dfa whose language is one of various simple functions of $L(M_1)$ and $L(M_2)$. These include $\neg L(M_1) := A^* \backslash L(M_1)$, $L(M_1) \cap L(M_2)$, $L(M_1) \cup L(M_2)$, $L(M_1)L(M_2)$ and $L(M_1)^*$. The last two of these are defined as follows.

DEFINITION 13.4 If $L, L_1, L_2 \subseteq A^*$, then

(i) $L_1 L_2 := \{ w_1 w_2 \mid w_1 \in L_1,\ w_2 \in L_2 \}$;

(ii) $L^0 := \{\varepsilon\}$ and $L^n := LL^{n-1}$ for $n > 0$;

(iii) $L^* := \bigcup_{n \geq 0} L^n$.

We shall merely explain how these algorithms work, without writing out detailed code for them. One reason for this is that the code in a practical implementation involves techniques such as hashing (see, for example the detailed treatment by Knuth in [Knu73]), which we do not want to discuss in detail here. We shall, however, make some general comments on implementation questions later on.

13.1.6.1   Making an fsa deterministic

Let us start with the algorithm for finding a dfa equivalent to a general fsa. Computations with dfa are generally much more efficient than those with equivalent fsa, and all of the other algorithms to be discussed require a dfa as input. Unfortunately this gain in efficiency comes at a price. For an fsa with state-set Σ, the state-set of the equivalent dfa that we shall construct is a subset of the power set $\mathcal{P}(\Sigma)$ of Σ. (By definition, $\mathcal{P}(\Sigma)$ is the set of all subsets of Σ.) So, if the original fsa has n states, then the equivalent dfa can have up to 2^n states, and there are examples (see exercise below) in which the minimal possible number of states of an equivalent dfa is 2^{n-1}. This exponential blowup in the number of states is the underlying reason why many fsa algorithms, including those for computing with automatic groups, can have very large memory requirements.

If the fsa M with state-set Σ has no ε-transitions, then we proceed as follows. We define M' with state-set $\mathcal{P}(\Sigma)$, put $S_{M'} := \{S_M\}$, and let $F_{M'}$ be the set of all elements of $\mathcal{P}(\Sigma)$ that contain a set in F_M. For $\Upsilon \in \mathcal{P}(\Sigma)$ and $a \in A$, we define

$$\Upsilon^a := \{\, \sigma_2 \in \Sigma \mid (\sigma_1, a, \sigma_2) \in \tau_M \text{ for some } \sigma_1 \in \Upsilon \,\}.$$

It is straightforward to show that $L(M') = L(M)$.

If M has ε-transitions, then we make the following modification. For each $\Upsilon \in \mathcal{P}(\Sigma)$, define the ε-closure $\varepsilon(\Upsilon)$ of Υ as follows. Let $\varepsilon_0(\Upsilon) := \Upsilon$ and, for $n > 0$, let

$$\varepsilon_n(\Upsilon) := \varepsilon_{n-1}(\Upsilon) \cup \{\, \sigma_2 \in \Sigma \mid (\sigma_1, \varepsilon, \sigma_2) \in \tau_M \text{ for some } \sigma_1 \in \varepsilon_{n-1}(\Upsilon) \,\}.$$

This is an ascending sequence of subsets of Σ, and we define $\varepsilon(\Upsilon)$ to be the largest subset in the sequence.

The state-set of M' is now defined to be $\{\, \varepsilon(\Upsilon) \mid \Upsilon \in \mathcal{P}(\Sigma) \,\}$, the start state of M' is set equal to $\varepsilon(S_M)$, and the transition Υ^a of M' is defined to be the ε-closure of Υ^a as defined above.

Allowing ε-transitions is an extra complication for the determinization algorithm, but it is worthwhile, because the constructions of the fsa $\mathsf{cat}(M_1, M_2)$ and M^* to be described in Subsubsection 13.1.6.5 below would be considerably more complicated without ε-transitions.

In practice, we would begin by constructing the start state of M', and then define new states as required as the targets of transitions from states already defined, using code like that in ORBIT in Section 4.1. Then we would only ever need to define and store the accessible states of M'; in many examples, including those arising in the automatic groups procedures, only a small number of the elements of $\mathcal{P}(\Sigma)$ are accessible; indeed, if that were not the case, then these methods would be totally impractical!

13.1.6.2 Minimizing an fsa

The next algorithm to be discussed is one for finding a dfa with a minimal number of states that is equivalent to a given dfa M. We shall see that this minimal dfa is unique up to the naming and ordering of its states. We start by summarizing the theory of the method. The reader could consult Section 4.4 of [HMU01] for a more detailed and leisurely treatment.

For a subset L of A^* and $w \in A^*$, define $L(w) := \{\, v \in A^* \mid wv \in L \,\}$. We say that $w_1, w_2 \in A^*$ are *L-equivalent* and write $w_1 \sim_L w_2$ if $L(w_1) = L(w_2)$. Notice that $w_1 \sim_L w_2 \Rightarrow w_1 a \sim_L w_2 a$ for all $a \in A$, so $\sim_L$ is a right congruence on A^*.

If there are only finitely many L-equivalence classes, then we can define a complete dfa M_L with language L, as follows. Let $[w]$ denote the $\sim_L$-class of $w \in A^*$. The states of M_L are the L-equivalence classes, the start state is $[\varepsilon]$, and $[w]$ is a final state if and only if $w \in L$ (note that $w \in L$ iff $\varepsilon \in L(w)$, so this is well-defined). The transitions are defined by $[w]^a = [wa]$; these are well-defined because $\sim_L$ is a right congruence. To see that $L(M_L) = L$ just note that $[\varepsilon]^w = [w]$ for all $w \in A^*$.

Suppose now that $M = (\Sigma, A, \tau, \{\sigma_0\}, F)$ is a complete accessible dfa with $L(M) = L$. For $\sigma \in \Sigma$, define $L(\sigma) := \{\, v \in A^* \mid \sigma^v \in F \,\}$. Then $L(\sigma) = L(w)$ for any $w \in A^*$ with $\sigma_0^w = \sigma$. Hence all such words w are L-equivalent. We say that $\sigma_1, \sigma_2 \in \Sigma$ are L-equivalent, if words w_1, w_2 with $\sigma_0^{w_i} = \sigma_i$ are L-equivalent (w_1, w_2 exist because we are assuming that M is accessible.) Then each L-equivalence class of states corresponds to an L-equivalence class of words in A^*, and all such L-equivalence classes of words occur in this way. Hence each state of M_L corresponds to an L-equivalence class of Σ. This shows that M_L is the unique (up to naming of states) complete dfa with a minimal number of states and language L.

Since M_L has at most one dead state, consisting of those w with $L(w)$ empty, we get the unique minimal partial dfa with language L by removing this dead state if it exists. We would do this if we wanted an equivalent trim dfa; as input for FSA-COUNT, for example.

We now describe an algorithm to construct M_L when $M = (\Sigma, A, \tau, \{\sigma_0\}, F)$ is a given dfa with $L(M) = L$. By adding a dead state if necessary, we can assume that M is complete and, by removing inaccessible states, we can assume that M is accessible. The problem is then to decide which states of M are L-equivalent. The cases $F = \phi$ and $F = \Sigma$, corresponding to $L = \phi$ and $L = A^*$ are easy: M_L has a single state in those cases. So let us assume that F is a proper nonempty subset of Σ.

If $\sigma_1 \sim_L \sigma_2$, then $\sigma_1^a \sim_L \sigma_2^a$ for all $a \in A$, so $\sim_L$ is a right congruence. Furthermore, if $\sigma_1 \in F$ and $\sigma_2 \notin F$, then $\sigma_1 \nsim_L \sigma_2$. In other words, $\sim_L$ is a refinement of the equivalence relation $\sim_F$ on Σ that has just the two classes F and $\Sigma \setminus F$.

Let $\sim$ be any right congruence on Σ that is a refinement of $\sim_F$. We claim that $\sim$ is a refinement of $\sim_L$. To see this, observe that $\sigma_1 \sim \sigma_2$ implies $\sigma_1^w \sim \sigma_2^w$ for any $w \in A^*$ and hence $\sigma_1^w \in F$ iff $\sigma_2^w \in F$, so $\sigma_1 \sim_L \sigma_2$.

So our task is to compute the coarsest (or largest) right congruence on Σ that is a refinement of $\sim_F$. Note that this is the dual of the problem considered in Sections 4.3 (finding block systems of permutation groups) and 5.1 (the coincidence routine in coset enumeration), where we wanted the smallest right congruence that contained a given equivalence relation.

The most straightforward way of computing $\sim_L$ is as follows. We start by putting $\sigma := \sigma_F$. The method consists of a sequence of passes. On each pass, we check whether σ is a right congruence. If not, we replace it by a proper refinement σ_R with the property that any right congruence refining σ must also refine σ_R, and perform another pass. This process must clearly terminate with $\sigma = \sigma_L$. The refinement σ_R is defined from σ as follows. For $\sigma_1, \sigma_2 \in \Sigma$, $\sigma_1 \sim_R \sigma_2$ if and only if $\sigma_1 \sim \sigma_2$ and $\sigma_1^a \sim \sigma_2^a$ for all $a \in A$.

Example 13.3

Let $A = [a, b]$, $\Sigma = \{1 .. 9\}$ $S = \{1\}$, $F = \{6, 7, 8, 9\}$, where the transitions of S are as follows:

σ:	1	2	3	4	5	6	7	8	9
σ^a:	2	2	5	6	8	6	7	8	9
σ^b:	3	4	3	7	9	6	7	8	9

Then $L(M)$ consists of all words that contain at least one of *aba*, *abb*, *baa*, *bab* as a substring. It has been constructed in a similar way to the construction of M_S as defined in Subsection 13.1.3. The states $1..9$ correspond to the word w read so far having the suffix ε, a, b, ab, ba, *aba*, *abb*, *baa*, *bab*, respectively, but we have put $\sigma^a = \sigma^b = \sigma$ for all $\sigma \in F$.

At the beginning of the first pass, we have two σ-classes, $\{1, 2, 3, 4, 5\}$ and $\{6, 7, 8, 9\}$. After the first pass, the first of these splits into $\{1, 2, 3\}$ and $\{4, 5\}$. During the second pass, the first of these splits into three separate classes, $\{1\}$, $\{2\}$ and $\{3\}$. The third pass reveals that we now have a right congruence, so the algorithm terminates after three passes, with 5 L-classes.

The resulting minimized dfa has 5 states $\{1..5\}$ corresponding to the classes $\{1\}$, $\{2\}$, $\{3\}$, $\{4, 5\}$, and $\{6, 7, 8, 9\}$ of M. We have $S = \{1\}$, $F = \{5\}$, and the transitions are:

σ:	1	2	3	4	5
σ^a:	2	2	4	5	5
σ^b:	3	4	3	5	5

☐

One problem with this algorithm is that its complexity is $\Theta(|A|n^2)$, where $n = |\Sigma|$. This is because each pass has complexity $\Theta(|A|n)$, and there are examples that require $n-1$ passes (see exercise below). It is, however, very efficient in terms of space requirements: the total memory used is typically less than twice that taken by the input dfa M and, since the transition table of M is always accessed sequentially during a given pass, it can be kept in a file on disk if necessary and read in as required. In applications to the automatic groups procedures, we need to minimize some very large fsa that have typically arisen through determinizing a nondeterministic fsa, and memory constraints tend to be more critical than time constraints. Furthermore, the number of passes required in typical applications is much less than n.

An $O(|A|n\log(n))$ algorithm for solving this problem is described in [Hop71] by J.E.Hopcroft, but its memory requirements are unclear to the author. Even a requirement of twice the space taken by the original fsa would be unacceptably large for the automatic groups applications.

13.1.6.3   Testing for language equality

We have just seen that, for a given language L of an fsa, there is a complete dfa M_L with a minimal number of states having that language, which is unique up to the labelling of the states. So, if we number the states $1..n$ with $n = |\Sigma|$, then M_L is unique up to the order of the states.

We now recall the procedure STANDARDIZE that we introduced in Subsection 5.2.3 to standardize the order of the cosets in a coset table. We can apply STANDARDIZE to the transition table of a dfa to permute the states so that they occur in a standard order, corresponding to the order in which they would occur if we applied ORBIT to the initial state of M. This produces a canonical complete dfa for L.

So, if we wish to test whether $L(M_1) = L(M_1)$ for two dfa M_1 and M_2, then we minimize and standardize both, and then just check whether the resulting dfa are identical.

13.1.6.4 Negation, union, and intersection

Given a complete dfa M, it is very easy indeed to define a dfa with language $\neg L(M) = A^* \setminus L(M)$. We simply replace F_M by $\Sigma_M \setminus F_M$.

Given dfa M_1 and M_2, we can define dfa with languages $L(M_1) \cup L(M_2)$ and $L(M_1) \cap L(M_2)$ as follows. In both cases, we define the state-set to be $\Sigma_{M_1} \times \Sigma_{M_2}$, the set of start states to be $S_{M_1} \times S_{M_2}$, and the set of transitions to be

$$\{ ((\sigma_1, \sigma_2), a, (\sigma_1^a, \sigma_2^a)) \mid \sigma_1 \in \Sigma_{M_1}, \sigma_2 \in \Sigma_{M_2} \}.$$

A state $(\sigma_1, \sigma_2) \in \Sigma_{M_1} \times \Sigma_{M_2}$ is defined to be a final state of $L(M_1) \cup L(M_2)$ if $\sigma_1 \in F_{M_1}$ or $\sigma_2 \in F_{M_2}$; and it is a final state of $L(M_1) \cap L(M_2)$ if $\sigma_1 \in F_{M_1}$ and $\sigma_2 \in F_{M_2}$. We leave readers to convince themselves that the dfa that we have defined have the required language.

As was the case for the subset construction of dfa from a general fsa defined in Subsubsection 13.1.6.1, when we construct the dfa for $L(M_1) \cup L(M_2)$ and $L(M_1) \cap L(M_2)$, in practice, we do not explicitly define the whole set $\Sigma_{M_1} \times \Sigma_{M_2}$ initially. Instead, we begin with just the start state and then construct those states in $\Sigma_{M_1} \times \Sigma_{M_2}$ that are accessible from the start state.

13.1.6.5 Concatenation and star

Let M_1 and M_2 be fsa. We define an fsa $\mathsf{cat}(M_1, M_2)$ with accepted language $L(M_1)L(M_2)$ as follows. By renaming the states of M_2 if necessary, we assume that Σ_{M_1} and Σ_{M_2} are disjoint, and define $\Sigma_{\mathsf{cat}(M_1,M_2)} := \Sigma_{M_1} \cup \Sigma_{M_2} \cup \{\sigma\}$, where $\sigma \notin \Sigma_{M_1} \cup \Sigma_{M_2}$. We define $S_{\mathsf{cat}(M_1,M_2)} := S_{M_1}$ and $F_{\mathsf{cat}(M_1,M_2)} := F_{M_2}$. The transitions of $\mathsf{cat}(M_1, M_2)$ consist of those of M_1 and M_2, together with ε-transitions from the final states of M_1 to σ, and ε-transitions from σ to the start states of M_2. That is:

$$\tau_{\mathsf{cat}(M_1,M_2)} := \tau_{M_1} \cup \tau_{M_2} \cup \{ (\rho, \varepsilon, \sigma) \mid \rho \in F_{M_1} \} \cup \{ (\sigma, \varepsilon, \rho) \mid \rho \in S_{M_2} \}.$$

So a path of arrows for a word accepted by $\mathsf{cat}(M_1, M_2)$ has the form $p_1 p_2 p_3 p_4$, where p_1 is a path of arrows for an accepted word through M_1, p_2 is an ε-arrow from a state in F_{M_1} to σ, p_3 is an ε-arrow from σ to a state in S_{M_2}, and p_4 is a path of arrows for an accepted word through M_2. Hence $L(\mathsf{cat}(M_1, M_2)) = L(M_1)L(M_2)$, as required.

Of course, if we want a dfa with language $L(M_1)L(M_2)$, then we will need to apply the algorithm described in Subsubsection 13.1.6.1 to $\mathsf{cat}(M_1, M_2)$.

Let M be a fsa. The construction of an fsa M^* with language $L(M^*)$ is of a similar nature to that for $L(M_1)L(M_2)$. The state-set of M^* is $\Sigma_M \cup \{\sigma\}$, where $\sigma \notin \Sigma_M$. The additional state σ is the only start state and the only final state; that is $S_{M^*} = F_{M^*} = \{\sigma\}$. The transitions of M^* include all transitions of M and, in addition, ε-transitions from σ to the start states of M and from the final states of M to σ. That is:

$$\tau_{M^*} := \tau_M \cup \{\, (\sigma, \varepsilon, \rho) \mid \rho \in S_M \,\} \cup \{\, (\rho, \varepsilon, \sigma) \mid \rho \in F_M \,\}.$$

13.1.7 Existential quantification

Let M be a 2-variable dfa over A. It is straightforward to construct an fsa $M_\exists$ over A with language $\{\, w \in A^* \mid \exists v \in A^* \times A^* : (w, v)^+ \in L(M) \,\}$. To do this, we let the state-set, the start states, and the final states be the same for $M_\exists$ as for M. We replace a transition $(\sigma_1, (\alpha, \beta), \sigma_2)$ of M by a transition $(\sigma_1, \alpha', \sigma_2)$ of $M_\exists$, where $\alpha' = \alpha$ if $\alpha \in A$ and $\alpha' = \varepsilon$ if $\alpha = \$$.

Then, for any accepting path of arrows of $M_\exists$ with label $\alpha_1' \cdots \alpha_r'$, there is a corresponding accepting path of M with label $(\alpha_1, \beta_1) \cdots (\alpha_r, \beta_r)$ for some $\beta_i \in A^+$. By definition of a 2-variable fsa, M accepts only words of the form $(w_1, w_2)^+$ for $w_1, w_2 \in A^*$, and so any occurrences of ε in the path $\alpha_1' \cdots \alpha_r'$ must occur at the end of the word. The corresponding accepted word of $M_\exists$ is then just $\alpha_1' \cdots \alpha_r'$ with any trailing ε symbols removed, which is precisely the word w_1 for which $(\alpha_1, \beta_1) \cdots (\alpha_r, \beta_r) = (w_1, w_2)^+$. Hence the language of $M_\exists$ is as required.

As was the case for $\mathsf{cat}(M_1, M_2)$ and for M^*, the fsa $M_\exists$ is nondeterministic in general, and so we need to apply the subset construction from Subsubsection 13.1.6.1 to get an equivalent dfa. This process often results in a substantial increase in the number of states. In practice, even the minimized version of an equivalent dfa for $M_\exists$ can have very many more states than the original nondeterministic version; see exercise below.

An fsa $M_\forall$ with language $\{\, w \in A^* \mid \forall v \in A^* \times A^* : (w, v)^+ \in L(M) \,\}$ could be constructed if required by using the fact that $L(M_\forall) = \neg L(M_\exists')$, where M' is a 2-variable fsa with language $\neg L(M)$.

In some applications, for a given 2-variable fsa M and a word $w \in A^*$, we need to test whether $w \in L(M_E)$ and, if so, to find a specific word v with $(w, v)^+ \in L(M)$. The following function achieves this aim. A brief explanation of the algorithm will follow the function itself, but the reader needs to be aware of one point of notation. A list $\mathcal{S}$ is constructed, where each $\mathcal{S}[i]$ is itself a list of triples (σ, a, ρ), where σ and ρ are states of M, and $a \in A \cup \{\varepsilon\}$. Within the function, the notation $\mathcal{S}[i][3]$ should be understood as meaning

$$\mathcal{S}[i][3] := \{\, \rho \mid \exists \sigma, a : (\sigma, a, \rho) \in \mathcal{S}[i] \,\}.$$

EXISTSWORD(M, w)

    Input: 2-variable dfa $M = (\Sigma, A, \tau, \{\sigma_0\}, F)$, $w = a_1 a_2 \cdots a_n \in A^*$
    Output: $v \in A^*$ with $(w, v)^+ \in L(M)$, or **false** if no such v exists

1   **for** $i \in [1 \mathinner{.\,.} n]$
2   **do if** $i = 1$ **then** $S := [\sigma_0]$; **else** $S := \mathcal{S}[i-1][3]$;
3          $\mathcal{S}[i] := []$;
4   **for** $(\sigma, a) \in S \times A^+$
5   **do if** $\sigma^{(a_i, a)} \notin \mathcal{S}[i][3]$
6   **then** APPEND$(\sim \mathcal{S}[i], (\sigma, a, \sigma^{(a_i, a)}))$;
7   **if** $\mathcal{S}[i] = []$ **then return false**;
8   $i := n$;
9   **if** $i = 0$ **then** $f := (\sigma_0 \in F)$; **else** $f := \exists \sigma \in \mathcal{S}[i][3] \cap F$;
10 **while** $\mathcal{S}[i] \neq []$ **and not** f
11   **do** $S := \mathcal{S}[i][3]$; $i := i+1$; $\mathcal{S}[i] := []$;
12   **for** $(\sigma, a) \in S \times A$
13   **do if** $\sigma^{(\$, a)} \notin \cup_{j=n}^{i} \mathcal{S}[j][3]$
14   **then** APPEND$(\sim \mathcal{S}[i], (\sigma, a, \sigma^{(\$, a)}))$;
15     $f := \exists \sigma \in \mathcal{S}[i][3] \cap F$;
16 **if not** f **then return false**;
17 Choose $\sigma \in \mathcal{S}[i][3] \cap F$;
18 $v := \varepsilon$;
19 **for** $j \in [i \mathinner{.\,.} 1$ **by** $-1]$
20   **do** Find $(\rho, a) \in \Sigma \times A$ with $(\rho, a, \sigma) \in \mathcal{S}[j]$;
21       $v := av$; $\sigma := \rho$;
22 **return** v;

In the 'For' loop of the function beginning at line 1, we read the input word $w = a_1 \cdots a_n$ and, for each prefix $w_i := a_1 \cdots a_i$, we store a list $\mathcal{S}[i]$ of triples (σ, a, ρ). The third components of these triples are those $\rho \in \Sigma_M$ for which $\sigma_0^{(w_i, w_i')} = \rho$ for some word $w_i' \in (A^+)^*$. The first two components specify the relevant transition of M of which ρ is the target, and enable the word w_i' to be reconstructed later.

At the end of the 'For' loop, if $\mathcal{S}[n]$ contains a success state, then there exists v with $(w, v) \in L(M)$ and we set the variable f to **true**. Otherwise it is still possible that there is $v \in A^*$ with $|v| > |w|$ and $(w, v)^+ \in L(M)$. In the 'While' loop at line 10, we test for this by appending the padding symbol to the end of w until no new states appear in the $\mathcal{S}$ list. (This is similar to the process of computing the ε-closure of a subset of states when we determinize a nondeterministic fsa.)

At the end of this loop, if we have found a success state as a third component of a triple in $\mathcal{S}[i]$ for some $i \geq n$, then there exists $v \in A^*$ with $(w, v)^+ \in L(M)$, and f is set to **true**. The word v is reconstructed in reversed order in the section of the function beginning at line 18, by using the triples in $\mathcal{S}[i]$ to trace back through the transitions in v in $\sigma_0^{(w, v)}$.

We shall also require a related but slightly more complicated construction. Let M_1 and M_2 be two 2-variable dfa over A. Then we define their *composite* M_{12} to be a 2-variable fsa with language

$$\{\,(w_1, w_2)^+ \in (A^+ \times A^+)^* \mid \exists w \in A^* : (w_1, w)^+ \in L(M_1) \text{ and } (w, w_2)^+ \in L(M_2)\,\}.$$

To construct M_{12}, we first construct a 3-variable fsa M'_{12} with language

$$\{\,(w_1, w, w_2)^+ \in (A^+ \times A^+ \times A^+)^* \mid$$
$$(w_1, w)^+ \in L(M_1) \text{ and } (w, w_2)^+ \in L(M_2)\,\}.$$

The state-set of M'_{12} is $\Sigma_{M_1} \times \Sigma_{M_2}$; the start state is (σ_1, σ_2), where σ_1 and σ_2 are the start states of M_1 and M_2, respectively; the final states are (σ_1, σ_2), where σ_1 and σ_2 are final states of M_1 and M_2, respectively; and the transitions are $(\sigma_1, \sigma_2)^{(a,b,c)} = (\sigma_1^{(a,b)}, \sigma_2^{(b,c)})$, where the transitions in the right-hand side are transitions of M_1 and M_2, respectively. There are some additional technicalities concerning the padding symbol that need to be got right, and which we leave to the reader! We then construct M_{12} from M'_{12}, by quantifying over the second variable, in the same way that we defined M_E from M.

Of course, we only actually define those states of M'_{12} that are accessible from the start state but, even so, it will generally have many more states than M_1 and M_2. The determinization of the nondeterministic fsa M_{12} resulting from the quantification can then result in the equivalent dfa having a very large number of states.

Exercises

1. If L is the language of an fsa, show that there is an fsa with the reversed language $L^R := \{\,a_n a_{n-1} \cdots a_1 \mid a_1 a_2 \cdots a_n \in L\,\}$.
   (*Hint*: Reverse the direction of all arrows, and interchange S and F.)

2. Let $A := \{0, 1\}$ and, for $n > 0$, define $L_n := A^* \{1\} A^{n-1}$. In other words, L_n consists of all words in A^* that have length at least n and have a 1 in the n-th position before the end. Show that there is an fsa M with $n+1$ states and $L(M) = L_n$, but that any dfa having language L_n has at least 2^n states.
   (*Hint*: For the second part, show that there are 2^n distinct sets $L_n(w) := \{\,v \in A^* \mid wv \in L_n\,\}$, one for each $w \in A^*$ with $|w| = n$.)

3. Let $n \in \mathbb{N}$ and define a dfa M_n with $A = \{a\}$, $\Sigma = [1 \mathinner{.\,.} n]$, $S = F = \{1\}$, and $i^a = i+1 \bmod n$ for $i \in \Sigma$. So $L(M) = (a^n)^*$. In fact M_n already has a minimal number of states, but show that the minimization procedure described in Subsubsection 13.1.6.2 takes $n-1$ passes to prove this.

4. Use the example in Exercise 2 to construct a 2-variable dfa M over $\{0, 1\}$ with $n+1$ states, with the property that any dfa with language equal to $L(M_\exists)$ has at least 2^n states.

13.2 Automatic groups

We are now ready to proceed to the main topic of this chapter, automatic groups.

13.2.1 Definitions, examples, and background

DEFINITION 13.5 Let G be a group that is generated as a monoid by the set A. Then G is said to be *automatic* (with respect to A), if there exist fsa W, and M_x for each $x \in A \cup \{\varepsilon\}$, such that:

(i) W has input alphabet A, and is a word-acceptor for G (cf Definition 13.3);

(ii) Each M_x is a 2-variable fsa over A and, for $v, w \in A^*$, $(v, w)^+ \in L(M_x)$ if and only if $v, w \in L(W)$ and $vx =_G w$.

The collection of automata W and M_x is known as an *automatic structure* for G. The M_x are known as the *multiplier automata*.

It is called an *automatic structure with uniqueness* if W is a unique word-acceptor for G.

The idea is that the fsa M_x recognizes multiplication on the right by x in $L(W)$. We have not insisted that W should be a unique word-acceptor for G; this is not necessary in general, because the fsa M_ε can be used to decide whether two words in $L(W)$ represent the same element of G. Note that W is a unique word-acceptor if and only if $L(M_\varepsilon) = \{\, (w, w) \mid w \in L(W) \,\}$.

There are a number of equivalent definitions of automatic groups, and some other related definitions such as *asynchronously automatic groups*, in which the two words in M_x may be read at different speeds, and *biautomatic groups* in which multiplication by a generator on the left in $L(W)$ can also be recognized by an fsa. See [ECH+92] for details.

It is proved in Theorem 2.4.1 of [ECH+92] that automaticity is a property of G and does not depend on A; that is, if G is automatic with respect to one monoid generating set, then it is automatic with respect to any other. We can therefore unambiguously say that G is an *automatic group*. We shall assume for the remainder of this section that A is chosen to be $X \cup X^{-1}$ for some set X of group generators of G.

The algorithms that we shall be describing are only able to compute shortlex automatic structures, which we shall now define.

DEFINITION 13.6   The group $G = \langle X \rangle$ is said to be *shortlex automatic* with respect to X, if it has a *shortlex automatic structure*. That is, an automatic structure W, $M_x (x \in A \cup \{\varepsilon\})$ with uniqueness in which W accepts precisely the minimal words for the group elements under some shortlex ordering of A^*. (See Definition 2.60. Such an ordering will of course depend on a given ordering of A.)

In other words,

$$L(W) = \{ w \in A^* \mid w \leq_S v \ \ \forall v \in A^* \text{ with } w =_G v \},$$

where $\leq_S$ is the shortlex ordering of A^* derived from the given ordering of A.

Unfortunately, for a given A, a group can be automatic, but not shortlex automatic with respect to any ordering of A; see Section 3.5 of [ECH+92] for an example. A group can also be shortlex automatic with respect to one ordering of A but not with respect to a different ordering; see exercise below. It is unknown whether every automatic group is shortlex automatic with respect to some ordering of some monoid generating set of G.

Here is a quick summary of some of the basic properties of automatic groups and some of the known classes of examples. See [ECH+92] for details unless otherwise indicated.

All automatic groups have finite presentations. The word problem is solvable in quadratic time in an automatic group. It is not known whether the conjugacy problem is solvable in automatic groups in general, but it is known to be solvable in biautomatic groups. It is also unknown whether all automatic groups are biautomatic.

The hyperbolic groups considered by Jim Cannon in [Can84] are automatic; in fact, *word-hyperbolic* groups, as defined by Gromov, are shortlex automatic and are biautomatic with any any ordering of any finite generating set. This class includes free groups, and various small cancellation groups. See the multiauthor paper [ABC+91] or the book by Ghys and de la Harpe [GdlH90] for further details on word-hyperbolic groups. Other groups satisfying various small cancellation hypotheses are also automatic; see the paper by Gersten and Short [GS91].

Euclidean groups (that is, extensions of torsion-free finitely generated abelian groups by finite groups) are biautomatic, but nonabelian torsion-free nilpotent groups are not automatic. This result has now been generalized in [Har], where it is proved that nonabelian torsion-free polycyclic groups are not automatic.

Braid groups, geometrically finite groups, and Artin groups of finite type (proved by Charney in [Cha95]) are biautomatic. It is proved by Brink and Howlett in [BH93] that Coxeter groups are shortlex automatic using the natural generating set, but it is still not known whether all Coxeter groups are biautomatic.

The class of automatic groups is closed under direct products, free products with finite amalgamated subgroup, and HNN-extensions with finite con-

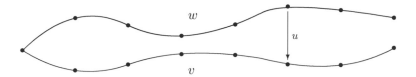

Figure 13.1: Word-differences.

jugated subgroup. Free factors of automatic groups are automatic, but it has not been proved that direct factors are.

If H has finite index in G, then G is automatic if and only if H is.

13.2.2 Word-differences and word-difference automata

The *word-differences* associated with two words $w, v \in A^*$ play a vital rôle in the algorithms for computing automatic structures. If $w = a_1 \ldots a_n$ with $a_i \in A$ then, for $0 \le i \le n$, we denote the prefix $a_1 \ldots a_i$ of length i of w by w_i, and we let $w_i = w$ for $i \ge n$.

DEFINITION 13.7 The set of word-differences associated with a pair of words $(w, v) \in A^* \times A^*$ is $\{\, w_i^{-1} v_i \mid i \in \mathbb{N}_0 \,\}$, regarded as a subset of G.

The set of word-differences associated with a set P of pairs of words $(w, v) \in A^* \times A^*$ is the union of the sets of word-differences associated with the pairs $(w, v) \in P$.

Although we appear to have defined infinitely many word-differences for a given pair (w, v), there are of course at most $\max(|w|, |v|) + 1$ distinct word-differences associated with (w, v).

The word-differences are the group elements represented by the paths joining corresponding vertices in the paths labelled w and v that start at the origin in the Cayley graph $\Gamma_X(G)$ of $G = \langle X \rangle$ (Definition 2.13). So, in Fig. 13.1, the group element represented by the path u is a word-difference.

The following result, which is equivalent to [ECH+92, Lemma 2.3.2] is fundamental in the theory of automatic groups. Its proof is easy, so we include it here.

THEOREM 13.8 *Suppose that the* fsa *W and M_x ($x \in A \cup \{\varepsilon\}$) form an automatic structure for G, and let D be the set of word-differences associated with the set of all $(w, v) \in A^* \times A^*$ for which $(w, v)^+ \in L(M_x)$ for some $x \in A \cup \{\varepsilon\}$. Then D is finite.*

PROOF   An element of D has the form $w_i^{-1}v_i$ for some $i \in \mathbb{N}_0$, where $(w, v)^+ \in L(M_x)$ for some $x \in A \cup \{\varepsilon\}$. Let $w', v' \in A^*$ satisfy $w = w_i w'$, $v = v_i v'$. Then $(w_i, v_i)^+(w', v')^+ \in L(M_x)$ so, if σ_0 is the start state of M_x, then there is a path of arrows in M_x from $\sigma_0^{(w_i, v_i)^+}$ to a success state. By removing closed circuits from this path, we can choose such a path of length at most the number of states in M_x. Hence, if K is the maximum number of states of any of the fsa M_x, then we can choose $w'', v'' \in A^*$ with $|w''|, |v''| \le K$ and $(w_i, v_i)^+(w'', v'')^+ \in L(M_x)$. Then, by definition of $L(M_x)$, we have $w_i w'' x =_G v_i v''$ and so $w_i^{-1} v_i =_G w'' x (v'')^{-1}$.

We have now proved that any word-difference in D is equal in G to a word of length at most $2K + 1$. Since there are only finitely many elements of G with that property, this proves the result.   ∎

Geometrically, this result says that the lengths of the paths u in the Cayley graph, as shown in Fig. 13.1, is uniformly bounded for pairs (w, v) such that $(w, v)^+ \in L(M_x)$ for some x, even though the words w and v themselves can be arbitrarily long. This property of the sets $L(M_x)$ is known as the *fellow-traveller* property. It is not hard to show (see exercises below) that this property can be used as an alternative definition of an automatic group.

DEFINITION 13.9   An accessible two-variable dfa Z with start state σ_0 is called a *word-difference automaton* for G, if there is a function $\delta : \Sigma_Z \to G$ such that

(i) $\delta(\sigma_0) = 1_G$;

(ii) For all $a, b \in A^+ := A \cup \{\$\}$ and $\sigma \in \Sigma_Z$ such that $\sigma^{(a,b)}$ is defined, we have $\delta(\sigma^{(a,b)}) =_G a^{-1}\delta(\sigma)b$.

Here and elsewhere, if we need to interpret the padding symbol $\$$ as an element of G, then $\$$ evaluates to 1_G.

Let Z be a word-difference machine with start state σ_0. Since we are assuming that Z is accessible, for all $\sigma \in \Sigma_Z$ there exist $v, w \in A^*$ with $\sigma_0^{(v,w)^+} = \sigma$, and then (i) and (ii) imply that $\delta(\sigma) =_G v^{-1}w$. Hence the map δ is determined by the group G and the transitions of Z.

The above definition says nothing about the accepting states of Z, but the following result follows immediately from the definition.

LEMMA 13.10   *If Z is a word-difference machine and there exists $g \in G$ such that $\delta(\sigma) = g$ for all accepting states σ of Z, then $wg =_G v$ for all $w, v \in A^*$ with $(w, v)^+ \in L(Z)$.*

For an automatic structure W, M_x $(x \in A \cup \{\varepsilon\})$ of G, we can define associated word-difference dfa Z_x as follows. There is one of these for each

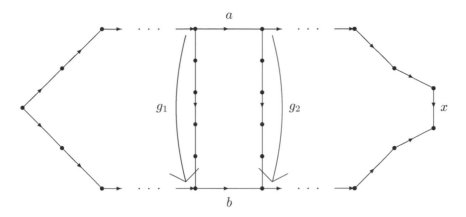

Figure 13.2: A transition in Z_x.

$x \in A \cup \{\varepsilon\}$, but they differ only in their accepting states. First we give a version that is not quite correct.

The state-set Σ_{Z_x} is the set D defined and proved to be finite in Theorem 13.8, δ is the identity map, and 1_G is the unique start state. For $\sigma \in D$ and $x, y \in A^+$, we define $\sigma^{(x,y)} = \rho$ if $x^{-1}\sigma y =_G \rho \in D$, and otherwise $\sigma^{(x,y)}$ is undefined. The unique final state is x.

The definition is not quite right as it stands, because the dfa defined might accept words in $(A^+ \times A^+)^*$ that are not padded words of the form $(w, v)^+$, and that would violate our definition of a 2-variable fsa. To remedy this, adjoin two extra copies D_L and D_R of D to Σ_{Z_x}. Transitions labelled $(a, \$)$ for $a \in A$ are only defined when the source state is in D or D_L and when the target is in D_L, and these are the only transitions defined with target in D_L. Similarly, transitions labelled $(\$, a)$ are only defined when the source state is in D or D_R and when the target is in D_R, and these are the only transitions defined with target in D_R. The element $1_G \in D$ remains the only start state of Z_x, and we make Z_x accessible by removing any inaccessible states. There are now (up to) three final states, namely the elements representing x in D, D_L and D_R.

PROPOSITION 13.11   *For each $x \in A \cup \{\varepsilon\}$, we have $wx =_G v$ for all $(w, v)^+ \in L(Z_x)$, and $(w, v)^+ \in L(Z_x)$ for all $(w, v)^+ \in L(M_x)$.*

PROOF   The first statement follows from Lemma 13.10, and the second statement follows from the definitions of D and of Z_x.   ∎

In Fig. 13.2, we illustrate Z_x accepting a pair $(w, v)^+$ with $(w, v) \in L(M_x)$.

The arrow labels g_1 and g_2 are states of Z_x, and there is a transition $g_1^{(a,b)} = g_2$ of Z_x, where $a^{-1}g_1 b =_G g_2$.

DEFINITION 13.12   A collection of word-difference dfa Z_x ($x \in A \cup \{\varepsilon\}$) for G is said to be *correct* for an automatic structure W, M_x of G, if the Z_x satisfy Proposition 13.11.

Constructing correct word-difference dfa forms an important component of our automatic structure algorithm. The following result follows immediately from the definitions of an automatic structure and of correctness. It will be used in the algorithm to construct the M_x from W and the Z_x.

PROPOSITION 13.13   *If the collection of word-difference* dfa Z_x *is correct for an automatic structure W, M_x of G, then, for each $x \in A \cup \{\varepsilon\}$, we have*

$$L(M_x) = \{\, (w,v)^+ \mid w,v \in L(W) \text{ and } (w,v) \in L(Z_x) \,\}.$$

13.3   The algorithm to compute the shortlex automatic structures

In this section, we shall describe a procedure for computing shortlex automatic structures for groups G defined by a finite presentation. As mentioned earlier, this has been implemented in the author's **KBMAG** package. Although it should, in principal, be possible to extend the procedure to handle general automatic structures, there are a number of features of the shortlex situation, principally the fact that words accepted by W are the shortest representatives of the associated group elements, which make this special case suitable for relatively efficient implementation. The only current extension of the methods to nonshortlex structures is described by Sarah Rees in [Ree98] and has been implemented by her as **GAP** procedures that call programs from **KBMAG**.

Let $G = \langle X \mid R \rangle$ with X, R finite, where a total order $<$ has been specified on $A := X \cup X^{-1}$. We shall use $<$ also to denote the corresponding shortlex ordering on A^*. If G is shortlex automatic with respect to this ordering then, given sufficient time and space, the procedure is guaranteed to compute the corresponding automatic structure and to verify its correctness. So we do not need to know in advance whether the group is shortlex automatic; indeed, these methods have proved most useful when applied to groups about which very little was known. If the group is not shortlex automatic, then the procedure will not complete.

There are five main steps in the procedure, which we shall describe in detail in separate subsections. For the remainder of this section, when we talk about the word-acceptor W or the multipliers M_x of G, we shall mean the corresponding fsa in the putative shortlex automatic structure of G. Of course, we do not know in advance whether G has such a structure. If not, then the procedures will fail and, in practice, they may fail as a result of insufficient memory even when G is shortlex automatic. We shall say that a candidate W' for W or M'_x for M_x is *correct* if $L(W') = L(W)$ or $L(M'_x) = L(M_x)$, respectively.

Step 1. Use Knuth-Bendix completion to compute word-difference dfa Z'_x ($x \in A \cup \{\varepsilon\}$) for G.

Step 2. Use Z'_ε to compute a candidate W' for the word acceptor W of G.

Step 3. Use the Z'_x and W' to compute candidates M'_x for the multiplier automata of G. If it transpires during the course of this construction that $L(W') \neq L(W)$, then redefine the dfa Z'_x and go back to Step 2.

Step 4. Carry out correctness tests on the M'_x. If it turns out that $L(M'_x) \neq L(M_x)$ for some x, then redefine the dfa Z'_x and go back to Step 3.

Step 5. Carry out full correctness tests on W' and M'_x to decide whether $L(W') = L(W)$ and $L(M'_x) = L(M_x)$ for all $x \in A \cup \{\varepsilon\}$.

We shall explain later exactly how the looping involved in Step 2 to Step 4 works.

The final step (Step 5) of the algorithm, which returns just `true` or `false`, attempts to verify that these candidates are all correct; that is, that they are equal to the fsa of a shortlex automatic structure of G. In practice, it has been observed that, unless one deliberately runs the procedure with very silly parameter settings, if the tests in Step 4 complete successfully, then so does the full test in Step 5. It is frustrating that Step 5 can often take up the bulk of the time and space requirements for the whole procedure, and at the end it merely reports `true`! In principle, if Step 5 returned `false`, then we could repeat the whole procedure again with larger parameter settings in Step 1.

13.3.1 Step 1

As already explained, we shall assume throughout (possibly erroneously!) that G is shortlex automatic, with automatic structure W, M_x ($x \in A \cup \{\varepsilon\}$). Let D be the finite set of word-differences defined in Theorem 13.8 for the shortlex automatic structure of G.

By Theorem 12.12, there is an associated monoid presentation $\mathrm{Mon}\langle A \mid \mathcal{I} \cup \mathcal{R} \rangle$ of G. As described in Section 12.3, we can use this monoid presentation to define a rewriting system $\mathcal{S}$, using our shortlex ordering $<$ on A^*, and we can run the Knuth-Bendix completion process on $\mathcal{S}$.

According to Definition 12.18, the $\mathcal{S}$-minimal strings in A^* are the minimal elements in their $\leftrightarrow^*$ equivalence classes, which are the minimal representatives of group elements under $<$. So $L(W)$ is equal to the set of $\mathcal{S}$-minimal strings in A^*.

In Definition 12.19, we defined an $\mathcal{S}$-essential rule (w, v) to be a pair of words $w, v \in A^*$ where $w \leftrightarrow^* v$, and v and all proper substrings of w are ($\mathcal{S}$-)minimal. Or, equivalently, $w =_G v$, and v and all proper substrings of w are in $L(W)$.

Let D_0 be the set of word-differences associated with the set of all $\mathcal{S}$-essential rules. If (w, v) is such a rule, then $w = ux$, where $x \in A$ and $u, v \in L(W)$. Hence, by the definition of an automatic structure, we have $(w, v) \in L(M_x)$. Since $|w| \geq |v|$, all word-differences associated with (ux, v) are also associated with (u, v), and hence $D_0 \subseteq D$, and D_0 is finite.

So, although there will normally be infinitely many $\mathcal{S}$-essential rules, there will only be finitely many associated word-differences.

We can now describe Step 1. We run Knuth-Bendix on $\mathcal{S}$, as already mentioned. At any stage in the procedure, for $w \in A^*$, let $r(w)$ be the current $\mathcal{S}$-reduction of w, which can be computed from $\mathcal{S}$ using the method described in Subsection 13.1.3. So $r(w) =_G w$. From time to time we interrupt the Knuth-Bendix procedure and, for each rule (w, v) in $\mathcal{S}$, we calculate

$$\{ r(w_i^{-1} v_i) \mid 0 \leq i \leq |w| \}$$

and let D_0' be the union of these sets over all rules currently in $\mathcal{S}$. (It is convenient to make this calculation immediately after we have carried out the simplification process on the current rule set, because that tends to avoid getting superfluous elements of D_0'.)

To be more precise, for $i \geq 1$, we compute $r(w_i^{-1} v_i)$ in the above expression as $r(a^{-1} r(w_{i-1}^{-1} v_{i-1}) b)$, where a and b are the i-th letters of w and v, respectively. The proof of the proposition below depends on this method of computing the $r(w_i^{-1} v_i)$; the observant reader will have noticed that, since $\mathcal{S}$ is not generally confluent, the value of $r(w)$ calculated can depend on the method used to $\mathcal{S}$-reduce w.

When D_0' appears to have become stable (more on this later!), we compute and output word-difference dfa Z_x' ($x \in A \cup \{\varepsilon\}$) as follows. The states of Z_x' are the elements of D_0', ε is the unique start state, and $r(x)$ is the unique accepting state of Z_x'. For $a, b \in A^+ := A \cup \{\$\}$ and $\sigma \in D_0'$, we define the transition $\sigma^{(a,b)} = r(a^{-1} \sigma b)$ if $r(a^{-1} \sigma b) \in D_0'$, and $\sigma^{(a,b)}$ is undefined otherwise. We also modify D_0' as described above for D to ensure that Z_x' accepts only padded strings. Then it is clear that Z_x' satisfies the requirements of Definition 13.9 for a word-difference fsa where, for $w \in D_0'$, $\delta(w)$ is defined to be the element of G represented by w.

PROPOSITION 13.14   *Provided that the Knuth-Bendix process is run for sufficiently long before computing and outputting the Z_x', we will have $(w, v)^+ \in L(Z_\varepsilon')$ for all $\mathcal{S}$-essential rules (w, v).*

PROOF   Since D_0 is finite, the elements of D_0 all occur as word-differences of a finite subset $\mathcal{T}$ of the set of all essential rules. By Proposition 12.20, if we run the Knuth-Bendix process for sufficiently long, then all rules in $\mathcal{T}$ will be rules in $\mathcal{S}$. Furthermore, if we run it for long enough then, for all $a, b \in A \cup \{\varepsilon\}$ and all minimal words w representing elements of D_0, the $\mathcal{S}$-reduction $r(w)$ of $a^{-1}wb$ will be equal to the minimal representative of $a^{-1}wb$.

At that stage, D'_0 will contain all minimal words representing elements of D_0, and also all possible transitions $\sigma^{(a,b)}$ for $a, b \in A \cup \{\$\}$. Since the word-differences associated with essential rules (w, v) all lie in D_0, we will then have $(w, v)^+ \in L(Z'_\varepsilon)$.   ∎

There remains the problem of deciding when to interrupt the Knuth-Bendix process or, in other words, how to judge when D'_0 has become stable. No completely satisfactory solution to this has been found to date. A strategy that works well on many straightforward examples is to stop when the number of rules in $\mathcal{S}$ is at least some minimum size, and has also doubled without D'_0 changing.

Unfortunately, in some difficult examples, a small number of elements of D'_0 take a very long time to appear, and we may decide to stop Step 1 before D'_0 is complete. But on the other hand this does not always matter, because the missing word-differences may be found during Step 2 to Step 4.

Another problem is that in some examples additional word-differences arise from rules that later turn out not to be reduced. Although superfluous word-differences do not prevent the procedure from functioning correctly, they are a nuisance, since they can make it difficult to decide, particularly mechanically, whether D'_0 has become stable.

One nice feature of the procedure is that in many examples that are not shortlex automatic, the number of elements in D'_0 increases steadily without stabilizing, so it becomes rapidly clear that the process is doomed to failure.

13.3.2   Step 2 and word reduction

The remaining steps of the procedure do not involve the Knuth-Bendix process, and consist entirely of constructions involving fsa. Since some of these fsa may have large numbers of states, it is important that for each fsa that we construct, we always compute an equivalent dfa if necessary, and that we always compute an equivalent dfa with a minimal number of states. We routinely standardize each dfa computed, as described in Subsubsection 13.1.6.3, so that we can easily test two dfa for equivalence.

For a shortlex ordering $<$ on A^*, it is straightforward to construct a 2-variable fsa over A with language $\{ (w, v)^+ \in (A^+ \times A^+)^* \mid v < w \}$; see exercise below. It is interesting to note, however, that for various other familiar orderings on A^*, such as the wreath-product orderings defined in Section 12.4, the set defined above is not the language of an fsa. This is the main reason why the methods for constructing shortlex automatic structures do not readily generalize to other orderings. A method of carrying out such generalizations was, however, proposed in [Ree98].

It follows from the above remark and the results and constructions described in Subsection 13.1.6 that we can construct a dfa W' with language $\neg(A^*LA^*)$, where

$$L := \{\, w \in A^* \mid \exists v \in A^* : (w,v)^+ \in L(Z'_\varepsilon) \text{ and } v < w \,\},$$

and Z'_ε was computed in Step 1. Step 2 consists of this construction of W'.

PROPOSITION 13.15 $L(W')$ *is prefix-closed with* $L(W) \subseteq L(W')$. *If the dfa* Z'_ε *returned in Step 1 accepts all essential rules, then* $L(W') = L(W)$.

PROOF   It follows immediately from $L(W') = \neg(A^*LA^*)$ that $L(W')$ is prefix-closed. If $u \in A^* \setminus L(W')$ then $u \in A^*LA^*$, so $u = u_1wu_2$ where there exists $v \in A^*$ with $v < w$ and $(w,v)^+ \in L(Z'_\varepsilon)$. But Proposition 13.10 implies that $w =_G v$, so $u =_G u_1vu_2$ with $u > u_1vu_2$, and hence $u \notin L(W)$. Thus $L(W) \subseteq L(W')$.

Conversely, suppose that $u \in A^* \setminus L(W)$. Then u is not minimal. Choose w to be a substring of u, which is shortest subject to being not minimal. Then there exists a minimal $v \in A^*$ with $w =_G v$, and all proper substrings of w are minimal, so (w,v) is an essential rule. Hence, if Z'_ε accepts all essential rules, then $(w,v)^+ \in L(Z'_\varepsilon)$ and so $u \notin L(W')$, which completes the proof. ∎

During the remaining steps of the procedure, we shall sometimes need to replace a word $u \in A^*$ by a word u' with $u =_G u'$ and $u' \in L(W')$. We can do that as follows.

If $u \notin L(W')$, then find a substring w of u of minimal length with $w \notin L(W')$. Then, from the way that W' was constructed, we know that there exists $v \in A^*$ with $(w,v)^+ \in L(Z'_\varepsilon)$. We can find such a word v by using the procedure EXISTSWORD described in Subsection 13.1.7. We then substitute v for w in u and repeat. Since each such substitution replaces u by a smaller word in the shortlex ordering, this process must eventually terminate with $u \in L(W')$.

In fact, if $(w,v)^+ \in L(Z'_\varepsilon)$, then $(u_1w, u_1v)^+ \in L(Z'_\varepsilon)$ for all $u_1 \in A^*$ (see exercise below). So it suffices to find a shortest prefix w of u with $w \notin L(W')$, which is easy: since $L(W')$ is prefix closed, when reading u into W', we will have read the shortest prefix not in $L(W')$ as soon as we reach a failure state.

13.3.3   Step 3

In Proposition 13.13, we saw that, if the dfa Z_x $(x \in A \cup \{\varepsilon\})$ are correct for the automatic structure as defined in Definition 13.12, then for each $x \in A \cup \{\varepsilon\}$, we have

$$L(M_x) = \{\, (w,v)^+ \mid w,v \in L(W) \text{ and } (w,v)^+ \in L(Z_x) \,\}.$$

In Step 3, we construct our candidate multiplier dfa M'_x with language equal to the right-hand side of this equality, but using our candidate dfa W' and Z'_x in place of W and Z_x. Of course we do not know yet whether the Z'_x really are correct and, as we shall explain later, they will not necessarily be correct after the first application of Step 2, even if $L(W') = L(W)$.

To be more precise, M_x is constructed as follows. (This construction is also described in Definition 2.3.3 of [ECH$^+$92].) The set Σ_{M_x} of states of M_x is

$$\Sigma_{M_x} := \{\, (\sigma_1, \sigma_2, \rho) \mid \sigma_1, \sigma_2 \in \Sigma_{W'},\ \rho \in \sigma_{Z'_x} \,\},$$

the start state is $(\sigma_0, \sigma_0, \rho_0)$, where σ_0 and ρ_0 are the start states of W' and Z'_x, and a state is accepting if σ_1 and σ_2 are accepting states of W' and ρ is an accepting state of Z'_x.

We need to extend the definition of W' to allow transitions labelled by $\$$. We define $\sigma^\$ = \sigma$ if σ is an accepting state of W', and $\sigma^\$$ is undefined otherwise. The transitions of M_x are derived from those of W' and Z'_x, and are

$$(\sigma_1, \sigma_2, \rho)^{(a,b)} = (\sigma_1^a, \sigma_2^b, \rho^{(a,b)})$$

for $a, b \in A^+$, where the transition is undefined if any of the components on the right-hand side are undefined.

It is easy to see, and left as an exercise to the reader, that

$$L(M'_x) = \{\, (w,v)^+ \mid w, v \in L(W') \text{ and } (w,v)^+ \in L(Z'_x) \,\},$$

as required.

We shall need the following straightforward properties of $L(M'_x)$ later.

PROPOSITION 13.16   (i) For all $w, v \in A^*$ with $(w,v)^+ \in L(M'_x)$, we have $w, v \in L(W')$ and $wx =_G v$.

(ii) If $a_1 a_2 \cdots a_n \in L(W')$ with $n > 0$, then $(a_1 a_2 \cdots a_{n-1}, a_1 a_2 \cdots a_n)^+ \in L(M'_{a_n})$.

PROOF   By definition of the word-difference dfa Z'_x, we have $\delta(\sigma) =_G x$ for all accepting states σ of Z'_x, so (i) follows from Lemma 13.10.

By Proposition 13.15, $a_1 a_2 \cdots a_n \in L(W')$ implies $a_1 a_2 \cdots a_{n-1} \in L(W')$, so to prove (ii) we just need to prove that $(a_1 a_2 \cdots a_{n-1}, a_1 a_2 \cdots a_n)^+ \in L(Z'_{a_n})$. If σ_0 is the start state of Z'_{a_n} then, by definition of Z'_{a_n}, we have $\sigma_0^{(a,a)} = \sigma_0$ for all $a \in A$. Hence

$$\sigma_0^{(a_1 a_2 \cdots a_{n-1},\, a_1 a_2 \cdots a_n)} = \sigma_0^{(\$, a_n)} = r(a_n),$$

so $(a_1 a_2 \cdots a_{n-1}, a_1 a_2 \cdots a_n)^+ \in L(Z'_{a_n})$.   ∎

As is usual with fsa constructions, when we construct M'_x in practice, we calculate the transitions beginning with the start state, and we only need to define those states that are accessible from the start state. As was the case with the Z'_x, we can store all of the M'_x as a single dfa, and just allow the final states to vary to provide the individual M'_x. It is even possible to do the dfa minimization using this single dfa, but we shall not go into details about that here.

Notice that, if $(w, v)^+ \in L(Z'_\varepsilon)$ with $w > v$, then w is in the language L defined in Step 2 and so, by definition of W', we cannot have $w \in L(W')$. Hence, if we apply Step 3 immediately after Step 2, then we cannot have $(w, v)^+ \in L(M'_\varepsilon)$ with $w > v$. However, as we shall see in the next subsection, we sometimes reapply Step 3 after a redefinition of the Z'_x in Step 4. When that happens, it is possible to find $(w, v)^+ \in L(M'_\varepsilon)$ with $w > v$ and indeed this occurs not infrequently in practice.

This tells us that $w \notin L(W)$ and so W' is incorrect, and typically there will be word-differences associated with the pair (w, v) that are not in the set D'_0 used to construct the Z'_x in Step 1. So we calculate the word-differences associated with (w, v) just as in Step 1, except that we now use the algorithm described at the end of Subsection 13.3.2 to reduce words to a word in $L(W')$. We adjoin these new word-differences to our set D'_0, recalculate the dfa Z'_x, and then repeat Step 2. It is possible to design the algorithm so that we find such pairs (w, v) as soon as they arise during the construction of the M_x, so we can then abort the construction.

13.3.4 Step 4

In general, the set D_0, defined in Subsection 13.3.1, of word-differences arising from the essential equations is a proper subset of the set D, defined in Theorem 13.8, of all word-differences arising from the sets $L(M_x)$. Proposition 13.14 guarantees that, if we run the Knuth-Bendix process in Step 1 for long enough, then the states of the Z'_x will contain minimal representatives of all elements of D_0. But they will not, in general, contain representatives of all elements in D, and so the Z'_x will not necessarily be correct after Step 3, and hence neither will the M'_x.

Let us assume for the moment that W' is correct; that is, $L(W) = L(W')$. From the construction of M'_x, we have $(w, v)^+ \in L(M'_x)$ implies $(w, v)^+ \in L(Z'_x)$, and hence $wx =_G v$. So, if some $L(M'_x)$ is not correct, then there exist $w, v \in L(W')$ with $wx =_G v$ and $(w, v)^+ \notin L(M'_x)$. But, by the construction of M'_x again, if $(w, v')^+ \in L(M'_x)$ for some $v' \in A^*$, then $w, v' \in L(W') = L(W)$ with $wx =_G v'$, and hence $v = v'$. So, if $L(M'_x)$ is not correct, then there exists $w \in L(W')$ such that there is no $v \in A^*$ with $(w, v)^+ \in L(M'_x)$.

In Step 4, we use the method described in Subsection 13.1.7 to construct, for each $x \in A \cup \{\varepsilon\}$, an fsa E_x with language

$$L(E_x) = \{ w \in A^* \mid \exists v \in A^* : (w, v)^+ \in L(M'_x) \}.$$

If $L(E_x) = A^*$ for all $x \in A \cup \{\varepsilon\}$ and $L(W) = L(W')$, then the argument above shows that $L(M_x) = L(M'_x)$ for all x, and so we have successfully calculated the automatic structure. But we do yet know that $L(W) = L(W')$, and we proceed to Step 5 for the final correctness verification.

Otherwise, $L(E_x) \neq A^*$ for some x, in which case we use the algorithm FSA-ENUMERATE presented in Subsection 13.1.2 to find some words $w \in \neg L(E_x)$. For each such w, we use the method described in Subsection 13.3.2 to find $v \in L(W')$ with $wx =_G v$. As in Steps 1 and 3, we then calculate words in $L(W')$ that represent the word-differences associated with the pairs (w, v), adjoin them to our set D_0', recalculate the fsa Z_x' (for all x), and return to Step 3.

We can now prove that, if G is shortlex automatic, then it is possible to calculate the shortlex automatic structure by Step 1 to Step 4.

THEOREM 13.17 *If the Knuth-Bendix process is run for sufficiently long in Step 1 and G is shortlex automatic, then the associated shortlex structure will be successfully computed in Step 2 to Step 4.*

PROOF By Propositions 13.14 and 13.15, if we run Knuth-Bendix for long enough, then we will have $L(W) = L(W')$ after Step 2, so let us assume that to be the case.

We also assume that we run Knuth-Bendix for long enough to ensure that, for all $a, b \in A \cup \{\varepsilon\}$ and all minimal words w representing elements of D, the $\mathcal{S}$-reduction $r(a^{-1}wb)$ of $a^{-1}wb$ is equal to the minimal representative of $a^{-1}wb$. (In the proof of Proposition 13.14 we just needed that property for D_0 rather than D.) Then, after Step 2, the minimal representative of each such $a^{-1}wb$ will be the unique representative of that word in $L(W') = L(W)$, and so the algorithm for finding representatives of words in $L(W')$ described at the end of Section 13.3.2 will successfully replace all of these words $a^{-1}wb$ by their minimal representatives.

So, if our set D_0' contains minimal representatives of all elements of D, then the fsa Z_x' computed will be correct and then, by Proposition 13.13, the M_x' computed in Step 3 will be correct, the sets E_x computed in Step 4 will all satisfy $L(E_x) = A^*$, and we will proceed to Step 5.

Otherwise, in Step 4 we will find pairs (w, v) with $w, v \in L(W')$, $wx =_G v$ and $(w, v)^+ \notin L(M_x')$ and hence $(w, v)^+ \notin L(Z_x')$. Since $(w, v)^+ \in L(M_x)$, its associated word-differences lie in D and so, by our assumption, they will all be correctly reduced to their minimal representatives. If these minimal representatives were all already in D_0', then we would have $(w, v)^+ \in L(M_x')$, which is not the case. So, when we redefine the fsa Z_x', D_0' will contain minimal representatives of more elements of D than previously. Since D is finite, the looping process must eventually terminate, at which stage the automatic structure will have been successfully calculated. ∎

As we saw in the proof, Propositions 13.14 and 13.15 ensure that, if we run Knuth-Bendix for long enough in Step 1, then we will get $L(W) = L(W')$

in Step 2, and then we will never need to repeat Step 2. It is important to stress once again, however, that even if we make a mistake and fail to get $L(W) = L(W')$ on the first application of Step 2, it is still possible that W' will be corrected after one or more applications of Steps 3 and 4. There is no obvious theoretical reason why this should happen, but in the more difficult examples, this feature has turned out to be essential in practice for the whole procedure to complete successfully.

13.3.5 Step 5

In Step 5, we decide whether W' and the M_x' constructed in Step 2 to Step 4 form a shortlex automatic structure for G. The method depends on the following theorem, which is Theorem 2.5 of [EHR91] and Theorem 6.1.2 of [ECH+92]. We recall from Theorem 12.12 that, for a group presentation $G = \langle\, X \mid R \,\rangle$, there is an associated monoid presentation $\mathrm{Mon}\langle\, A \mid \mathcal{I} \cup \mathcal{R} \,\rangle$ of G, where

$$\mathcal{I} = \{\, (xx^{-1}, \varepsilon) \mid x \in X \,\} \cup \{\, (x^{-1}x, \varepsilon) \mid x \in X \,\} \text{ and } \mathcal{R} = \{\, (r, \varepsilon) \mid r \in R \,\}.$$

THEOREM 13.18   *Let $G = \langle\, X \mid R \,\rangle$ be a finite group presentation and let $\mathrm{Mon}\langle\, A \mid \mathcal{I} \cup \mathcal{R} \,\rangle$ be the associated monoid presentation of G. Suppose that W is a* fsa *over A and M_x $(x \in A \cup \{\varepsilon\})$ are 2-variable* fsa *over A, which satisfy the following hypotheses:*

(i) If $(w, v)^+ \in L(M_x)$ for some $w, v \in A^*$ and $x \in A \cup \{\varepsilon\}$, then $w, v \in L(W)$ and $wx =_G v$.

(ii) $L(W)$ is nonempty and, if $a_1 a_2 \cdots a_n \in L(W)$ with $n > 0$, then $a_1 a_2 \cdots a_{n-1} \in L(W)$ and $(a_1 a_2 \cdots a_{n-1}, a_1 a_2 \cdots a_n)^+ \in L(M_{a_n})$.

(iii) Let $w = w_0 \in L(W)$ and $(a_1 a_2 \cdots a_n, \varepsilon) \in \mathcal{I} \cup \mathcal{R}$. Then, for a word $w_n \in L(W)$, there exist elements $w_1, w_2, \ldots, w_{n-1} \in L(W)$ satisfying $(w_{i-1}, w_i)^+ \in L(M_{a_i})$ for $1 \le i \le n$ if and only if $w = w_n$.

Then W and the M_x form an automatic structure for G with uniqueness.

PROOF   Suppose that the three hypotheses hold. Let $x \in A$ and $w \in L(W)$. Then, by hypothesis, we have $(xx^{-1}, \varepsilon), (x^{-1}x, \varepsilon) \in \mathcal{I}$. By applying hypothesis (iii) to w and xx^{-1}, we deduce that there exists $w' \in L(W)$ with $(w, w')^+ \in L(M_x)$ and $(w', w)^+ \in L(M_{x^{-1}})$. By applying hypothesis (iii) to w' and $x^{-1}x$, we find that w' is the unique word with this property. Hence we can define a map $\varphi(x) : L(W) \to L(W)$ that maps w to w' and, since $\varphi(x)$ and $\varphi(x^{-1})$ are mutually-inverse maps, $\varphi(x)$ is a bijection; that is, $\varphi(x) \in \mathrm{Sym}((L(W))$. Furthermore, hypothesis (iii) implies that, for any $(a_1 a_2 \cdots a_n, \varepsilon) \in \mathcal{I} \cup \mathcal{R}$, the permutation $\varphi(a_1)\varphi(a_2)\cdots\varphi(a_n)$ of $L(W)$ is the identity, and so by Theorem 12.11, φ extends to a monoid (and hence also a group) homomorphism from G to $\mathrm{Sym}(L(W))$.

Now let $g \in G$ and let $a_1 a_2 \cdots a_n \in A^*$ represent g. Note that $\varepsilon \in L(W)$ by hypothesis (ii), and hypothesis (i) implies that the image of ε under the permutation $\varphi(a_1)\varphi(a_2)\cdots\varphi(a_n)$ also represents g. So $L(W)$ contains at least one word representing g, which we may as well assume to be $a_1 a_2 \cdots a_n$. Let $b_1 b_2 \cdots b_m \in L(W)$ represent g. Then, since φ is a homomorphism, we have

$$\varphi(a_1)\varphi(a_2)\cdots\varphi(a_n) = \varphi(a_1 a_2 \cdots a_n) = \varphi(b_1 b_2 \cdots b_m) = \varphi(b_1)\varphi(b_2)\cdots\varphi(b_m).$$

But hypothesis (ii) implies that $\varphi(a_1)\varphi(a_2)\cdots\varphi(a_n)$ and $\varphi(b_1)\varphi(b_2)\cdots\varphi(b_m)$ map ε to $a_1 a_2 \cdots a_n$ and $b_1 b_2 \cdots b_m$, respectively, so $a_1 a_2 \cdots a_n = b_1 b_2 \cdots b_m$. This proves that W is a unique word-acceptor for G.

We know from hypothesis (i) that, for $x \in A \cup \{\varepsilon\}$, $(w,v)^+ \in L(M_x)$ implies $w, v \in L(W)$ and $wx =_G v$. Conversely, for $w \in L(W)$ and $x \in A \cup \{\varepsilon\}$, we saw above that there is a unique $v \in A^*$ with $(w,v)^+ \in L(M_x)$, so v must be the unique element of $L(W)$ with $wx =_G v$. Hence $w, v \in L(W)$ and $wx =_G v$ implies that $(w,v)^+ \in L(M_x)$, so W and the M_x form an automatic structure for G with uniqueness. ∎

We wish to apply this theorem to W' and the M'_x computed in Step 2 to Step 4. By Proposition 13.15, $L(W')$ contains the shortlex minimal representative of each group element so, if the hypotheses of the above theorem holds, then W' and the M'_x form the shortlex automatic structure for G.

Now hypothesis (i) holds by Proposition 13.16 (i). By Proposition 13.15 $L(W')$ is prefix-closed and contains the shortlex minimal representative of each group element, so we certainly have $\varepsilon \in L(W')$. Also, if $a_1 \cdots a_n \in L(W')$ with $n > 0$, then, by Proposition 13.16 (ii), we have $(a_1 a_2 \cdots a_{n-1}, a_1 a_2 \cdots a_n)^+ \in L(M'_{a_n})$.

So hypotheses (i) and (ii) are definitely satisfied, and we just have to check whether hypothesis (iii) holds. (We leave it to the reader to verify that, if W' and M'_x form the shortlex automatic structure for G, then hypothesis (iii) must hold.)

In order to check hypothesis (iii), we use the construction of the composite fsa defined in Subsection 13.1.7. The composite operation is easily seen to be associative so, for a word $u = a_1 a_2 \cdots a_n \in A^*$, we can unambiguously define a multiplier fsa M_u to be the composite of $M'_{a_1}, M'_{a_2}, \ldots, M'_{a_n}$. Then

$$L(M_u) = \{\, (w,v)^+ \mid \exists w = w_0, w_1, \ldots, w_n = v \in A^* : \\ (w_{i-1}, w_i)^+ \in L(M'_{a_i}) \text{ for } 1 \le i \le n \,\}.$$

So, hypothesis (iii) says exactly that $L(M_u) = \{\, (w,w) \mid w \in L(W') \,\}$ for all $(u, \varepsilon) \in \mathcal{I} \cup \mathcal{R}$. The right-hand side of this equation is equal to the language of a 2-variable fsa that can easily be constructed from W', so we can carry out these checks as required.

13.3.6   Comments on the implementation and examples

We start with a few extra comments on various steps in the the above procedure, some of which relate to the implementation in KBMAG.

Although we have described the fsa constructions in Step 2 to Step 5 in terms of the standard operations introduced in Sections 13.1.6 and 13.1.7, in the implementation in KBMAG, we have written separate stand-alone programs for each of the main processes; this has resulted in faster running times and smaller memory requirements than we would get by using combinations of the basic operations.

Since $(w, v)^+ \in L(M_x) \Leftrightarrow (v, w)^+ \in L(M_{x^{-1}})$, the set D of word-differences is closed under inversion. The subset D_0 is not necessarily inverse-closed but, since we eventually need to extend D_0 to D, we can substantially reduce the amount of looping required in Step 2 to Step 4 by adjoining $r(w^{-1})$ to D_0' for each $w \in D_0'$ in Step 1. We do this by default in the implementation and, whenever we need to extend the set D_0' in Step 3 or Step 4, we always adjoin word-differences $r(w)$ and $r(w^{-1})$ together.

On a related point, we said above that, in Step 3, we abort and return to Step 2 if we find $(w, v)^+ \in L(M_\varepsilon')$ with $w > v$. In fact we do this if we find any $(w, v)^+ \in L(M_\varepsilon')$ with $w \neq v$ and hence, if we complete Step 3 without aborting, then it is guaranteed that $L(M_\varepsilon') = \{ (w, w) \mid w \in L(W') \}$. Thus, in the testing of hypothesis (iii) in Step 5, we can use M_ε' for the language equality tests.

In principal, if the test in hypothesis (iii) failed, then we could find $(w, v)^+ \in L(M_u)$ for some u with $(u, \varepsilon) \in \mathcal{I} \cup \mathcal{R}$ and $w \neq v$. Then we would have $w =_G v$ and we could adjoin the word-differences associated with (w, v) to D_0, redefine the Z_x' and go back to Step 2. In fact, we have not implemented this process, mainly because, in all examples that we have run in practice, the test in Step 5 has never failed, except when we have deliberately run the Knuth-Bendix process in Step 1 for an absurdly short time.

In difficult examples, the construction of the composite dfa in Step 5 is the most expensive in memory requirements, because the dfa M_u can have large numbers of states. Of course, we always minimize them immediately after constructing them. Generally, the longer the word u, the larger will M_u be. We always start by checking hypothesis (iii) for the short words $u = xx^{-1}$ with $x \in A$. Once we have done this, if $u = u_1 u_2 \in R$, then checking that $L(M_u) = L(M_\varepsilon)$ is easily seen to be equivalent to checking that $L(M_{u_1}) = L(M_{u_2^{-1}})$, and we can improve performance somewhat by doing this with u_1 chosen to have about half the length of u.

To give the reader an idea of the difficulty of the procedure, we shall now present some statistics on successful computations of automatic structures using KBMAG, on a collection of examples of increasing difficulty. The programs were run on a Sun Ultra 80 Workstation with 4 GigaBytes RAM.

There are some very straightforward examples, such as the triangle groups $\langle x, y \mid x^l, y^m, (xy)^n \rangle$, for which the complete procedure typically takes less than a second.

The following moderately easy example is a symmetry group of a certain tesselation of 3-dimensional hyperbolic space into dodecahedra. As in previous examples, we use the case-change for inverses convention, $A = a^{-1}$, despite the clash with our notation for the alphabet!

Example 13.4

$$G = \langle a, b, c, d, e, f \mid a^4, b^4, c^4, d^4, e^4, f^4, abAe, bcBf, cdCa, deDb, efEc, faFd \rangle.$$

In Step 1, Knuth-Bendix was interrupted with 2000 rewrite-rules and $|D'_0| = 49$, after $|D'_0|$ had remained constant while the final 1000 rules were found. After closing under inversion, $|D'_0| = 53$. In Step 2, W' had 438 and 47 states before and after minimization. Steps 3 and 4 were executed three times each. On the third and final iteration, the multipliers M'_a had 2293 and 385 states before and after minimization, with $|D'| = 75$, after which the correctness test in Step 4 was passed. In fact $|D_0| = 49$ and $|D| = 75$, so the computed sets are exactly right. Since all the group relators have length 4, it was only necessary to compute M_w for various words w with $|w| = 2$. For example, M_{ab} had 1264 and 291 states before and after minimization. The complete procedure took just over 4 seconds to run. ▯

Example 13.5

$$G = \langle x, y, z \mid [x, [x, y]] = z, [y, [y, z]] = x, [z, [z, x]] = y \rangle.$$

This example, due to Heineken, was mentioned earlier, in Section 5.4. It was proposed as a possible candidate for 3-generator, 3-relator finite group, and was motivated by the fact that the group with the simpler presentation $\langle x, y, z \mid [x, y] = z, [y, z] = x, [z, x] = y \rangle$ was known to be trivial. By means of the low-index subgroup and p-quotient algorithms, it was established relatively easily that G has a finite quotient of order 60.2^{24}, but for many years nothing further was discovered about it.

It was finally proved to be automatic and infinite using KBMAG. It is even word-hyperbolic (see Section 13.4 below). In Step 1, Knuth-Bendix was interrupted with about 40000 rewrite-rules and $|D'_0| = 71$, after $|D'_0|$ had remained constant while the final 20000 rules were found. After closing under inversion, we still had $|D'_0| = 171$. In Step 2, W' had 33114 and 1106 states before and after minimization. Steps 3 and 4 were executed only once, and the multipliers M'_a had 49158 and 2428 states before and after minimization. Step 5 was also reasonably straightforward, and the complete computation took about 110 seconds to run. ▯

Example 13.6

$$G = \langle\ a, b, c, d, e, f, g, h, i\ |$$
$$ab = c,\ bc = d,\ cd = e,\ de = f,\ ef = g,\ fg = h,\ gh = i,\ hi = a,\ ia = b \rangle.$$

This is the Fibonacci group $F(2, 9)$, and is a difficult example. Step 1 halted at 500 000 equations after 9574 seconds, with $|D_0'| = 541$ and fluctuating around 541 for the final 120 000 equations.

On the first run of Step 2, W' had 124 003 states initially, which minimized to 8543. But W' was not correct at this stage. Step 2 was executed three times in all, and on the third time W' had 129 169 states initially, minimizing to 3251. Steps 3 and 4 were executed 7 times altogether. On the final run of Step 3, the multipliers M_a' had 872 309 and 25 741 states before and after minimization, with $|D'| = 665$, after which the correctness test in Step 4 was passed. Steps 2 to 4 took about 7568 seconds altogether.

Step 5 took 3042 seconds to verify the correctness of the final versions of D', W', and the M_a'. All the group relators have length 3 so, as with the previous example, it was only necessary to compute M_w for various words w with $|w| = 2$. For example, M_{ab} had 1 137 648 and 25 555 states before and after minimization. In fact Step 5 is relatively undemanding in this particular case. In examples with long relators, Step 5 can dominate the total time and memory requirements.

After the automatic structure had been successfully computed, it was not difficult to use it to calculate the correct value of $|D_0|$, which is 563. So we were missing 22 word-differences when we halted Step 1, but these were found without undue difficulty during Steps 2 to 4. $\Box$

13.4   Related algorithms

There are a number of other procedures that are related to and extend the methods that we have described for automatic groups. We shall mention them briefly here, and refer the reader elsewhere for details.

A group is defined to be *word-hyperbolic* if geodesic triangles in its Cayley graph are uniformly thin. See [ABC+91] or the book by Ghys and de la Harpe [GdlH90] for details. All such groups are shortlex automatic with any choice of generators, so we can use the procedures described above to find their automatic structures.

If a group is word-hyperbolic, then the geodesics between any two vertices in the Cayley graph have the fellow-traveller property; that is, they remain within a bounded distance of each other. It has been proved by Papasoglu in [Pap95] that the converse is true; that is, if the geodesics in the Cayley graph have the fellow-traveller property, then the group is word-hyperbolic.

Starting from the shortlex automatic structure, it is possible and not difficult to attempt to verify that the geodesics in the Cayley graph have the fellow-traveller property. If we succeed then, by Papasoglu's result, we will have proved that the group is word-hyperbolic. As usual in this area, if we do not succeed, then we gain no further information. (As far as we know, it is not known whether it is decidable whether a group known to be shortlex automatic is word-hyperbolic.) This procedure is described by Epstein and Holt in [EH01b] and is implemented in **KBMAG**. For example, Examples 13.4 and 13.5 above above word-hyperbolic. It takes only a few seconds to verify this in the first of these examples, and to calculate a word-acceptor with 63 states that accepts all geodesic words, and a geodesic word-difference machine with 91 states (= word-differences). More recently, in a computation taking about two weeks of cpu-time, we have verified that Examples 13.6 is word-hyperbolic. The geodesic word-acceptor and word-difference machine have 7688 and 1730 states, respectively.

One can also attempt to compute the thinness constant for hyperbolic triangles in a word-hyperbolic group, but that is significantly more difficult. An algorithm for this is also described in [EH01b]. This, in turn, gives rise to algorithms for solving the conjugacy problem in these groups.

It is possible to generalize the notion of an automatic group to a group that is automatic with respect to a subgroup. Some procedures for constructing the related coset automatic structures are described in detail by Holt and Hurt in [HH99]. They follow the same lines as the procedure for finding standard shortlex automatic structures. For example, in Step 1, the Knuth-Bendix process is run for the group relative to the subgroup. The remaining steps involve the manipulation of an interesting class of fsa that may have more than one start state, but are otherwise deterministic.

When successful, these procedures provide the possibility of proving that a subgroup H has infinite index in G and, in some examples, they represent the only currently known method of doing this. They may also be used to compute a finite presentation of H. These methods are implemented in **KBMAG**.

13.5 Applications

A number of applications of the automatic group programs within group theory have been mentioned already. They were used to prove the infiniteness of various groups, including the Heineken group (Example 13.5), and to prove that the generators of the Fibonacci group $F(2, 9)$ (Example 13.6) have infinite order.

As was mentioned in Subsection 13.1.4, the use of the programs to construct unique word-acceptors for various groups has led to applications outside of group theory. These applications are based on the simple fact that a unique word-acceptor allows one to use FSA-ENUMERATE to rapidly enumerate unique shortest representatives of the elements of the group.

The simplest applications of this type involve drawing pictures of geometric objects. Some of the scenes in the Geometry Center movie "Not Knot" [GM92] were made with the help of automatic groups software: Example 13.4 above was one of the groups used. One of the posters depicting a frame from the fly-through of hyperbolic space from "Not Knot" used a list of over 100,000 group elements; this list took only a few minutes to compute using the word-acceptor. Without it, the same list would have required hours of computation. Another advantage of the method is its extremely low memory requirements; without it, the full list of group elements would have to be kept in memory, as matrices, throughout the computation.

For a more sophisticated application, to the drawing of limit sets of Kleinian groups, see the article by McShane, Parker, and Redfern [MPR94].

Exercises

1. Let $G := \langle\, x, y, a \mid a^2 = 1,\ xy = yx,\ a^{-1}xa = y \,\rangle$. Show that G is shortlex automatic with respect to the ordering $x < x^{-1} < y < y^{-1} < a < a^{-1}$ of A, but not with respect to the ordering $a < a^{-1} < x < x^{-1} < y < y^{-1}$.

2. Suppose that W is a word-acceptor for the group G and that the set of word-differences associated with the union of the sets

$$L_x := \{\, (w, v)^+ \mid w, v \in L(W) \text{ and } wx =_G v \,\}$$

for $x \in A \cup \{\varepsilon\}$ is finite. Prove that each L_x is the language of an fsa, and hence that G is automatic.

3. Let $<$ be a shortlex ordering on A^*. Show how to construct a 2-variable dfa with four states and language $\{\, (w, v)^+ \in (A^+ \times A^+)^* \mid v < w \,\}$.

4. In the construction of W' described in Subsection 13.3.2 show that, if $(w, v) \in L$, then $(u_1 w, u_1 v) \in L$ for any $u_1 \in A^*$. Deduce that $L(W') = \neg(LA^*)$. (This fact enables us to construct W' slightly more efficiently.)

5. Show that the fsa M'_x constructed in Subsection 13.3.3 have the language claimed.

References

[ABC⁺91] J.M. Alonso, T. Brady, D. Cooper, V. Ferlini, M. Lustig, M. Mihalik, M. Shapiro, and H. Short. Notes on word hyperbolic groups. In A. Haefliger É. Ghys and A. Verjovsky, editors, *Group Theory from a Geometrical Viewpoint* (Trieste, 1990), pages 3–63. World Sci. Publishing Co., Inc., River Edge, NJ, 1991.

[Adi58] S.I. Adian. On algorithmic problems in effectively complete classes of groups. *Doklady Akad. Nauk. SSR*, 123:13–16, 1958.

[Adi79] S. I. Adian. *The Burnside Problem and Identities in Groups*. Springer-Verlag, Berlin, 1979. Translated from the Russian by John Lennox and James Wiegold.

[AHU74] A.V. Aho, A.V. Hopcroft, and J.D. Ullman. *The Design and Analysis of Computer Algorithms*. Reading: Addison-Wesley, 1974.

[AK92] E.F. Assmus, Jr. and Jenny Key. *Designs and Their Codes*, volume 103 of *CUP Tracts in Mathematics*. Cambridge University Press, 1992.

[AMW82] D.G. Arrell, S. Manrai, and M.F. Worboys. A procedure for obtaining simplified defining relations for a subgroup. In C.M. Campbell and E.F. Robertson, editors, *Groups — St Andrews 1981*, volume 71 of *London Math. Soc. Lecture Note Ser.*, pages 155–159, Cambridge, 1982. Cambridge University Press.

[Arc02] Claude Archer. Classification of group extensions. PhD thesis, Université Libre de Bruxelles, 2002.

[Art55] E. Artin. The orders of the classical simple groups. *Comm. Pure Appl. Math.*, 8:455–472, 1955.

[Asc84] M. Aschbacher. On the maximal subgroups of the finite classical groups. *Invent. Math.*, 76:469–514, 1984.

[Atk75] M.D. Atkinson. An algorithm for finding the blocks of a permutation group. *Math. Comp.*, 29:911–913, 1975.

[Atk84] Michael D. Atkinson, editor. *Computational Group Theory*, London, New York, 1984. (Durham, 1982), Academic Press.

[ATL] ATLAS of Finite Group Representations.
http://web.mat.bham.ac.uk/atlas/index.html.

[Bab91] László Babai. Local expansion of vertex-transitive graphs and random generation in finite groups. In *Theory of Computing*, pages 164–174, New York, 1991. (Los Angeles, 1991), Association for Computing Machinery.

[BB99] L. Babai and R. Beals. A polynomial-time theory for black box groups I. In *Groups St Andrews* 1997 *in Bath*, volume 261 of *London Math. Soc. Lecture Note Ser.*, pages 30–64, Cambridge, 1999. Cambridge University Press.

[BBN+78] Harold Brown, Rolf Bülow, Joachim Neubüser, Hans Wondratschek, and Hans Zassenhaus. *Crystallographic Groups of Four-Dimensional Space*. Wiley-Interscience, New York, Chicester, Brisbane, Toronto, 1978.

[BBR93] László Babai, Robert Beals, and Daniel Rockmore. Deciding finiteness of matrix groups in deterministic polynomial time. In *Proc. of International Symposium on Symbolic and Algebraic Computation* ISSAC '93, pages 117–126. (Kiev), ACM Press, 1993.

[BC82] Gregory Butler and John J. Cannon. Computing in permutation and matrix groups I: Normal closure, commutator subgroups, series. *Math. Comp.*, 39:663–670, 1982.

[BC89] Gregory Butler and John J. Cannon. Computing in permutation and matrix groups III: Sylow subgroups. *J. Symbolic Comput.*, 8:241–252, 1989.

[BC91] Gregory Butler and John J. Cannon. Computing Sylow subgroups using homomorphic images of centralizers. *J. Symbolic Comput.*, 12:443–458, 1991.

[BC93] Wieb Bosma and John Cannon. *Handbook of* MAGMA *Functions*. School of Mathematics and Statistics, University of Sydney, 1993.

[BCFS91] László Babai, Gene Cooperman, Larry Finkelstein, and Ákos Seress. Nearly linear time algorithms for permutation groups with a small base. In *Proc. of International Symposium on Symbolic and Algebraic Computation* ISSAC '91, pages 200–209. (Bonn), ACM Press, 1991.

[BCM81a] G. Baumslag, F.B. Cannonito, and C.F. Miller III. Computable algebra and group embeddings. *J. Algebra*, 69:186–212, 1981.

[BCM81b] G. Baumslag, F.B. Cannonito, and C.F. Miller III. Some recognizable properties of solvable groups. *Math. Z.*, 178:289–295, 1981.

[BE99a] Hans Ulrich Besche and Bettina Eick. Construction of finite groups. *J. Symbolic Comput.*, 27:387–404, 1999.

[BE99b] Hans Ulrich Besche and Bettina Eick. The groups of order at most 1000 except 512 and 768. *J. Symbolic Comput.*, 27:405–413, 1999.

[BE01] Hans Ulrich Besche and Bettina Eick. The groups of order $q^n \cdot p$. *Comm. Algebra*, 29:1759–1772, 2001.

[Bea93a] Robert M. Beals. Computing blocks of imprimitivity for small-base groups in nearly linear time. In Finkelstein and Kantor [FK93], pages 17–26.

[Bea93b] Robert M. Beals. An elementary algorithm for computing the composition factors of a permutation group. In *Proc. of International Symposium on Symbolic and Algebraic Computation ISSAC '93*, pages 127–134. Kiev, ACM Press, 1993.

[Bee84] M.J. Beetham. Space saving in coset enumeration. In Atkinson [Atk84], pages 19–25.

[BEO01] Hans Ulrich Besche, Bettina Eick, and E. A. O'Brien. The groups of order at most 2000. *Electronic Research Announcements of the AMS*, 7:1–4, 2001.

[BEO02] Hans Ulrich Besche, Bettina Eick, and E. A. O'Brien. A millennium project: constructing small groups. *Internat. J. Algebra Comput.*, 12:623–644, 2002.

[Ber67] E.R. Berlekamp. Factoring polynomials over finite fields. *Bell System Technical Journal*, 46:1853–1859, 1967.

[Ber70] E.R. Berlekamp. Factoring polynomials over large finite fields. *Math. Comp.*, 24:713–735, 1970.

[BH93] Brigitte Brink and Robert B. Howlett. A finiteness property and an automatic structure for Coxeter groups. *Math. Ann.*, 296:179–190, 1993.

[BK01] P.A. Brooksbank and W.M. Kantor. On constructive recognition of a black box PSL(d, q). In Kantor and Seress [KS97], pages 95–111.

[BKPS02] L. Babai, W.M. Kantor, P. Pálfy, and Á. Seress. Black-box recognition of finite simple groups of Lie type by statistics of element orders. *J. Group Theory*, 5:383–401, 2002.

[BKW74] A.J. Bayes, J. Kautsky, and J.W. Wamsley. Computation in nilpotent groups (application). In *Proc. Second Internat. Conf. Theory of Groups*, volume 372 of *Lecture Notes in Math.*, pages 82–89, Berlin, Heidelberg, New York, 1974. (Canberra, 1973), Springer-Verlag.

[BLGN+02] R. Beals, C.R. Leedham-Green, A.C. Niemeyer, C.E. Praeger, and A. Seress. Permutations with restricted cycle structure and an algorithmic application. *Combin. Probab. Comput.*, 11:447–464, 2002.

[BM83] G. Butler and J. McKay. The transitive groups of degree up to 11. *Comm. Algebra*, 11:863–911, 1983.

[Boo59] W.W. Boone. The word problem. *Annals of Mathematics*, 70:207–265, 1959.

[BP00] S. Bratus and I. Pak. Fast constructive recognition of a black box group isomorphic to S_n or A_n using Goldbach's conjecture. *J. Symbolic Comput.*, 29:33–57, 2000.

[Bro69] Harold Brown. An algorithm for the determination of space groups. *Math. Comp.*, 23:499–514, 1969.

[Bro03] P.A. Brooksbank. Constructive recognition of classical groups in their natural representation. *J. Symbolic Comput.*, 35:195–239, 2003.

[BS84] László Babai and Endre Szemerédi. On the complexity of matrix group problems, I. In *Proc. 25th IEEE Sympos. Foundations Comp. Sci.*, pages 229–240, 1984.

[Buc87] B. Buchberger. The history and basic features of the critical-pair/completion procedure. *J. Symbolic Comput.*, 3:3–38, 1987.

[Bur02] W. Burnside. On an unsettled question in the theory of discontinuous groups. *Quart. J. Pure Appl. Math*, 33:230–238, 1902.

[Bur11] W. Burnside. *Theory of Groups of Finite Order*. Cambridge University Press. Reprinted by Dover 1955, New York, 2nd edition, 1911.

[But76] Gregory Butler. The Schreier algorithm for matrix groups. In SYMSAC '76, *Proc. ACM Sympos. Symbolic and Algebraic Computation*, pages 167–170, New York, 1976. (New York, 1976), Association for Computing Machinery.

[But79] Gregory Butler. *Computational Approaches to Certain Problems in the Theory of Finite Groups*. PhD thesis, University of Sydney, 1979.

[But83] Gregory Butler. Computing normalizers in permutation groups. *J. Algorithms*, 4:163–175, 1983.

[But84] Gregory Butler. On computing double coset representatives in permutation groups. In Atkinson [Atk84], pages 283–290.

[But91] Gregory Butler. *Fundamental Algorithms for Permutation Groups*, volume 559 of *Lecture Notes in Comput. Sci.* Springer-Verlag, Berlin, Heidelberg, New York, 1991.

[But93] Gregory Butler. The transitive groups of degree fourteen and fifteen. *J. Symbolic Comput.*, 16:413–422, 1993.

[Cam81] P.J. Cameron. Finite permutation groups and finite simple groups. *Bull. Lon. Math. Soc.*, 13:1–22, 1981.

[Cam99] Peter J. Cameron. *Permutation Groups*, volume 45 of *London Math. Soc. Stud. Texts*. Cambridge University Press, Cambridge, 1999.

[Can72] John J. Cannon. Graphs and defining relations. In *Proc. First Australian Conf. on Combinatorial Mathematics*, pages 215–233. (Newcastle, 1972), 1972.

[Can73] John J. Cannon. Construction of defining relators for finite groups. *Discrete Math.*, 5:105–129, 1973.

[Can84] James W. Cannon. The combinatorial structure of cocompact discrete hyperbolic group. *Geom. Dedicata*, 296:123–148, 1984.

[Car72] Roger W. Carter. *Simple Groups of Lie Type*. John Wiley & Sons, London, New York, Sydney, Toronto, 1972.

[CCH97a] J.J. Cannon, B.C. Cox, and D.F. Holt. Computing chief series, composition series and socles in large permutation groups. *J. Symbolic Comput.*, 24:285–301, 1997.

[CCH97b] J.J. Cannon, B.C. Cox, and D.F. Holt. Computing Sylow subgroups of permutation groups. *J. Symbolic Comput.*, 24:303–316, 1997.

[CCH01] J.J. Cannon, B.C. Cox, and D.F. Holt. Computing the subgroups of a permutation group. *J. Symbolic Comput.*, 31:149–161, 2001.

[CCN+85] J.H. Conway, R.T. Curtis, S.P. Norton, R.A. Parker, and R.A. Wilson. *Atlas of Finite Groups*. Clarendon Press, Oxford, 1985.

[CD02] M. Conder and P. Dobcsányi. Trivalent symmetric graphs on up to 768 vertices. *J. Combin. Math. Combin. Comput.*, 40:41–63, 2002.

[CD05] M. Conder and P. Dobcsányi. Applications and adaptations of the low index subgroups procedure. *Math. Comp.*, 74:485–497, 2005.

[CDHW73] John J. Cannon, Lucien A. Dimino, George Havas, and Jane M. Watson. Implementation and analysis of the Todd-Coxeter algorithm. *Math. Comp.*, 27:463–490, 1973.

[CELG04] John Cannon, Bettina Eick, and Charles Leedham-Green. Special polycyclic generating sequences for finite soluble groups. To appear in *J. Symbolic Comput.*, 2004.

[CF92] Gene Cooperman and Larry Finkelstein. A fast cyclic base change for permutation groups. In *Proceedings of International Symposium on Symbolic and Algebraic Computation* IS-SAC '92, pages 224–232. ACM Press, 1992.

[CFJR01] T. Chinburg, E. Friedman, K.N. Jones, and A.W. Reid. The arithmetic hyperbolic 3-manifold of smallest volume. *Ann. Scuola Norm. Sup. Pisa Cl. Sci. (4)*, 30:1–40, 2001.

[CFS90] G. Cooperman, L. Finkelstein, and N. Sarawagi. A random base change algorithm for permutation groups. In *Proc. of International Symposium on Symbolic and Algebraic Computation* ISSAC '90, pages 161–168. (Kiev), ACM Press, 1990.

[CH03] John J. Cannon and Derek F. Holt. Automorphism group computation and isomorphism testing in finite groups. *J. Symbolic Comput.*, 35:241–267, 2003.

[CH04] John J. Cannon and Derek F. Holt. Computing maximal subgroups of finite groups. *J. Symbolic Comput.*, 37:589–609, 2004.

[Cha95] R. Charney. Geodesic automation and growth functions for Artin groups of finite type. *Math. Ann.*, 301:307–324, 1995.

[Che] The CHEVIE Homepage. http://www.math.rwth-aachen.de/homes/CHEVIE/.

[CLG97] F. Celler and C.R. Leedham-Green. Calculating the order of an invertible matrix. In Finkelstein and Kantor [FK97], pages 55–60.

[CLG98] F. Celler and C.R. Leedham-Green. A constructive recognition algorithm for the special linear group. In *The Atlas of Finite Groups: Ten Years On*, volume 249 of *London Mathemtical Society Lecture Note Series*, pages 11–26, 1998.

[CLGM+95] Frank Celler, Charles R. Leedham-Green, Scott H. Murray, Alice C. Niemeyer, and E.A. O'Brien. Generating random elements of a finite group. *Comm. Algebra*, 23:4931–4948, 1995.

[CLGO] M.D.E. Conder, C.R. Leedham-Green, and E.A. O'Brien. Constructive recognition of PSL(2, q). To appear in *Trans. Amer. Math. Soc.*

[CLRS02] Thomas H. Cormen, Charles E. Leiserson, Ronald L. Rivest, and Clifford Stein. *Introduction to Algorithms.* MIT Press, Cambridge, MA, London, 2nd edition, 2002.

[CMT04] Arjeh M. Cohen, Scott H. Murray, and D.E. Taylor. Computing in groups of Lie type. *Math. Comp.*, 73:1477–1498, 2004.

[CNW90] Frank Celler, Joachim Neubüser, and Charles R. B. Wright. Some remarks on the computation of complements and normalizers in soluble groups. *Acta Applicandae Mathematicae*, 21:57–76, 1990.

[Coh73] Henri Cohen. *A Course in Computational Algebraic Number Theory*, volume 138 of *Graduate Texts in Math.* Springer-Verlag Inc., New York, 1973.

[Con] Conway polynomials for finite fields. http://www.math.rwth-aachen.de/~Frank.Luebeck/ConwayPol/.

[Cor] The clrscode Package for LaTeX2e. http://www.cs.dartmouth.edu/~thc/clrscode/.

[CP93] John Cannon and Catherine Playoust. *An Introduction to* MAGMA. School of Mathematics and Statistics, University of Sydney, 1993.

[CR80] C.M. Campbell and E.F. Robertson. A deficiency zero presentation for SL(2, p). *Bull. Lon. Math. Soc.*, 12:17–20, 1980.

[CS97] J.J. Cannon and B. Souvignier. On the computation of conjugacy classes in permutation groups. In W. Küchlin, editor, *Proceedings of International Symposium on Symbolic and Algebraic Computation, Maui, July 21–23, 1997*, pages 392–399. Association for Computing Machinery, 1997.

[CS98] J.J. Cannon and B. Souvignier. On the computation of normal subgroups in permutation groups. unpublished, 1998.

[CW90] D. Coppersmith and S. Winograd. Matrix multiplication via arithmetic progressions. *J. Symbolic Comput.*, 8:251–280, 1990.

[CZ81] D.G. Cantor and H. Zassenhaus. A new algorithm for factoring polynomials over a finite field. *Math. Comp*, 36:587–592, 1981.

[Deh11] M. Dehn. Über unendliche diskontinuierliche Gruppen. *Math. Ann.*, 71:116–144, 1911.

[Dek96] Karel Dekimpe. *Almost-Bieberbach Groups: Affine and Polynomial Structures*, volume 1639 of *Lecture notes in Math.* Springer, 1996.

[DF87] J.R. Driscoll and M.L. Furst. Computing short generating sequences. *Info. and Comput.*, 72:117–132, 1987.

[DGK83] P. Diaconis, R.L. Graham, and W.M. Kantor. The mathematics of perfect shuffles. *Adv. in Appl. Math.*, 4:175–196, 1983.

[DH92] Klaus Doerk and Trevor Hawkes. *Finite Solvable Groups.* De Gruyter, 1992.

[Dix67] John D. Dixon. High speed computation of group characters. *Numer. Math.*, 10:446–450, 1967.

[Dix68] John D. Dixon. The solvable length of a solvable linear group. *Math. Z.*, 107:151–158, 1968.

[DM88] John D. Dixon and Brian Mortimer. The primitive permutation groups of degree less than 1000. *Math. Proc. Camb. Phil. Soc.*, 103:213–238, 1988.

[DM96] John D. Dixon and Brian Mortimer. *Permutation Groups*, volume 163 of *Graduate Texts in Math.* Springer-Verlag, New York, Heidelberg, Berlin, 1996.

[DS74] Anke Dietze and Mary Schaps. Determining subgroups of a given finite index in a finitely presented group. *Canad. J. Math.*, 26:769–782, 1974.

[DT03] N. M. Dunfield and W. P. Thurston. The virtual Haken conjecture: experiments and examples. *Geom. Topol.*, 7:399–441, 2003.

[ECH+92] David B.A. Epstein, J.W. Cannon, D.F. Holt, S.V.F. Levy, M.S. Paterson, and W.P Thurston. *Word Processing in Groups.* Jones and Bartlett, Boston, 1992.

[EG81] S. Even and O. Goldreich. The minimum length generator problem is NP-hard. *J. Algorithms*, 2:311–313, 1981.

[EGN97] Bettina Eick, F. Gähler, and W. Nickel. Computing maximal subgroups and Wyckoff positions of space groups. *Acta Cryst. A*, 53:467–474, 1997.

[EH01a] Bettina Eick and A. Hulpke. Computing the maximal subgroups of a permutation group I. In Kantor and Seress [KS97], pages 155–168.

[EH01b] D.B.A. Epstein and D.F. Holt. Computation in word-hyperbolic groups. *Internat. J. Algebra Comput.*, 11:467–487, 2001.

[EH03]   Bettina Eick and Burkhard Höfling. The solvable primitive
          permutation groups of degree at most 6560. *LMS J. Comput.
          Math.*, 6:29–39, 2003.

[EHR91]   D.B.A. Epstein, D.F. Holt, and S.E. Rees. The use of Knuth-
          Bendix methods to solve the word problem in automatic
          groups. *J. Symbolic Comput.*, 12:397–414, 1991.

[Eic93]   Bettina Eick. *Spezielle PAG-Systeme im Computeralgebrasys-
          tem* GAP. Diplomarbeit, RWTH Aachen, 1993.

[Eic97]   Bettina Eick. Special presentations for finite soluble groups
          and computing (pre-)Frattini subgroups. In Finkelstein and
          Kantor [FK97], pages 101–112.

[Eic01a] Bettina Eick. Algorithms for polycyclic groups. *Habilitation-
          sschrift*, Universität Kassel, 2001.

[Eic01b] Bettina Eick. Computing with infinite polycyclic groups. In
          Kantor and Seress [KS97], pages 139–153.

[Eic02]   Bettina Eick. Orbit-stabilizer problems and computing nor-
          malizers for polycyclic groups. *J. Symbolic Comput.*, 34:1–19,
          2002.

[EIFZ96] D.B.A. Epstein, A.R. Iano-Fletcher, and U. Zwick. Growth
          functions and automatic groups. *Experimental Math.*, 5:297–
          315, 1996.

[ELGO02] Bettina Eick, C.R. Leedham-Green, and E.A. O'Brien. Con-
          structing automorphism groups of p-groups. *Comm. Algebra*,
          30:2271–2295, 2002.

[ENP]   Bettina Eick, A.C. Niemeyer, and O. Panaia. An infinite poly-
          cyclic quotient algorithm. Submitted.

[EO99]   Bettina Eick and E.A. O'Brien. Enumerating p-groups. *J.
          Austral. Math. Soc. Ser. A*, 67:191–205, 1999.

[EO03]   Bettina Eick and Gretchen Ostheimer. On the orbit stabilizer
          problem for integral matrix actions of polycyclic groups. *Math.
          Comp.*, 72:1511–1529, 2003.

[EP88]   D. Easdown and C.E. Praeger. On minimal faithful permuta-
          tion representations of finite groups. *Bull. Aust. Math. Soc.*,
          38:207–220, 1988.

[EW02]   Bettina Eick and Charles R.B. Wright. Computing subgroups
          by exhibition in finite solvable groups. *J. Symbolic Comput.*,
          33:129–143, 2002.

[Fel61] H. Felsch. Programmierung der Restklassenabzählung einer Gruppe nach Untergruppen. *Numer. Math.*, 3:250–256, 1961.

[FHL80] M. Furst, J. Hopcroft, and E. Luks. Polynomial-time algorithms for permutation groups. In *Proc. 21st IEEE Sympos. Foundations Comp. Sci.*, pages 36–41, 1980.

[FK93] Larry Finkelstein and William M. Kantor, editors. *Groups and Computation*, volume 11 of *Amer. Math. Soc. DIMACS Series*. (DIMACS, 1991), 1993.

[FK97] Larry Finkelstein and William M. Kantor, editors. *Groups and Computation II*, volume 28 of *Amer. Math. Soc. DIMACS Series*. (DIMACS, 1995), 1997.

[FN68] V. Felsch and J. Neubüser. Über ein Programm zur Berechnung der Automorphismengruppe einer endlichen Gruppe. *Numer. Math.*, 11:277–292, 1968.

[FN70] V. Felsch and J. Neubüser. On a programme for the determination of the automorphism group of a finite group. In *Computational Problems in Abstract Algebra*, pages 59–60, Oxford, London, Edinburgh, 1970. (Oxford, 1967), Pergamon Press.

[FN79] V. Felsch and J. Neubüser. An algorithm for the computation of conjugacy classes and centralizers in p-groups. In Edward W. Ng, editor, *Symbolic and Algebraic Computation*, volume 72 of *Lecture Notes in Comput. Sci.*, pages 452–465, Berlin, Heidelberg, New York, 1979. (Marseille, 1979), Springer-Verlag.

[GAP] The **GAP** Character Table Library. `http://www.math.rwth-aachen.de/~Thomas.Breuer/ctbllib/`.

[GAP04] The GAP Group. *GAP — Groups, Algorithms, and Programming, Version 4.4*, 2004. (`http://www.gap-system.org`).

[Gas53] Wolfgang Gaschütz. Über die ϕ-Untergruppe endlicher Gruppen. *Math. Z.*, 58:160–170, 1953.

[GCL92] K.O. Geddes, S.R. Czapor, and G. Labahn. *Algorithms for Computer Algebra*. Kluwer Academic Publishers, 1992.

[GdlH90] E. Ghys and P. de la Harpe., editors. *Sur les groupes hyperboliques d'après Mikhael Gromov*, volume 83 of *Progress in Mathematics*. Birkhäuser Boston, Inc., Boston, MA, 1990.

[GH97] S.P. Glasby and R.B. Howlett. Writing representations over minimal fields. *Comm. Algebra*, 25:1703–1711, 1997.

[Gil79] R.H. Gilman. Presentations of groups and monoids. *J. Algebra*, 57:544–554, 1979.

[Gil84] R.H. Gilman. Enumerating infinitely many cosets. In Atkinson [Atk84], pages 267–274.

[GM92] C. Gunn and D. Maxwell. Not knot. Video, ISBN 0-86720-240-8, published by AKPeters, Natick, MA, 1992.

[Gor82] Daniel Gorenstein. *Finite Simple Groups. An Introduction to Their Classification.* The University Series in Mathematics. Plenum Press, New York and London, 1982.

[Gri82] R.L. Griess. The friendly giant. *Invent. Math.*, 69:1–102, 1982.

[GS90] S.P. Glasby and Michael C. Slattery. Computing intersections and normalizers in soluble groups. *J. Symbolic Comput.*, 9:637–651, 1990.

[GS91] S.M. Gersten and H. Short. Small cancellation theory and automatic groups, II. *Invent. Math.*, 105(3):641–662, 1991.

[Hal59] Marshall Hall, Jr. *The Theory of Groups.* Macmillan Co., New York, 1959.

[Har] A. Harkins. Polycyclic groups are not automatic. Submitted.

[Hav74] George Havas. A Reidemeister-Schreier program. In *Proceedings of the Second Internat. Conf. Theory of Groups*, volume 322 of *Lecture Notes in Mathematics*, pages 347–356, Berlin, Heidelberg, New York, 1974. (Canberra, 1973), Springer-Verlag.

[Hei61] Hermann Heineken. Engelsche Elemente der Länge drei. *Illinois J. Math.*, 5:681–707, 1961.

[Her94] Susan M. Hermiller. Rewriting systems for Coxeter groups. *J. of Pure and Applied Algebra*, 92:137–148, 1994.

[Hey97] M-C. Heydemann. Cayley graphs and interconnection networks. In *Graph Symmetry (Montreal 1996)*, volume 497 of *NATO ASI Ser. C, Math. Phys. Sci.*, pages 167–224. Kluwer Academic Publishers, 1997.

[HH56] P. Hall and Graham Higman. On the p-length of p-soluble groups and reduction theorems for Burnside's problem. *Proc. London Math. Soc.* (3), 6:1–42, 1956.

[HH99] Derek F. Holt and Darren F. Hurt. Computing automatic coset systems and subgroup presentations. *J. Symbolic Comput.*, 27:1–19, 1999.

[HHN01] G. Havas, D.F. Holt, and M.F. Newman. Certain cyclically presented groups are infinite. *Comm. Algebra*, 29:5175–5178, 2001.

[HHR93]   George Havas, Derek F. Holt, and Sarah Rees. Recognizing badly presented *Z*-modules. *Linear Algebra Appl.*, 192:137–163, 1993.

[Hig59]   Graham Higman. Some remarks on varieties of groups. *Quart. J. Math. Oxford Ser.* (2), 10:165–178, 1959.

[Hil02]   J.A. Hillman. *Four-Manifolds, Geometries and Knots*, volume 5 of *Geometry and Topology Monographs*. Geometry and Topology Publications, Mathematics Institute, University of Warwick, 2002.

[His03]   G. Hiss. Algorithms of representation theory. In *Computer Algebra Handbook: Foundations, Applications, Systems*, pages 84–88. Springer-Verlag, 2003.

[HK84]   George Havas and L.G. Kovács. Distinguishing eleven crossing knots. In Atkinson [Atk84], pages 367–373.

[HL89]   G. Hiss and K. Lux. *Brauer Trees of Sporadic Groups*. Oxford University Press, 1989.

[HL04]   L.S. Heath and N.A. Loehr. New algorithms for generating Conway polynomials over finite fields. *J. Symbolic Comput.*, 38:1003–1024, 2004.

[HLGOR96a]   Derek F. Holt, C.R. Leedham-Green, E.A. O'Brien, and Sarah Rees. Computing matrix group decompositions with respect to a normal subgroup. *J. Algebra*, 184:818–838, 1996.

[HLGOR96b]   Derek F. Holt, C.R. Leedham-Green, E.A. O'Brien, and Sarah Rees. Testing matrix groups for primitivity. *J. Algebra*, 184:795–817, 1996.

[HLO+04]   P.E. Holmes, S.A. Linton, E.A. O'Brien, A.J.E. Ryba, and R. A. Wilson. Constructive membership testing in black-box groups. Preprint, 2004.

[HM97]   George Havas and Bohdan S. Majewski. Integer matrix diagonalization. *J. Symbolic Comput.*, 24:399–408, 1997.

[HMM98]   G. Havas, Bohdan S. Majewski, and K.R. Matthews. Extended GCD and Hermite normal form algorithms via lattice basis reduction. *Experimental Math.*, 7:125–136, 1998.

[HMM99]   G. Havas, Bohdan S. Majewski, and K.R. Matthews. Extended GCD and Hermite normal form algorithms via lattice basis reduction (addenda and errata). *Experimental Math.*, 8:205, 1999.

[HMU01] J.E. Hopcroft, R. Motwani, and J.D. Ullman. *Introduction to Automata Theory, Languages, and Computation.* Addison-Wesley, 2nd edition, 2001.

[HN80] George Havas and M.F. Newman. Application of computers to questions like those of Burnside. In *Burnside Groups*, volume 806 of *Lecture Notes in Math.*, pages 211–230, Berlin, Heidelberg, New York, 1980. (Bielefeld, 1977), Springer-Verlag.

[HNO95] George Havas, M.F. Newman, and E.A. O'Brien. Groups of deficiency zero. In G. Baumslag et al., editors, *Geometric and Computational Perspectives on Infinite Groups*, volume 25 of *Amer. Math. Soc. DIMACS Series*, pages 53–67. (DIMACS, 1994), 1995.

[Hol84] D.F. Holt. The calculation of the Schur multiplier of a permutation group. In Atkinson [Atk84], pages 307–318.

[Hol85a] D.F. Holt. A computer program for the calculation of a covering group of a finite group. *J. Pure Appl. Algebra*, 35:287–295, 1985.

[Hol85b] D.F. Holt. The mechanical computation of first and second cohomology groups. *J. Symbolic Comput.*, 1:351–361, 1985.

[Hol91] D.F. Holt. The computation of normalizers in permutation groups. *J. Symbolic Comput.*, 12:499–516, 1991.

[Hol95] Derek F. Holt. The Warwick automatic group software. In G. Baumslag et al., editors, *Geometric and Computational Perspectives on Infinite Groups*, volume 25 of *Amer. Math. Soc. DIMACS Series*, pages 69–82. (DIMACS, 1994), 1995.

[Hol97] Derek F. Holt. Representing quotients of permutation groups. *Quart. J. Math. (Oxford)*, 48:347–350, 1997.

[Hop71] J.E. Hopcroft. An $n \log n$ algorithm for minimizing the states in a finite automaton. In Z. Kohavi, editor, *The Theory of Machines and Computations*, pages 189–196, New York, 1971. Academic Press.

[HP89] Derek F. Holt and W. Plesken. *Perfect Groups*. Clarendon Press, Oxford, 1989.

[HR92] Derek F. Holt and Sarah Rees. Testing for isomorphism between finitely presented groups. In *Groups, Combinatorics and Geometry*, volume 165 of *London Math. Soc. Lecture Note Ser.*, pages 459–475, London, 1992. Durham, 1989, Cambridge University Press.

[HR93] Derek F. Holt and Sarah Rees. A graphics system for displaying finite quotients of finitely presented groups. In Finkelstein and Kantor [FK93], pages 113–126.

[HR94] Derek F. Holt and Sarah Rees. Testing modules for irreducibility. *J. Austral. Math. Soc. Ser. A*, 57:1–16, 1994.

[HR99] George Havas and Colin Ramsay. *Coset enumeration: ACE version 3.* (http://www.csee.uq.edu.au/~cram/ce.html), 1999.

[HS79] George Havas and L.S. Sterling. Integer matrices and abelian groups. In *Symbolic and Algebraic Computation*, volume 72 of *Lecture Notes in Comput. Sci.*, pages 431–451, Berlin, Heidelberg, New York, 1979. (Marseille, 1979), Springer-Verlag.

[Hul] Alexander Hulpke. Constructing transitive permutation groups. To appear in *J. Symbolic Comput.*

[Hul96] Alexander Hulpke. *Konstruktion transitiver Permutationsgruppen*. PhD thesis, RWTH Aachen, 1996.

[Hul99a] A. Hulpke. Computing subgroups invariant under a set of automorphisms. *J. Symbolic Comput.*, 27:415–427, 1999.

[Hul99b] A. Hulpke. Galois groups through invariant relations. In *Groups St Andrews 1997 in Bath*, volume 261 of *London Math. Soc. Lecture Note Ser.*, pages 379–393, Cambridge, 1999. Cambridge University Press.

[Hup67] B. Huppert. *Endliche Gruppen I*, volume 134 of *Grundlehren Math. Wiss.* Springer-Verlag, Berlin, Heidelberg, New York, 1967.

[HWW74] George Havas, G.E. Wall, and J.W. Wamsley. The two generator restricted Burnside group of exponent five. *Bull. Austral. Math. Soc.*, 10:459–470, 1974.

[IL00] Gábor Ivanyos and Klaus Lux. Treating the exceptional cases of the Meataxe. *Experimental Math.*, 9:373–381, 2000.

[Isa76] I.M. Isaacs. *Character Theory of Finite Groups*, volume 69 of *Pure and Applied Mathematics*. Academic Press, Berlin, Heidelberg, New York, 1976.

[Iva94] Sergei V. Ivanov. The free Burnside groups of sufficiently large exponents. *Internat. J. Algebra Comput.*, 4(1-2):ii+308, 1994.

[Jer82] M. Jerrum. A compact representation for permutation groups. In *Proc. 23rd IEEE Sympos. Foundations Comp. Sci.*, pages 126–133, 1982.

[Jer85] M. Jerrum. The complexity of finding minimum length generator sequences. *Theoretical Computer Science*, 36:256–289, 1985.

[JKO99] D.L. Johnson, A.C. Kim, and E.A. O'Brien. Certain cyclically presented groups are isomorphic. *Comm. Algebra*, 27:3531–3536, 1999.

[JLPW95] Christoph Jansen, Klaux Lux, Richard Parker, and Robert Wilson. *An Atlas of Brauer Characters*, volume 11 of *London Math. Soc. Monographs (New Series)*. Clarendon Press, Oxford, 1995.

[Joh98] D.L. Johnson. *Presentations of Groups, Second Edition*, volume 15 of *London Math. Soc. Stud. Texts*. Cambridge University Press, Cambridge, 1998.

[Jor73] C. Jordan. Sur la limite de transitivité des groupes non alternés. *Bull. Soc. Math. France*, 1:40–71, 1873.

[Jür70] H. Jürgensen. Calculation with the elements of a finite group given by generators and defining relations. In Leech [Lee70], pages 47–57.

[Kan85] William M. Kantor. Sylow's theorem in polynomial time. *J. Comp. Syst. Sci.*, 30:359–394, 1985.

[Kan90] William M. Kantor. Finding Sylow normalizers in polynomial time. *J. Algorithms*, 11:523–563, 1990.

[Kan91] William M. Kantor. Finding composition factors of permutation groups of degree $n \leq 10^6$. *J. Symbolic Comput.*, 12:517–526, 1991.

[KB70] D.E. Knuth and P.B. Bendix. Simple word problems in universal algebras. In Leech [Lee70], pages 263–297.

[KB79] R. Kannan and A. Bachem. Polynomial algorithms for computing the Smith and Hermite normal forms of an integer matrix. *SIAM J. Computing*, 9:499–507, 1979.

[Ker99] Adalbert Kerber. *Applied finite group actions*, volume 19 of *Algorithms and Combinatorics*. Springer-Verlag, New York, Heidelberg, Berlin, 2nd edition, 1999.

[KM02] J.D. Key and J. Moori. Codes, designs and graphs from the Janko groups J_1 and J_2. *J. Combin. Math. Combin. Comput.*, 40:143–159, 2002.

[KMR03] J.D. Key, J. Moori, and B.G. Rodrigues. On some designs and codes from primitive representations of some finite simple groups. *J. Combin. Math. Combin. Comput.*, 45:3–19, 2003.

[Knu69] Donald E. Knuth. *The Art of Computer Programming. Volume 2: Seminumerical Algorithms.* Addison-Wesley, Massachusetts, 1969.

[Knu73] Donald E. Knuth. *The Art of Computer Programming. Volume 3: Sorting and Searching.* Addison-Wesley, Massachusetts, 1973.

[Knu81] D. Knuth. Notes on efficient representation of perm groups. 1981.

[Kol82] G. Kolata. Perfect shuffles and their relation to math. *Science*, 216:505–506, 1982.

[KS97] William M. Kantor and Ákos Seress, editors. *Groups and Computation III*, volume 8 of *Ohio State University Research Institute Publications*, Berlin, New York, 1997. (DIMACS, 1999), Walter de Gruyter.

[KS01] W.M. Kantor and Á. Seress. Black box classical groups. *Memoirs Amer. Math. Soc.*, 149, No. 708, 2001.

[Lau82] Reinhard Laue. Computing double coset representatives for the generation of solvable groups. In Jacques Calmet, editor, *Computer algebra : EUROCAM '82, European Computer Algebra Conference, Marseille, France, 5-7 April 1982*, volume 144 of *Lecture Notes in Comput. Sci.*, pages 65–70, Berlin, 1982. Springer.

[LeC86] P. LeChenadec. A catalogue of complete presentations. *J. Symbolic Comput.*, 2:363–381, 1986.

[Lee63] John Leech. Coset enumeration on digital computers. *Proc. Camb. Phil. Soc.*, 59:257–267, 1963.

[Lee70] J. Leech, editor. *Computational problems in abstract algebra*, Oxford, 1970. (Oxford, 1967), Pergamon Press.

[Leo80a] Jeffrey S. Leon. Finding the order of a permutation group. In Bruce Cooperstein and Geoffrey Mason, editors, *Finite Groups*, volume 37 of *Proc. Sympos. Pure Math.*, pages 511–517, Providence, RI, 1980. (Santa Cruz, 1979), Amer. Math. Soc.

[Leo80b] Jeffrey S. Leon. On an algorithm for finding a base and strong generating set for a group given by generating permutations. *Math. Comp.*, 20:941–974, 1980.

[Leo84] Jeffrey S. Leon. Computing automorphism groups of combinatorial objects. In Atkinson [Atk84], pages 321–335.

[Leo91] Jeffrey S. Leon. Permutation group algorithms based on partitions, I: Theory and algorithms. *J. Symbolic Comput.*, 12:533–583, 1991.

[LG84] C.R. Leedham-Green. A soluble group algorithm. In Atkinson [Atk84], pages 85–101.

[LGN80] C.R. Leedham-Green and M.F. Newman. Space groups and groups of prime-power order I. *Arch. Math. (Basel)*, 35:193–202, 1980.

[LGNW69] C.R. Leedham-Green, P.M. Neumann, and J. Wiegold. The breadth and the class of a finite p-group. *J. London Math. Soc.* (2), 1:409–420, 1969.

[LGO97a] C.R. Leedham-Green and E.A. O'Brien. Recognising tensor products of matrix groups. *Internat. J. Algebra Comput.*, 7:541–559, 1997.

[LGO97b] C.R. Leedham-Green and E.A. O'Brien. Tensor products are projective geometries. *J. Algebra*, 189:514–528, 1997.

[LGO02] C.R. Leedham-Green and E.A. O'Brien. Recognising tensor-induced matrix groups. *J. Algebra*, 253:14–30, 2002.

[LGS90] C.R. Leedham-Green and L.H. Soicher. Collection from the left and other strategies. *J. Symbolic Comput.*, 9:665–675, 1990.

[LGS98] C. R. Leedham-Green and L.H. Soicher. Symbolic collection using Deep Thought. *LMS J. Comput. Math.*, 1:9–24, 1998.

[Lie85] Martin W. Liebeck. On the orders of maximal subgroups of the finite classical groups. *Proc. London Math. Soc.* (3), 50:426–446, 1985.

[LMR94] Klaus Lux, Jürgen Müller, and Michael Ringe. Peakword condensation and submodule lattices: an application of the Meat-Axe. *J. Symbolic Comput.*, 17:529–544, 1994.

[LNS84] R. Laue, J. Neubüser, and U. Schoenwaelder. Algorithms for finite soluble groups and the SOGOS system. In Atkinson [Atk84], pages 105–135.

[Lo98] Eddie H. Lo. A polycyclic quotient algorithm. *J. Symbolic Comput.*, 25:61–97, 1998.

[LPS88] Martin W. Liebeck, Cheryl E. Praeger, and Jan Saxl. On the O'Nan-Scott theorem for finite primitive permutation groups. *J. Austral. Math. Soc. (Series A)*, 44:389–396, 1988.

[Lüb02] Frank Lübeck. Computation of elementary divisors of integer matrices. *J. Symbolic Comput.*, 33:57–65, 2002.

[Luk82] Eugene M. Luks. Isomorphism of graphs of bounded valence can be tested in polynomial time. *J. Comp. Syst. Sci.*, 25:42–65, 1982.

[Luk92] Eugene M. Luks. Computing in solvable matrix groups. In *Proc. 33rd IEEE Sympos. Foundations Comp. Sci.*, pages 111–120, 1992.

[Luk93] Eugene M. Luks. Permutation groups and polynomial-time computation. In Finkelstein and Kantor [FK93], pages 139–175.

[Lys96] I. G. Lysënok. Infinite Burnside groups of even period. *Izv. Ross. Akad. Nauk Ser. Mat.*, 60(3):3–224, 1996.

[Mac74] I.D. Macdonald. A computer application to finite p-groups. *J. Austral. Math. Soc. Ser. A*, 17:102–112, 1974.

[Mac87] I.D. Macdonald. Nilpotent quotient algorithms. In *Proceedings of Groups — St Andrews* 1985, volume 121 of *London Math. Soc. Lecture Note Ser.*, pages 268–272, Cambridge, 1987. Cambridge University Press.

[Mag] New York Group Theory Cooperative at CCNY. http://www.grouptheory.org/.

[Mag04] Computational Algebra Group, School of Mathematics and Statistics, University of Sydney. *The Magma Computational Algebra System for Algebra, Number Theory and Geometry*, 2004. (http://magma.maths.usyd.edu.au/magma/).

[Mas91] William S. Massey. *A Basic Course in Algebraic Topology*, volume 127 of *Graduate Texts in Math.* Springer-Verlag Inc., New York, 1991.

[McK70] John McKay. The construction of the character table of a finite group from generators and relations. In Leech [Lee70], pages 89–100.

[Mil92] C.F. Miller III. Decision problems for groups — survey and reflections. In G. Baumslag and C.F. Miller III, editors, *Algorithms and Classification in Combinatorial Group Theory*, volume 23 of *MSRI publications*, pages 1–60. Springer-Verlag, 1992.

[Min98] Torsten Minkwitz. An algorithm for solving the factorization problem in permutation groups. *J. Symbolic Comput.*, 26:89–95, 1998.

[MM87] S. Medvedoff and K. Morrison. Groups of perfect shuffles. *Math. Mag.*, 60:3–14, 1987.

[MN89] M. Mecky and J. Neubüser. Some remarks on the computation of conjugacy classes of soluble groups. *Bull. Austral. Math. Soc.*, 40:281–292, 1989.

[MNRW02] Jürgen Müller, Max Neunhöffer, Frank Röhr, and Robert Wilson. Completing the Brauer trees for the sporadic simple Lyons group. *LMS J. Comput. Math.*, 5:18–33, 2002.

[MO95] Scott H. Murray and E.A. O'Brien. Selecting base points for the Schreier-Sims algorithm for matrix groups. *J. Symbolic Comput.*, 19:577–584, 1995.

[MPR94] G. McShane, J. Parker, and I. Redfern. Drawing limit sets of Kleinian groups using finite state automata. *Experimental Math.*, 3:153–172, 1994.

[Neb95] Gabrielle Nebe. *Endliche rationale Matrixgruppen vom Grad 24*, volume 12 of *Aachener Beiträge zur Mathematik*. RWTH Aachen, 1995.

[Neu60] J. Neubüser. Untersuchungen des Untergruppenverbandes endlicher Gruppen auf einer programmgesteuerten elektronischen Dualmaschine. *Numer. Math.*, 2:280–292, 1960.

[Neu61] J. Neubüser. Bestimmung der Untergruppenverbände endlicher *p*-Gruppen auf einer programmgesteuerten elektronischen Dualmaschine. *Numer. Math.*, 3:271–278, 1961.

[Neu70] J. Neubüser. Investigations of groups on computers. In Leech [Lee70], pages 1–19.

[Neu82] J. Neubüser. An elementary introduction to coset table methods in computational group theory. In *Groups — St Andrews 1981*, volume 71 of *London Math. Soc. Lecture Note Ser.*, pages 1–45, Cambridge, 1982. Cambridge University Press.

[Neu87] Peter M. Neumann. Some algorithms for computing with finite permutation groups. In E.F. Robertson and C.M. Campbell, editors, *Proceedings of Groups — St Andrews 1985*, volume 121 of *London Math. Soc. Lecture Note Ser.*, pages 59–92, Cambridge, 1987. Cambridge University Press.

[Neu95] J. Neubüser. An invitation to computational group theory. In C.M. Campbell, T.C. Hurley, E.F. Robertson, S.J. Tobin, and J.J. Ward, editors, *Groups'93 — Galway/St. Andrews*, volume 212 of *London Math. Soc. Lecture Note Ser.*, pages 457–475. Cambridge University Press, 1995.

[New51] M.H.A. Newman. The influence of automatic computers on mathematical methods. In *Manchester University Computer Inaugural Conference*, pages 13–15, 1951.

[New76] M.F. Newman. Calculating presentations for certain kinds of quotient groups. In SYMSAC '76, *Proc. ACM Sympos. Symbolic and Algebraic Computation*, pages 2–8, New York, 1976. (New York, 1976), Association for Computing Machinery.

[New77] M.F. Newman. Determination of groups of prime-power order. In *Group Theory*, volume 573 of *Lecture Notes in Math.*, pages 73–84, Berlin, Heidelberg, New York, 1977. (Canberra, 1975), Springer-Verlag.

[New90] M.F. Newman. Proving a group infinite. *Arch. Math.*, 54:209–211, 1990.

[Nic88] Werner Nickel. *Endliche Körper in dem Gruppentheoretischen Programmsystem* GAP. Diplomarbeit, RWTH, Aachen, 1988.

[Nic93] Werner Nickel. *Central extensions of polycyclic groups.* PhD thesis, Australian National University, Australian National University, 1993.

[Nic95] Werner Nickel. Computing nilpotent quotients of finitely presented groups. In G. Baumslag et al., editors, *Geometric and Computational Perspectives on Infinite Groups*, volume 25 of *Amer. Math. Soc. DIMACS Series*, pages 175–191. (DIMACS, 1994), 1995.

[Nie] A.C. Niemeyer. Constructive recognition of normalisers of small extra-special matrix groups. To appear in *Internat. J. Algebra Comput.*

[Nie93] Alice C. Niemeyer. *Computing Presentations for Soluble Groups.* PhD thesis, Australian National University, 1993.

[NNN98] M.F. Newman, Werner Nickel, and Alice C. Niemeyer. Descriptions of groups of prime-power order. *J. Symbolic Comput.*, 25:665–682, 1998.

[NO96] M.F. Newman and E.A. O'Brien. Application of computers to questions like those of Burnside, II. *Internat. J. Algebra Comput.*, 6:593–605, 1996.

[NO99] M.F. Newman and E.A. O'Brien. Classifying 2-groups by coclass. *Trans. Amer. Math. Soc.*, 351:131–169, 1999.

[Nor80] Simon Norton. The construction of J_4. In Bruce Cooperstein and Geoffrey Mason, editors, *Proc. Santa Cruz Conference on Finite Groups*, pages 271–277. Amer. Math. Soc., 1980.

[Nov55] P.S. Novikov. On the algorithmic unsolvability of the word problem in group theory. *Trudy Math. Inst. im. Stevlov*, 44:1–143, 1955.

[NOVL04] M.F. Newman, E.A. O'Brien, and M.R. Vaughan-Lee. Groups and nilpotent Lie rings whose order is the sixth power of a prime. *J. Algebra*, 278:383–401, 2004.

[NP92] Peter M. Neumann and Cheryl E. Praeger. A recognition algorithm for special linear groups. *Proc. London Math. Soc.* (3), 65:555–603, 1992.

[NP95] G. Nebe and W. Plesken. Finite rational matrix groups. *Memoirs Amer. Math. Soc.*, 116, No. 556, 1995.

[NP98] Alice C. Niemeyer and Cheryl E. Praeger. A recognition algorithm for classical groups over finite fields. *Proc. London Math. Soc.*, 77:117–169, 1998.

[NPP84] J. Neubüser, H. Pahlings, and W. Plesken. CAS; design and use of a system for the handling of characters of finite groups. In Atkinson [Atk84], pages 145–183.

[O'B89] E.A. O'Brien. The groups of order dividing 256. *Bull. Austral. Math. Soc.*, 39:159–160, 1989.

[O'B90] E.A. O'Brien. The *p*-group generation algorithm. *J. Symbolic Comput.*, 9:677–698, 1990.

[O'B91] E.A. O'Brien. The groups of order 256. *J. Algebra*, 143(1):219–235, 1991.

[O'B94] E.A. O'Brien. Isomorphism testing for *p*-groups. *J. Symbolic Comput.*, 17:133–147, 1994.

[OPS98] J. Opgenorth, W. Plesken, and T. Schulz. Crystallographic algorithms and tables. *Acta. Cryst.*, A54:517–531, 1998.

[OVL02] E.A. O'Brien and M.R. Vaughan-Lee. The 2-generator restricted Burnside group of exponent 7. *Internat. J. Algebra Comput.*, 12:575–592, 2002.

[Pak01] I. Pak. What do we know about the product replacement algorithm? In Kantor and Seress [KS97], pages 301–347.

[Pap95] P. Papasoglu. Strongly geodesically automatic groups are hyperbolic. *Invent. Math.*, 121(2):323–334, 1995.

[Par84] R.A. Parker. The computer calculation of modular characters (the Meat-Axe). In Atkinson [Atk84], pages 267–274.

[Pas68] D. Passman. *Permutation Groups*. W.A. Benjamin, New York, Amsterdam, 1968.

[Ple81] Vera Pless. *Introduction to the Theory of Error-Correcting Codes*. Wiley-Interscience Series in Discrete Mathematics. John Wiley & Sons Inc, 1981.

[Ple87]   W. Plesken. Towards a soluble quotient algorithm. *J. Symbolic Comput.*, 4:111–122, 1987.

[Rab58]   M.O. Rabin. Recursive unsolvability of group theoretic problems. *Ann. Math.*, 67:172–194, 1958.

[RD04]   C.M. Roney-Dougal. Conjugacy of subgroups of the general linear group. *Experimental Math.*, 13:151–163, 2004.

[RDU03]   C.M. Roney-Dougal and W.R. Unger. The affine primitive permutation groups of degree less than 1000. *J. Symbolic Comput.*, 35:421–439, 2003.

[Ree98]   Sarah Rees. Automatic groups associated with word orders other than shortlex. *Internat. J. Algebra Comput.*, 8:575–598, 1998.

[Rei95]   Alan W. Reid. A non-Haken hyperbolic 3-manifold covered by a surface bundle. *Pacific J. Math.*, 167:163–182, 1995.

[Rin92]   Michael Ringe. *The C* MeatAxe. Lehrstuhl D für Mathematik, RWTH, Aachen, 1992.

[RM90]   G.F. Royle and B.D. McKay. The transitive graphs with at most 26 vertices. *Ars Combin.*, 30:161–176, 1990.

[Rob81]   Derek J. Robinson. Applications of cohomology to the theory of groups. In C.M. Campbell and E.F. Robertson, editors, *Groups — St Andrews 1981*, volume 71 of *London Math. Soc. Lecture Note Ser.*, pages 46–80. Cambridge University Press, 1981.

[Rob82]   Derek J. Robinson. *A Course in the Theory of Groups*, volume 80 of *Graduate Texts in Math.* Springer-Verlag, New York, Heidelberg, Berlin, 1982.

[Rón90]   Lajos Rónyai. Computing the structure of finite algebras. *J. Symbolic Comput.*, 9:355–373, 1990.

[Rot94]   Joseph J. Rotman. *An Introduction to the Theory of Groups*. Springer-Verlag, Berlin and Heidelberg, 4th edition, 1994.

[Rot02]   Joseph J. Rotman. *Advanced Modern Algebra*. Prentice-Hall, Pearson Education, NJ, 2002.

[Roy87]   Gordon F. Royle. The transitive groups of degree twelve. *J. Symbolic Comput.*, 4:255–268, 1987.

[RP89]   G.F. Royle and C.E. Praeger. Constructing the vertex-transitive graphs of order 24. *J. Symbolic Comput.*, 8:309–326, 1989.

[Ryb90]   A.J.E. Ryba. Computer condensation of modular representations. *J. Symbolic Comput.*, 9:591–600, 1990.

[Ryb01] A.J.E. Ryba. Condensation of symmetrized tensor powers. *J. Symbolic Comput.*, 32:273–289, 2001.

[San81] G. Sandlöbes. Perfect groups of order less than 10^4. *Comm. Algebra*, 9:477–490, 1981.

[Sch90a] Gerhard J.A. Schneider. Computing with endomorphism rings of modular representations. *J. Symbolic Comput.*, 9:607–636, 1990.

[Sch90b] Gerhard J.A. Schneider. Dixon's character table algorithm revisited. *J. Symbolic Comput.*, 9:601–606, 1990.

[Sco64] W.R. Scott. *Group Theory*. Dover Publications, Inc., Mineola, NY, 1964.

[Sco80] L.L. Scott. Representations in characteristic p. In Bruce Cooperstein and Geoffrey Mason, editors, *Finite Groups*, volume 37 of *Proc. Sympos. Pure Math.*, pages 319–331, Providence, RI, 1980. (Santa Cruz, 1979), Amer. Math. Soc.

[Seg83] Daniel Segal. *Polycyclic Groups*. Cambridge University Press, Cambridge, 1983.

[Seg90] Dan Segal. Decidable properties of polycyclic groups. *Proc. London Math. Soc.* (3), 61:497–528, 1990.

[Ser97] Ákos Seress. An introduction to computational group theory. *Notices Amer. Math. Soc.*, 44:671–679, 1997.

[Ser03] Ákos Seress. *Permutation Group Algorithms*. Cambridge Tracts in Mathematics 152. Cambridge University Press, 2003.

[Sho92] M.W. Short. *The Primitive Soluble Permutation Groups of Degree Less than* 256, volume 1519 of *Lecture Notes in Math.* Springer-Verlag, Berlin, Heidelberg, New York, 1992.

[Shp92] Igor Shparlinski. *Computational and Algorithmic Problems in Finite Fields*, volume 1519. Kluwer Academic Publishers, 1992.

[Sim70] Charles C. Sims. Computational methods in the study of permutation groups. In Leech [Lee70], pages 169–183.

[Sim71a] Charles C. Sims. Computation with permutation groups. In *Proc. Second Symp. on Symbolic and Algebraic Manipulation*. ACM Press, 1971.

[Sim71b] Charles C. Sims. Determining the conjugacy classes of permutation groups. In Garret Birkhoff and Marshall Hall, Jr., editors, *Computers in Algebra and Number Theory*, volume 4 of *Proc. Amer. Math. Soc.*, pages 191–195, Providence, RI, 1971. (New York, 1970), Amer. Math. Soc.

[Sim73] Charles C. Sims. The existence and uniqueness of Lyons' group. In T. Hagen, M.P. Hale, and E.E. Shult, editors, *Finite Groups '72*, pages 138–141. North Holland, 1973.

[Sim74] Charles C. Sims. Some algorithms based on coset enumeration. Unpublished notes, Rutgers University, New Brunswick, NJ, 1974.

[Sim80] Charles C. Sims. How to construct a baby monster. In M.J. Collins, editor, *Finite Simple Groups II*, pages 339–345. (Durham 1978), Academic Press, 1980.

[Sim87] Charles C. Sims. Verifying nilpotence. *J. Symbolic Comput.*, 3:231–247, 1987.

[Sim90a] Charles C. Sims. Computing the order of a solvable permutation group. *J. Symbolic Comput.*, 9:699–705, 1990.

[Sim90b] Charles C. Sims. Implementing the Baumslag-Cannonito-Miller polycyclic quotient algorithm. *J. Symbolic Comput.*, 9:707–723, 1990.

[Sim94] Charles C. Sims. *Computation with Finitely Presented Groups*. Cambridge University Press, 1994.

[Sim03] Charles C. Sims. Computational group theory. In *Computer Algebra Handbook: Foundations, Applications, Systems*, pages 65–83. Springer-Verlag, 2003.

[SL96] A. Storjohann and G. Labahn. Asymptotically fast computation of Hermite normal forms of integer matrices. In *Proc. of International Symposium on Symbolic and Algebraic Computation* ISSAC '96, pages 259–266. (Zürich), ACM Press, 1996.

[SM85] L.H. Soicher and J. McKay. Computing Galois groups over the rationals. *J. Number Theory*, 20:273–281, 1985.

[Smi94] Michael J. Smith. *Computing automorphisms of finite soluble groups*. PhD thesis, Australian National University, 1994.

[SS92] S.T. Schibell and R.M. Stafford. Processor interconnection networks and Cayley graphs. *Discrete Applied Mathematics*, 40:337–357, 1992.

[SS94] M. Schönert and Á. Seress. Finding blocks of imprimitivity in small-base groups in nearly linear time. In *Proc. of International Symposium on Symbolic and Algebraic Computation* ISSAC '94, pages 154–157. (Oxford), ACM Press, 1994.

[ST02] Ian Stewart and David Tall. *Algebraic Number Theory and Fermat's Last Theorem*. A.K. Peters, 2002.

[Sto98] Arne Storjohann. Computing Hermite and Smith normal forms of triangular integer matrices. *Linear Algebra Appl.*, 282:25–45, 1998.

[SWD96] O. Schirokauer, D. Weber, and Th. F. Denny. Discrete logarithms: the effectiveness of the index calculus method. In *Algorithmic Number Theory — ANTS II*, volume 1122 of *Lecture Notes in Comput. Sci.*, pages 337–361, Berlin, 1996. Springer.

[Syl72] L. Sylow. Théorèmes sur les groupes de substitutions. *Math. Ann.*, 5:584–594, 1872.

[TC36] J.A. Todd and H.S.M. Coxeter. A practical method for enumerating cosets of a finite abstract group. *Proc. Edinburgh Math. Soc.*, 5:26–34, 1936.

[The97] Heiko Theißen. *Eine Methode zur Normalisatorberechnung in Permutationsgruppen mit Anwendungen in der Konstruktion primitiver Gruppen.* PhD thesis, RWTH Aachen, 1997.

[Tie08] H. Tietze. Über die topologischen Invarianten mehrdimensionalen Mannigfaltigkeiten. *Monatsh. f. Math. u. Physik*, 19:1–118, 1908.

[Tro64] H.F. Trotter. A machine program for coset enumeration. *Canad. Math. Bull.*, 7:357–368, 1964.

[Ung02] W.R. Unger. Computing the solvable radical of a permutation group. In preparation, 2002.

[VL84] M.R. Vaughan-Lee. An aspect of the nilpotent quotient algorithm. In Atkinson [Atk84], pages 76–83.

[VL93] M.R. Vaughan-Lee. *The Restricted Burnside Problem*, volume 5 of *London Math. Soc. Monographs (New Series)*. Oxford University Press, Oxford, 2nd edition, 1993.

[VLZ99] M.R. Vaughan-Lee and E.I. Zel'manov. Bounds in the restricted Burnside problem. *J. Austral. Math. Soc. Ser. A*, 67:261–271, 1999.

[Wam70] J.W. Wamsley. The deficiency of metacyclic groups. *Proc. Amer. Math. Soc.*, 24:724–726, 1970.

[Wam74] J.W. Wamsley. Computation in nilpotent groups (theory). In *Proc. Second Internat. Conf. Theory of Groups*, volume 372 of *Lecture Notes in Math.*, pages 691–700, Berlin, Heidelberg, New York, 1974. (Canberra, 1973), Springer-Verlag.

[Wam77] J.W. Wamsley. Computing soluble groups. In A. Dold and B. Eckmann, editors, *Group Theory*, volume 573 of *Lecture Notes in Math.*, pages 118–125, Berlin, Heidelberg, New York, 1977. (Canberra, 1975), Springer-Verlag.

[War74] J.N. Ward. A note on the Todd-Coxeter algorithm. In R.A. Bryce, J. Cossey, and M.F. Newman, editors, *Group Theory*, volume 573 of *Lecture Notes in Mathematics*, pages 126–129, Berlin, 1974. Springer.

[Wie64] Helmut Wielandt. *Finite Permutation Groups*. Academic Press, New York, 1964.

[Wil96] Robert A. Wilson. Standard generators for sporadic simple groups. *J. Algebra*, 184:505–515, 1996.

[Wur93] Martin Wursthorn. Isomorphisms of modular group algebras: an algorithm and its application to groups of order 2^6. *J. Symbolic Comput.*, 15:211–227, 1993.

[Zas48] H. Zassenhaus. Über einen Algorithmus zur Bestimmung der Raumgruppen. *Comment. Math. Helv.*, 21:117–141, 1948.

[Zel91a] E.I. Zel'manov. Solution of the restricted Burnside problem for 2-groups (Russian). *Math. Sb.*, 182:568–592, 1991.

[Zel91b] E.I. Zel'manov. Solution of the restricted Burnside problem for groups of odd exponent. *Math. USSR-Izv.*, 36:41–60, 1991.

Index of Displayed Procedures

Author Index

Subject Index

Printed and bound by CPI Group (UK) Ltd, Croydon, CR0 4YY

28/10/2024

01780263-0003